Principles of Ocean Physics

JOHN R. APEL, PH.D.

Chief Scientist,
Milton S. Eisenhower Research Center
Applied Physics Laboratory
The Johns Hopkins University
Laurel, Maryland 20207 USA

D1394261

ACADEMIC PRESS

Harcourt Brace & Company, Publishers

London   San Diego   New York

Boston   Sydney   Tokyo

ACADEMIC PRESS LIMITED
24–28 Oval Road, London NW1 7DX

United States Edition published by
ACADEMIC PRESS INC.
San Diego, CA 92101

Second printing with corrections 1988
Third printing 1990
Fourth printing 1995

British Library Cataloguing in Publication Data

Apel, J. R.
  Principles of ocean physics. – (International
  geophysics series, ISSN 0074-6142)
  1. Oceanography
  I. Title   II. Series
  551.46'01   GC150.5

  ISBN 0-12-058865-X
  ISBN 0-12-058866-8 (Pbk)

LCCCN 87-70490
LCCCN 87-70489 (Pbk)

Printed in Great Britain by
St Edmundsbury Press Limited, Bury St Edmunds, Suffolk

Preface

The study of the physics of the sea has been somewhat detached from the study of physics *per se;* even its common name of physical oceanography reveals an alliance with geography that is as strong as that with physics. In recent years, however, significant advances in both the theoretical and observational sides of the discipline have allowed more quantitative descriptions of the physical behavior of the ocean to be derived. As a result, it becomes possible to carry on much discourse in physical oceanography in the traditional language of physics, that is, in terms of processes and mechanisms clothed in the idealized garb of equations, numbers, and graphs.

This is not to say that all is known of the physics of the sea, or that descriptive physical oceanography no longer has a role – far from it. The real world is much too complicated for either of these statements to be true. It is simply that the subject has happily progressed to a point where the unification and synthesis resulting from a deepened physical understanding no longer have the associated risk of simplification *ad absurdum* that they once might have had.

The preparation of this book was originally undertaken in the hope that it might stimulate a few physics graduate students to the serious study of the oceans, perhaps even as a life's work. To the extent that it has provided any such migration across disciplines, the book will have served its purpose. In writing it, I have tried to address the nonexpert but reasonably mature physical scientist who might like to know something of the fundamentals of the quantitative physics of the sea without being put off by having to read introductory student texts, or diverted by the focus and detail of the research literature. In this audience I would count working physicists, graduate students, and perhaps even an occasional scientist from an allied discipline (i.e., dynamics, electromagnetics, optics, and the like) who is interested in the marine environment. In doing so, however, I have tried to present a viewpoint that a physical oceanographer might not object to, and have attempted to use his or her language alongside our common tongue of physics. In this way the book may also serve as a lexicon, giving the reader access to the information shrouded by such mysteries as "quasi-geostrophic motions in baroclinic instability" or "conservation of potential vorticity." In discussing a process or a mechanism, I have first tried to describe it and then to account for it theoretically, if possible. Purely deductive reasoning from equations is often not rewarding in a subject as complicated as this. I have also freely used results from numerical models, laboratory experiments, and remote measurements, in addition to the classical sources of observations made in the water and calculations made with theory. The text has not attempted to be exhaustive, but rather to present the fundamentals with what is hoped to be a balanced perspective. Enough access is provided to more advanced or specialized texts so that the reader can eventually find a way into the research literature, if desired.

The untimely deaths of two colleagues occurred while I was readying this manuscript for publication, and their passings have deprived the field of two of its most able and productive scientists. Those familiar with Prof. Adrian Gill's excellent monograph, *Atmosphere–Ocean Dynamics,* will recognize the large intellectual debt owed to that work by the sections of this book concerned with geophysical fluid dynamics. In quite another discipline, Dr. Rudolph Preisendorfer put the field of optical oceanography on a firm mathematical basis with the publication of his four-volume work, *Hydrologic Optics,* on which Chapter 9 draws so heavily. Because of the inspiration that their research and writing have provided to the community, and their importance in my own efforts, I dedicate this text to A. E. Gill and R. W. Preisendorfer. I also dedicate it to those five fathers of modern physical oceanography whose names appear so frequently here: H. U. Sverdrup, C.-G. Rossby, V. W. Ekman, H. M. Stommel, and W. H. Munk, all teachers as well as researchers.

In the Revised Edition, a number of errors have been corrected, notation has been made more uniform, and a few recent references have been added.

J.R.A.
Manor Park, Maryland

Acknowledgments

In a book of this length and novelty, it is inevitable that errors—be they conceptual, calculational, or typographical—will be present. I am grateful to colleagues who reviewed portions of the manuscript in response to appeals for assistance in reducing its error rate to acceptable levels; they included Roswell Austin, Robert Beal, Jack Calman, Janet Campbell, Nicholas Fofonoff, Richard Gasparovic, Howard Gordon, Jimmy Larsen, Lawrence McGoldrick, Owen Phillips, Charles Schemm, Morris Schulkin, Donald Thompson, Kenneth Voss, Robert Winokur, Warren Wooster, and Charles Yentsch. The remaining errors (ever too numerous) are clearly my responsibility. It is also with thanks that I acknowledge the patience of my students, who were exposed to the early drafts of the text and who suffered in unknown ways as a result of its imperfections.

I am especially appreciative of the support given by The Johns Hopkins University Applied Physics Laboratory in the award of a Stewart S. Janney Fellowship, which allowed me to convert a draft manuscript into a finished publication; and in the editorial, illustrations, and typesetting functions carried out by the Laboratory staff. A number of associates assisted in this regard, and special thanks go to my editor, Al Brogdon; my editorial assistants, Linda Muegge and Jacqueline Apel; artists Mary O'Toole and Stephen Smith; secretaries Anne Landry, Barbara Goldsmith, and Jeaneen Jernigan; and compositors Veronica Lorentz, Patrice Zurvalec, Barbara Bankert, Nancy Zepp, Sandy Bridges, Barbara Northrop, and Brenda Laub. The encouragement and assistance of Academic Press, especially of Conrad Guettler, are gratefully acknowledged.

JOHN R. APEL

Publisher's Credits

The following publishers have kindly granted permission to reprint or adapt the materials cited; in all cases listed, attempts have been made to reach the senior author or his survivors to obtain their permission as well. We apologize to those whom we have not reached, for whatever reason. Complete citations of the sources are found in the Bibliography at the end of the chapter indicated by the figure or table number. All such material is copyrighted by the publisher as of the date listed, and each has granted nonexclusive world rights in English and, in some cases, in all languages.

Aberdeen University Press, Aberdeen, Scotland: 9.27
Academic Press, Inc., Orlando, Fla.: Figs. 2.5, 4.3, 4.13, 5.32, 6.11, 6.48a, 6.48b, 6.52, 9.17b, and 9.50; Table 9.5
Addison–Wesley Publishing Co., Inc., Reading, Mass.: Figs. 5.23, 6.8, 6.15, 6.19, 6.25, 6.44, and 6.45
American Association for the Advancement of Science, Washington, D.C.: Figs. 8.33, 8.34, and 8.46
American Geophysical Union, Washington, D.C.: Figs. 2.32, 3.7, 6.40, 6.41, 6.42, 7.2, 8.22, 8.35, and 9.57
American Institute of Physics, New York, N.Y.: Figs. 7.11, 7.24, and 7.35
American Meteorological Society, Boston, Mass.: Figs. 2.7, 5.2, 6.33, 6.34, 9.12, 9.13, and 9.14
American Physical Society, New York, N.Y.: Figs. 2.3 and 2.12
American Society of Limnology and Oceanography, Grafton, Wisc.: Fig. 9.29
Artech House, Inc., Norwood, Mass.: Figs. 8.14 and 8.16
Belknap Press of Harvard University Press, Cambridge, Mass.: Fig. 1.1
C. Boysen, Hamburg, Federal Republic of Germany: Fig. 2.24
Cahners Publishing Co., Newton, Mass.: Fig. 2.31
Cambridge University Press, New York, N.Y.: Figs. 3.14, 5.4, 5.6, 9.6, 9.7, and 9.20; Table 9.6
Columbia University, Palisades, N.Y.: Fig. 3.8
Deutches Hydrographisches Institut, Hamburg, Federal Republic of Germany: Figs. 5.14, 5.15, and 6.22
Dodd, Mead and Co., New York, N.Y.: Fig. 5.5
Dover Publications, Inc., New York, N.Y.: Fig. 5.13
Elsevier Scientific Publishing Company, Amsterdam, Netherlands: Figs. 9.5, 9.8, 9.12, 9.22, 9.25, 9.33, 9.36, 9.38, 9.42, 9.44, 9.45, and 9.47
Fisheries and Marine Service, Ottawa, Ontario, Canada: Fig. 5.17, 5.18, and 5.19
Florida State University, Tallahassee, Fla.: Fig. 2.15a
Gebruder Bornstraeger, Stuttgart, Federal Republic of Germany: Figs. 2.16, 2.26, and 4.9
Geological Society of America, Boulder, Colo.: Figs. 2.29 and 7.8
Harper and Row Publishers, New York, N.Y.: Figs. 1.2 and 1.3
Imperial College of Science and Technology, London, England: Fig. 6.47
Institute of Electrical and Electronic Engineers, New York, N.Y.: Figs. 8.11, 8.12, 8.15, and 8.26
International Union of Geodesy and Geophysics, Paris, France: Fig. 9.10
Jet Propulsion Laboratory, Pasadena, Calif.: Figs. 2.15b, 2.32, 3.7, 5.20, 5.24, 6.40, 6.41, 6.42, 7.2, 8.22, 8.35, 9.54, and 9.57
The Johns Hopkins Applied Physics Laboratory, Laurel, Md.: Figs. 8.36 and 8.37
The Johns Hopkins University Press, Baltimore, Md.: Figs. 6.35, 6.55, and 6.56
Journal of Marine Research, New Haven, Conn.: Figs. 6.18, 7.23, and 9.24
Macmillan Journals Limited, Washington, D.C.: Fig. 5.16
Marine Technical Society, Washington, D.C.: Fig. 8.25 and 9.53
Massachusetts Institute of Technology, Cambridge, Mass.: Fig. 4.12
McGraw–Hill Book Co., New York, N.Y.: Figs. 7.9, 7.10, and 7.32
National Academy Press, Washington, D.C.: Figs. 6.51 and 6.57
NOVA N.Y.I.T. University Press, Ft. Lauderdale, Fla.: Fig. 6.53
Optical Society of America, Washington, D.C.: Figs. 8.2, 8.39, 9.11, 9.19, 9.28, and 9.32; Table 9.4
Peninsula Publishing, Los Altos, Calif.: Figs. 7.14, 7.16, 7.17, 7.19, 7.21, and 7.26
Pergamon Press, New York, N.Y.: Figs. 2.23, 3.11, 5.26, 5.27, 6.43, 6.54, and 7.3
Plenum Press, New York, N.Y.: Fig. 7.25; Table 7.1
Prentice-Hall, Inc., Englewood Cliffs, N.J.: Figs. 4.4, 4.8, 5.10, and 5.12
D. Reidel Publishing Co., Dordrecht, Netherlands: Figs. 8.7, 8.13, 8.24, 8.40, and 8.41
Royal Society of London, London, England: Fig. 6.49
Royal Swedish Academy of Sciences, Stockholm, Sweden: Fig. 2.19
Society of Photo-Optical Instrumentation Engineers, Bellingham, Wash.: Figs. 9.26, 9.31, 9.35, and 9.43
Springer-Verlag, New York, N.Y.: Figs. 6.2, 6.3, 7.33 7.34, 9.51, and 9.52
Tetra Tech., Inc., Pasadena, Calif.: Fig. 9.34
Trewin Copplestone Books, Ltd., London, England: Fig. 2.2
United States Naval Institute, Annapolis, Md.: Fig. 9.18
University of California Press, Berkeley, Calif.: Figs. 6.20, 6.21, 6.23, 6.26, 6.27, 6.28, 6.29, 6.30, 6.31, 6.32, 6.38, 8.19, 8.27, 8.31, and 8.38
University of Göteborg, Göteborg, Sweden: Figs. 9.55 and 9.56
University of Washington Press, Seattle, Wash.: Fig. 6.58
John Wiley and Sons, New York, N.Y.: Figs. 7.22, 7.30, 8.4a, and 8.4b
Woods Hole Oceanographic Institution Atlas Series, Woods Hole, Mass.: Fig. 2.25

Contents

Chapter Three   **Hydrodynamic Equations of the Sea**

Chapter Four   **Thermodynamics and Energy Relations**

Chapter Five   **Geophysical Fluid Dynamics I:**
                Waves and Tides

Chapter Six   **Geophysical Fluid Dynamics II:**
               Currents and Circulation

Chapter Seven   **Acoustical Oceanography**

Chapter Eight   **Electromagnetics and the Sea**

Chapter Nine   **Optics of the Sea**

Why did the old Persians hold the sea holy?
Why did the Greeks give it a separate deity, an own brother of Jove?
Surely all this is not without meaning.

Melville, *Moby Dick*

Chapter One

Physical Oceanography: An Overview

The history of science is science itself; the history of the individual, the individual.

Goethe, *Mineralogy and Geology*

1.1 Introduction

The study of the physics of the sea is the study of a broad range of classical and modern physics as applied to a complicated and pervasive medium covering much of the earth. The subject is properly a branch of geophysics, although that term is often applied in a more narrow sense to the physical characteristics of the solid earth; we shall not so restrict it here.

Ocean physics, which grew out of both the physical geography of the sea and meteorology, appears to have diverged from the mainstream of the discipline sometime around the turn of this century, when the attention of basic researchers in physics became increasingly directed toward the atomic and subatomic nature of matter. However, in the hands of northern European scientists who, to some extent, were conditioned by the need for solving maritime and fisheries problems, the subject of physical oceanography led a limited but vigorous existence under the *nom de choix* of hydrography, as it is known in those quarters even today. In North America the more common name of physical oceanography has been used, while in the USSR the preferred terminology is physical oceanology.

The growth of ocean physics has been stimulated not only by its intellectual character and by the practical needs of maritime affairs, but also by the slowly growing realization that the ocean plays an enormously important role in conditioning both weather and climate, over land and sea alike. The exploitation of this awareness, however, has at the same time been severely limited by the lack of an adequate data base that would allow one to construct a dynamical picture of physical events in the sea.

Atmospheric science has made significant contributions to the subdiscipline of fluid dynamics of the sea. The exigencies of producing a daily forecast of weather long ago drove scientists and nations to erect what has become a synoptic scale (approximately 250 km) weather-observing network over essentially all of the land areas of the Northern Hemisphere, and a considerably more rarefied system over the Southern Hemisphere and the oceans. More importantly, this network has provided data leading to dynamical concepts that appear to explain atmospheric processes on an extremely wide range of space and time scales. There currently exist reasonably quantitative theoretical and numerical explanations of events in the atmosphere ranging from tornadoes to the general circulation of the earth's atmosphere. The scales of these events span dimensions of the order of 5 km at the small end to 5×10^5 km at the large.

The realization that the dynamics of the atmosphere might be translated into the dynamics of the ocean with appropriate changes in scaling factors occurred over a period of years before, during, and after World War II. While the data base to verify the validity of these similarity transformations and changes in boundary conditions is as of now only partially assembled, enough has been done to give one confidence that the subject of geophysical fluid dynamics, as it has come to be known, is on firm ground for both the atmosphere and the ocean. Indeed, it has even been applied with varying degrees of success to the motion of the atmospheres of Venus and Jupiter. This success does not imply that all is known about ocean dynamics, but only that the basic framework is intact. The problems arise because many processes taking place in the ocean are so complicated and span such a range of scales that their reasonably rigorous description in terms of first principles exceeds all available computational power. The analyst is then forced to *parameterize* many of the processes, i.e., replace their rigorous physical descriptions with a simplified model whose adequacy is often difficult to assay. (An example of a widely used parameterization is the treatment of large-scale turbulence in terms of a relatively simple eddy diffusion process; this is discussed in the chapters on hydrodynamics and geophysical fluid dynamics.) Parameterization is essential for real progress to be made in geophysics.

While fluid dynamics was probably the branch of physics first finding significant application to the sea (and it is still the dominant one), it is by no means the sole subdiscipline. Thermodynamics and its characterization of matter by bulk properties such as an equation of state was also an early entry. Underwater acoustics is relevant because it is the only known form of signaling that can carry information through large distances in the sea—a fact of much interest to submarine warfare, to the study of marine life, and to charting the ocean depths. Electromagnetism finds application through the measurement of voltages induced by the conducting sea water moving in the earth's magnetic field, as well as through the techniques of remote measurement of sea surface properties via radio and microwave probing. The study of optics of the sea has been stimulated by the need to understand absorption or emission of infrared, visible, and ultraviolet radiation, which in turn are relevant to the life cycle of flora and fauna and to the interchange of heat and water vapor between the sea and the atmosphere. The detailed description of the earth's gravity and geoidal fields finds application in relatively subtle ways in the establishment of mean sea level and in the dynamics of large-scale currents. Geology and solid earth geophysics concern themselves with the basins that contain the oceans, with the crust moving on global but geological time scales that imply that ancient oceans were quite different in many ways from the present ones. The atomic and molecular physics of the constituents of seawater largely determine the macroscopic properties of the medium with which the oceanographer so unconcernedly deals. At a smaller level still, nuclear physics in the sea is in part occupied with the migration of radioactive materials, with nuclear radiation, and with the behavior of reactor neutrons from ships.

These, then, are the subject areas of the physics of the sea.

1.2 The Evolution of Modern Physical Oceanography

It is abundantly clear that the origins of the scientific study of the ocean are to be found in the sailing and navigational needs of the sea-going commerce that took place among the Mediterranean countries of the ancient world and which later included those of the Far East. Piloting and navigation as arts (rather than sciences) have been practiced by sailors for some 6000 to 8000 years, using the accumulated knowledge of the sea and its winds, waves, and currents. This understanding was passed on by a combination of (1) tutelage before the mast, where its imperfections were often punctuated by finding one's vessel *in extremis* due to poor navigation or sudden bad weather, and (2) in maritime schools, where master seamen and scholars perpetuated what was then known—or thought—about the sea and its navigation. Even at that stage, there were close interactions between sailors returning from a voyage and reporting their observations to their sponsors, and scholars who, with varying degrees of success, would attempt to fit those observations together with others to form some sort of world view. On occasion a scholar would even go to sea and experience its wonders personally, a practice that is not followed frequently enough even today.

In the Greco–Roman world, maritime charts often contained summaries of what was known about both the geography of the Mediterranean region and its navigation. The earliest surviving map of the region showing quantitative character is due to Eratosthenes (ca. 250 B.C.), who used lines of latitude and longitude and the 16-point wind rose on his chart of the then-known world. He also determined the circumference of the earth with considerable accuracy by comparing the length of a shadow in Alexandria at high noon on the summer solstice with the observation, on the same day, of the exact vertical illumination of a well located on the Tropic of Cancer; the value he obtained was about 43,000 km. In the second century A.D., Claudius Ptolmey published an outstanding world map that used a conic projection; however, Ptolmey apparently ignored Eratosthenes' value for the circumference of the earth and instead used a lesser value of approximately 32,000 km. So influential was Ptolmey's cartography that, upon the rediscovery and translation from Greek to Latin of his *Cosmographia,* the Rennaissance navigators (including Columbus) easily became convinced that the distance around the world was only some 75% of its actual circumference. This error was not generally rectified until the late 1600's.

Along with the charts and maps were produced *sailing directions* containing information on distances, aids and dangers, and characteristics of the sea, the earliest of which dates from sometime between the sixth and fourth centuries B.C. The practice of publishing these volumes has continued to the present, with considerable improvements being made by 16th century navigators and scholars. For a time both charts and sailing directions were bound together, but increasing knowledge as time went on made these documents so bulky that they eventually came to be published as separate entities. These directions contained much of what was known of the physical oceanography of the day. It is fair to say, though, that there was no organized body of knowledge known by that name until the 19th century, and such knowledge as existed was in the province of the navigator as much as the scholar.

The beginnings of modern navigation and scientific study of the sea are probably marked by the three Pacific voyages of Captain James Cook of the British Royal Navy between 1768 and 1779. Among the array of other scientific instruments, his parties carried chronometers that allowed the precision determination of time and longitude and, hence, the positions of land masses, reasonable estimates of ship's positions, and the influence of currents on those positions. With the completion of Cook's voyages and the publication of the findings in several volumes and journals, the major dimensions of the world's oceans were known and the first glimmerings of the science of the sea were apparent.

During the same general time period, Benjamin Franklin, then Postmaster General of the United States, prepared a surprisingly accurate chart of the path of the Gulf Stream by compiling reports of the drift of ships plying the waters between Britain and the United States. This chart enabled American ships to reduce the round-trip time to England over their British competitors by several days, by riding the Gulf Stream while going east and avoiding it when returning. Modern vessels make use of updated, more accurate versions of the same information.

The H.M.S. *Beagle* expedition of 1831–36 with Charles Darwin as naturalist is considered the next major oceangoing campaign to intensively study environmental sciences, this time with a concentration on regions of the Southern Hemisphere. While more directed toward biological and geological studies than physics as such, its scientific orientation and the breadth of its scientific party ensured that there would be contributions to a range of disciplines. Darwin's conclusions on species variability are well known, but his first-hand observations of mountain building in the Andes and his theories on coral reef formation and seamount subsidence laid the groundwork for much later concepts of the dynamics of the earth's crust. The publication of his *Journal of Researches* in 1845 predates his *Origin of Species* by 14 years.

While organized knowledge derived from expeditions of the types just described was highly valuable to the advance of science, the example of Franklin's Gulf Stream chart nevertheless demonstrated the utility of ships' navigational and meteorological data in ocean studies. In the hands of Lt. Matthew F. Maury, an American naval officer, this type of analysis reached its pinnacle. Confined to limited duty by an injury, he was placed in charge of the Naval Depot of Charts and Instruments in 1842, where he had access to ships' log books and other records. He quickly realized the enormous value of the information in them and, over the next few years, subjected the data to analyses that yielded currents, winds, and shipping routes throughout much of the world's oceans. His publication of charts and sailing directions starting in 1849 caught the attention of seafarers throughout the world and led to an international cooperative effort in collecting marine data at regular intervals that persists today with some modification. Chapter I of Maury's 1855 book entitled *The Physical Geography of the Sea* opens with the paragraph:

> THERE is a river in the ocean. In the severest droughts it never fails, and in the mightiest floods it never overflows. Its banks and its bottom are of cold water, while its current is of warm. The Gulf of Mexico is its fountain, and its mouth is in the Arctic Seas. It is the Gulf Stream.

Figure 1.1 is a reproduction of Maury's chart of "Gulf Stream and Drift," which is both a scientific and an artistic masterpiece. Later work has added greatly to the understanding of this "river" (see Chapter 6), but none has come close to having the impact on ocean science of his pioneering work.

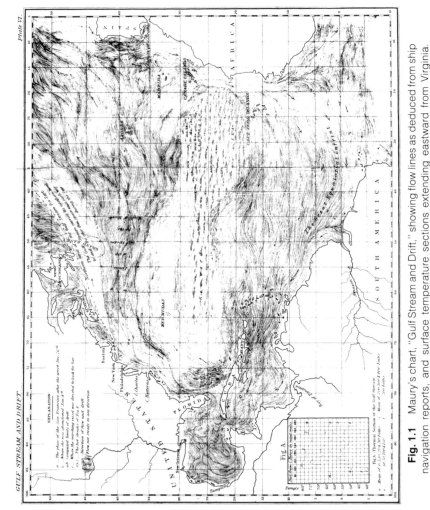

Fig. 1.1 Maury's chart, "Gulf Stream and Drift," showing flow lines as deduced from ship navigation reports, and surface temperature sections extending eastward from Virginia. [From Maury, M. F., *The Physical Geography of the Sea* (1855).]

Fig. 1.2 Drawing of H.M.S. *Challenger* [From Thompson, C.W., *Voyage of the "Challenger"* (1878).]

Today, the analysis of century-old data from the "Marine Deck" continues to yield long-term trends in oceanography and meteorology. The collecting of global marine weather information at 6 hour intervals continues as well, now under the auspices of the World Meteorological Organization, and constitutes an essential data source for real-time numerical weather forecasts.

By the last third of the 19th century, a new force had come into play in the science of the sea: the transoceanic telegraph. The requirement for laying cable along the sea floor to join continents via communications links led nations to the realization that the physical, chemical, and biological conditions of the deep sea floor might well determine the success or failure of such ventures. The British Royal Society recommended that an expedition should survey the physical and biological state of the sea along certain sections, and H.M.S. *Challenger* (Fig. 1.2) was assigned to the task, with C. Wyville Thomson as director of the scientific staff. From 1872 to 1876, the ship surveyed major lines in three oceans in a series of research cruises that constituted

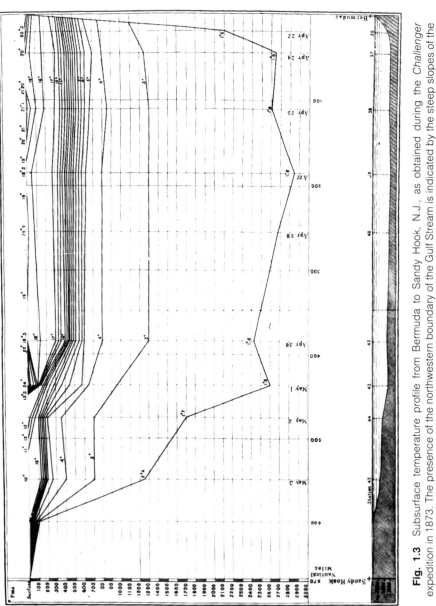

Fig. 1.3 Subsurface temperature profile from Bermuda to Sandy Hook, N.J., as obtained during the *Challenger* expedition in 1873. The presence of the northwestern boundary of the Gulf Stream is indicated by the steep slopes of the isotherms in the vicinity of stations 42-44. [From Thompson, C. W., *The Voyage of the "Challenger."* (1878).]

the first modern oceanographic expedition. Among the extensive new information developed and ultimately published in some 50 volumes were temperature and depth measurements and bottom samples that revealed oddities of the deep. A temperature section from Bermuda to Sandy Hook, N.J., is shown in Fig. 1.3, with the slopes of the isotherms showing the northwest boundary of the Gulf Stream near Stations 42 and 43. The dynamical significance of these data was probably not appreciated at the time, but such information would later allow oceanographers to compute the variation of current speeds with depth using the "dynamic method." It is probably fair to say that the *Challenger* expedition marked the beginnings of oceanography as a systematic scientific enterprise.

The one-third century between the *Challenger* expedition and the onset of World War I saw a flowering of both exploration and the academic, basic research phases of ocean science. It was a time of great public faith in the benefits of science and technology and in the ultimate "perfectibility of man," and the general scientific advances in theoretical understanding and in improved measurement capabilities led to concomitant advances in field oceanography. Exploration of the Arctic and Antarctic continued apace, with scientists as often as military officers leading exploration parties; in these, Scandinavian and Russian investigators figured prominently, with the most famous of the scientific polar expeditions being headed by F. Nansen during the three-year drift of the Norwegian vessel *Fram* through the Arctic Basin. On the academic side, basic developments in classical fluid dynamics were soon followed by their application to both meteorology and physical oceanography; for a time, the practitioners of the analytical parts of those sciences were generally theoretical physicists and applied mathematicians, with the names V. F. K. Bjerknes, V. W. Ekman, and V. J. Boussinesq frequently occurring in the literature. In chemistry, the physicochemical properties of seawater, especially its thermodynamics, were established with accuracies that allowed the quantitative interpretation of field data, an effort in which M. Knudsen figured prominently. In general, it was an era when the basics of the science were being uncovered through the interactions of exploration, field and laboratory measurement, and theoretical formulation. These methods continue to be the *modi operandi* of ocean science even today, but with the addition of large-scale numerical modeling as a major new ingredient not available to the earlier investigators.

During this time, there was an increasing understanding of the interrelationships among the physics, chemistry, and biology of the sea and of the importance of these sciences in the very practical requirements of commercial fisheries. Since edible fish are generally among the climax species in the oceanic food chain, the impacts of far-removed physical and chemical factors such as wind-driven surface currents, upper ocean temperatures, dissolved oxygen, and suspended particulates upon higher trophic levels in the marine ecosystem were only suspected to exist. The conviction that such factors are relevant but not well enough understood led to the establishment, in Copenhagen in 1902, of the International Council for the Exploration of the Sea, a unique treaty/scientific organization of North Atlantic states that continues today to foster research in these subjects.

World War I provided two additional stimuli for the study of the sea: the submarine, and the asdic, almost the sole means of detecting the submerged submarine. (Asdic evolved into the more familiar sonar, which reached operational use after the war.) Experiments by the British showed that the detectability of submarines by sonar was largely conditioned by the temperature and density structure of the upper ocean, and that correctly acquired measurements of thermal profiles were of considerable assistance in understanding and predicting the detection process. As with so many other inventions of engines of destruction, developments in antisubmarine warfare have had beneficial influences on nonmilitary oceanography, in that ocean instrumentation and theoretical comprehension have been significantly advanced as consequences of submarine-oriented studies. Today, a major portion of the basic oceanic research sponsored by the navies of the world has as its end objective the improved forecasting of marine environmental conditions for both submarine and surface ship operations.

The period between the two World Wars saw the growth of ocean research institutions and programs in almost all of the technologically advanced nations. New research vessels were commissioned and built by many, including *Meteor* by the German Navy in 1924, H.M.S. *Discovery* by the British in 1925, and R.V. *Atlantis* by the newly founded Woods Hole Oceanographic Institution in 1930. The Scripps Institution of Oceanography, while first organized in 1903, acquired its present name and was dedicated to research in 1925. The exponential-like growth that characterized research in other fields and which extended approximately from those years through to the 1960's also affected oceanography, although that subject started from a much smaller base. The discipline is still miniscule compared with, say, chemistry or physics (the *U.S. Directory of Marine Scientists* for 1987 listed about 6600 names active in all areas of marine science, compared with the approximately 60,000 members of the American Institute of Physics societies alone).

Stimulated by the events of World War II and the technologies that grew from it, the post-war oceanic sciences expanded greatly in breadth and depth, both literally and figuratively. The global nature of the conflict, which involved naval operations in waters around the world, led to increased demands for mapping and marine forecasts. Acoustic bathymetry (depth sounding) allowed rapid charting of regions untouched by the sounding lead line; ultimately these soundings, along with other geophysical measurements (e.g., of magnetics and gravity) gave rise to the theories of sea floor spreading and provided a mechanistic underpinning for the hypothesis of continental drift put forth by A. Wegener in 1915. By the mid-1960's, the plate tectonic theory was generally (but not universally) accepted by marine geologists and geophysicists (see Chapter 2). The submarine problem, too, had oceanographic by-products of importance in the understanding of upper ocean dynamics. The invention of the mechanical bathythermograph by Spilhaus and the follow-on development of expendable temperature, conductivity, and current probes significantly extended measurement capabilities in the upper several hundred meters of the sea. The installation of precision loran navigation near shore, and then the invention of satellite Doppler navigation by The Johns Hopkins University Applied Physics Laboratory allowed positioning accuracies of features at sea in the range of tens of meters. The successful deployment of deep water current meter and thermistor arrays for months on end became feasible because of the efforts of Woods Hole scientists, among others. Since oceanography has always been a data-starved science, these extensions of measurement capability and efficiency had great impact on the understanding of physical processes in the sea. A new round of ship construction after the war placed improved sea-going platforms at the disposal of the wet-deck oceanographer, although in the United States many of these ships are currently nearing the ends of their useful lives. An entirely different kind of ocean-viewing platform, the spacecraft, returned quite useful information via sensors not originally intended for observing the sea. Recent developments in dedicated satellite remote measurement, such as the determination of surface temperature, current velocity, wind stress, wave spectra, and the visualization of flow patterns with color images of the sea, are just now making serious impacts on the data-scarcity problem, and are probably still only in their embryonic state. Moored and drifting buoys carrying surface and near-surface instrumentation can forward their data and position through satellite communication links, returning information on the oceans with much efficiency.

Theoretical developments have accelerated as well, driven in part by the expanding data base resulting from such instrumentation, in part by the increased understanding of atmospheric dynamics that has diffused over into ocean dynamics, and to a very great degree by the availability of large computing resources. The last quarter-century has seen a significant movement of physical oceanography away from its origins in the physical geography of the sea toward a quantitative, hypothesis-testing kind of activity, in considerable degree because of the unifying theoretical concepts that have grown out of the accumulated body of information derived from the ocean research enterprise to date.

It is thus to a major degree that this book is directed toward the basic the-

Chapter Two

Forcing Functions and Responses

*First, they should have right ideas of things,
ideas that are based on careful observation,
and understand causes and effects and their
significance correctly.*

The Teaching of Buddha, Book 3,
"The Way of Practice."

2.1 Introduction

In order to help the understanding of the more quantitative material found later in the book, this chapter is devoted to a description, in minimally mathematical format, of the major forces and constraints acting on the ocean, as well as the responses with which the sea acknowledges those nudgings and pullings. This description will make somewhat more clear why *dynamics* (i.e., the study of oceanic motions) forms such a major portion of the physics of the sea. This is not to give short shrift to the other subdisciplines of acoustics, optics, etc., for indeed, these are treated somewhat more fully than is usual in introductory texts on physical oceanography; it is only that the dynamics condition much of the state of the medium in which higher frequency wave propagation—be it acoustic, electromagnetic, or optic—takes place.

2.2 Forcing Functions On and In the Sea

It is convenient to divide the forces acting on the sea into a few classes, although as with all categorizations, there is more than one way to do this, and the categories are not necessarily mutually exclusive. In order of their scale and their usual appearance in the Navier–Stokes dynamical equations, they are: (1) gravitational and rotational forces, which permeate the entire fluid and which have large scales compared with most other forces; (2) thermodynamic forces, such as radiative transfer, heating, cooling, precipitation, and evaporation; (3) mechanical forcing, such as surface wind stress, atmospheric pressure variations, seismic sea floor motions, and other mechanical perturbations; (4) internal forces – pressure and viscosity – exerted by one portion of the fluid on its other parts, and which serve to make fluid dynamics much more complicated than single particle dynamics. In a somewhat separate category but overlapping the others to a degree is (5) boundary forcing, which on the bottom usually acts statically as a simple containment force, but which is generally formulated as boundary conditions on the equations. The fluid invariably acts to develop boundary layers in response to these – for example, the Ekman layer at the ocean surface, the benthic boundary layer at the bottom, and western and eastern boundary currents along the margins of the sea near continents.

2.3 Gravitational and Rotational Forces

If the volume of ocean water, V, of approximately 1.370×10^9 km^3 were uniformly distributed over a smooth, spherical earth with a mean radius, R_e, of 6371 km, it would have a mean depth of approximately 2700 m; since only 70.8% of the earth's surface is covered with water, however, the actual mean depth, H, is closer to 3800 m. The aspect ratio of the ocean is $H/2\pi R_e \sqrt{0.708} \approx 1 \times 10^{-4}$, which is approximately the same as the thickness-to-width ratio of the paper on which a typical chart of the world's oceans is printed. Relative to its breadth, then, the ocean is very shallow, and the force of gravity that maintains this thin sheet in place is very nearly, but not exactly, the value of the gravity at the solid surface of the earth.

The subject of marine gravity has some important implications not only for solid earth geophysics, but also for physical oceanography, as we shall see ahead, since small-scale variations in the magnitude and direction of the gravitational acceleration have observable consequences on the equipotential surface of the sea. In addition to the earth's gravity, clearly the moon and sun exert gravitational effects on the sea, these (plus spin, as will be discussed later) being the major tide-producing forces on both the solid earth and its fluid envelopes of air and water. Figure 2.1 is a schematic diagram of these gravitational and rotational forces. Such time-dependent forces are driven at exact astronomical frequencies and can be predicted with considerable precision for years in advance; tidal height forecasts are among the most successful operational predictions available for the sea.

Rotational forces also play a role in geophysical fluid dynamics, with three

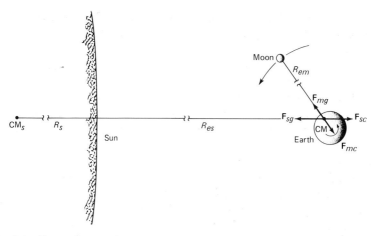

Fig. 2.1 Force diagram for the earth–moon–sun system. Tidal forces arise from both gravitational and centrifugal forces caused by the moon and the sun.

distinct motions contributing: (1) the earth's daily spin; (2) the rotation of the moon on a near-monthly period; and (3) the annual rotation of the earth-moon system about the common center of mass of the solar system. The centripetal acceleration and the concomitant flattening of the earth due to the first of these motions results in a variation of the effective surface gravity by approximately 0.52% from pole to equator. The other two rotations also produce centripetal accelerations that contribute to the tide-producing forces in a fashion that will be discussed in Chapters 3 and 5. There the gravitational and rotational forces will be mathematically summarized in terms of a potential function, Φ, whose negative gradient gives both forces.

2.4 Radiative, Thermodynamic, and Related Forces

The radiant energy that drives the earth's fluid envelopes comes from only two major sources: (1) the sun, as an external source; and (2) the decay of radioactive matter within the earth's interior, which maintains the material below the surface crust in either a plastic or a fluid condition (except for the inner, solid core). The latter energy influences, to a degree, the bottom temperature of the deep sea but is not an important source for dynamical effects.

The sun is approximately a graybody with an emissivity, e_λ, of 1 or less, characterized by an emission temperature, T_s, of about 5900 K (Fig. 2.2); the blackbody radiant intensity curve is modulated by the solar Fraunhofer lines in the visible wavelength region. In this band, the radiative spectral intensity of a 5900 K blackbody has a maximum at a wavelength, λ, in the blue at about $\lambda = 491$ nm. The total solar radiative emission per unit area, M, or exitance (see Chapter 8), is

$$M = \sigma_{SB} T_s^4 \qquad \text{W m}^{-2} , \tag{2.1}$$

where the Stefan–Boltzmann constant is $\sigma_{SB} = 5.67 \times 10^{-8}$ W $(\text{m}^2 \text{ K}^4)^{-1}$. At the earth's surface, at a radius of one astronomical unit ($R_{es} = 1.49598 \times 10^{11}$ m), the mean theoretical solar irradiance, $\langle S_{th} \rangle$, is, according to the numbers above, $\sigma_{SB} T_s^4 (R_s/R_{es})^2 = 1.487$ kW m^{-2}, where R_s is the radius of the sun ($R_s = 6.960 \times 10^8$ m). The correct value at the top of the atmosphere, i.e., the solar "constant" is, from observation,

$$\langle S \rangle = 1.376 \text{ kW m}^{-2} , \tag{2.2}$$

which is somewhat lower than the theoretical value because of the gray-bodiness of the sun. In addition, this *insolation* is not constant but varies by $\pm 3.2\%$ annually because of the ellipticity of the earth's orbit. The projection of the disk of the earth in the sun's direction receives an amount of solar power, P_s, of

$$P_s = \pi R_e^2 \langle S \rangle , \tag{2.3}$$

of which a fraction, $\langle \alpha \rangle$, the mean *albedo* or reflected power, is reflected back to space, chiefly by clouds, snow, ice, and desert surface. Since the total area of the earth's surface is $4\pi R_e^2$, the average power flux absorbed per unit area is

$$\tfrac{1}{4} \langle S \rangle (1 - \langle \alpha \rangle) = 240.8 \text{ W m}^{-2} . \tag{2.4}$$

It is this energy that is responsible for the mean temperature of the earth, and since the planet is neither heating up nor cooling down (at least not on decadal or shorter time scales), the same amount of energy must be re-radiated to space as infrared radiation. If the earth, too, were a blackbody, its equilibrium emission temperature, T_e, would be found from the equation: re-radiation = insolation. Thus

$$4\pi R_e^2 \sigma_{SB} T_e^4 = \pi R_e^2 (1 - \langle \alpha \rangle) \sigma_{SB} T_s^4 (R_s/R_{es})^2 . \tag{2.5}$$

For an average albedo, $\langle \alpha \rangle = 0.3$, as measured from spacecraft, Eq. 2.5 yields a theoretical value for T_e of 260 K. However, the re-absorption of emitted infrared energy by radiatively active constituents in the atmosphere (the greenhouse effect), coupled with convection, raises the observed value for T_e to about 288 K. Figure 2.2 shows the spectral exitance of a 300 K blackbody as an envelope for the observed IR emission from the earth. By the Wien displacement law, the wavelength, λ_m, of the maximum of the Planck distribution of blackbody radiation at temperature T is

$$\lambda_m = \frac{\alpha_r}{T} \, , \tag{2.6}$$

where $\alpha_r = 2897.8\ \mu m$ K. For the earth at $T_e = 288$ K, the maximum occurs at

$$\lambda_m = 10.1\ \mu m \ . \tag{2.7}$$

This region of infrared emission is termed the *thermal infrared,* and it is coincidently a region of relatively high transparency of the atmosphere (see Fig. 2.2). One would then expect that the land and water would have a globally and seasonally averaged equilibrium temperature, in degrees Celsius, of

$$T_{eq} = T_e - 273.15 \simeq 15°C \ . \tag{2.8}$$

This temperature is not far from the observed average surface temperature of the oceans, which have extreme values of approximately $-2°C$ to $+35°C$ and which have by far the largest heat capacity of the earth's surface. Thus the processes of insolation, reflection, absorption, and re-radiation establish the gross temperature balance of the earth.

Figure 2.3 is a schematic diagram of the flow of energy in the geophysical system consisting of atmosphere, hydrosphere (ocean, lakes, and other waters), lithosphere (land), and cryosphere (ice). Of the totality of incoming radiation, 30% is reflected by clouds, dust, and the earth's surface (the albedo); 70% is re-radiated as IR radiation, which in turn is divided into 64% emitted by the radiatively active gases and clouds in the atmosphere, and 6% by the earth's surface, both land and water. Before this re-emission occurs, many transmutations of the incident energy have taken place, as the diagram suggests, with 20% of the total incident radiation being directly absorbed by the atmosphere, and 50% entering the nonatmospheric portions of the system – mainly the oceans, where it heats the upper levels of the sea and directly contributes to the oceanic *thermohaline circulation* (temperature- and salt-driven). As will be seen below, it also indirectly contributes to the wind-driven circulation via evaporation of water in the tropics and subtropics. Water has an enormous specific heat, the second highest of any

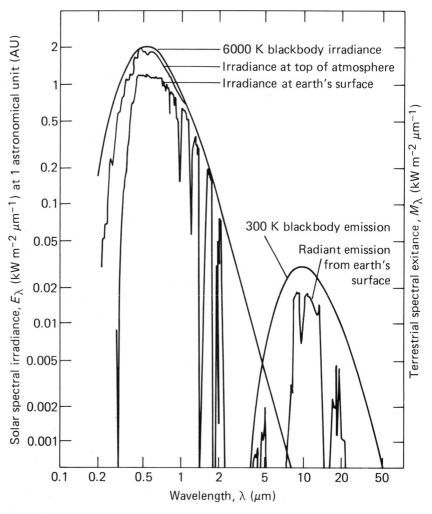

Fig. 2.2 Incident solar spectral irradiances at top of atmosphere and at the earth's surface, compared with that from a 6000 K blackbody (left). Spectral emittance from the earth's surface, compared with that from a 300 K blackbody (right). [Adapted from Smith, D. G., Ed., *The Cambridge Encyclopedia of Earth Sciences* (1981).]

known substance; for seawater, the specific heat at constant pressure, C_p, is 0.955 cal (g °C)$^{-1}$, or 3998 J (kg °C)$^{-1}$ at a temperature, T, of 20°C, a surface pressure, p, of 1.024×10^5 Pa, and a salinity, s, of 35 practical salinity units (psu, or 35 parts per thousand). It also has a very large *latent heat of vaporization, L_v*:

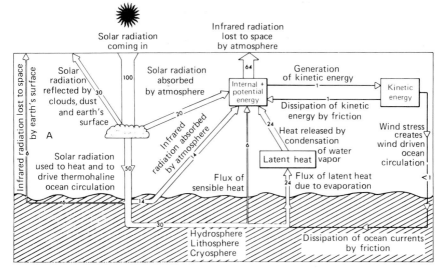

Fig. 2.3 Schematic diagram of the flow of energy in the components of the climate system. Numbers show the percentages of incoming solar radiation entering into the energy transformation processes illustrated. [From Piexóto, J. P., and A. H. Oort, *Rev. Mod. Phys.* (1984).]

$$L_v = (2.501 - 0.00239\ T) \times 10^6$$

$$= 2.453 \times 10^6\ \text{J kg}^{-1}\ \text{at } 20°\text{C}$$

$$= 586\ \text{cal g}^{-1}. \tag{2.9}$$

These two seemingly simple facts have important consequences for the entire ocean–ice–atmosphere system on both short and long time scales, as much of the subsequent discussion in this book will reveal. The thermodynamic properties of water in its solid, liquid, and vaporous states establish the thermal inertia of the climate system and the ability of that system to redistribute and transform heat, energy, and momentum.

Returning to Fig. 2.3, our ledger showing the energy budget of the system: It is seen that 50% of the incoming solar radiation is deposited in the surface, mostly via warming of the surface waters of the sea. A quantity of 6% + 14% is re-radiated as infrared radiation, with the former percentage escaping to space and the latter being absorbed by the atmosphere; 30% is transported by ocean currents to areas removed from the immediate regions of deposition. Essentially all of the latter is eventually surrendered to the atmosphere, 24% as latent heat of vaporization and 6% as *sensible heat,* i.e., direct thermal conduction across the air/sea interface. Thus, on the average, 83 W m^{-2}, or nearly one-third of the incoming solar radiation of 241 W m^{-2}, is deposited within the atmosphere via the agent of water vapor. In the tropics, the actual amounts are nearer 120 W m^{-2} of evaporative flux, which is four times more effective in air/sea energy interchange than sensible heat conduction. The heat of vaporization is surrendered to the atmosphere as internal thermal energy upon condensation, i.e., cloud formation and rainfall, whereupon it is immediately available to drive the atmospheric circulation.

The discussion to this point has been along the lines of globally averaged quantities. The obliqueness of the polar regions to solar radiation, the seasonally varying tilt of the spin axis, the interruption of flow by the continental land masses, and the large asymmetry in the land/water ratio between the Northern and Southern Hemispheres also have profound influences on both oceanic and atmospheric dynamics. We shall now consider some of these.

The solar energy absorbed in the ocean is deposited in the upper levels of the sea and is mixed downward by wind waves, convective overturning, and turbulent processes, as well as being forced downward by Ekman pumping in certain regions, a process to be considered later. Figure 2.4 shows six typical vertical profiles of temperature for Northern Hemisphere summer and winter, in the tropics (15°N), at mid-latitudes (40°N), and in the polar North Atlantic (75°N). The three vertical regimes are: the near-surface *mixed*

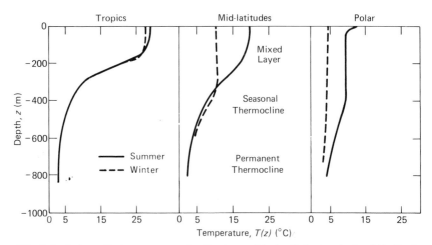

Fig. 2.4 Schematic temperature distributions with depth for tropical, mid-latitude, and polar regions, for summer (solid) and winter (dashed) conditions. The major regimes in the vertical are termed the near-surface mixed layer, the seasonal thermocline, and the deep, permanent thermocline.

layer, which is isothermal (or nearly so); the *seasonal thermocline,* where the temperature changes relatively rapidly below the base of the mixed layer; and the deep *permanent thermocline,* where the temperature changes are more gradual with depth and exceedingly slow with time. While there are only small variations between the summer and winter thermoclines in the tropics, there are very large alterations in the subpolar and even the mid-latitude profiles. It is in this upper region of the sea that the main portion of the incoming solar energy is stored and redistributed by currents and air/sea interchanges.

2.5   Zonal and Meridional Variations

If the spin axis of the earth were normal to the ecliptic, one would expect a total extinction of the mean insolation as the oblique angles of illumination at the poles are approached, but the seasonal variation due to the 23.5° tilt of the earth results in a much smoother mean north–south distribution of the bath of solar radiation reaching the ground than otherwise. This *mean meridional variation* is shown in Fig. 2.5. If there were no tilt, the radiation per unit area received would vary from zero at the poles to $\langle S \rangle / \pi$ at the equator. However, the tilt, plus the eccentricity of the earth's orbit, result in the distribution of solar radiation shown in Fig. 2.6. It can be seen that the radiation varies from over 500 W m^{-2} in the Austral summer to zero

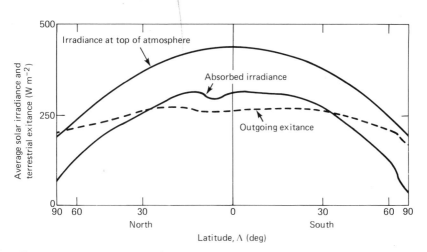

Fig. 2.5   Zonal distribution of insolation and terrestrial reradiation. The upper curve shows the incident flux of solar energy; the middle, the flux absorbed by the earth and atmosphere; and the lower (dashed) curve, the outgoing reradiated power. [Adapted from Gill, A. E., *Atmosphere-Ocean Dynamics* (1982).]

at the poles in the winter, with the the north–south asymmetry arising from the ± 3.2% variation in total radiation due to the ellipticity of the orbit. The outgoing radiation in Fig. 2.5, on the other hand, is much more uniform in latitude than is the incident radiation, which implies that the equator-to-pole temperature gradient has acted to smooth out the distribution of heat resulting from the radiation deposited in the atmosphere. The heat transport required to reduce the gradient is carried by both ocean and atmosphere, and it is this driving force of north–south temperature gradient that is the major source of the dynamical motions of both geophysical fluids. Figure 2.7 shows, for the Northern Hemisphere, the poleward transport of heat due to both fluids. As we shall see, the heat is carried mainly by meridional (north–south) circulation in the atmosphere and the oceanic equivalent, i.e., boundary currents in the sea, with both the average circulations and their fluctuating components playing roles in the transports. These motions constitute the great atmospheric and oceanic circulation systems, but the multiple processes that establish and maintain them are indirect, complicated, and

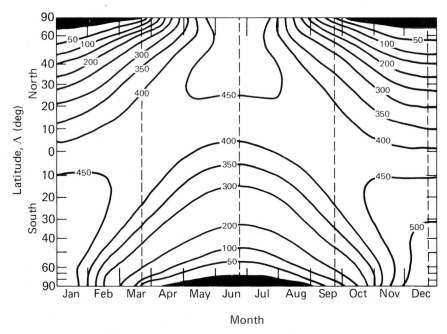

Month

Fig. 2.6 Average daily solar irradiance in W m⁻² for a unit horizontal surface at the top of the atmosphere, versus latitude and month. The dotted vertical lines show the equinoxes and solstices; the dark regions represent the polar zones during absence of sunshine. The north–south asymmetry is largely due to the eccentricity of the earth's orbit, with the perihelion occurring during the Southern Hemisphere summer. [Adapted from Milankovitch, M., in *Handbuch der Klimatologie* (1930).]

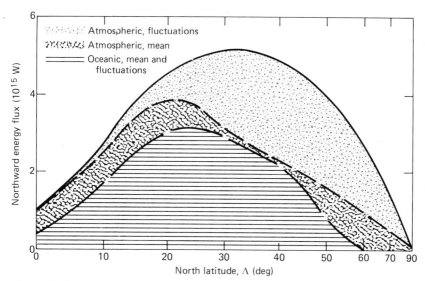

Fig. 2.7   Rate of poleward (meridional) transport of energy by the atmosphere and the ocean as a function of latitude. The outer curve is net transport deduced from radiation measurements from satellites; the middle curve derives from atmospheric mean circulation; lower curve is for atmospheric eddy transport. The remainder is attributed to oceanic mean and eddy transports, but direct measurements to support this are not generally available. [Adapted from Vonder Haar, T. H., and A. H. Oort, *J. Phys. Oceanogr.* (1973).]

tightly coupled to one another. It is these processes that we shall discuss in the remainder of this section.

According to the diagram of the atmospheric heat engine (Fig. 2.3), about 50% of the incoming solar radiation is absorbed by the earth's surface, most of it through the large heat capacity of the ocean, where it is stored for redistribution to the atmosphere. The single most important mechanism in the redistribution is evaporation, which is active where there are high temperatures and copious water supplies. The major regions of the earth where this occurs are the tropical oceans, the Amazon and Congo River basins, and the so-called "Far Eastern Island Continent," the myriad of islands and archipelagos in the tropical Far East. Here the absorption of solar radiation by land and sea, and the attendant high daily and average temperatures, result in large evaporation rates, rising air, strongly buoyant cumulonimbus cloud formation, and heavy rainfall. Because of this, the ocean has its warmest surface waters in the tropical Eastern Indian Ocean and Western Pacific. This rising, unstable *convective motion* carries water vapor high into the atmosphere, with the maximum release occurring near mid-altitudes (approximately 500 mbar pressure height), but with the tops of the convective systems often reaching altitudes of 12 to 15 km (40,000 to 50,000 ft), even injecting some water vapor into the stratosphere. The condensation of the vapor into liquid water and the subsequent rainfall releases the latent heat of evaporation, L_v, to the surrounding air and warms it even more, frequently leading to superbuoyant conditions and convective instabilities.

Air must flow into these buoyant regions at low levels to replace the upwardly displaced moist air, which departs from the unstable convective "chimneys" at mid to high altitudes. A return flow at height then occurs, along with a degree of poleward migration of air and vapor caused by vortex-line stretching from Coriolis forces (which will be explained more fully in Chapter 6). The air finally sinks toward the surface near the cooler western coasts of continents to complete the circulation. The *zonal* (east–west) cellular patterns thereby established, called *Walker cells,* are shown schematically in Fig. 2.8.

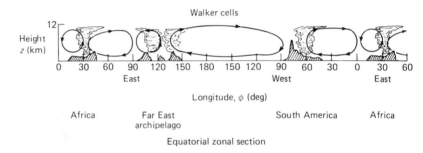

Fig. 2.8 Schematic diagram of Walker cell circulation. Heating and evaporation of water are greatest over the Amazon and Congo River basins and the Far Eastern Island Continent. Convection carries water vapor to high altitudes; surface flows occur to replace rising air. The circulation is closed in the upper atmosphere.

Thus the process of *convective motion,* with its release of latent heat, is the fundamental means by which the ocean communicates a flux of moisture, momentum, and energy to the atmosphere. It is active in the formation of tropical cyclones (called hurricanes in the Atlantic and Eastern Pacific, typhoons in the Western Pacific, and cyclones in the Indian Ocean), but is not confined to the tropics, although it is most pronounced there. As an example of temperate zone convection, Fig. 2.9 shows an image made from the geosynchronous satellite GOES at thermal infrared wavelengths, illustrating cold polar air progressing out across a much warmer ocean and resulting in very large evaporation and formation of linear *cloud streets* over the Gulf Stream and Gulf of Mexico. Thus strong exchanges between sea and air are not confined to the tropics, but occur when there is a negative difference between the temperature of the air, T_a, and of the sea, T_s. The rate of transfer also depends on surface wind stress, as will be discussed further in Chapter 3.

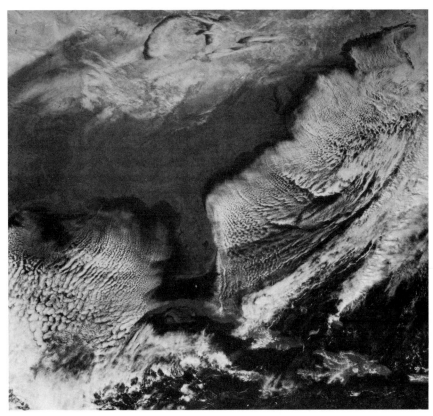

Fig. 2.9 Image from geosynchronous satellite, GOES, at infrared wavelengths, showing large evaporative fluxes occurring over the warm ocean during a cold, polar air outbreak. [Figure courtesy of National Oceanic and Atmospheric Administration.]

Returning to the atmospheric circulation: The idealized east–west Walker cell is modified by north–south and low–high circulations induced by continuity of flow and by the Coriolis force. The rising air leaving the rotating surface near the equator must conserve its absolute angular momentum and, finding itself at a larger distance from the surface of the earth at altitude, moves poleward. At high altitudes, it radiatively cools by emitting long-wave infrared radiation and thereby becomes denser. By the time the poleward-flowing air has reached 25° to 30° latitude, it has acquired an appreciable easterly or zonal component, has dried considerably through rainfall, and begins to sink. The zone of dry, sinking air is a desert region globally, the atmospheric moisture having been wrung out of the air along the way. This meridional cell is called a *Hadley cell,* and is a region of net energy release into the atmosphere. In the simple Hadley cell picture, the return surface winds would have only an equatorward component, but in actuality are modified by interacting with the Walker circulation to take on strong zonal components as well. Over the oceans, these combined lower level Walker–Hadley wind components form the *trade winds,* and blow from approximately the east-northeast in the Northern Hemisphere and the east-southeast in the Southern Hemisphere, to converge just north of the equator in the *Intertropical Convergence Zone* (ITCZ). Because of the asymmetrical distribution of land masses and attendant heat storage between the hemispheres, the ITCZ is actually located at an average of several degrees north of the equator, and is effectively the thermal and meteorological equator. It is a region of persistent cloudiness and rainfall and is known to mariners as the *doldrums.* Figure 2.10 sketches the Hadley cells and the surface winds for the Northern Hemisphere; the upper level winds tend to be oppositely directed.

   Poleward of the zones of dry, sinking air, a second set of meridional cells develops called *Ferrel cells,* and in the Northern Hemisphere extend from approximately 25° or 30°N to 60°N; these cells are much weaker than the directly driven Hadley cells. In the polar regions is a third set known appropriately as *polar cells.* In the Ferrel cell there is a rising of relatively cold air in high latitudes and a sinking of relatively warm air in lower middle lati-

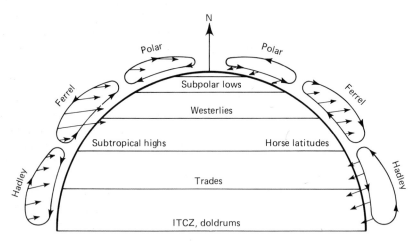

Fig. 2.10   Diagram of Northern Hemisphere meridional cellular circulation, showing the tropical Hadley cells, the mid-latitude Ferrel cells, and the polar cells. Upper and lower level winds have directions suggested by the top and bottom arrows attached to the cells.

tudes, thereby requiring net work. It is the region of high meteorological variability, strong westerlies, the subpolar jet streams, and the production of mid-latitude *cyclones,* or low-pressure storm systems. In the Southern Hemisphere, the latitudinal center of the Ferrel cell falls over the infamous "roaring forties" of maritime lore. The three-cell regime results from zonal radiation and temperature gradients combined with the Coriolis force, and is a consequence of the relatively high spin rate of the earth. If the earth were spinning significantly less rapidly, there would be only one cell, the Hadley cell, rather than the three. Figure 2.11, a schematic of an average vertical section of the atmosphere from north to south, illustrates how tropical convection elevates the *tropopause,* or the top of the lower atmosphere; the three-cell circulation is suggested as well.

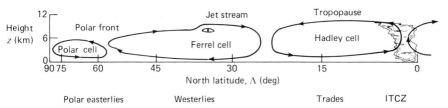

Fig. 2.11 Vertical section of three-cell circulation from equator to pole, showing elevation of tropopause in the tropics relative to polar regions.

Figure 2.12 shows north–south cross sections of mean zonal and meridional wind speeds, $\langle u \rangle$ and $\langle v \rangle$, respectively, in meters per second, and the transport rate of atmospheric mass, in units of 10^{10} kilograms per second. The vertical coordinate is atmospheric pressure in decibars or 0.1 bar (1 bar $= 10^5$ N m^{-2} = 10^5 Pa). Figure 2.13 gives the mean global distribution of surface pressure in millibars ($= 10^{-3}$ bar), surface streamlines, and mean surface wind velocity, as indicated by arrows; a constant pressure of 1000 mbar must be added to the numbers shown to obtain the actual pressure. It can be seen that the ITCZ is a region of lower average pressure, while the "horse latitudes" near 30°N and 30°S (cf. Fig. 2.10) average near 1020 mbar. These surface pressure and wind distributions provide the basic mechanical forcing functions for the upper ocean, and we will return to study their effects in more detail at a later time.

The sum total of the mean general circulation of the troposphere is a complicated, three-dimensional flow field; Fig. 2.14 (Plate 1) attempts to illustrate this. While it is only the surface of the sea that is in intimate contact with the atmosphere and which exchanges fluxes of moisture, energy, and momentum with it, one also sees that the entire air mass up to the tropopause is strongly conditioned by the ocean on all time scales ranging through daily, seasonal, and climatological periods. Conversely, the ocean is in turn condi-

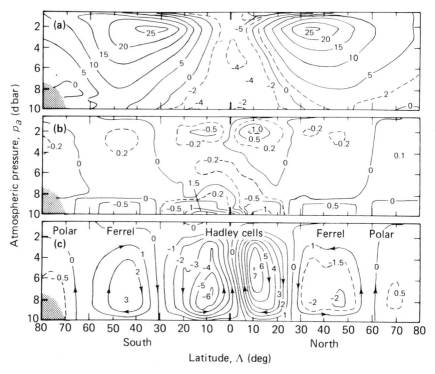

Fig. 2.12　Cross sections of (a) zonal wind component, $\langle u \rangle$, and (b) meridional wind component, $\langle v \rangle$, in m s^{-1}, as averaged over the year and zonally around the globe. (c) Mass transport in 10^{10} kg s^{-1}. Jet stream axes are near 250 mbar in both hemispheres. [From Piexóto, J. P., and A. H. Oort, *Rev. Mod. Phys.* (1984).]

tioned by atmospheric radiation, wind stress, temperature, pressure, and rainfall, although its response times are much longer than those of the atmosphere. The ocean thus is the flywheel or the inertia on the global heat engine and serves to smooth out the high-frequency fluctuations of the atmosphere. The two fluids are strongly coupled, having both positive and negative feedback loops and greatly differing time constants.

2.6　Wind Stress

While Fig. 2.13 shows the mean global wind velocity and surface pressure distributions, these have been greatly averaged in both space and time and present a relatively bland picture of the spatial patterns. To illustrate shorter-time variability, Fig. 2.15a demonstrates the distribution of *surface wind stress*

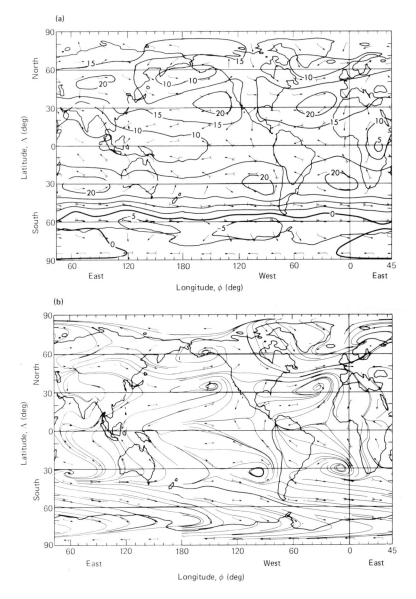

Fig. 2.13 (a) Global distribution of averaged surface pressure, p_a, reduced to sea level; units are $p_a - 1000$ mbar. Also shown are arrows indicating average velocity of wind, with each barb on tail representing a speed of 5 m s^{-1}. (b) Surface streamlines and wind barbs. [From Oort, A. H., *Global Atmospheric Circulation Statistics* (1982).]

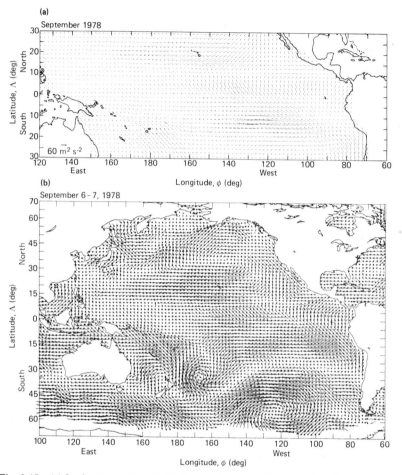

Fig. 2.15 (a) Surface wind stress for the tropical Pacific, proportional to the square of the wind velocity, for the month of September, 1978. Dominant features are the northeast and southeast trades, and the strong coastal winds along South America. [From Legler, D. M., and J. J. O'Brien, *Atlas of Tropical Pacific Wind-Stress Climatology 1971–1980* (1984).] (b) Surface wind velocity obtained from radar scatterometer on the Seasat spacecraft for September 6–7, 1978. Large Southern Hemisphere storm systems are major features. [Courtesy of P. M. Woiceshyn and G. F. Cunningham, Jet Propulsion Laboratory (1986).]

(proportional to the square of wind speed) over the Tropical Pacific as averaged over September 1978, this month being one of both well-developed trade winds and cyclonic storm activity. However, the winds due to the latter have been smoothed out because of the movement of storms. On an even shorter time scale, Fig. 2.15b is a "snapshot" of surface wind velocity for most of

the Pacific for a two-day period in September 1978 that shows several large
cyclonic systems. The data were obtained by a satellite radar scatterometer.
It is these winds that dominate the transfer of horizontal momentum to the
ocean and which produce the wind-driven ocean currents, these representing the most energetic component of the oceanic circulation.

The elevation at which the measurement of the "surface" wind is made
is an important quantity to specify, since near the surface of land, water,
or ice, the winds vary sharply in speed and direction throughout the near-
surface *planetary boundary layer* or the so-called *atmospheric Ekman layer*
(named after the Swedish meteorologist/oceanographer, V. W. Ekman). The
usual meteorological heights for specifying winds at sea are either 10 m or
19.5 m, set chiefly by the mast height of vessels. At a solid boundary, the
wind speed must be zero just at the surface; for water, wind-induced motion
means that the wind and water velocities must match just at the boundary.
Thus the transition regime in the vertical, i.e., the turbulent boundary layer,
is one in which the *wind shear,* or vertical derivative of velocity, is large and
its direction variable; because of the latter effect, the wind is said to "veer
aloft." The mean wind speed profile in the lower few hundred meters of
atmospheric height is approximately logarithmic in the vertical coordinate, z.

In Fig. 2.15a, the *stress*, τ, is approximately quadratically dependent on
wind velocity. The stress is the aerodynamic force per unit area exerted by
the wind on the sea surface and may be written as

$$\tau = c_d \rho_a |\mathbf{u}_w| \, \mathbf{u}_w \qquad \text{N m}^{-2} , \qquad (2.10)$$

where $\mathbf{u}_w$ is the wind velocity; ρ_a, the density of air; and c_d is a dimensionless quantity called the *aerodynamic drag coefficient*. The latter is itself a
function of wind speed, being approximately constant at low speeds, but increasing linearly above some 10 m s^{-1}. The stress directly induces surface
waves and horizontal momentum, i.e., horizontal currents. Because of the
more-than-quadratic dependence of the stress on $\mathbf{u}_w$, it seems likely that
much momentum is conveyed to the ocean during brief intense storms (i.e.,
is episodic); thus, the type of data in Fig. 2.15b becomes highly relevant.
However, storms are usually separated by long intervals of lower winds that
can do work on the ocean for extended times and which are therefore effective in driving currents. Thus for an accurate description of wind-induced
currents, it is necessary to know both the storm and the steady winds.

Not only the wind velocity but also the heat, temperature, and humidity
profiles vary sharply in the lower few hundred meters of the atmosphere.
The turbulence in the lowest 10 m or so is largely driven by wind shear; at
slightly higher levels, heat and moisture are carried upward by interchanges
of warm, moist lower air with cooler, drier upper air, and depend on buoyan-

cy differences as well. It is these interchanges that bring about the turbulence in the planetary boundary layer; many of the properties of the layer depend on the sign of the air–sea temperature difference, $T_a - T_s$, which, when negative (colder air over warmer water), leads to strong interchange, considerable turbulence, and roughened seas.

By definition, the currents induced by the wind stress make up the *wind-driven circulation* of the ocean. While the direct effects of the stress are felt close to the ocean surface (mainly within the so-called *oceanic Ekman layer* extending over the upper 10 to 100 m), it is the deeper components of circulation brought about by the surface motions that bring into play a significant volume of the waters of the sea. Furthermore, it is not only the stress itself, but also its *curl, divergence,* and *gradient* that drive the circulation; the division of the wind effects into various portions is based partly on mathematical convenience and partly on physical response; this will be discussed in more detail in Chapters 3 and 6.

2.7 The General Circulation of the Ocean

For purposes of setting the stage for the mathematics to follow, a qualitative description of oceanic response will suffice, although it is clear that the justification for and quantitative descriptions of the wind-driven circulation must follow a more mathematical approach.

In the absence of rotation, it would be expected that the wind would drive currents in the direction of the stress exerted. However, the Coriolis force diverts the ocean waters at some angle to the wind as time progresses, so that the response is more complicated than expected. Furthermore, the continental boundaries interrupt and also divert the otherwise purely wind-driven currents. Thermal and wind forcing also lead to the slow vertical motions called the thermohaline circulation (from the Greek: heat plus salt), so that the *general circulation* is a very complicated flow field conditioned by many factors. Figure 2.16 is a schematic rendition of the long-term-average surface current distribution during the Northern Hemisphere winter. Amidst the plethora of detail in Fig. 2.16 are some characteristic surface flow patterns whose summation makes up the general circulation; these are sketched in Fig. 2.17. The associated mean surface temperatures are shown in Fig. 2.18, a long-term average based on nearly 1.6 million observations.

The most ubiquitous features of the flow patterns are the very large *subtropical gyres* that span the entire east–west dimension of each ocean basin in both the Northern and Southern Hemispheres. The most familiar example is probably the North Atlantic gyre, whose western leg is made up of the Florida Current–Gulf Stream system, which subsequently becomes the

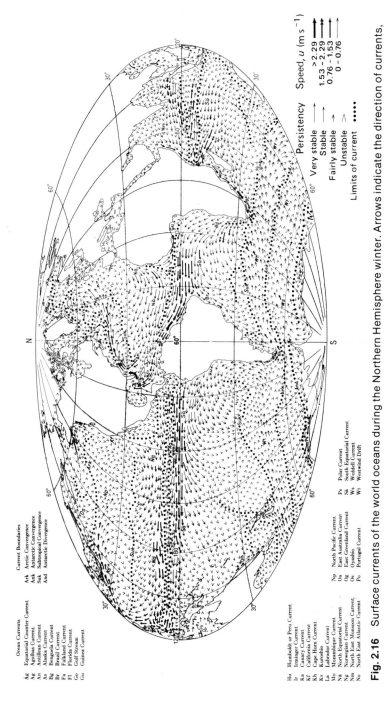

Fig. 2.16 Surface currents of the world oceans during the Northern Hemisphere winter. Arrows indicate the direction of currents, while their lengths indicate persistency. [From Dietrich, G., *General Oceanography* (1963); originally due to G. Schott (1943).]

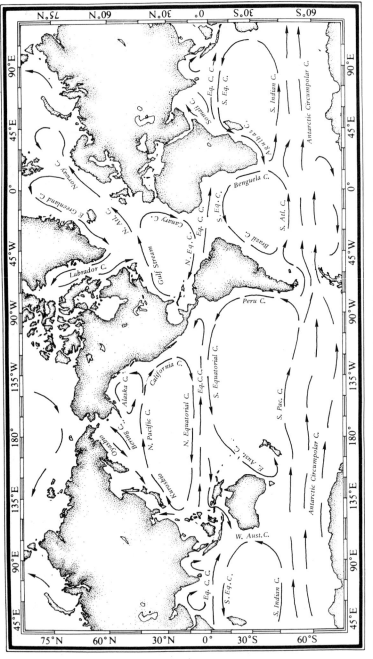

Fig. 2.17 The general surface circulation of the ocean in schematic form. The dominant features are the large, anticyclonic sub-tropical gyres in each ocean basin, the equatorial current systems, and the Antarctic Circumpolar Current.

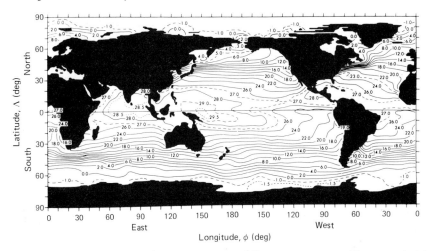

Fig. 2.18 The general surface temperature of the ocean, averaged from nearly 1.6 million observations. The warmest major ocean area is near the Far Eastern Island Continent. [From Levitus, S., *Climatological Atlas of the World Ocean* (1982).]

North Atlantic Current as it crosses that ocean in an easterly direction at latitudes of 40° to 50°N. Nearing Europe, it divides into a stronger Canary Current directed toward the south, and the Norwegian Current headed toward the north. The southern leg merges with the North Equatorial Current off Africa to return across the Atlantic near 20° to 30°N and flow on through the Caribbean to complete the circuit. This large, clockwise gyre carries a total flow of the order of 10^8 m^3 s^{-1} in its most intense regions, with the Gulf Stream portion transporting perhaps 60×10^6 m^3 s^{-1}. (The name given to the unit representing a volume transport of 10^6 m^3 s^{-1} is the *sverdrup,* abbreviated Sv, after the Norwegian oceanographer.) The average surface flow speed is 1 to 2 m s^{-1}, although in its most intense portions the Florida Current may reach speeds of 2.5 to 3.0 m s^{-1}. This gyre constitutes a very large oceanic high pressure system whose surface elevation extends perhaps a meter or more above the equipotential surface and whose western interior constitutes the Sargasso Sea, a region also known as the Atlantic Subtropical Convergence.

Every ocean except the Arctic Ocean has such a basinwide subtropical gyre, although there are important modifications brought about by geography and winds in the Indian Ocean and in the Antarctic Circumpolar current. It is apparent from the latitudes of flow cited above that the eastward ocean currents occur where the winds are westerlies, to use the meteorological convention (i.e., from the west and directed toward the east), while the regions of flow toward the west are in the easterly trade wind zones. In the Northern

Hemisphere, the gyres are clockwise; in the Southern, counterclockwise. In either event, they are termed *anticyclonic,* reflecting their high pressure character. It is the combined forces of surface wind stress and Coriolis acceleration coupled with continental boundaries that establish the anticyclonic flow patterns.

The Antarctic Circumpolar Current lies between 40° and 60°S in the zone of Southern Hemisphere westerlies and in a region of ocean uninterrupted by land masses for the entire circuit around the globe. This current, whose surface speeds are of order 0.5 to 1.5 m s^{-1} and whose volume transport is of order 200×10^6 m^3 s^{-1}, is the oceanic analog of the jet streams in the atmosphere.

The Indian Ocean, which in some ways is only half an ocean, has a current system whose northern part is dominated by the variable *monsoon* winds. During the Northern Hemisphere summer, the southwest wet monsoon drives surface currents north and east between Africa and India. In the winter, when the dry, northeast monsoon occurs, both wind and current reverse direction to flow south and west in the northern part of the basin. The South Indian Ocean, which is probably the least-well understood major basin, also possesses a subtropical gyre.

In the equatorial zone, separating the oppositely rotating major gyres is a somewhat complicated system of currents flowing mainly in the zonal direction. The westward-directed *North and South Equatorial Currents* are the equatorward arms of the gyres, but these are separated by an *Equatorial Countercurrent* and a near-surface *Equatorial Undercurrent,* both flowing east.

Branching from and merging with these components of the general circulation are numbers of secondary flows known by a variety of geographically oriented names. The major secondary systems in the north are, for example, the Norway, Greenland, and Labrador Currents in the Atlantic, and the Alaska and Oyashio Currents in the Pacific. These, plus other similar currents, comprise the secondary oceanic surface and near-surface circulation features.

From their latitudinal positions, it is clear that the transoceanic arms of the gyres are closely linked to the surface wind stress (in fact, to the curl of the stress, as will be shown in Chapter 6), with Coriolis effects playing a role. We will now attempt to describe how the wind stress couples into the water and how it responds over very large areas in a way that makes up the surface circulation patterns shown in Fig. 2.16. We will see that the subsurface circulation is also strongly conditioned by the surface flows, but that cooling and sinking of water in the polar regions (i.e., the thermohaline circulation) also contribute to the subsurface flow in important ways.

2.8   The Wind-Driven Oceanic Circulation

The response of the ocean to the wind stress defined by Eq. 2.10 is not at all in accord with intuition, for the intuitive response – motion along the direction of the forcing – is modified, or rather, significantly altered by the Coriolis force, the stratification of the ocean, and the redirection of flow due to its boundaries. The Coriolis force per unit mass, f_c (which is a pseudoforce introduced so that we may consider the rotating earth as an inertial system in which Newton's equation holds), depends on the cross-product of the earth's angular velocity, Ω, which is parallel to its spin axis, and the fluid velocity, u, as follows:

$$f_c = -2\Omega \times u . \tag{2.11}$$

Thus the surface velocity, which is primarily due to the wind stress, τ, provokes an additional force that is at right angles to both Ω and u. A more detailed consideration of the variations in the cross-product of Eq. 2.11 over the surface of the spherical earth leads to the conclusion that the *horizontal* component of f_c (which is the component active in diverting horizontal flows) varies from zero at the equator to a maximum at the poles. At all latitudes, however, this component acts at right angles to the current. In the Northern Hemisphere, the left-hand cross-product rule (which is due to the minus sign in Eq. 2.11) states that the Coriolis force will act to the *right of the flow,* and as time goes on, will cause a parcel of moving fluid to orbit in clockwise circles, in the absence of other forces. In the Southern Hemisphere, the sense of rotation is opposite.

These circular motions are called *inertial oscillations* and have a period of the order of one day at mid-latitudes; however, the local *inertial periods* range from 12 hours at the poles to infinity at the equator. This type of motion, with some modification, is often observed in the ocean following impulsive wind events; its characteristics are discussed more completely in Chapters 3 and 6.

We must also consider the effects of friction on the flow, because it is always present. The damping that friction causes slowly reduces the amplitude of the inertial oscillations and in addition, allows the parcel of fluid to take up a component of motion along the direction of the stress, so that the net, long-term surface flow is to the right of the wind stress (in the Northern Hemisphere) at some angle, typically 10° to 45°. Such flow is termed *Ekman wind drift.* Now the near-surface Ekman flow in turn exerts a frictional stress on the layer of fluid immediately below, which also responds by moving to the right of the surface flow, but at a somewhat reduced velocity because of friction. This veering effect continues to migrate downward,

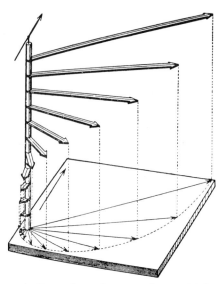

Fig. 2.19 Rotation and attenuation of near-surface velocity vector with depth through the surface Ekman layer of the ocean. Wind direction is indicated by topmost vane. [From Ekman, V. W., *Ark. f. Mat., Astron. och Fysik* (1905).]

so that the local velocity vector of the oceanic current rotates continuously to the right with depth, while decaying exponentially with a scale of order 10 to 20 m. Figure 2.19 illustrates Ekman's original diagram of the behavior of the current velocity vector, a representation called a *hodograph*. If the wind drift current is integrated over all depths, it is found that the *net transport* of water is at 90° to the direction of wind stress, and is to the right of it in the Northern Hemisphere and to the left in the Southern Hemisphere. This volume flow rate per unit horizontal distance is termed *Ekman transport*, with units of square meters per second. On the equatorward flanks of the easterly trade winds, the right-angle forcing moves surface water poleward away from the equatorial region and results in cold, subsurface water flowing upward to replace the missing surface water, a process called *equatorial upwelling*.

Ekman transport represents the first step in the formation of the major subtropical and subpolar oceanic gyres as well as the equatorial current systems. In a typical ocean basin (Fig. 2.20), both the southeasterly trade winds and the mid-latitude westerlies force surface water toward the gyre interior because of Ekman transport. The surface convergence causes an accumulation there of warmer, lighter water, and results in a small elevation (of order 1 m or less) of the surface above the equipotential. It also causes a much larger deepening of the thermocline, which is found down to depths of several

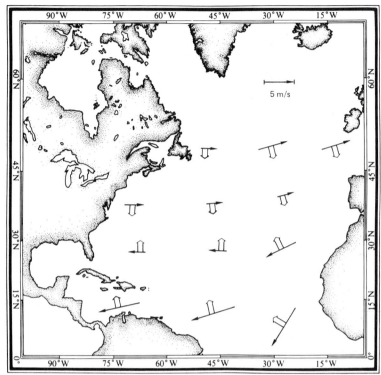

Fig. 2.20 Convergence of surface water toward the interior of the North Atlantic Gyre under the influence of Ekman transport. Both trade winds and westerlies contribute to accumulation and deepening of the warm water pool.

hundred meters in the western Sargasso Sea, for example. Thus Ekman inflow into the subtropic convergence region (i.e., the interior of the gyre) results in a *downwelling* of the surface waters and a decrease in their angular momentum and *vorticity* by way of their change in height. Vorticity is a measure of the rotation rate of a small fluid element, and is equal to twice the angular velocity of the element. It is termed *cyclonic* if it has the same sense of rotation as the earth, and *anticyclonic* if opposite. Since rotating fluids must move so as to maintain constant their total angular momentum per unit mass and per unit area, the loss of vorticity caused by sinking must be made up by an increase in vorticity caused by the earth's rotation. As will be shown in Chapter 6, this may be accomplished by the movement of an extensive mass of the interior, deeper fluid toward the equator. Thus in the Northern Hemisphere gyres, water at depth moves southward, a motion called *Sverdrup interior flow*.

The small surface elevation is termed *setup*, and has associated with it a hydraulic head and a horizontal pressure gradient, $-\nabla_h p$, that is opposed by the horizontal component of Coriolis force, $\rho(2\Omega \times \mathbf{u})_h$. The currents resulting from this balance of forces are called *geostrophic* (earth-turning) flows, and in the absence of time variations and friction, move approximately along surfaces of constant elevation. Thus the geostrophic equation is

$$\nabla_h p = -\rho(2\Omega \times \mathbf{u})_h , \qquad (2.12)$$

where ρ is the density of seawater and h denotes the horizontal component. Geostrophic balance is not confined to surface currents, but may exist at all depths. This equation for balanced flow forms an important tool in the analytical tool kit of the oceanographer.

As the geostrophically balanced deeper waters vacate the northern regions of the gyre as Sverdrup flow, they must be replaced by a return flow to the north in such a manner as to ensure continuity of the flow and local conservation of vorticity, i.e., angular momentum. Additionally, the northward-flowing water will become concentrated along the western boundary of the basin (the eastern edge of the continent) because of reasons of the *overall* vorticity balance of the gyre; since, in the longer term, the gyre is neither increasing nor decreasing its rotation rate, approximate equilibrium must exist between sources and sinks of vorticity on each side of the ocean basin. This circumstance, which is somewhat difficult to grasp conceptually without a more extensive discourse on its causes and effects, results in the formation of intense, narrow current systems off the east coasts of continents everywhere in the world, which are called, appropriately enough, *western boundary currents*. The Gulf Stream is the best-known example of such a vorticity-balancing flow. Along with the concentration of the flow, there is a concomitant large increase in fluid friction along the western, inshore side of the Stream as a result of flow instability and eddy production. This "eddy viscosity" is an important source of positive, *cyclonic* vorticity in the overall balance. In a sense, a western boundary current in the Northern Hemisphere will produce vorticity by "rubbing its left shoulder" against the continental shelf; in the Southern Hemisphere it is the right shoulder that is rubbed. At some point along the continental margin, the boundary currents leave the land mass and turn seaward to complete the flow circuit by rejoining the eastward-flowing wind-driven current. Thus the overall gyre has the appearance of the surface flow shown in Fig. 2.21, which gives surface streamlines of an idealized subtropical gyre. It is estimated that the time required for a typical water parcel to circulate around the entire gyre is of order five years. It should be noted that the subsurface flow is generally quite different from this (see below), and that the overall circulation is a complicated function of all three spatial coordinates. Additionally, there are major

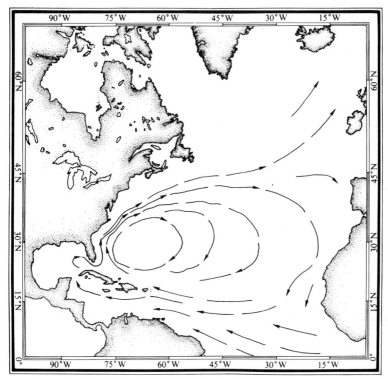

Fig. 2.21 Surface streamlines of an idealized subtropical gyre. Intensification of flow along the western boundary is due to the overall vorticity balance of the gyre.

time variations in the currents caused by wind stress variations and large-scale flow instabilities.

2.9 Pressure Forcing

Another mechanical force that also drives the upper ocean is *pressure forcing*, also called the *inverse barometer effect*. Atmospheric high and low pressure systems exert vertical forces on the sea that result in elevation changes of the sea surface, and associated flows. A high pressure system will depress the sea surface and a low will elevate it, with the changes in surface height being surprisingly large. For example, in Fig. 2.13, the difference between the Azores High with an average pressure of about 1020 mbar, and the Icelandic Low with an average of about 1005 mbar, is approximately 15 mbar. From the hydrostatic equation (Eq. 5.25) it is simple to derive the rule of thumb that the change in ocean surface elevation, $\delta\xi$ (in centimeters), is numerically the same as but opposite in sign to the change in atmospheric pressure, δp_a (in millibars):

$$\delta \xi \text{ (cm)} = -\delta p_a \text{ (mbar)} . \tag{2.13}$$

Thus sea level setup near Iceland is of order 15 cm higher than near the Azores, compared with the constant pressure equilibrium. For example, corrections must be applied to bottom-mounted pressure gauges that measure the depth-integrated density of the overhead water column, in order to remove the effects of atmospheric pressure fluctuations. Such gauges can measure mean water elevations with a precision greater than 1 cm, with surface waves being effectively filtered out if the water is deeper than a few wavelengths.

2.10   The Thermohaline Circulation

Poleward of perhaps 30° to 35°N or S, the wind-driven surface currents begin to give up rather than to receive net thermal energy from the atmosphere via the processes of evaporation, radiation, and conduction, thereby warming the air and cooling the surface layers of the water. By the time these currents have made their way to the polar regions, they have lost much heat and have greatly reduced their temperature and have thereby increased their density. In the regions near the Norwegian and Greenland Seas in the Northern Hemisphere, and again in the Antarctic Circumpolar Current near that continent's borders, the surface temperatures have dropped to the vicinity of 0.5° to $-2.0°C$ (cf. Fig. 2.18). The densities have increased to values where the surface waters become heavier than the subsurface waters, and sinking and convective overturning occur. In the Norwegian Sea, this sinking water flows slowly south along the bottom, crossing over the relatively shallow underwater sills connecting Greenland, Iceland, and the British Isles, where it undergoes a certain amount of mixing with warmer water above and thereby raises its temperature slightly. From there it continues to flow south along the bottom in a very indistinct, density-driven circulation that creeps at velocities too slow to be directly measured. This water mass is called North Atlantic Deep Water (NADW) and is shown schematically in Fig. 2.22 as extending as far south as 45° to 50°S. Such flows are termed *thermohaline circulations* and can usually be identified by the distributions of certain water type properties, particularly by their dissolved oxygen (DO) and salinity–temperature–depth (STD) correlations. The bottom temperature of NADW is near 2°C and salinity 34.95 psu (*practical salinity units*, formerly parts per thousand, or ‰).

In the Antarctic near the sea ice boundaries, especially in the Weddell Sea, a similar sinking process occurs, now aided and abetted by the formation of the ice, which rejects much salt upon change of phase from liquid to crystalline and which leaves behind water that is more saline, and thus more dense because of both decreased temperature and increased salt content. This water also sinks to the bottom to become Antarctic Bottom Water (AABW). While not initially as dense as NADW because of its lower surface salinity, it undergoes no appreciable mixing across sills as does the North Atlantic Deep Water, and thus it maintains its low temperature and high density. It apparently underflows the NADW and extends as far north as the equator in the Atlantic (see Fig. 2.22). In the Pacific, the analogous northward flow may be traced all the way to 40° to 50°N.

A third, somewhat different process contributes to another deep water mass globally, the Antarctic Intermediate Water (AAIW). In a region near the edge of the Antarctic Current termed the *Antarctic Convergence*, the surface waters flow together and are forced downward due to both increased density and the requirement for continuity of flow, there to migrate north at intermediate depths near 1 km and overlie the other water masses. This region is a source of intermediate water in all three major oceans.

The picture of the thermohaline circulation as a deep, broad, and weak flow extending over much of the oceans is probably not altogether correct, since the dynamics of these flows must obey the same laws as their surface counterparts, although with quite different magnitudes. Analytical and numerical models suggest a distinctly altered picture from Fig. 2.22; in particular, the deep flow away from the polar regions must occur as a western boundary current in each basin, as suggested in Fig. 2.23. The return flow is thought to be eastward and poleward everywhere and is much more diffuse, but in theory cannot cross the equator. Also a large eastward flow at depth accompanies the surface portion of the Antarctic Circumpolar Current. The models suggest a total deep western boundary transport of about 10 Sv. However, the picture is relatively uncertain and the subject remains an active research area.

One other less important but nevertheless interesting thermohaline flow occurs in the Atlantic at the mouth of the Mediterranean. This sea, which because of low rainfall, small riverine input, and limited communication with the open ocean, has a high salinity—about 39.1 psu in its eastern portion. The dense, salty water makes its way out of the Strait of Gibraltar at about the depth of the sill (450 to 500 m) and protrudes into the Atlantic at depths near 1000 to 1200 m (Fig. 2.24). It can be identified across the entire width of the Atlantic via its properties (see Fig. 2.25) and is denoted in Fig. 2.22 as "MED." The high salinity water gradually spreads and mixes throughout the basin, and is thought to be the reason that the North Atlantic is,

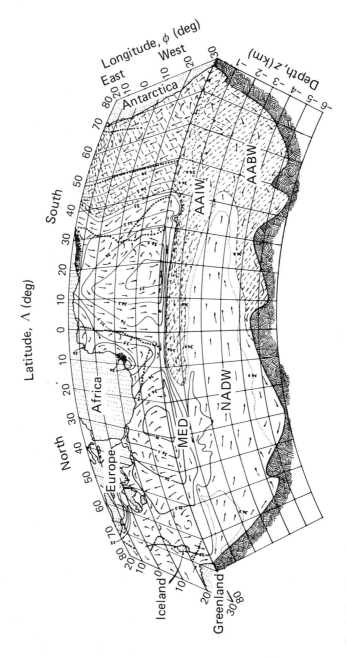

Fig. 2.22 Schematic cross section of surface and subsurface currents in the North and South Atlantic. [Adapted from Wüst, G., *Kieler Meeresforschungen* (1950).]

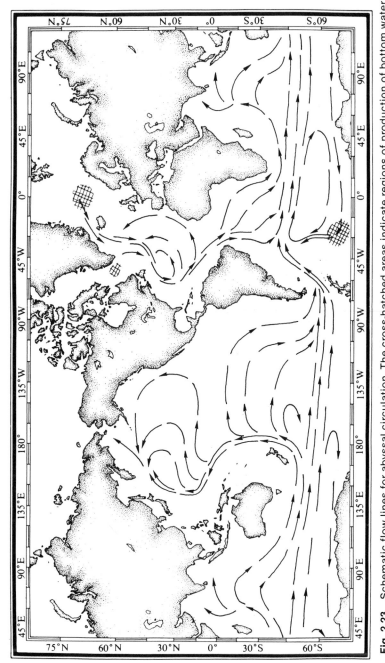

Fig. 2.23 Schematic flow lines for abyssal circulation. The cross-hatched areas indicate regions of production of bottom water. [Adapted from Stommel, H., *Deep Sea Research* (1958).]

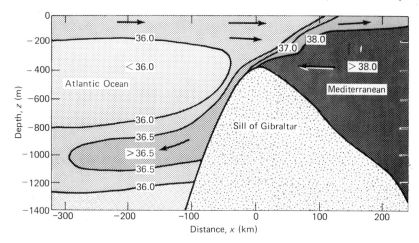

Fig. 2.24 Vertical cross section of the Atlantic and the Mediterranean, illustrating the high salinity tongue of water flowing over the sill at the Strait of Gibraltar and sinking to depths of order 1000 m. [Adapted from Schott, G., *Geographie des Atlantischen Ozeans* (1942).]

on the average, saltier than the other oceans: a mean surface salinity of approximately 36.5 psu, compared with perhaps 35.0 psu for the Pacific and Indian Oceans.

As these multiple interleaving flows make their way on a slow but ponderous scale, they gradually mix to form a composite water mass constituting perhaps 40% of the volume of the sea.

The sinking motion in the limited areas of the polar regions must be compensated by rising motions elsewhere, and it is currently thought that these may take place over relatively large areas of the ocean. The cold, deep water is very slowly mixed upward into the warmer, shallow water, thereby in considerable degree establishing the permanent thermocline. The upward velocities are not at all measurable, but the numerical models give estimates of vertical velocities of perhaps 10^{-7} m s^{-1} (about 3 m yr^{-1}). As a consequence, the residence time of deep waters—i.e., the time for them to sink, flow, and be mixed upward again—may be as great as 1000 years.

This thermohaline circulation also explains the oceanic thermocline in a qualitative way. If there were no deep water formation in cold regions, the ocean would gradually warm up throughout its vertical extent due to the heat sources cited earlier and, under the influence of mixing and diffusion, become more or less isothermal in depth. The presence of cold waters with clearly identifiable properties attests to the continued renewal of the deep via sinking in cold regions, offset by widespread upward flow elsewhere at a rate

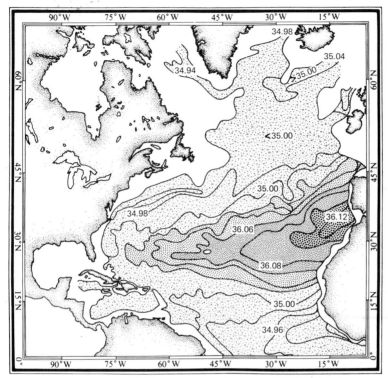

Fig. 2.25 Spreading of high-salinity Mediterranean water across the Atlantic at intermediate depths. [Adapted from Worthington, L. V., and W. R. Wright, *North Atlantic Ocean Atlas* (1970).]

great enough to balance the downward diffusion of heat and thereby maintain the vertical temperature gradients.

This sinking of cold water also aids in the removal of heat from the tropics by providing a conduit for surface waters, their thermal free energy now exhausted to their surroundings, to return to the subsurface system and thus complete the cycle started in the equatorial regions under the forcing of winds that the currents themselves have played a major role in establishing.

2.11 The Sea Floor and Its Dynamics

Having in this discussion reached to the oceanic deep and its circulation, we next encounter the sea floor, which is obviously the containment vessel for the oceans, among many other things. While from the viewpoint of the hydrodynamicist, the ocean bottom may simply be a surface over which flows

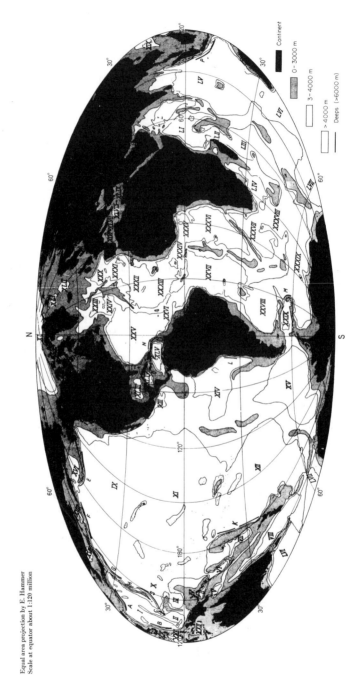

Equal area projection by E. Hammer
Scale at equator about 1:120 million

Fig. 2.26 The gross depth structure of the world's oceans, showing three depth ranges and the axes of deep sea trenches. [From Dietrich, G., *General Oceanography* (1980); originally due to Wüst (1940).]

traverse and on which the fluid boundary conditions must be satisfied, to the solid earth geophysicist or marine geologist, the sea floor is an immensely rich and dynamic region whose study has yielded one of the major intellectual advances in geophysics of the century. These advances are summarized in the topics of *sea floor spreading, plate tectonics,* and *continental drift.* Before launching into their study, however, a short diversion into a view of the morphology of the ocean floor is in order.

Figure 2.26 is a bathymetric map of the world's oceans showing broad depth categories. A study of the oceanic relief shows several distinct regimes of the sea floor. The *continental shelves* are gently sloping seaward projections of the continents themselves, and extend into the oceans for distances that range from tens to a few hundreds of kilometers; the shelf is conventionally considered to end at a depth of approximately 200 m, which contour forms the *shelf break.* Still further at sea, the *shelf slope* or *continental margin* exists, often cut through by submarine canyons and drowned river beds; the slope is the site of *turbidity currents,* mixtures of sediment and water that occasionally tumble down the slope, sometimes with enough force to sever submarine communication cables. The margin generally terminates at the *abyssal deep,* the flat, sediment-filled plain that occupies much of the sea floor. The accumulation of the organic detritus that constantly rains down onto the sea floor from above at rates of millimeters per year has filled the abyss with sediments whose depths may approach several kilometers. Located out in the breadth of the sea floor but sometimes intersecting the continents and islands are the more shallow *mid-ocean ridges,* which are regions of mountains, volcanoes, seismic activity, and bottom heat flow, and which are also the sites of new sea floor production. The ridges generally lie at depths near 1 to 2 km, but actually broach the surface in a few places such as Iceland. Scattered at various locations throughout the breadth of the sea are volcanic islands, atolls, and seamounts produced by *vulcanism,* wherein heated material from the interior of the earth has been transported to the surface by internal forces. Closer to the continents are islands and undersea structures that are not volcanic in their origin but rather are *continental fragments* that have broken off from the main land masses. Also near the continents are the *deep ocean* or *submarine trenches,* regions where the ocean bottom reaches its greatest depths—the deepest being more than 11 km— under the influence of subduction forces occurring between the sea floor and the adjacent continents. The trenches are the sites of numerous deep earthquakes that extend to perhaps 700 km beneath the surface.

In order to understand the theoretical origin of the motions of the solid upper layers of the earth implied by the submarine features of Fig. 2.26, it is necessary to know something of the hypothesized interior distribution and state of matter of the planet. Figure 2.27 represents a vertical wedge of the interior showing divisions into solid and liquid cores, a solid but somewhat plastic mantle, and a crustal region consisting of several distinct layers whose thicknesses in the figure have been somewhat exaggerated. Pressures within the interior of the earth are largely hydrostatic and the approach of the crust

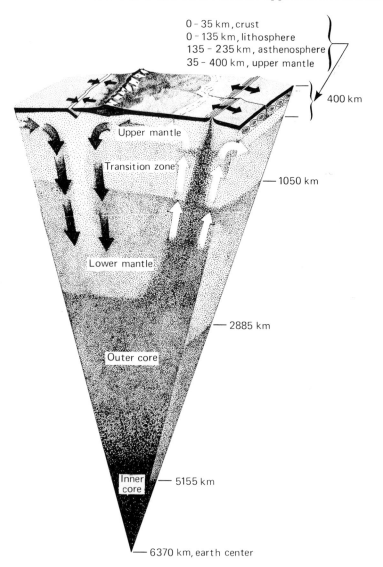

0 – 35 km, crust
0 – 135 km, lithosphere
135 – 235 km, asthenosphere
35 – 400 km, upper mantle

400 km

Upper mantle

Transition zone

1050 km

Lower mantle

2885 km

Outer core

Inner core — 5155 km

— 6370 km, earth center

Fig. 2.27 A wedge-shaped sector of the earth's interior, illustrating solid and liquid cores and near-surface crustal differentiations. [Adapted from *Geopotential Research Mission*, National Aeronautics and Space Administration (1984).]

to hydrostatic equilibrium is termed *isostatic adjustment*; it is not perfect, but results in gravity anomalies and dynamical motions. The shallow layer termed the *lithosphere* can be considered as supporting both continental rock masses and the ocean floor, all viewed as floating on the upper mantle. Although the lithosphere cannot support vertical stresses well, it nevertheless transmits horizontal stresses readily, somewhat as a sheet of thin ice might.

Radioactive decay within the earth is thought to provide the heat necessary to keep much of the interior plastic or liquid, as well as the potential and kinetic energy needed to drive the very slow but extensive motions within the liquid core and mantle. Enormous hydrostatic pressure is thought to have solidified the inner core, in spite of its high temperature. These motions, which must also obey the laws of fluid dynamics, are exceedingly viscous (in fluid terms), but are also conditioned by rotational and convective forces, as are the oceans and atmosphere—although the time scales for interior motions are millions of times longer than for the surface fluids. The existence of cellular convection currents within the mantle has been hypothesized, of somewhat the same genre as the combined Walker–Hadley cells in the atmosphere, and it is thought that these convection cells bring heat and molten material from the deep interior upward to the crustal complex, and exert horizontal and vertical forces on the athenosphere and lithosphere, as suggested in Fig. 2.28. The molten interior material broaches the crust along the *rift valleys* centered on the mid-ocean ridges, and in *convective plumes*, or "hot spots," where undersea mountain building, volcanoes, and shallow

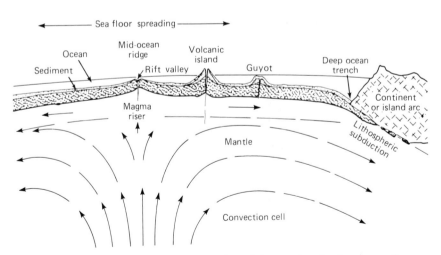

Fig. 2.28 Schematic vertical cross section of the crust under the ocean. Convection cells advect the lithosphere away from mid-ocean ridges as on a conveyor belt. The crust is consumed under the continental edges beneath deep ocean trenches.

earthquakes (depths of 1 to 3 km) occur. New sea floor is created at the mid-ocean ridges and is slowly carried away from the regions of formation, called *spreading centers*, by the motion of the lithosphere, somewhat as on a conveyor belt. Speeds range from a very few to perhaps 10 cm yr^{-1} (10 m per century). This is the process of sea floor spreading, and takes place along ridges such as the Mid-Atlantic Ridge and the East Pacific Rise, for example (see Fig. 2.26). These ridges comprise the oceanic legs of the major crustal *fault zones* of the earth; together with their landward extensions (such as the San Andreas Fault and the East African Rift Valley), they physically split the surface of the globe into perhaps 14 major continental plates (Fig. 2.29).

On a spherical earth, the motion of the rigid lithospheric plates away from the spreading centers must be considered, by Euler's theorem on the general displacement of a rigid body with one point fixed, to be a rotation about a fixed axis called the *Euler pole*, as suggested in Fig. 2.30. The Euler pole does not generally coincide with the spin axis. Plate elements close to the pole move with lower azimuthal velocities than do those closer to the Euler equator, varying as the sine of the colatitude, θ. Under these differential velocities, the spreading centers are thought to split zonally into a large number of *ridge–ridge transform faults*, which then become secondary sites of mountain building, volcanoes, and earthquakes.

During the process of sea floor spreading, undersea volcanoes and volcanic islands are also carried away from the spreading centers or hot spots by the motion of the lithosphere, cooling in the meantime. Over a period measured in a few tens of millions of years, the newly formed sea floor and volcanoes are conveyed away from their sources over convective plumes, usually to become inactive, then erode, then accumulate sediments from the rain of ocean biogenic and terrigenous (land-originated) material, and sink into the underlying mantle. This progressive sinking of the older bottom is due to the increased weight from accumulated benthic material, the cooling of the extruded matter, and the isostatic adjustment process. Often the volcanoes in the tropics that slowly sink beneath the sea surface along with the sea floor become *atolls* as corals build around their lips, or become flat-topped seamounts called *guyots*.

If the sub-crustal heat source happens to be a localized hot spot fixed in the mantle, the plate motion past the spot may result in a long series of volcanic islands and sea mounts; such is the case of the combined Hawaiian–Midway–Emperor Seamount chain extending northwest from the big island of Hawaii. This chain has formed over a period of approximately 70 million years as the Pacific Plate drifted away from the East Pacific Rise at a mean velocity of some 6 cm yr^{-1}. It is clear that the island building has been episodic and quasi-random, but has taken place over this entire time, continuing on to the present. This same plate motion is expected to carry Baja

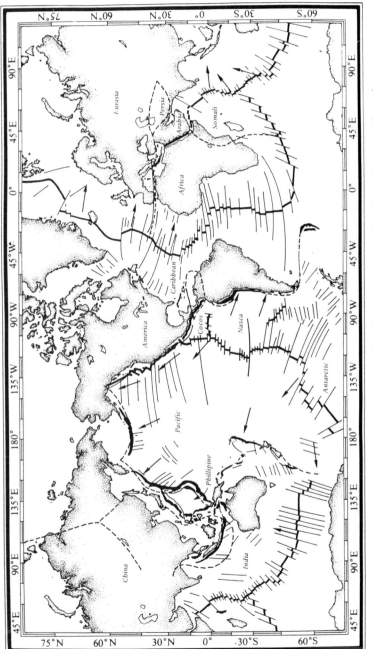

Fig. 2.29 Major crustal plates and directions of motion. Plate boundaries are defined by mid-ocean ridges, fracture zones, and subduction zones. [Adapted from Morgan, W. J., *Studies in Earth and Space Sciences* (1972).]

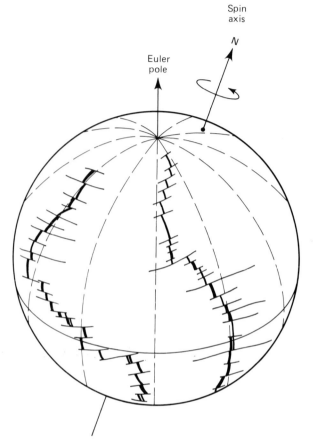

Fig. 2.30   Motion of a rigid lithospheric plate about the Euler pole. The variation in azimuthal velocity is thought to create ridge–ridge transform faults.

California and Southern California northwest at a rate such that in approximately 50 million years, that region will be located off the Pacific Northwest (Fig. 2.31).

The linear regions where oceanic plates thrust up against or otherwise intersect continental plates are areas of active destruction of the lithosphere, deep earthquakes, and mountain building (Fig. 2.28). Such a region extends around a major portion of the entire Pacific from the tip of South America, through North America, Northeastern Asia, Japan, on to the Marianas and Tonga Trenches, and past New Zealand; it is sometimes referred to as the Pacific "Rim of Fire." *Seismic sea waves* or *tsunami* are produced by earthquakes near these trenches and often propagate across the entire Pacific.

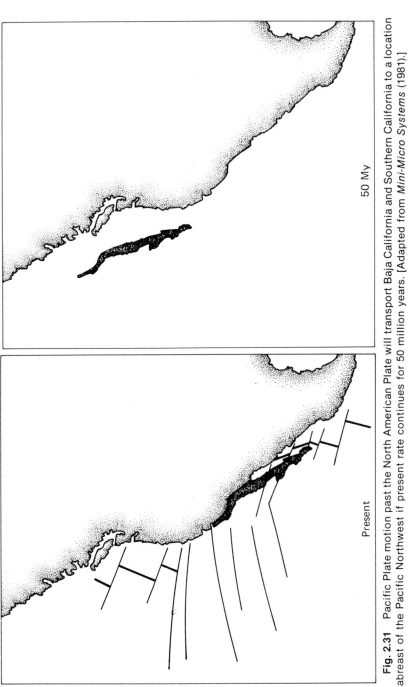

Present

50 My

Fig. 2.31 Pacific Plate motion past the North American Plate will transport Baja California and Southern California to a location abreast of the Pacific Northwest if present rate continues for 50 million years. [Adapted from *Mini-Micro Systems* (1981).]

The collision of the sea floor with the lighter continental land mass generally results in underthrusting and consumption of the lithosphere, a process termed *subduction*, and is accompanied by the return of oceanic sediments to the continental roots and the up-thrusting of continental material, often producing coastal mountain ranges thereby. The Peru–Chile Trench and the Andes Mountains are examples of this type of tectonic activity. Plate collision and *orogeny* (mountain building) over geological time also explain why erosion and weathering have not long since worn down the earth's surface to a smooth, water-covered sphere, and also explain the appearance of oceanic fossils high atop mountains.

In other regions where the oceanic plate moves more or less parallel to the continental plate, a *strike–slip fault* is said to occur, with the San Andreas Fault being the best known example. The relative motion here is usually episodic, with long periods during which no slippage occurs but during which considerable strain is accumulated along the fault. The sudden release of this strain via differential motion between segments of the fault results in earthquakes and sharp elevation changes. In some regions, the friction produced may result in remelting of subsurface material, with the associated faulting allowing magma to flow to the surface. The chain of volcanoes in the Cascade Mountains have their origin in this type of motion.

Figure 2.29 is a map that summarizes the principal plate tectonic features and their manifestations on a global scale. Over geological time, the plate tectonic process appears capable of transporting entire continental and oceanic masses about the earth, a mechanism that now gives credence to the earlier continental drift theory advanced by Wegener in 1915, based on the geometric fit of the Atlantic coastlines of South America and Africa; however, Wegener's hypothesis was rejected by his contemporaries for lack of any known process by which the continents could move. Since then, by linking his idea to the concepts of sea floor spreading and plate tectonics, geologists and geophysicists have been able to legitimize the hypothesis and even to reconstruct the probable distribution of the crustal plates in the past, back to approximately 200 million years before present (Mybp). Figure 2.32 shows such a reconstruction, with South America and Africa fitting together in the Southern Hemisphere along the lines of the continental shelf break, and North America, Africa, and Eurasia nestling together in the Northern Hemisphere. Near the south pole, one sees that Antarctica, India, and Australia may also have been conjoined with South America and Africa to form a supercontinent named *Pangaea* (all earth), with a superocean called *Panthalassa* (all ocean) that would have occupied some 60% of the earth's surface. The climate interior to the continents was probably exceedingly dry and the oceanic circulation quite different from that of today. As time went on, under the influence of the conjectured mantle convective cells, rifts developed that

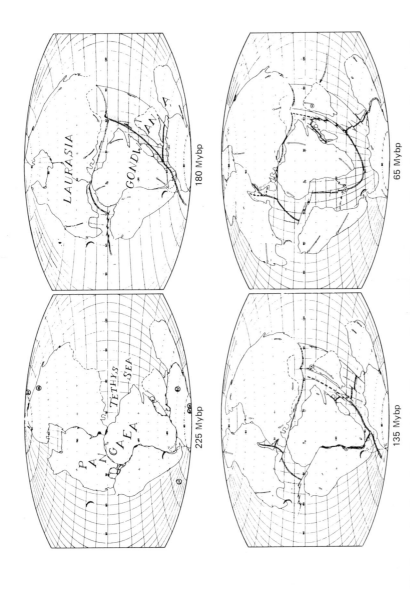

Fig. 2.32 Reconstruction of continental positions over the last 200 million years. The proto-continent is named *Pangaea* and the ocean *Panthalassa*. [From Dietz, R., and J. Holden, *J. Geophys. Res.* (1970).]

split the continents along lines that are still traceable today, and the subsequent spreading of the sea floor carried them apart at speeds up to the order of 0.1 km per millennium. By 65 Mybp, the oceans as we know them had more or less opened up, while Panthalassa had shrunk to become the proto-Pacific. To some extent, the changing oceanic circulation patterns of the past can be discerned in the deposition and type of sediments laid down on the sea floor during that time.

The process of plate motion continues, and in some areas its drift rates have been directly measured over a decade or more by geodetic and satellite means. Projections of the future distributions of land masses have been made based on extrapolations of the present motions. While these projections are no more certain than the reconstructions of the past, what is certain is that in another 100 million years, the surface of the earth and the configuration of the oceans will be quite different from what they are today.

Bibliography

General References

Defant, A., *Physical Oceanography*, Vols. I and II. Pergamon Press, New York, N.Y. (1961).

Dietrich, G., *General Oceanography: An Introduction*, 2nd ed., Interscience, New York, N.Y. (1980).

Emiliani, C., Ed., *The Sea,* Vol. 7, *The Oceanic Lithosphere*, John Wiley & Sons, Inc., New York, N.Y. (1981).

Fleagle, R. C., and J. A. Businger, *An Introduction to Atmospheric Physics*, 2nd ed., Academic Press, New York, N.Y. (1982).

Gill, A. E., *Atmosphere-Ocean Dynamics*, Academic Press, New York, N.Y. (1982).

The Global Climate System, World Meteorological Organization, Geneva, Switzerland (1985).

Goldberg, E. D., Ed., *The Sea,* Vol. 5, *Marine Chemistry*, John Wiley & Sons, New York, N.Y. (1974).

Goldberg, E. D., I. N. McCave, J. J. O'Brien, and J. H. Steele, Eds., *The Sea,* Vol. 6, *Marine Modeling*, John Wiley & Sons, New York, N.Y. (1977).

Gross, M. G., *Oceanography: A View of the Earth*, 3rd ed., Prentice-Hall, Inc., Englewood Cliffs, N.J. (1982).

Hill, M. N., Ed., *The Sea:* Vol. 1, *Physical Oceanography,* 2nd ed. (1982). Vol. 2, *Composition of Seawater; Comparative and Descriptive Oceanography* (1963). Vol. 3, *The Earth Beneath the Sea; History* (1981), John Wiley & Sons, New York, N.Y.

Holton, J. R., *An Introduction to Dynamic Meteorology*, 2nd ed., Academic Press, New York, N.Y. (1979).

Kamenkovich, V. M., and A. S. Monin, Eds., *Ocean Physics*, Vols. 1 & 2, NAUKA Publishing House, Moscow, U.S.S.R. (1978) (in Russian).

Knauss, J. A., *Introduction to Physical Oceanography*, Prentice-Hall, Englewood Cliffs, N.J. (1978).

Kraus, E. B., *Atmosphere–Ocean Interaction*, Oxford University Press, London, England (1972).

Maxwell, A. E., Ed., *The Sea*, Vol. 4, *New Concepts of Sea Floor Evolution,* Parts 1 and 2, John Wiley & Sons, New York, N.Y. (1970).

Menard, H. W., *Geology, Resources, and Society*, W. H. Freeman & Co., San Francisco, Calif. (1974).

Menard, H. W., "Introduction," *Ocean Science,* W. H. Freeman & Co., San Francisco, Calif. (1977).

Officer, C. B., *Introduction to Theoretical Geophysics,* Springer-Verlag, New York, N.Y. (1974).

Perry, A. H., and J. M. Walker, *The Ocean–Atmosphere System*, Longman, London, England (1977).

Pickard, G. L., and W. J. Emery, *Descriptive Physical Oceanography: An Introduction,* 4th ed., Pergamon Press, Oxford, England (1982).

Pickard, G. L., and S. Pond, *Introductory Dynamic Oceanography,* 2nd ed., Pergamon Press, Oxford, England (1978).

Schott, G., *Geographie des Atlantischen Ozeans*, 3rd ed., C. Boysen, Hamburg, Germany (1942).

Shepard, F. P., *Submarine Geology,* 3rd ed., Harper and Row, New York, N.Y. (1973).

Smith, D. G., Ed., *The Cambridge Encyclopedia of Earth Sciences*, Crown Publishers, Inc./Cambridge University Press, New York, N.Y. (1981).

Sverdrup, H. O., M. W. Johnson, and R. H. Fleming, *The Oceans: Their Physics, Chemistry, and General Biology*, Prentice-Hall, Inc., Englewood Cliffs, N.J. (1963).

Wallace, J. M., and P. V. Hobbs, *Atmospheric Science: An Introductory Survey*, Academic Press, New York, N.Y. (1977).

Warren, B. A., and C. Wunsch, *Evolution of Physical Oceanography*, MIT Press, Boston, Mass. (1981).

Weyl, P. K., *Oceanography: An Introduction to the Marine Environment*, John Wiley & Sons, New York, N.Y. (1970).

Atlas Guides and Lexicons

Stommel, H., and M. Fieux, *Oceanographic Atlases: A Guide to their Geographic Coverage and Contents,* Woods Hole Press, Woods Hole, Mass. (1978).

Glossary of Oceanographic Terms, U.S. Naval Oceanographic Office, U.S. Government Printing Office, Washington, D.C. (1966).

Atlases

Oceanographic Atlas of the North Atlantic Ocean, U.S. Naval Oceanographic Office, Washington, D.C.:
  Section I, *Tides and Currents* (1969)
  Section II, *Physical Properties* (1967)
  Section III, *Ice* (1968)
  Section IV, *Sea and Swell* (1963)
  Section V, *Marine Geology* (1970)
  Section VI, *Sound Velocity* (1974)

The Times Atlas of the Oceans, Van Nostrand, New York, N.Y. (1983).
The Rand McNally Atlas of the Oceans, Rand McNally, New York, N.Y. (1977).
Fuglister, F. C., *Atlantic Ocean Atlas*, Woods Hole Oceanographic Institution, Woods Hole, Mass. (1960).
Legler, D. M., and J. J. O'Brien, *Atlas of Tropical Pacific Wind Stress Climatology 1971–1980*, Florida State Univ., Tallahassee, Fla. (1984).
Levitus, S., *Climatological Atlas of the World Ocean*, National Oceanic and Atmospheric Administration, Washington, D.C. (1982).
Oort, A. H., *Global Atmospheric Circulation Statistics*, National Oceanic and Atmospheric Administration, Washington, D.C. (1983).
Worthington, L. V., and W. R. Wright, *North Atlantic Ocean Atlas of Potential Temperature and Salinity in the Deep Water*, Vol. 2, Woods Hole Oceanographic Institution Atlas Series (1970).

Journal Articles and Reports

Dietz, R., and J. Holden, "Reconstruction of Pangaea: Breakup and Dispersion of Continents, Permian to Present," *J. Geophys. Res.,* Vol. 75, p. 4939 (1970).
Ekman, V. W., "On the Influence of the Earth's Rotation on Ocean Currents," *Ark. f. Mat., Astron. och Fysik,* Vol. 2, p. 1 (1905).
Geopotential Research Mission, National Aeronautics and Space Administration, Washington, D.C. (1984).
Milankovitch, M., "Mathematische Klimalehre und astronomische Theorie der Klimaschwankungen," *Handbuch der Klimatologie*, Bd. 1, Teil A, p. 1, N. Köppen and R. Geiger, Eds., Borntraeger, Berlin, Germany (1930).
Morgan, W. J., "Plate Motions and Deep Mantle Convection," in *Studies in Earth and Space Sciences*, p. 7, R. Shagam, Ed., Geological Society of America, New York, N.Y., p. 7 (1972).
Piexóto, J. P., and A. H. Oort, "Physics of Climate," *Rev. Mod. Phys.,* Vol. 56, p. 365 (1984).
Stewart, R. W., "The Atmosphere and the Ocean," in *Ocean Science*, W. H. Freeman & Co., San Francisco, Calif. (1977).
Stommel, H., "The Abyssal Circulation," in *Deep Sea Res.,* Vol. 5, p. 80 (1958).
Vonder Haar, T. H., and A. H. Oort, "New Estimate of Annual Poleward Energy Transport by Northern Hemisphere Oceans," *J. Phys. Ocean.,* Vol. 3, p. 169 (1973).
Wüst, G., "Blockdiagramme der Atlantischen Zirkulation auf Grund der 'Meteor-' Ergebnisse," *Kieler Meeresforschungen*, Vol. 7, p. 24 (1950).

Chapter Three

Hydrodynamic Equations of the Sea

Given for one instant an intelligence which could compre-
hend all the forces by which nature is animated,...to it
nothing would be uncertain, and the future as the past
would be present to its eyes.

Laplace, *Oeuvres*

So what is the incomprehensible secret force driving
me,...carried from sea to sea through your entire expanse?

Gogol, *Dead Souls*

3.1 Introduction

In the previous chapters we have set the stage for a more quantitative description of the myriad motions of the sea. The task now at hand is to derive the equations governing those motions, along with appropriate initial and boundary conditions. We shall find in the course of this program that the mechanical equations alone contain more dependent variables than equations, and that they must be rounded out with thermodynamic statements. In addition, the concept of eddy diffusion coefficients will be introduced as a means of mathematically modeling the always difficult problems of turbulence, friction, and dissipation. For these reasons, as well as others, we will be led inexorably to the subject of thermodynamics as a companion topic to hydrodynamics; indeed, the need for introducing thermodynamics was already clear in our earlier discussions of oceanic heat absorption and transport.

Without much doubt, the hydrodynamic equations were originally derived for smallish samples of fluids, not the vast expanse of the sea, and it is something of a practice of faith to apply them to the motions of a fluid that covers 71% of the globe and then to expect predictive capability in the dynamics of the ocean. However, the physics of the small has been notoriously successful in predicting the characteristics of the physics of the large, astrophysics being the most conspicuous example. Even so, occasional doubt has been expressed as to whether the properties of water derived from measurements on minuscule samples would continue to govern on large scales, or whether there might be some effects—overlooked in the small, but ponderous on the scale of global oceans—that would render the hydrodynamic equations as we know them less than satisfactory. The ocean is a very noisy medium at all frequencies and it is usually quite difficult to do quantitative calculations that reproduce the jumbled measurements with the precision required to test such a hypothesis. To the extent that one can tell, though, the predictive skill of the equations seems to exceed the calculational skill of the manipulators of those equations, so that we may proceed with some tentative confidence toward our mathematical objective, until otherwise diverted.

3.2 The Convective Derivative and the Momentum Equation

The equations of motion for a fluid are the momentum equation (quite simply, Newton's Law applied to a unit mass of fluid) and the equation expressing conservation of mass per unit volume—the continuity equation. From these, equations can be derived for energy flux and associated quantities. Our derivation of these equations will *per force* introduce additional variables requiring additional equations; also, certain coefficients or parameters (e.g., sound speed or thermal expansion coefficient) must be considered as given, rather than as being solvable within the context of the present theory. To try to calculate these *constitutive parameters* from first principles would take us far afield into the physics of the liquid state, to statistical mechanics, and ultimately to the molecular and atomic physics of H_2O and other constituents. We will not tread that path here.

A word about notation is in order. We will use for a three-vector quantity a boldface letter coinciding with the first letter of its triad of scalar components; the second and third of the triad will be the letters immediately following in the alphabet. Where this does not work, we will use scalar subscripts; the latter will never indicate partial differentiation, which is occasionally abbreviated as ∂_x, ∂_y, etc. Thus the spatial Cartesian coordinates are written as

$$\mathbf{x} \equiv (x,y,z) \ , \tag{3.1}$$

the components of the fluid velocity field, $\mathbf{u}$, are

$$\mathbf{u} \equiv (u,v,w) \ , \tag{3.2}$$

and the components of the wave vector, **k**, of a harmonically varying quantity are likewise

$$\mathbf{k} \equiv (k,l,m) \ . \tag{3.3}$$

Consider now the concept of an infinitesimal *material element* or *parcel of fluid* located in a right-handed coordinate system at position **x** at time *t*. Actually, the parcel must be tiny enough to be mathematically infinitely small, but large enough to contain ample water molecules and dissolved materials to be physically representative of seawater in the large, to avoid explicitly considering intermolecular forces and voids in what for our subject should be an ideal continuum. Then any arbitrary *property,* γ (e.g., temperature, entropy), of the fluid element may be considered as a continuously differentiable function of space and time, or a smooth field:

$$\gamma = \gamma(x,y,z,t) \equiv \gamma(\mathbf{x},t) \ . \tag{3.4}$$

The material element has position $\mathbf{x}(t)$ at time *t* and its property γ will therefore vary with time according to

$$\gamma[x(t),y(t),z(t),t] \equiv \gamma[\mathbf{x}(t),t] \ . \tag{3.5}$$

Then the rate at which the property changes is, according to the chain rule of calculus,

$$
\begin{aligned}
\frac{d\gamma}{dt} &= \left(\frac{\partial\gamma}{\partial t}\right)_{xyz} + \left(\frac{\partial\gamma}{\partial x}\right)_{yzt}\frac{dx}{dt} + \left(\frac{\partial\gamma}{\partial y}\right)_{ztx}\frac{dy}{dt} + \left(\frac{\partial\gamma}{\partial z}\right)_{txy}\frac{dz}{dt} \\
&\equiv \frac{\partial\gamma}{\partial t} + \frac{d\mathbf{x}}{dt}\cdot\nabla\gamma \ .
\end{aligned}
\tag{3.6}
$$

(We will henceforth drop the subscript notation for those independent variables held constant in the partial differentiation.) Now $d\mathbf{x}/dt$ is clearly the rate of change of position of the fluid element, i.e., the fluid velocity, **u**, following the element, a quantity that is also considered as a field:

$$\frac{d\mathbf{x}}{dt} \equiv \mathbf{u}(\mathbf{x},t) \ . \tag{3.7}$$

We now introduce notation for the operator D/Dt, known variously as the convective, advective, material, Stokes, substantive derivative, or the derivative following a fluid parcel:

$$\frac{D}{Dt} \equiv \frac{\partial}{\partial t} + \mathbf{u} \cdot \nabla \ . \tag{3.8}$$

This equation, when applied to the property γ, gives the total rate of change of γ following the motion, and states that the total rate is the sum of an intrinsic local rate that is observed at a fixed point, plus the change due to advection of the quantity at the velocity $\mathbf{u}$ while moving by $d\mathbf{x}$ in time dt. It is different from the operator d/dt, which is henceforth considered to operate on functions of time only. When applied to the velocity field itself, the advective rate of change is the acceleration of the fluid parcel, $\mathbf{a}$:

$$\mathbf{a} = \frac{D\mathbf{u}}{Dt} = \frac{\partial \mathbf{u}}{\partial t} + \mathbf{u} \cdot \nabla \mathbf{u} \ . \tag{3.9}$$

Let $\rho = \rho(\mathbf{x}, t)$ be the fluid density; then the momentum equation is simply Newton's First Law applied to a small mass of fluid, $d^3m = \rho \, dx \, dy \, dz$. Thus

$$\rho \frac{D\mathbf{u}}{Dt} \, dx \, dy \, dz = \sum_i \rho \, \mathbf{f}_i \, dx \, dy \, dz \ , \tag{3.10}$$

where the $\mathbf{f}_i$ are forces per unit mass (or body forces) acting on the fluid element, and the sum is over all such forces present (not integration over the volume), including externally imposed and internally generated fluid forces.

3.3 Gravitational Forces

Gravity is clearly the strongest and most pervasive force acting on the sea, and it manifests itself in a number of obvious ways. It establishes an equipotential surface to which sea level closely, but not exactly, conforms; it supplies the restoring force for surface and internal gravity waves; it compresses (to a small degree) the water at depth via the weight of the overlying fluid column; it establishes the sun–moon–earth tidal forces; and it provides the energy for the subtle but important density-driven circulations at depth, i.e., the thermohaline currents previously mentioned.

The gravitational force on a parcel of fluid at radius r from the center of mass of a uniform, nonrotating, spherical earth is

$$d^3\mathbf{F}_g = - \hat{r} \, \frac{Gm_e}{r^2} \, \rho \, d^3\mathbf{x} \,, \tag{3.11}$$

where $\hat{r}$ is a unit vector directed from the center of mass to the fluid parcel, G is the universal gravitational constant (6.673×10^{-11} N m^2 kg^{-2}); m_e is the mass of the earth, including water and air (5.973×10^{24} kg); and $d^3\mathbf{x}$ is an element of volume in whatever coordinate system is used to express r. The force per unit mass is the local gravitational acceleration, $\mathbf{g}_0$:

$$\frac{d^3\mathbf{F}}{\rho \, d^3\mathbf{x}} = \mathbf{g}_0 = - \frac{Gm_e}{r^2} \, \hat{r} \quad \text{m s}^{-2} \,. \tag{3.12}$$

This body force can be derived from a scalar potential function, Φ_0, the geopotential, by taking the negative gradient:

$$\mathbf{g}_0 = - \nabla \Phi_0 \,. \tag{3.13}$$

For a spherical earth, the gravitational potential is radially symmetric and $\nabla = \hat{r}(\partial/\partial r)$; hence

$$\Phi_0(r) = - \frac{Gm_e}{r} \quad \text{m}^2 \text{ s}^{-2} \,. \tag{3.14}$$

The *geoid* is defined as a surface on which the geopotential is constant, which is clearly a sphere for the nonrotating spherical earth model. The *marine geoid* is further defined as the surface assumed by a uniform, motionless ocean under the influence of terrestrial gravitational forces alone, again assuming (for the moment) no rotation. However, the ocean is not uniform, is not motionless, and is not driven by earth-derived gravity alone. Furthermore, the actual gravitational forces are not spherically symmetric, but vary in significant ways because of local mass concentrations and anomalies, with the greatest departure from spherical symmetry coming from the flattening of the earth at the poles caused by its spinning about its north–south axis.

In order to introduce these complicating factors, we shall next derive expressions for the dynamical effects on a spinning earth, and then return to complete the story on gravity.

3.4 Effects of Spin

There are two primary effects of the earth's spin, both classified as apparent or fictitious forces—the *Coriolis force* and the *centrifugal force*. Here we shall use *spin* to describe the earth's daily motion about its axis, and *rotation* to describe its annual motion about the sun. In later chapters, however, we shall speak more loosely about rotations, meaning spin all the while. We will also neglect other very small accelerations caused by nonuniform spin rates brought about by redistributions of mass and hence moments of inertia, both interior to the earth and within the atmosphere. As an aside, it is interesting to note that observed small variations in the length of day are well correlated with seasonal redistributions of the earth's atmosphere as a result of changes in solar heating and orbital eccentricity throughout the year.

To derive the fictitious forces, let Ω be the angular velocity of the earth's spin relative to the "fixed" stars, i.e., inertial space. This vector points north along the north–south axis, and the earth's spin is thus right-handed, having positive helicity. Its scalar value, $\Omega = |\mathbf{\Omega}|$, is

$$\Omega = 2\pi(1 + 1/365.24) \quad \text{rad day}^{-1}$$

$$= 7.292 \times 10^{-5} \quad \text{rad s}^{-1} . \tag{3.15}$$

The correction term in parentheses represents the fact that during a complete solar day, in order to bring the sun overhead again, the earth must spin through an additional angle approximately equal to the angular change incurred during that day's rotation about the sun. This rotation rate is therefore measured with respect to inertial space. Motions taking place on the spinning earth, when viewed from a coordinate system fixed on the earth, will clearly be different than when viewed from inertial space. We will therefore require relationships between derivatives in inertial and rotating coordinate systems.

Let $\mathbf{A}$ be an arbitrary vector, with Cartesian components A_x, A_y, and A_z in an inertial frame and A_x', A_y', and A_z' in the spinning frame (which is the one from which we shall view our dynamics). Then in the inertial frame we may write

$$\mathbf{A} = \hat{i}A_x + \hat{j}A_y + \hat{k}A_z , \tag{3.16}$$

while the same vector, when viewed from the frame spinning at angular velocity Ω has the representation

$$\mathbf{A} = \hat{i}'A_x' + \hat{j}'A_y' + \hat{k}'A_z' . \tag{3.17}$$

Let $(d\mathbf{A}/dt)_i$ be the total time derivative in the unprimed (inertial) frame, in which Newton's equation holds; then

$$\left(\frac{d\mathbf{A}}{dt}\right)_i = \hat{i}\,\frac{dA_x}{dt} + \hat{j}\,\frac{dA_y}{dt} + \hat{k}\,\frac{dA_z}{dt}$$

$$= \hat{i}'\,\frac{dA_x'}{dt} + \hat{j}'\,\frac{dA_y'}{dt} + \hat{k}'\,\frac{dA_z'}{dt}$$

$$+ \frac{d\hat{i}'}{dt}\,A_x' + \frac{d\hat{j}'}{dt}\,A_y' + \frac{d\hat{k}'}{dt}\,A_z' , \qquad (3.18)$$

where the second expression accounts for the fact that the primed unit vectors are rotating in the coordinate system as well. Note that $d\hat{i}'/dt$ is the velocity of $\hat{i}'$ due to its spin; from the diagram of Fig. 3.1, one concludes

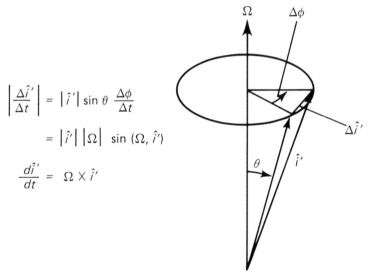

$$\left|\frac{\Delta\hat{i}'}{\Delta t}\right| = |\hat{i}'|\sin\theta\,\frac{\Delta\phi}{\Delta t}$$

$$= |\hat{i}'|\,|\Omega|\,\sin(\Omega, \hat{i}')$$

$$\frac{d\hat{i}'}{dt} = \Omega \times \hat{i}'$$

Fig. 3.1 Diagram for derivation of the rate of change of a unit vector in a rotating coordinate system.

that $d\hat{i}'/dt = \Omega \times \hat{i}'$, and similarly for $d\hat{j}'/dt$ and $d\hat{k}'/dt$. The first three terms in Eq. 3.18 give just the rate of change of $\mathbf{A}$ as observed in the rotating coordinate system. Thus we may write

$$\left(\frac{d\mathbf{A}}{dt}\right)_i = \left(\frac{d\mathbf{A}}{dt}\right)_r + \Omega \times \mathbf{A} , \qquad (3.19)$$

or in general, $(d/dt)_i = (d/dt)_r + \Omega \times$, which is a kinematic relationship giving an inertial derivative in terms of a derivative observed in the rotating frame, plus a temporal change induced by spin. Thus if $\mathbf{A} = \mathbf{x}$, the position vector of a fluid element, then

$$\left(\frac{d\mathbf{x}}{dt}\right)_i = \left(\frac{d\mathbf{x}}{dt}\right)_r + \Omega \times \mathbf{x} \ . \tag{3.20}$$

The left side is the velocity in inertial space while on the right is the velocity as observed on the spinning earth. Thus

$$\mathbf{u}_i = \mathbf{u}_r + \Omega \times \mathbf{x} \ . \tag{3.21}$$

If we apply the kinematic relationship to the velocity $\mathbf{u}_i$ itself, we obtain

$$\left(\frac{d\mathbf{u}_i}{dt}\right)_i = \left(\frac{d\mathbf{u}_i}{dt}\right)_r + \Omega \times \mathbf{u}_i \ , \tag{3.22}$$

which, upon substituting Eq. 3.21 into Eq. 3.22, yields

$$\left(\frac{d\mathbf{u}_i}{dt}\right)_i = \left(\frac{d}{dt}\right)_r \left[\mathbf{u}_r + \Omega \times \mathbf{x}\right] + \Omega \times \left[\mathbf{u}_r + \Omega \times \mathbf{x}\right]$$

$$= \left(\frac{d\mathbf{u}_r}{dt}\right)_r + \Omega \times \left(\frac{d\mathbf{x}}{dt}\right)_r + \Omega \times \mathbf{u}_r + \Omega \times \left(\Omega \times \mathbf{x}\right)$$

$$= \left(\frac{d\mathbf{u}_r}{dt}\right)_r + 2\Omega \times \mathbf{u}_r + \Omega \times \left(\Omega \times \mathbf{x}\right) \ . \tag{3.23}$$

The term on the left-hand side of Eq. 3.23 is the acceleration in the inertial frame; on the right side are the accelerations observed in the rotating frame— the apparent acceleration, $(d\mathbf{u}_r/dt)_r$, the Coriolis acceleration (which acts at right angles to both spin and velocity), and the centripetal acceleration (which acts outward at right angles to the spin axis at a rate quadratic in Ω). We may rewrite the latter by inspecting Fig. 3.2, which represents the gravitational and rotational forces acting on an ellipsoidal earth whose radius, $R_e(\theta)$, varies with colatitude, θ. The negative of the double cross-product in Eq. 3.23 points in a direction perpendicular to the spin axis, as indicated by the unit vector, $\hat{\rho}$, in that direction, and the resultant centripetal acceleration has the effect of changing the gravity vector $\mathbf{g}_0$ to a new effective gravity, $\mathbf{g}_{eff}$, as indicated by the force parallelogram in Fig. 3.2:

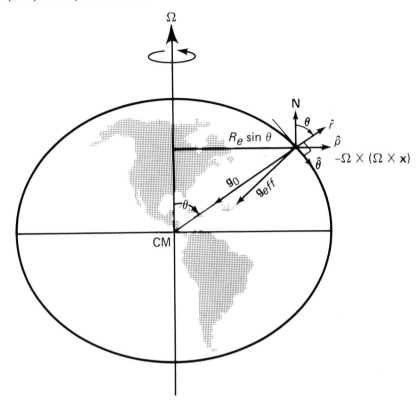

Fig. 3.2 Force diagram for the effective gravity, $\mathbf{g}_{eff}$, on an ellipsoidal earth.

$$\mathbf{g}_{eff} = \mathbf{g}_0 + \Omega^2 R_e \sin \theta \, \hat{\rho} \, , \tag{3.24}$$

where, in terms of unit vectors in the directions of increasing r and θ,

$$\hat{\rho} = \hat{r} \sin \theta + \hat{\theta} \cos \theta \, . \tag{3.25}$$

At the equator, this value of effective gravity is less than the mean gravitational acceleration, $g_0 = 9.80 \text{ m s}^{-2}$, by approximately 0.35% for an average earth radius, R_e, of 6.371×10^6 m. The equatorial flattening of $1/298.257$ implies that the equatorial radius of the earth is about 21 km larger than the polar radius. By the process of *isostatic compensation*, the plastic spinning earth has apparently adjusted itself to the centrifugal force over geological times by assuming the shape of an approximate ellipsoid of revolution in which g_{eff} is very close to the perpendicular at the local surface. This figure of revolution is termed the *reference ellipsoid* and is used as the basis for measuring anomalies in both gravity and geoidal heights.

The centripetal acceleration can be derived from a potential function, $\Phi(r,\theta) = -(\Omega^2 r^2/2)\sin^2\theta$, which can be added to the spherical potential of Eq. 3.14 to form an effective potential, Φ_{eff}:

$$\Phi_{eff} = -\frac{Gm_e}{r} - \frac{\Omega^2 r^2}{2}\sin^2\theta . \tag{3.26}$$

Clearly the centripetal acceleration and the ellipticity of the earth must be included in the specification of the equipotential surface that defines the marine geoid. However, higher order terms must be incorporated into the geopotential for precision work in marine geodesy (cf. Section 3.6).

During the manipulations involving equations written in inertial and rotating frames, the meaning of the derivative, d/dt, as applied to $x(t)$ did not require that a distinction be made between the ordinary and the convective derivative, since x is a function only of time. However, the total derivative of Eq. 3.22, in which the velocity is a function of space and time, requires that distinction to be made. The proper writing of Eq. 3.23 for the acceleration is then

$$\left(\frac{Du_i}{Dt}\right)_i = \left(\frac{\partial u_r}{\partial t}\right)_r + u_r \cdot \nabla u_r + 2\Omega \times u_r + \Omega \times (\Omega \times x) , \tag{3.27}$$

where all quantities on the right-hand side are specified in the rotating frame. This form, in which the time derivative contains four terms describing distinct physical effects, will henceforth be used in the fluid dynamics of the ocean.

3.5 The Coriolis Force

Although the Coriolis force is often described as a fictitious force, its effects are nevertheless demonstrably real; for oceanic and atmospheric motions, they are, in fact, both subtle and profound. An intuitive understanding of the origins and impacts of rotational forces is of considerable value in the analytical work to come.

In Eq. 3.23, the transformation to the rotating frame of reference generates two terms, the Coriolis force and the centrifugal force. What is called the Coriolis effect should to some degree be considered as the combined action of both terms. Behind their dynamics are three distinguishable but closely interrelated physical phenomena: (1) the rotation of the earth under a moving particle during its time of flight; (2) changes in the centripetal acceleration with variation in the particle's distance from the spin axis; and (3) conservation of the particle's absolute angular momentum, i.e., the sum of its planetary-induced momentum and its relative fluid angular momentum. We shall discuss each of these in turn.

Consider a freely rotating, frictionless pendulum of the type developed by Foucault, swinging from a suspension at the North Pole; take the plane of the swing to coincide initially with the Greenwich Meridian, ($\phi = 0°$) at $t = 0$ (see Fig. 3.3). During its rotation at the rate Ω (which is $15°$ h^{-1} as measured in solar time), the earth will turn under the pendulum, so that after time t, the reference meridian will have advanced through an *hour angle,* Ωt. An observer in inertial space will see the pendulum continuing to swing in the same plane in which it started, but to an observer on the rotating earth near the pole it will appear that the plane of the pendulum's motion has rotated to the right, i.e., clockwise. After 12 hours, the plane of rotation will have rotated through $180°$ and will once again coincide with the original plane of oscillation, passing through the Greenwich Meridian. Thus the spin of the earth apparently displaces a particle in the clockwise direction (in the Northern Hemisphere) for this purely meridional motion, and in the counterclockwise direction in the Southern Hemisphere. The period for $180°$ of rotation (not oscillation) of the pendulum is termed the *half-pendulum day,* and varies with latitude, as will be discussed below, but in this case is one-half a sidereal (or stellar) day (approximately 11 h 58 min of solar time).

A somewhat different *gedanke* (thought) experiment will demonstrate the Coriolis effect away from the Pole. Consider a circular pan of water, steadily rotating at angular speed Ω_z after the transients associated with its spin-up have damped out (Figs. 3.4a and 3.4b). Under the combined influences of gravity and centripetal acceleration, the surface of the water will assume the shape of a paraboloid of revolution, which for small displacements from equilibrium will be almost spherical, with the downward normal to the surface lying along the resultant of the vectors $\mathbf{g}$ and $\Omega_z^2 R\hat{\rho}$. At the point O_1 on the surface of the water at a distance R from the axis (Fig. 3.4b), allow a small floating object initially to co-rotate with the water, but to move across the surface without friction. Let the direction from O_1 to the axis of rotation be called *north*. The forces at O_1 are in equilibrium, with the poleward slope of the surface just balancing the outward centripetal acceleration. Starting at O_1 (Fig. 3.4a), if we now impart a small southward (outward) impulsive velocity, δu_r, to the particle in the purely radial direction, it will initially move in that direction. However, upon arriving at O_2, one-eighth of a period later, it will have a larger radius, $R + r_i$, from the spin axis than

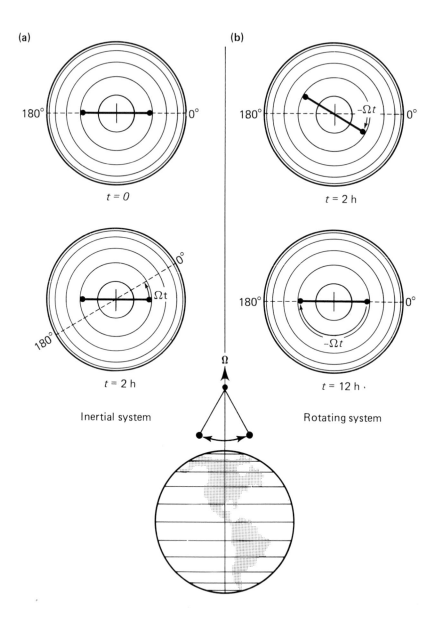

Fig. 3.3 Rotations of a Foucault pendulum at the North Pole as viewed (a) from inertial space and (b) from an earthbound coordinate system.

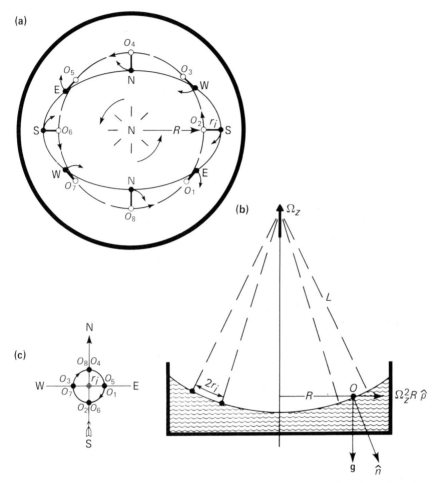

Fig. 3.4 (a), (b). Frictionless motion of a particle moving over the surface of a spinning pan of water, demonstrating the Coriolis effect. The particle is arranged on a very long pendulum of length L, and co-rotates with the pan at an angular speed of Ω_z. Under the combined influences of gravity and centripetal acceleration, the water's surface is nearly a spheroid of revolution, across which the pendulum swings tangentially. A small radial velocity, δu_r, is imparted impulsively to the particle at point O_1, where an observer is located. The particle then describes an elliptical path as seen from inertial space, but appears to an observer on the spinning pan to rotate clockwise in an inertial circle of radius r_i, lying first to the east of O_1, then south at O_2, west at O_3, and north at O_4. As viewed by the observer, the particle makes *two* complete revolutions in the inertial circle about the points O_i for one revolution of the pan, thereby giving rise to the factor of 2 in the Coriolis acceleration. (c) Positions of the particle relative to the observer located at the center of the compass rose. As applied to the earth, Ω_z is the local vertical component of the earth's angular velocity, and R the distance from the spin axis.

at equilibrium; however, it will have retained its azimuthal velocity, $u_\phi = R\Omega_z$, which is lower than that of its surroundings, $u'_\phi = (R + r_i)\Omega_z$. Thus the particle will fall behind the rotating water, i.e., it will appear to be deflected clockwise to the right. At this point, the particle lies to the *south* of the observer at O_2 and its linear velocity, $u_\phi = (R + r_i)\Omega_z - \delta u$, is purely azimuthal; however, its effective angular velocity around the spin axis has been reduced from Ω_z to $\Omega_z - \delta u/(R + r_i)$. As a consequence of this lowered spin rate, the centripetal acceleration at O_2 will be reduced from $\Omega_z^2(R + r_i)$ to

$$a_c = \left(\Omega_z - \frac{\delta u}{R + r_i}\right)^2 (R + r_i) \ . \tag{3.28}$$

The difference between the centripetal acceleration at radius $R + r_i$ for the water and the particle is

$$\Delta a_c = 2\Omega_z \delta u + \delta u^2/(R + r_i) \ . \tag{3.29}$$

It will be recognized that the first term is the Coriolis acceleration, which is, after all, the source of the deflections to the right. The remaining term is of second order and is negligible for velocities in the ocean.

By the time the particle has advanced with the earth to O_3, the clockwise deflection has resulted in a northward radial velocity, so that the particle now appears to be *west* of the observer at O_3. The motion, however, carries it inward past its equilibrium radius, R, and this will again result in deflection to the right, since the particle's azimuthal component of total velocity for $r_i < R$ will be greater than the nearby water and it will overtake its local environment. At position O_4, the particle is to the *north* and its increased angular velocity will result in an increased centripetal acceleration and a southward deflection. At position O_5 (one half-period from the start), the velocity will have become purely radial again with the conditions the same as at O_1 and with the particle once again located to the *east* of the observer.

Thus, in a half-period of rotation of the pan of water, the particle, as seen by an observer in the rotating coordinate system, will appear to have traversed a full circle whose radius, r_i, is called the *inertial circle*. The associated period is also a half-pendulum day, and is termed an *inertial period, T_i*. Its value ranges from approximately 12 hours at the pole, through 24 hours at about 30°, to infinity at the equator. The relationship to the pendulum comes from the fact that a pendulum whose geometry is the same as that shown in Fig. 3.4 (which is not the same one as the Foucault pendulum of Fig. 3.3) will oscillate with the inertial period. To show this, recall that the period of a pendulum, T_p, of length L is

$$T_p = 2\pi (L/g)^{1/2} \ . \tag{3.30}$$

From Fig. 3.4, the balance of forces and the geometry give, for small angles from the vertical,

$$L/R \simeq g/\Omega_z^2 R \ ,$$

or

$$\Omega_z \simeq (g/L)^{1/2} \ ,$$

and thus the period of the rotation is equal to the period of the pendulum:

$$T = 2\pi/\Omega_z = 2\pi (L/g)^{1/2} = T_p$$

$$= 2T_i \ . \tag{3.31}$$

Therefore, an imaginary pendulum of length L that undergoes small oscillations of radius r_i about O as an equilibrium point would have a period given by twice the inertial period. For the earth, the length of the pendulum is very large, depending on the local value of the inertial period; at the pole, this length is approximately 1.84×10^6 km.

The radius of oscillation, r_i, can be calculated by equating the Coriolis force to the centrifugal force that maintains the motion in the small circle. Since the small velocity, $\delta u = \delta u_r$, is measured with respect to the rotating system, this results in

$$(\delta u^2/r_i) = 2\Omega_z \delta u \ ,$$

or

$$r_i = (\delta u/2\Omega_z) \tag{3.32}$$

as the radius of the inertial circle. Thus the radius is proportional to the velocity imparted to the particle.

Returning to the upper diagram of Fig. 3.4, the path of the particle viewed from inertial space is an ellipse, while that of the observer is a circle of radius R, about which the ellipse is centered. It is clear that the observer will see the particle rotating clockwise about him in the inertial circle, first to the south, then west, north, and east, to return to the south within one-half a rotation period, $2\pi/\Omega_z$: this is suggested by the small compass rose diagram of Fig. 3.4c. This doubling of the apparent rotation rate is the origin of the factor of 2 in the formula for the Coriolis frequency (Eq. 3.23).

If small to moderate friction exists, the radius of oscillation of the particle in the inertial circle will gradually be reduced as it loses its initial velocity, δu, but the period will remain nearly unchanged. In the rotating coordinate system, the particle's path will be viewed as an inward spiral that converges on the observer's position.

An alternative view of the dynamics is that expressed by conservation of angular momentum. The particle's total angular momentum, L, can be considered as the sum of its momentum about the spin axis and the angular momentum of its small oscillation. The movement in the inertial circle can be thought of as the particle's attempt to conserve its angular momentum as it oscillates between the extremes of its moment arm, $R \pm r_i$.

On a spherical earth, the situation is more complicated and more subtle. Now Ω and g are no longer antiparallel, and the spin vector at any colatitude, θ, has both horizontal and vertical components (Fig. 3.5). As a consequence, any large-scale motions that involve appreciable north–south excursions will see a *varying* component of Coriolis frequency and force. This will turn out to have far-reaching implications for mesoscale and planetary-scale motions, a fact that has been mentioned earlier and whose consequences will be developed in Chapter 6.

For the present, let us consider the Coriolis force as viewed in a local right-handed Cartesian coordinate system called the *tangent plane system* (Fig. 3.5), which touches the earth's surface at polar angles θ and ϕ, with x to the east, y to the north, and z upward; such a coordinate system will be used extensively in this book. An expansion of the Coriolis term in this coordinate system is

$$2\Omega \times \mathbf{u} = 2 \begin{bmatrix} \hat{i} & \hat{j} & \hat{k} \\ 0 & \Omega \sin \theta & \Omega \cos \theta \\ u & v & w \end{bmatrix}$$

$$= \hat{i}(2w\Omega \sin \theta - 2v\Omega \cos \theta)$$

$$+ \hat{j}(2u\Omega \cos \theta) - \hat{k}(2u\Omega \sin \theta) , \qquad (3.33)$$

where $(\hat{i}, \hat{j}, \hat{k})$ are unit vectors in the tangent plane system. Of these four terms, only two are important in geophysical fluids. First, the vertical component of velocity, w, is generally very small compared with the horizontal

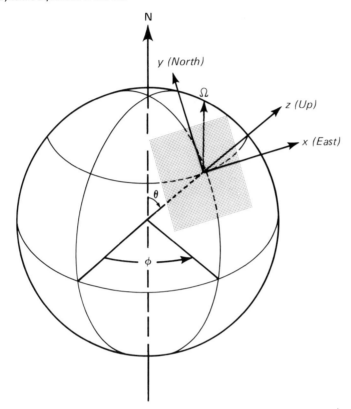

Fig. 3.5   Tangent plane coordinate system touches the earth at the point at which the radius vector at (θ,ϕ) intersects the surface; Cartesian coordinates constitute a right-handed system. The local Coriolis frequency varies linearly as $f = f_0 + \beta y$ in the north–south direction.

components, u and v, and the term involving w is therefore negligible. Next, the vertical or $\hat{z}$ component of Eq. 3.33 (which is called the Eötvös correction in marine gravity) is many orders of magnitude smaller than its competing vertical forces, i.e., gravity and fluid buoyancy, and can also be neglected. This leaves as the effective components of Coriolis acceleration the quantities

$$- (2\mathbf{\Omega} \times \mathbf{u})_x \simeq 2v\Omega \cos \theta \qquad (3.34)$$

and

$$- (2\mathbf{\Omega} \times \mathbf{u})_y = - 2u\Omega \cos \theta \, , \qquad (3.35)$$

where we have written the Coriolis term as it will appear on the right-hand side of the momentum equation ahead, Eq. 3.83, so that it now acts as a

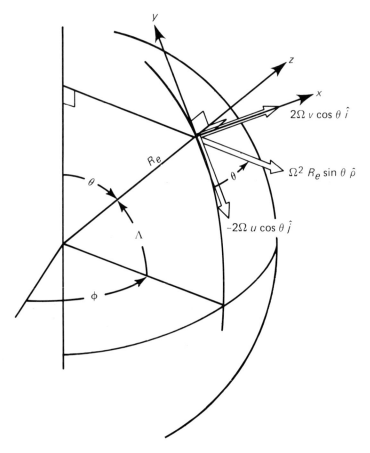

Fig. 3.6 Force diagram showing the components of Coriolis and centrifugal forces at the earth's surface. These behave as real forces to an observer on the rotating earth.

forcing term. In Fig. 3.6, this redefinition in terms of horizontal force components makes clear the origin of the inertial motion: In the Northern Hemisphere, an initially northward-directed particle ($v > 0$) is deflected eastward by the x component of Coriolis force, and thereby acquires an eastward velocity component ($u > 0$). This component then interacts with southward-directed meridional component of Coriolis force to continue the clockwise deflection. In the Southern Hemisphere, $\cos \theta < 0$ and the direction of the force is opposite: The inertial circle rotates counterclockwise.

We next expand the centripetal term of Eq. 3.27 evaluated for $r = R_e$, and also consider it as a forcing function on the right-hand side of the momentum equation:

$$-\mathbf{\Omega} \times (\mathbf{\Omega} \times \mathbf{x}) = -\Omega^2 R_e \sin\theta(\hat{j}\cos\theta - \hat{k}\sin\theta) , \tag{3.36}$$

$$= \Omega^2 R_e \sin\theta \, \hat{\rho}$$

which has both vertical and horizontal components in the tangent-plane system. Here the unit vector, $\hat{\rho}$, is directed outward at right angles from the spin axis and is, from Eq. 3.36,

$$\hat{\rho} = -(\hat{j}\cos\theta - \hat{k}\sin\theta) = \hat{\theta}\cos\theta + \hat{r}\sin\theta , \tag{3.37}$$

where $\hat{\theta}$ and $\hat{r}$ are unit vectors in the θ and r directions, respectively.

It is now clear how to interpret the motion shown in Fig. 3.4 on a rotating planet. The local vertical component of the planet's angular velocity vector, $\Omega_z = \Omega\cos\theta$, interacts with the horizontal components of particle velocity, $\mathbf{u}$, to form the fictitious horizontal Coriolis force at right angles to both. The horizontal component of planetary angular velocity causes small, essentially negligible vertical forces. Thus the angular velocity, Ω_z, in Fig. 3.4 is to be re-written as $\Omega\cos\theta$, and the lever arm, R, becomes $R_e\sin\theta$. The local vertical component of centripetal acceleration, $\Omega^2 R_e \sin^2\theta$, changes the value of $\mathbf{g}$ by only a small percentage, while the local horizontal component, $\Omega^2 R_e \sin\theta\cos\theta$, provides the same restoring force as for the rotating disk. We can then write for the local inertial period on an approximately spherical earth,

$$T_i = \frac{2\pi}{2\Omega\cos\theta} , \tag{3.38}$$

and for the radius of the inertial circle,

$$r_i = \frac{u}{2\Omega\cos\theta} . \tag{3.39}$$

If the north–south particle motions are extensive enough to range through significant changes in latitude, Λ, the effective value of the earth's angular velocity also changes. In the tangent plane approximation, the angular velocity is expanded in a series about some value f_0 and only linear terms are retained. Denote the important *Coriolis parameter* (also called the planetary vorticity) by f:

$$f \equiv 2\Omega\cos\theta = 2\Omega\sin\Lambda , \tag{3.40}$$

where $\Lambda = (\pi/2) - \theta$ is the latitude. For small changes in θ (or Λ) about a point where $f = f_0$, one writes

$$f = f_0 + \frac{\partial f}{\partial y} y + \ldots$$

$$\simeq f_0 + \beta y . \tag{3.41}$$

Here the tangent plane touches the earth where the Coriolis parameter has the value f_0. Denote by β the derivative at that point, which then becomes

$$\beta \equiv \frac{df}{dy}\bigg|_{f_0} = \frac{df}{d\Lambda}\frac{d\Lambda}{dy}\bigg|_{\Lambda_0} = \frac{2\Omega}{R_e} \cos \Lambda_0 . \tag{3.42}$$

At mid-latitudes,. say $\Lambda_0 = 45°$, numerical values of f and β are:

$$f = 2\Omega \sin \Lambda_0 = 1.031 \times 10^{-4} \text{ s}^{-1} , \tag{3.43}$$

and

$$\beta = 1.619 \times 10^{-11} \text{ (m s)}^{-1} . \tag{3.44}$$

The interpretation of these values is as follows. The Coriolis parameter is a measure of the *planetary vorticity,* or the earth's rotation rate, and is approximately 10^{-4} rad s^{-1} in mid-latitudes. Fluids with relative intrinsic vorticities of the same order or smaller will be strongly affected by rotation, all else being equal. For example, if some meridional motion carries the fluid through a north–south excursion of one-tenth of the earth's radius (637 km), the Coriolis force will change roughly by 10% because of the β term, and that change will give rise to a type of planetary-scale wave termed a *Rossby wave* or a *potential vorticity wave*, whose dynamics we shall study in Chapter 6. The tangent-plane approximation with $f = f_0$ constant is called an *f plane*; with *f* varying linearly, it is called a *β plane*. These two coordinate approximations are convenient and are widely used in oceanography and meteorology, with the β plane being of particular interest for the description of Rossby waves.

3.6 More Gravity

Beyond the 21 km ellipsoidal flattening of the shape of the earth, other gravitational deformations exist on smaller scales, and these have the effect of introducing further departures of the geoid from the ellipsoid of revolution by amounts up to the order of ± 100 m. The departures are due to concentrations of mass, other density anomalies, and surface topographic variations, and can be determined accurately via satellite techniques. To outline the method, we expand the geopotential, $\Phi_g(r,\theta,\phi)$, in spherical coordinates, (r,θ,ϕ), with origin at the center of mass and the z axis coincident with the spin axis. The basis functions for the expansion are associated Legendre polynomials, $P_n^m(\cos \theta)\, e^{im\phi}$, which form a complete, orthonormal set defined on a sphere. Thus we may write the geopotential as

$$\Phi_g(r,\theta,\phi) = -\frac{Gm_e}{r}\left[1 + \sum_{n=2}^{\infty}\sum_{m=0}^{n} a_{nm}\left(\frac{R_e}{r}\right)^n P_n^m(\cos\theta)\, e^{im\phi}\right]. \quad (3.45)$$

The complex coefficients are a_{nm}, where $a_{nm} = \mathrm{Re}\,(a_{nm}) + i\,\mathrm{Im}\,(a_{nm})$, and may be evaluated from a large number of satellite orbit and radar altimeter measurements; a rather complete description of the gravity field may so be obtained. The gravity is then calculated as the gradient of the geopotential,

$$\mathbf{g} = -\nabla\Phi_g = -\left(\hat{r}\frac{\partial}{\partial r} + \frac{\hat{\theta}}{r}\frac{\partial}{\partial\theta} + \frac{\hat{\phi}}{r\sin\theta}\frac{\partial}{\partial\phi}\right)\Phi_g , \quad (3.46)$$

where $\hat{r}$, $\hat{\theta}$, and $\hat{\phi}$ are unit vectors, as before. With this definition of geopotential, terms are automatically included that describe both the spherical earth as the Gm_e/r term, and the ellipsoidal flattening or "equatorial bulge" as $(1/2)J_2(R_e/r)^2(1 - 3\cos^2\theta)$, where $J_2 = -[5\,\mathrm{Re}(a_{20})]^{1/2} = 1.0826 \times 10^{-3}$ is the harmonic that gives the gravity change due to the bulge. Hence the total gravitational acceleration can be derived from Eqs. 3.45 and 3.46, with the latter being a much improved estimate replacing the simpler definition of Eq. 3.12. The centripetal acceleration, as incorporated in Eqs. 3.24 and 3.25, must still be added to Eq. 3.46 to obtain the effective gravity.

The utility of the somewhat elaborate descriptions (Eqs. 3.45 and 3.46) lies in the fact that the mean topography of the sea surface conforms rather closely to the actual marine geoid, and not to the simple ellipsoidal model. Later in this book, such questions as mean sea level, the level-of-no-motion, geostrophic setup, and satellite altimeter measurements of currents and tides will require an understanding of the geopotential and geoid. On the other hand, most dynamical effects appearing in the equations of motion (where **g** represents an inertial term) will be adequately modeled by gravity plus centripetal acceleration (Eq. 3.24), or even the spherical approximation (Eq. 3.12).

Figure 3.7 is a contour map of the mean sea surface topography, with elevations given relative to an ellipsoid of revolution having the mean radius and flattening described previously. The contour interval is 5 m; the contours have been derived from a combination of satellite altimeter and orbit perturbation data. A remarkable range of features is visible, including the large gravity depressions in the Indian Ocean and the Eastern Pacific, and the high in the North Atlantic off Europe; these represent examples of the anomalies mentioned previously, with extreme ranges of approximately -100 to $+65$ m. Also visible are sharp depressions over the deep ocean trenches where sea floor spreading and lithospheric subduction have caused negative geoidal anomalies of order -15 m, as well as geoidal highs associated with mid-ocean ridges and island arcs having anomalies of order $+4$ m. Figure 3.8 illustrates the marine gravity anomaly as derived purely from satellite altimetry data; the anomaly represents the departure of the observed scalar values of gravity from the values given by the ellipsoidal approximation carried out to the J_2 term, and is given in units of milligals (mGal), with 1 mGal equal to an acceleration of 10^{-5} m s^{-2}. Although the total range of gravity anomalies existing is near 100 mGal, only a range of ± 60 mGal has been displayed, in order to emphasize the small-size anomalies.

Figure 3.9 is a map of the land topography and ocean bathymetry (*bathos* = deep; *metry* = measure), or bottom topography of the sea floor, based on acoustic fathometer observations. A comparison between Figs. 3.8 and 3.9 reveals that a high degree of correlation exists between the gravity anomaly measured at the surface of the sea and the bottom contours measured at depth, which illustrates the effectiveness of local bathymetry in establishing small-scale gravity anomalies. However, not all bathymetric features have surface gravity expressions (e.g., the East Pacific Rise off South America); the lack of such a relationship may be caused by isostatic adjustment, or perhaps other geophysical mechanisms.

3.7 Tidal Forces

In principle, fluids respond to all gravitational fields exerted on them, but in practice, except for the earth's own gravity, only the sun and the moon exert influences strong enough to induce detectable tidal excursions in the sea. The orbits of these two bodies carry them through a very complex sequence of motions that vary the frequency, phase, and amplitude of the tidal response; the most important tidal forcings have frequencies ranging from approximately twice a day to twice a year.

Consider again the three-body configuration shown in Fig. 2.1, in which the sun, moon, and earth are all assumed to lie approximately in the plane of the ecliptic and to be rotating about their common center of mass. Because of the enormous mass of the sun—approximately 1.99×10^{30} kg— the center of mass of the system lies only 0.06% of the solar radius from the center of the sun. On the other hand, for the earth–moon system, the system center of mass lies within the earth about 4660 km (some 73% of its radius) from its center. It will be seen that not only gravitational but also centrifugal forces are important in the tidal problem. We shall delay the development of the tidal potential function until Chapter 5, but for now it may be said that this function, expressed in spherical coordinates and evaluated on the surface of the earth, is a complicated one that reflects the astronomical motions of the three bodies. The dominant periods fall in three groups— *semidiurnal* (half-day), *diurnal* (daily), and *long period* (14, 28, and 180 days and longer). The 11 most important periods are listed in Table 3.1 along with their commonly accepted nomenclature: M_2, K_1, etc., with the subscript 2 denoting semidiurnal and 1, diurnal periods.

TABLE 3.1
Major Tidal Modes

Tidal Mode		Period, T (h)
Semidiurnal Tides		
M_2	principal lunar	12.421
S_2	principal solar	12.000
N_2	elliptical lunar	12.658
K_2	declination luni–solar	11.967
Diurnal Tides		
K_1	declination luni–solar	23.935
O_1	principal lunar	25.819
P_1	principal solar	24.066
Q_1	elliptical lunar	26.868
Long-Period Tides		
Mf	fortnightly lunar	327.86
Mm	monthly lunar	661.31
Ssa	semiannual solar	4383.04

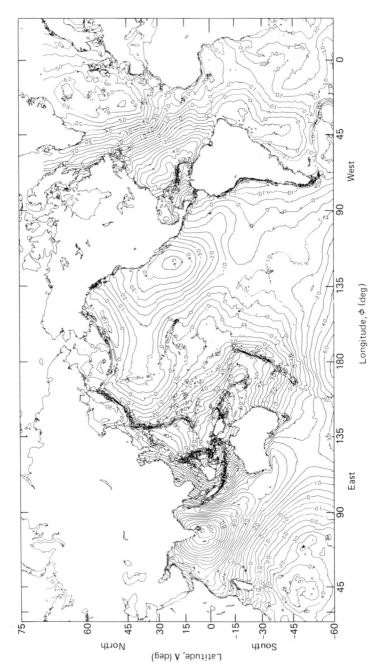

Fig. 3.7 Mean sea surface topography as derived from the radar altimeter on the Seasat satellite. Contours represent elevations and depressions as departures from an ellipsoid of revolution having an equatorial radius, R_e, of 6378.137 km, and a flattening of 1/298.257, and are spaced at 5 m intervals. Extremes of elevation are near 65 m above the ellipsoid; the extreme depression is near 100 m below it. [From Marsh, J. G., et al., *J. Geophys. Res.* (1986).]

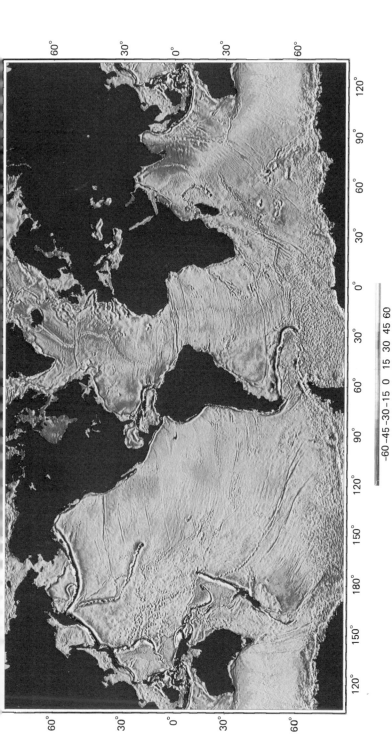

Fig. 3.8 Surface marine gravity anomaly from Seasat altimetry, derived from the gradient of a surface geoid similar to that of Fig. 3.7. Anomaly range shown covers ±60 mGal. A high correlation exists between bathymetry (Fig. 3.9) and gravity anomaly. [From Haxby, W. F., *1982–83 Lamont-Doherty Geological Observatory Yearbook.*]

The periods of the diurnal constituents clearly support the rule of thumb that "the tides come one hour later each day" (i.e., have roughly a 25 h period) although this neglects the intervening semidiurnal cycle; however, the importance of this rule varies greatly from place to place. The maximum of the fortnightly modulation is said to be a "spring tide" (which, however, has nothing to do with the spring season), and occurs approximately at the time that the sun and moon are in either conjunction or anticonjunction. The minimum tide over 14 days is termed a "neap tide" and happens approximately when the two celestial objects are in quadrature. The graphs in Fig. 3.10 show tidal elevations at four near-shore locations that exhibit, respectively, semidiurnal, mixed/dominant semidiurnal, mixed/dominant diurnal, and full diurnal types, with the mix at any location due in part to the

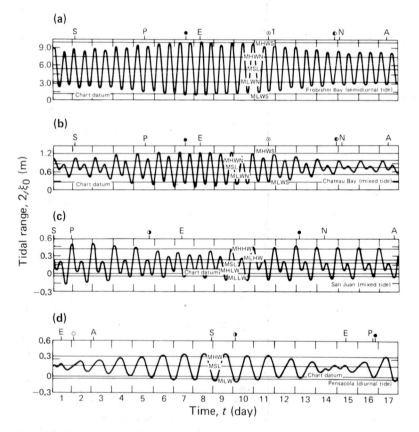

Fig. 3.10 Time series of tidal amplitudes showing four characteristic types of tides: (a) semidiurnal, (b) mixed, dominantly semidiurnal, (c) mixed, dominantly diurnal, (d) full diurnal. [Adapted from *Oceanographic Atlas of the North Atlantic Ocean* (1969).]

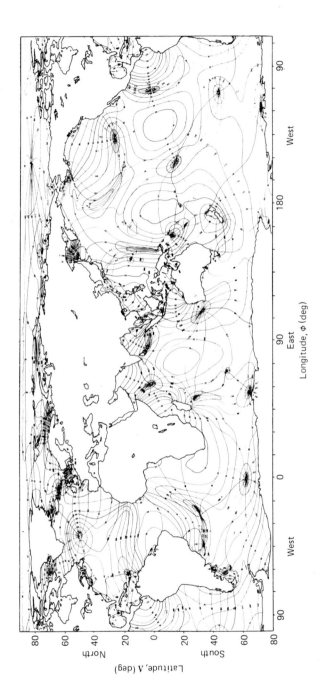

Fig. 3.12 Chart of tidal response characteristics for the M_2 tide in the world's oceans. Amphidromes are points of zero amplitude about which tidal currents rotate; co-range lines are contours of constant elevations (in meters); co-phase lines are contours of constant phase in degrees relative to passage of the moon over the Greenwich meridian. [From Schwerdski, E. W., *Global Ocean Tides, Part II* (1979).]

3.8 Total Potential Caused by Gravity and Rotation

Collecting now the potentials due to the earth's gravity (Eq. 3.45), the centrifugal potential (the rightmost term of Eq. 3.26) and the tidal potential (Eq. 3.47), we have for the total potential, $\Phi(r,\theta,\phi,t)$:

$$\Phi(r,\theta,\phi,t) \ = \ \Phi_g \ + \ \Phi_c \ + \ \Phi_t \ , \tag{3.49}$$

and the associated body force on a unit mass,

$$\mathbf{f} \ = \ -\nabla\Phi \ . \tag{3.50}$$

It is this quantity that will appear in the momentum equation to follow.

3.9 Internal Forces in a Viscous Fluid

The earlier chapters identified many, but not all of the forces acting on a fluid, including those exerted on or near the fluid boundaries. Missing in that tabulation were the *internal* forces that the fluid exerts on itself, i.e., the pressure gradient and viscous or pseudoviscous (eddy) forces. We shall complete that task before returning to the momentum equation.

Pressure Forces

We first derive the pressure force exerted on an infinitesimal material element of fluid by its surroundings. Again consider the volume element, *dx dy dz*, with its center at **x** and with the pressure, *p*, at that point, as shown in Fig. 3.13. Pressure is the force per unit area normal to the face of the element. First we expand the pressure in a Taylor series about the point **x**, and then vectorially sum all of the pressure forces acting on each element. The net force in the $+x$ direction acting on the faces of area *dy dz* is approx-

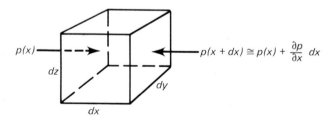

Fig. 3.13 Pressure forces acting on a differential element of fluid.

imately $-(\partial p/\partial x)\ dx\ dy\ dz$, with the minus sign indicating that the force is directed toward decreasing x (i.e., toward lower pressure). Similar arguments apply to the other faces, so that the vector sum of force per unit volume over the entire element is

$$-\left(\frac{\partial p}{\partial x}\ \hat{i} + \frac{\partial p}{\partial y}\ \hat{j} + \frac{\partial p}{\partial z}\ \hat{k}\right) = -\nabla p\ , \qquad (3.51)$$

where, as before, $\hat{i}$, $\hat{j}$, and $\hat{k}$ are unit vectors in the x, y, and z directions, respectively. Hence the pressure forces on a fluid are given by the negative gradient of the pressure, which acts in the down-gradient direction much like a potential function.

Viscous Forces

Next we derive the term in the momentum equation describing viscous forces, which generally act to dissipate and disperse quantities and to dampen or otherwise modify instabilities. If μ is the molecular viscosity of water (with a numerical value of 1.075×10^{-3} kg (m s)$^{-1}$ or 0.01075 poise at $T = 20°C$), then the form of the term due to molecular viscosity will be shown to be $\mu \nabla^2 \mathbf{u}$. Now the damping due to molecular viscosity is very small and its direct effect on larger-scale motions of the sea is essentially negligible; however, on very small scales of motion (e.g., capillary waves, acoustic waves, and small-scale turbulence), its effects become appreciable. There exist unresolved theoretical problems in large-scale turbulence in both oceans and atmospheres, in that most motions are far too complicated and detailed to be described on all scales at once. The first-order effect of instabilities and turbulence on the larger motions is to provide a sink, i.e., a mechanism for dissipation and diffusion of large-scale momentum and energy. These effects are often modeled by the inclusion of an artificially large *anisotropic eddy viscosity* in the momentum equation, whose numerical value depends on the scales in question, but whose functional form is analogous to that of molecular viscosity. Similarly, *eddy diffusivities* for heat and salinity will be used in the governing equations that will be introduced later in order to model those processes on larger scales. Such terms are said to *parameterize* the very complicated processes occurring on scales smaller than the ones of interest, and they have proven to be both useful and moderately effective, if treated with care. We include these terms in the equations that will follow, briefly justifying their use by invoking *Reynolds stresses* and *mixing length theory* in Section 3.12 and referring the reader to the references for additional information.

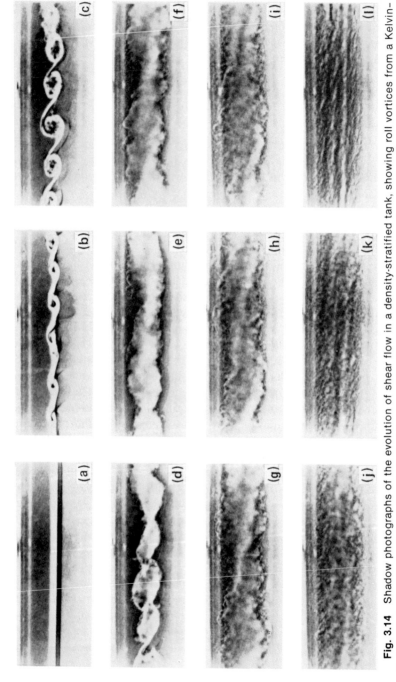

Fig. 3.14 Shadow photographs of the evolution of shear flow in a density-stratified tank, showing roll vortices from a Kelvin–Helmholtz instability evolving into a turbulent layer and finally diffusing out to reduce the gradient. [From Thorpe, S. A., *J. Fluid Mech.* (1971).]

Figure 3.14 is intended to provide an intuitive understanding of the effects of turbulent eddy diffusion. These photographs show how a series of small-scale nonlinear waves on a fluid density interface are driven to a turbulent state by flow instability, which then acts to diffuse the interface and weaken the vertical density and velocity gradients. Thus diffusion of larger-scale density and momentum occurs during such exchanges, but with numerical values of the eddy diffusivity that are orders of magnitude larger than the molecular value. It is clear that this process is very complicated in detail and that a more simplified description of the effect on larger-scale dynamics is required—unless, of course, the object of interest is the instability process itself.

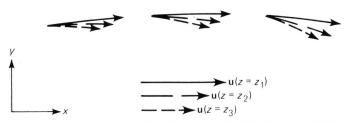

Fig. 3.15 Schematic of changes in fluid velocity field in the horizontal plane, at various depths. The fluid exhibits both shear and divergence.

To derive the term for viscous body forces, consider a fluid flow having the general properties shown in Fig. 3.15, with a velocity $\mathbf{u}(\mathbf{x},t)$ (as indicated by arrows) varying in all three dimensions in both strength and direction. The flow field shown exhibits both velocity *divergence* (or dilitation) and *shear* in both the horizontal (x,y) and vertical (z) directions, where the divergence is proportional to $\partial u_i/\partial x_i$, and the shear to $\partial u_i/\partial x_j$ (with $i \neq j$). Thus there are nine partial derivatives, $\partial u_i/\partial x_j$ (with $i,j = x,y,z$) describing the spatial variations in current velocity. These nine derivatives are proportional to the components of the *stress tensor*, τ_{ij} (of dimensions N m^{-2}), describing viscous, expansion, and pressure forces acting on a unit area within the body of the fluid. For the viscous forces, the proportionality factors are the coefficients of viscosity, A_i. Here we are allowing for the fact that, in some fluids, the effective viscosity may be anisotropic (i.e., it may vary with direction). Such is the case in the ocean and atmosphere, although for those fluids, the pure molecular viscosity, μ, is isotropic. We will assume that A_i includes the scalar molecular viscosity, μ, in addition to the turbulent viscosity. To derive the relationship between τ_{ij} and $\partial u_i/\partial x_j$, consider the x component of current, u, to be stressed from above (by wind, for example) and thus to be varying in the vertical (Fig. 3.16). The shear stress is transmitted downward by viscous and pseudoviscous forces that have the net result of transferring *horizontal* momentum in the *vertical* direction. For small distances, dz, it is found experimentally that the stress is proportional

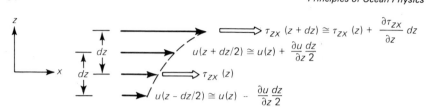

Fig. 3.16 Schematic of vertically sheared flow showing changes of horizontal velocity and of internal shear stress with depth.

to the velocity difference existing between $z + (dz/2)$ and $z - (dz/2)$; inversely to the distance, dz; and to increase directly with the area, $dx\,dy$. Upon expansion of the speed and stress elements in Taylor series, the net stress component, τ_{zx}, can be written as $\tau_{zx} = A_z(\partial u/\partial z)$ (see Fig. 3.16). However, in addition to the vertical variation of horizontal velocity, there is another contribution to τ_{zx} from the *horizontal* variation of *vertical* velocity; this term transmits a net stress either if the fluid is compressible or if the viscosity varies in space (see Eq. 3.57 ahead); this is required so that a fluid that is in uniform rotation at angular velocity ω, and whose velocity is thus $\mathbf{u} = \omega \times \mathbf{r}$, is subject to no viscous body forces. We therefore add a term given by $A_x \partial w/\partial z$ to the stress element to obtain[1]

$$\tau_{zx} = A_z\,\frac{\partial u}{\partial z} + A_x\,\frac{\partial w}{\partial x}\ . \tag{3.52}$$

From the symmetry of this element, it is clear that $\tau_{zx} = \tau_{xz}$.

---

[1]Here we have discussed only the stress due to velocity variations in the presence of viscous shear forces. In fact, the total stress tensor for a fluid is much more complicated than described to this point, and has terms due to the scalar pressure, p, the velocity divergence, $\nabla \cdot \mathbf{u}$, as well as the shear, and is given by

$$\tau_{ij} = (-p + \mu''\,\nabla \cdot \mathbf{u})\,\delta_{ij} + \mu\left(\frac{\partial u_i}{\partial x_j} + \frac{\partial u_j}{\partial x_i}\right)\ .$$

Thus the pressure and the velocity divergence are diagonal, as represented by the Kronecker delta, δ_{ij}, while the shear contributes off-diagonal terms as well. We have derived p as a distinct force, as given by Eq. 3.51, but it is part of the stress tensor nevertheless. The quantity $\mu'' + \tfrac{2}{3}\mu$ is called the *volume viscosity*, μ_v; its correct representation is not altogether clear for fluids. It is ordinarily assumed that $\mu_v = 0$, so that $\mu'' = -\tfrac{2}{3}\mu$. Then the viscous forces may be represented entirely by μ and its generalization for eddy motions, $\mathbf{A}$. See the more advanced texts in the Bibliography for a discussion of this arcane subject.

Note the indices for τ: the first index, z, denotes the fact that the velocity change occurs over the coordinate z, and that the stress is exerted over an area perpendicular to z, i.e., an element in the x,y plane. The second index denotes the current component and hence stress direction in question. The viscosity coefficient A_z indicates that the stress due to this element is transmitted in the z direction. Similar reasoning can be applied to the remaining term and indeed, all eight elements of the stress tensor.

Consider again our small rectangular parallelepiped containing the parcel of fluid immersed in the flow of Fig. 3.15. At time $t = t_0$, the nine stresses, τ_{ij}, are the forces per unit area exerted over each face in the directions indicated in Fig. 3.17. At a small time, Δt, later, the stresses have deformed the cube somewhat, although by assumption it still retains its original mass, $d^3m = \rho\, dx\, dy\, dz$. If the fluid is also incompressible, ρ is a constant and hence the volume, $d^3\mathbf{x} = dx\, dy\, dz$, must also be constant, if slightly distorted geometrically.

By noting that the nine elements of stress shown in Fig. 3.17 are functionally related to the velocity components in a form similar to Eq. 3.52, we can write the stress relationship in matrix form as

$$
\tau = \begin{bmatrix} \tau_{xx} & \tau_{xy} & \tau_{xz} \\ \tau_{yx} & \tau_{yy} & \tau_{yz} \\ \tau_{zx} & \tau_{zy} & \tau_{zz} \end{bmatrix}
$$

$$
= \begin{bmatrix} A_x \dfrac{\partial u}{\partial x} + A_x \dfrac{\partial u}{\partial x} & A_x \dfrac{\partial v}{\partial x} + A_y \dfrac{\partial u}{\partial y} & A_x \dfrac{\partial w}{\partial x} + A_z \dfrac{\partial u}{\partial z} \\[2ex] A_y \dfrac{\partial u}{\partial y} + A_x \dfrac{\partial v}{\partial x} & A_y \dfrac{\partial v}{\partial y} + A_y \dfrac{\partial v}{\partial y} & A_y \dfrac{\partial w}{\partial y} + A_z \dfrac{\partial v}{\partial z} \\[2ex] A_z \dfrac{\partial u}{\partial z} + A_x \dfrac{\partial w}{\partial x} & A_z \dfrac{\partial v}{\partial z} + A_y \dfrac{\partial w}{\partial y} & A_z \dfrac{\partial w}{\partial z} + A_z \dfrac{\partial w}{\partial z} \end{bmatrix} \,. \qquad (3.53)
$$

Now the stress itself can vary throughout the fluid and can similarly be expanded in a series, so that the change in τ_{zx} across dz is $(\partial \tau_{zx}/\partial z)\, dz$, as in Fig. 3.16, for instance. By considering the fluid element of Fig. 3.17, the

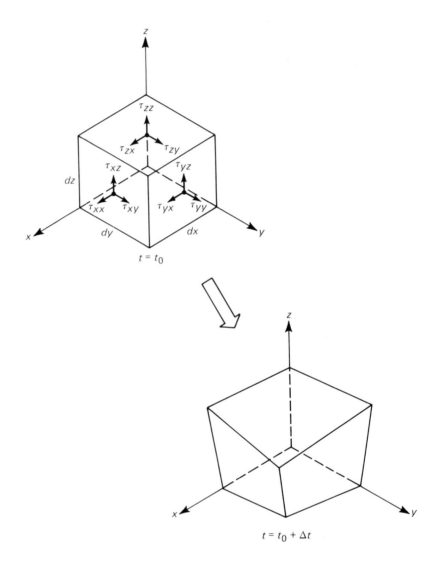

Fig. 3.17 Deformation of elemental fluid volume with time under shear stresses, shown schematically. For incompressible flow, both the mass and the volume of the element remain constant during the deformation.

net viscous force across all six faces in the x direction is seen to be given by the sum

$$d^3F_x = \frac{\partial \tau_{xx}}{\partial x} \, dx \, dy \, dz + \frac{\partial \tau_{yx}}{\partial y} \, dy \, dz \, dx + \frac{\partial \tau_{zx}}{\partial z} \, dz \, dx \, dy \,, \quad (3.54)$$

and the force per unit mass, $d^3F_x/\rho \, dx \, dy \, dz$, is

$$f_x = \frac{1}{\rho} \left(\frac{\partial \tau_{xx}}{\partial x} + \frac{\partial \tau_{yx}}{\partial y} + \frac{\partial \tau_{zx}}{\partial z} \right) . \quad (3.55)$$

Note that this summation proceeds down columns in Eq. 3.53. Repeating the sums for the y and z components, we obtain the vector form for the viscous body force:

$$\mathbf{f}_{vis} = \frac{1}{\rho} \nabla \cdot \boldsymbol{\tau} \,. \quad (3.56)$$

Here the divergence of the stress tensor has the matrix representation as:

$$\nabla \cdot \boldsymbol{\tau} = [\partial_x \partial_y \partial_z] \begin{bmatrix} \tau_{xx} & \tau_{xy} & \tau_{xz} \\[6pt] \tau_{yx} & \tau_{yy} & \tau_{yz} \\[6pt] \tau_{zx} & \tau_{zy} & \tau_{zz} \end{bmatrix} = \begin{bmatrix} \dfrac{\partial \tau_{xx}}{\partial x} + \dfrac{\partial \tau_{yx}}{\partial y} + \dfrac{\partial \tau_{zx}}{\partial z} \\[10pt] \dfrac{\partial \tau_{xy}}{\partial x} + \dfrac{\partial \tau_{yy}}{\partial y} + \dfrac{\partial \tau_{zy}}{\partial z} \\[10pt] \dfrac{\partial \tau_{xz}}{\partial x} + \dfrac{\partial \tau_{yz}}{\partial y} + \dfrac{\partial \tau_{zz}}{\partial z} \end{bmatrix}$$

$$= (\nabla \cdot \mathbf{A} \cdot \nabla)\mathbf{u} + (\mathbf{A} \cdot \nabla) \nabla \cdot \mathbf{u}$$

$$+ [(\partial_x + \partial_y + \partial_z) \, \mathrm{Tr}(\mathbf{A})] \nabla (u + v + w) \,. \quad (3.57)$$

The meaning of the various differential operators in Eq. 3.57 may be derived by taking the gradient of the matrix elements of Eq. 3.53, and then grouping the results into three expressions, as shown in Eq. 3.57. The first of these arises from the left-hand term in the matrix elements given by Eq.

3.52, and the remaining two come for the right-hand term. Discussing them in that order, they represent (1) the effects of eddy viscosity interacting with fluid shear, $(\nabla \cdot \mathbf{A} \cdot \nabla)\mathbf{u}$; (2) the effects of compressibility, as represented by $(\mathbf{A} \cdot \nabla)\nabla \cdot \mathbf{u}$ (which we will discuss in more detail ahead); and (3) the effects of spatial variations in the eddy viscosity as manifested by the expression $[(\partial_x + \partial_y + \partial_z)\mathrm{Tr}(\mathbf{A})]\nabla(u + v + w)$, where Tr $(\mathbf{A})$ is the trace, or sum of diagonal elements of $\mathbf{A}$. The first of these is by far the most important in geophysical fluids, and in fact, for an incompressible, uniform fluid, the last two are zero. For the remainder of this book (except for the chapter on acoustics), we shall assume that the fluid is incompressible, except for the static compression resulting from the weight of the overlying water. We shall also assume that the eddy viscosity does not vary in space. Then the sole remaining contribution to the viscous body force is

$$\mathbf{f}_{vis} = \frac{1}{\rho} \nabla \cdot \mathbf{A} \cdot \nabla \mathbf{u}$$

$$= \frac{1}{\rho} [\partial_x A_x \partial_x + \partial_y A_y \partial_y + \partial_z A_z \partial_z]\begin{bmatrix} u \\ v \\ w \end{bmatrix} . \tag{3.58}$$

With these assumptions, the complex sequence of differentiations of a tensor has contracted to a comparatively simple scalar operator working on each component of velocity, the expression for which can be written in the succinct form of Eq. 3.58. It is clear that the viscous forces involve second derivatives of velocity; to the extent that the coefficients A_i do not vary, those forces are proportional to the *curvature* of the velocity field and to the magnitude of the coefficients of viscosity. Thus viscosities are most important where the velocity field varies most rapidly in its second derivatives, which generally occurs in boundary layers.

In the ocean, the eddy coefficients for horizontal motions are much greater than for vertical movement. In the horizontal they are often but not always isotropic, although in general they are dependent on horizontal position. If we call the *horizontal* and *vertical* eddy coefficients A_h and A_v, respectively, then the *total* viscosities may be written as

$$A_x = A_y = A_h + \mu \tag{3.59a}$$

and

$$A_z = A_v + \mu . \tag{3.59b}$$

Thus the tensor form of the eddy and molecular viscosities becomes

$$\mathbf{A} = \begin{bmatrix} A_h & 0 & 0 \\ 0 & A_h & 0 \\ 0 & 0 & A_v \end{bmatrix} + \mu \mathbf{I}, \tag{3.60}$$

where $\mathbf{I}$ is the unit 3×3 diagonal matrix. If, in addition, the eddy coefficients are independent of position, the viscous operator simplifies to

$$\frac{1}{\rho} \nabla \cdot \mathbf{A} \cdot \nabla = K_h \nabla_h^2 + K_v \frac{\partial^2}{\partial z^2} + \nu_m \nabla^2, \tag{3.61}$$

where the horizontal Laplacian is

$$\nabla_h^2 \equiv \frac{\partial^2}{\partial x^2} + \frac{\partial^2}{\partial y^2}, \tag{3.62}$$

and the quantity

$$\nu_m = \frac{\mu}{\rho} \tag{3.63}$$

is termed the *kinematic viscosity,* which may be regarded as the molecular momentum diffusivity of the fluid. Its value in seawater at 20°C and 35 psu is 1.049×10^{-6} m^2 s^{-1}. The new quantities K_h and K_v are analogous to the molecular kinematic viscosity and are termed the *eddy diffusivities* for momentum. Thus

$$K_h \equiv A_h / \rho \tag{3.64a}$$

and

$$K_v \equiv A_v / \rho. \tag{3.64b}$$

Some representative values for A_h / ρ and A_v / ρ are listed in Table 3.2; the lower values go with smaller-scale motions, and conversely.

The dimensions of the diffusion coefficients (square meters per second) reveal something of the physics they describe, i.e., the mean-square step size of random diffusion and mixing processes per unit of time. It will become clear later that the horizontal dimensions of most orderly large-scale motions of the ocean (10 to 1000 km) are much greater than the vertical motions (0.01 to 100 m), and that the random motions must follow suit. The reasons for choosing anisotropic eddy diffusion coefficients follow from two of the properties of the sea discussed earlier: (1) its essential shallowness when compared with the horizontal scales of many of its motions, and (2) the strong vertical stability that the ocean generally possesses. Both of these characteristics make mixing and diffusion in the vertical much smaller than in the horizontal.

TABLE 3.2
Typical Values of Eddy Diffusion Coefficients

$$K_v = A_v/\rho = 3 \times 10^{-5} \text{ to } 2 \times 10^{-2} \text{ m}^2 \text{ s}^{-1}$$

$$K_h = A_h/\rho = 10^2 \text{ to } 10^5 \text{ m}^2 \text{ s}^{-1}$$

In summary, the internal body forces in the ocean are modeled by pressure and eddy viscosity, viz:

$$\mathbf{f}_{int} = -\frac{1}{\rho} \nabla p + \frac{1}{\rho} \nabla \cdot \mathbf{A} \cdot \nabla \mathbf{u} , \tag{3.65}$$

where the second term is given by Eq. 3.61.

3.10 Conservation Equations for Mass and Salinity

The Continuity Equation

The continuity equation is a statement of conservation of mass in a unit volume of a moving fluid. In words, it says that in the absence of sources or sinks of matter, a temporal change of the fluid density, $\partial\rho/\partial t$, within the parcel must result from an accumulation of the flux of mass, $\rho\mathbf{u}$; therefore, a nonzero divergence through the boundaries of the fluid volume must exist.

For an infinitesimal volume element of the type that we have been considering, the mass flux in the x direction into the left-hand face (of area $dy\,dz$) is $\rho u\,dy\,dz$; the flux out of the right-hand face is $[\rho u + \partial_x(\rho u)\,dx]\,dy\,dz$ because of mass accumulation within the volume, so that the net flux in the x direction is

$$\frac{\partial}{\partial x} (\rho u) \ dx \ dy \ dz \ .$$

Similar expressions for the other two pairs of faces give an overall net convergence (negative divergence) into the parcel of

$$-\left[\frac{\partial}{\partial x} (\rho u) + \frac{\partial}{\partial y} (\rho v) + \frac{\partial}{\partial z} (\rho w)\right] dx \ dy \ dz = - \nabla \cdot \rho \mathbf{u} \ d^3 \mathbf{x} \ . \quad (3.66)$$

This must obviously equal the rate at which mass is accumulating within the fixed volume, which is just the local rate of density change, $\partial \rho / \partial t$, times $d^3 \mathbf{x}$. Equating these rates, we obtain the *continuity equation*, one of the basic equations of fluid dynamics:

$$\frac{\partial \rho}{\partial t} + \nabla \cdot \rho \mathbf{u} = 0 \ . \quad (3.67)$$

An alternative form of the continuity equation can be derived using the convective derivative notation (Eq. 3.8). By expanding Eq. 3.67, one obtains

$$\frac{\partial \rho}{\partial t} + \mathbf{u} \cdot \nabla \rho + \rho \nabla \cdot \mathbf{u} = \frac{D\rho}{Dt} + \rho \nabla \cdot \mathbf{u} = 0 \ , \quad (3.68)$$

which ascribes the rate of change of density following the fluid motion to the velocity divergence.

If the fluid is incompressible, the density of a fluid element remains constant and $D\rho / Dt = 0;$ then the continuity equation requires that through any closed surface the inward flux be equal to the outward flux. That equation reduces to

$$\nabla \cdot \mathbf{u} = \frac{\partial u}{\partial x} + \frac{\partial v}{\partial y} + \frac{\partial w}{\partial z} = 0 \ . \quad (3.69)$$

In the ocean, the density of seawater varies slowly in space and time, having a typical surface value in temperate zones of $\rho \approx 1.026 \times 10^3$ kg m^{-3}, depending on salinity and temperature. Seawater is slightly compressible, with an isothermal compressibility, $a_p = (1/\rho)(\partial \rho / \partial p)_T = 4.3 \times 10^{-10}$ Pa^{-1} $= 43 \times 10^{-6}$ bar^{-1} at 2000 m depth and 4°C (see Chapter 4). This leads to an average fractional increase in density with depth, assuming only hydrostatic pressure, of

$$\frac{1}{\rho} \left(\frac{\partial \rho}{\partial p} \right)_T \left(\frac{\partial p}{\partial z} \right) = a_p \rho g \simeq 4.3 \times 10^{-6} \text{ m}^{-1} . \tag{3.70}$$

Thus seawater increases its density due to the weight of overlying fluid at a rate of approximately 4.3 parts per million (ppm) per meter. The reciprocal of this quantity is called the *scale height, H_s*, and has a value of approximately 230 km, which is the depth at which the ocean would have increased its density by a factor of $e = 2.718$ because of isothermal compressibility alone. Since the average depth of the sea is approximately 5 km, this means that the density of the ocean varies only slightly, even in the deepest portions of the sea. However, it does not mean that this variation is negligible; if seawater were truly incompressible, the surface of the ocean would be perhaps 30 m higher than it is. Futhermore, the compressibility must be taken into account in temperature and salinity measurements made in the deep ocean. However, for many calculations, especially those involving accelerations, the density can be assumed constant and the resultant equations considerably simplified; this forms the basis of the *Boussinesq approximation*, to be discussed in Chapter 5.

Returning to Eq. 3.67: For acoustic waves, the continuity equation as shown must be used. However, for slower fluid motions, even those with time scales characteristic of surface gravity waves, say, the incompressible version (Eq. 3.69) may generally be used. This forms the basis of one method of estimating the very slow vertical velocities of flow in the sea. If measurements can be made of the divergence of the *horizontal* flow, $\nabla_h \cdot \mathbf{u}$, over some area as a function of depth, these can be integrated to arrive at an estimate of the *vertical* flow speed, w, between any two depths, z_1 and z_2:

$$w(z_2) - w(z_1) = \int_{z_1}^{z_2} \frac{\partial w}{\partial z} \, dz$$

$$= -\int_{z_1}^{z_2} \left(\frac{\partial u}{\partial x} + \frac{\partial v}{\partial y} \right) dz . \tag{3.71}$$

Thus *upwelling* ($w > 0$) or *downwelling* ($w < 0$) speeds can be derived from an array of current meter moorings measuring horizontal currents alone, simply by assuming divergence-free flow. Such speeds are typically of order 10^{-4} to 10^{-6} m s^{-1} and are essentially impossible to determine by other means.

The Salinity Equation

The continuity equation is an example of a class of conservation equations for the fluxes of various properties, $\gamma(\mathbf{x}, t)$, which, unlike mass, may have sources and sinks. Thus in near-surface waters, oceanic salinity ($\gamma = s$) may increase or decrease due to riverine input, evaporation, rainfall, or freezing and thawing. An equation for the rate of change of salinity may then be written:

$$\frac{Ds}{Dt} = \frac{\partial s}{\partial t} + \mathbf{u} \cdot \nabla s = S_{in} - S_{out} , \qquad (3.72)$$

where the right-hand side represents the sources and sinks of salinity. In the Mediterranean Sea, for example, the evaporation excess has important consequences for the circulation of both surface and subsurface waters (cf. Chapter 2), and has led to the accumulation of massive quantities of salt on the floor of that Sea over geological times, when the Strait of Gibraltar was closed.

An explicit form for the balance of constitutive quantities such as salinity, heat, humidity, or energy may be derived from conservation arguments, and from these, equations describing advective and diffusive processes may be obtained. As a generic example, let $\mathbf{J}$ be the *advective flux per unit area* of some property such as salinity or density, and C be the quantity of that property per unit volume, or *concentration*. The advective flux and concentration are clearly related by

$$\mathbf{J} = C\mathbf{u} . \qquad (3.73)$$

Now the flux through some elemental area, dA, in the fluid is

$$\mathbf{J} \cdot \hat{n} \, dA , \qquad (3.74)$$

where $\hat{n}$ is the unit normal to the area dA. Since C can change within the volume, the advective flux must also change with a loss rate, or divergence per unit volume, of

$$\nabla \cdot \mathbf{J} = \frac{\partial J_x}{\partial x} + \frac{\partial J_y}{\partial y} + \frac{\partial J_z}{\partial z} , \qquad (3.75)$$

as obtained by a series expansion, as previously (see Fig. 3.13, for example). Then the rate of change of concentration due to advection alone is

$$\frac{\partial C}{\partial t} = -\nabla \cdot \mathbf{J} = -\nabla \cdot C\mathbf{u} \ . \tag{3.76}$$

In addition to a convergence due to advection, the concentration can change due to *diffusive processes*, both molecular and eddy-like. The *diffusive flux,* $\mathbf{J}_d$, is usually found to be proportional to the gradient of the concentration, ∇C, in the lowest approximation, with the constant of proportionality being the *diffusion coefficient*. As with the diffusion of momentum (cf. Section 3.9) this quantity may be anisotropic, so that the diffusive flux, which may not be in the same direction as the concentration gradient, may be written

$$\mathbf{J}_d = -\kappa \cdot \nabla C \ , \tag{3.77}$$

where the negative sign indicates that diffusion proceeds from regions of high concentration to regions of low concentration. For the ocean, we take the diffusion tensor, κ, to be diagonal but differing in the horizontal and vertical, as was the case with the eddy viscosity:

$$\kappa = \begin{bmatrix} \kappa_h & 0 & 0 \\ 0 & \kappa_h & 0 \\ 0 & 0 & \kappa_v \end{bmatrix} + \kappa_m \mathbf{I} \ . \tag{3.78}$$

Here κ_m is an isotropic molecular diffusion coefficient, as before. The rate at which the diffusive flux changes within the elemental volume is similar in functional form to the advective flux, so that the accumulation is proportional to the *convergence* of the total flux, $\mathbf{J} + \mathbf{J}_d$. The rate equation for the concentration is thus

$$\frac{\partial C}{\partial t} + \nabla \cdot (C\mathbf{u} - \kappa \cdot \nabla C) = C_{in} - C_{out} \ . \tag{3.79}$$

To apply this to salinity, let s be the fraction, by mass, of dissolved salts, multiplied by 1000; this unit is termed the *practical salinity unit* (psu), and replaces the former units of ‰, ppt, or parts per thousand. Then the concentration of salt per unit mass, C_s, is

$$C_s = \rho s \tag{3.80}$$

and the conservation equation for salinity becomes

$$\frac{\partial (\rho s)}{\partial t} + \nabla \cdot [\rho s \mathbf{u} - \kappa_s \cdot \nabla (\rho s)] = (E - P) + (F - \Theta) - R . \quad (3.81)$$

Here the *diffusivity of salt* in seawater, κ_s, depends on the thermodynamic state of the fluid, i.e., temperature, pressure, etc. For isotropic molecular diffusion alone, $\kappa_s \approx 1.5 \times 10^{-9}$ m^2 s^{-1} at $T = 25°C$, and at normal oceanic salinities. As with eddy viscosity, the eddy diffusion of salt is much larger than molecular diffusion, of course, but less is known about numerical values for salinity eddy diffusion than for momentum. However, estimates from the Mediterranean salt tongue of Figs. 2.24 and 2.25 give values of

$$\kappa_{sh} \approx 3 \times 10^3 \text{ m}^2 \text{ s}^{-1}$$

and

$$\kappa_{sv} \approx 5 \times 10^{-5} \text{ m}^2 \text{ s}^{-1} , \quad (3.82)$$

which are somewhat smaller than the momentum diffusivities. The source/sink terms represent either surface or volumetric rates for evaporation, E, minus precipitation, P, and freezing, F, minus thawing, Θ, minus R, the input of fresh water from riverine sources. This equation, together with analogous ones for momentum and heat, can be used to describe thermohaline circulations, salt-fingering and other density-driven motions.

3.11 The Momentum Equation

We have by now developed expressions for the major forces acting on an elemental volume of the ocean and we can collect these together into the momentum equation as outlined by Eq. 3.10. On the left-hand side will be the acceleration term in the rotating system (Eq. 3.27) but with the rotational terms transposed to the right-hand side to appear as fictitious forces; on the right will also be the gravitational forces (Eq. 3.50) with the centripetal term absorbed therein, and the pressure and viscous forces (Eq. 3.65). Gathering these together and dividing through by the density, we obtain from Newton's equation:

$$\frac{\partial \mathbf{u}}{\partial t} + \mathbf{u} \cdot \nabla \mathbf{u} = -2\Omega \times \mathbf{u} - \nabla \Phi - \frac{1}{\rho} \nabla p + \frac{1}{\rho} \nabla \cdot \mathbf{A} \cdot \nabla \mathbf{u} . \quad (3.83)$$

This is our required equation of motion, or momentum equation, written with all of the major terms displayed explicitly except for the centripetal contribution.

To recapitulate Eq. 3.83 term by term: The first is the acceleration at a fixed point, the so-called *inertial term*; the second is the *advective nonlinear term* giving the velocity change along a trajectory; the third is the Coriolis acceleration; the fourth is the combined gravity and centripetal accelerations; the fifth is the pressure force per unit mass; and the sixth is the combined eddy and viscous decelerations. Collectively they describe most of the known volume forces acting on a unit mass of ocean. Other forces may also be included as boundary forcing terms, or implicit thermodynamic effects, and these will be introduced as the case requires. This form of the momentum equation with the molecular viscosity included but without the eddy viscosity was originally due to Navier and Stokes and is known by those names even today.

It is instructive to write the momentum equation in component form and we will do so in Cartesian coordinates, for use in the tangent-plane geometry. In spherical coordinates, which must be used in planetary-scale numerical models, the equations are much more complicated. In Cartesian coordinates, we have defined *u, v,* and *w* to be the *x, y,* and *z* components of current velocity, respectively; in spherical coordinates, however, they become the east (ϕ, or zonal), north (Λ or meridional), and vertical (*r*, or up). Note that we have shifted from using the colatitude, θ, to the latitude, Λ, so that the direction of "north" and the associated unit vectors in both the spherical and tangent-plane systems have the same sense, i.e., toward the pole, and so that the tangent-plane system may be a slightly better approximation to the spherical system. This means that in order to maintain a right-handed coordinate system, the ordered triads of vectors in the two systems have different sequences than the usual one of *r*, θ, and ϕ. This system is then one of geographical latitude and longitude (Λ,ϕ), rather than the usual mathematical spherical coordinates (θ,ϕ), although only the transformation $\theta = \pi/2 - \Lambda$ has been made. It should be noted that *geographical coordinates* (as distinct from spherical) use ϕ for the latitude and λ for the longitude. In spite of their widespread use in geophysics, we have chosen not to write the salient equations in geographical coordinates because of the confusion that might result. The transformations from spherical to geographical coordinates are, respectively,

$$\theta_{sph} \rightarrow \pi/2 - \Lambda_{sph} \equiv \pi/2 - \phi_{geo}$$

and

$$\phi_{sph} \rightarrow \lambda_{geo} \; .$$

Momentum Equations in Cartesian Coordinates

(1) x component:

$$\frac{\partial u}{\partial t} + u \frac{\partial u}{\partial x} + v \frac{\partial u}{\partial y} + w \frac{\partial u}{\partial z} = 2v\Omega \sin\Lambda - 2w\Omega \cos\Lambda$$

$$- \frac{\partial \Phi}{\partial x} - \frac{1}{\rho} \frac{\partial p}{\partial x} + \frac{1}{\rho} \nabla \cdot \mathbf{A} \cdot \nabla u \; . \tag{3.84}$$

(2) y component:

$$\frac{\partial v}{\partial t} + u \frac{\partial v}{\partial x} + v \frac{\partial v}{\partial y} + w \frac{\partial v}{\partial z} = -2u\Omega \sin\Lambda$$

$$- \frac{\partial \Phi}{\partial y} - \frac{1}{\rho} \frac{\partial p}{\partial y} + \frac{1}{\rho} \nabla \cdot \mathbf{A} \cdot \nabla v \; . \tag{3.85}$$

(3) z component:

$$\frac{\partial w}{\partial t} + u \frac{\partial w}{\partial x} + v \frac{\partial w}{\partial y} + w \frac{\partial w}{\partial z} = 2u\Omega \cos\Lambda$$

$$- \frac{\partial \Phi}{\partial z} - \frac{1}{\rho} \frac{\partial p}{\partial z} + \frac{1}{\rho} \nabla \cdot \mathbf{A} \cdot \nabla w \; , \tag{3.86}$$

where the viscous operator is (from Eqs. 3.57 and 3.60)

$$\nabla \cdot \mathbf{A} \cdot \nabla = \frac{\partial}{\partial x}\left(A_h \frac{\partial}{\partial x} \right) + \frac{\partial}{\partial y}\left(A_h \frac{\partial}{\partial y} \right)$$

$$+ \frac{\partial}{\partial z}\left(A_v \frac{\partial}{\partial z} \right) + \mu \nabla^2 \; . \tag{3.87}$$

Momentum Equations in Spherical Coordinates

The momentum equations in spherical coordinates are much more complicated than in x, y, z coordinates, and even more so if the oblate spheroidal coordinates appropriate to Fig. 3.2 are used. An additional complication of the latter system arises because gravity is such a dominant force in the equations that even a very small but erroneous component of $-\nabla\Phi$ in the local horizontal direction, arising from an incorrect specification of a coordinate system, would constitute an important forcing term. For this reason an ideal if complex coordinate system could be defined by equipotential surfaces such as shown in Fig. 3.7, with z locally perpendicular to that surface. A. E. Gill discusses this as applied to an oblate spheroid system that approximates the geoid to within 0.2%. Here we shall simply assume that the system is spherical, then write appropriate expressions for the various operators in Eq. 3.83, and apply those to the momentum equation for this case. The other fluid equations will follow from the application of the operators listed below.

(1) Convective derivative:

$$\frac{D\gamma}{Dt} = \frac{\partial\gamma}{\partial t} + \frac{u}{r\cos\Lambda}\frac{\partial\gamma}{\partial\phi} + \frac{v}{r}\frac{\partial\gamma}{\partial\Lambda} + w\frac{\partial\gamma}{\partial r} \ . \tag{3.88}$$

(2) Gradient operator:

$$\nabla\gamma = \hat{\phi}\frac{1}{r\cos\Lambda}\frac{\partial\gamma}{\partial\phi} + \hat{\Lambda}\frac{1}{r}\frac{\partial\gamma}{\partial\Lambda} + \hat{r}\frac{\partial\gamma}{\partial r} \ . \tag{3.89}$$

(3) Divergence operator:

$$\nabla\cdot\mathbf{F} = \frac{1}{r\cos\Lambda}\left[\frac{\partial F_{\phi}}{\partial\phi} + \frac{\partial}{\partial\Lambda}(F_{\Lambda}\cos\Lambda)\right] + \frac{1}{r^{2}}\frac{\partial}{\partial r}(r^{2}F_{r}) \ . \tag{3.90}$$

(4) Viscosity operator (assuming $\mathbf{A}$ constant and isotropic):

$$(A\nabla^{2}\mathbf{u})_{\phi} = A\left\{\Delta u - \frac{1}{r^{2}\cos^{2}\Lambda}\right.$$
$$\left.\times\left[u + 2\frac{\partial}{\partial\phi}(v\sin\Lambda - w\cos\Lambda)\right]\right\} \ , \tag{3.91}$$

$$(A\nabla^{2}\mathbf{u})_{\Lambda} = A\left\{\Delta v - \frac{v}{r^{2}\cos^{2}\Lambda}\right.$$

$$+ \frac{2 \sin \Lambda}{r^2 \cos^2 \Lambda} \frac{\partial u}{\partial \phi} + \frac{2}{r^2} \frac{\partial w}{\partial \Lambda} \Big\} , \tag{3.92}$$

$$(A \nabla^2 \mathbf{u})_r = A \Big\{ \Delta w - \frac{2w}{r^2} - \frac{2}{r^2 \cos \Lambda} \frac{\partial u}{\partial \phi}$$

$$- \frac{2}{r^2 \cos \Lambda} \frac{\partial}{\partial \Lambda} (v \cos \Lambda) \Big\} , \tag{3.93}$$

$$\Delta \gamma = \frac{1}{r^2 \cos^2 \Lambda} \frac{\partial^2 \gamma}{\partial \phi^2} + \frac{1}{r^2 \cos \Lambda} \frac{\partial}{\partial \Lambda} \left(\cos \Lambda \frac{\partial \gamma}{\partial \Lambda} \right)$$

$$+ \frac{1}{r^2} \frac{\partial}{\partial r} \left(r^2 \frac{\partial \gamma}{\partial r} \right) . \tag{3.94}$$

(5) ϕ component:

$$\frac{Du}{Dt} = \left(2\Omega + \frac{u}{r \cos \Lambda} \right) (v \sin \Lambda - w \cos \Lambda)$$

$$- \frac{1}{r \cos \Lambda} \left(\frac{\partial \Phi}{\partial \phi} + \frac{1}{\rho} \frac{\partial p}{\partial \phi} \right) + \frac{1}{\rho} (A \nabla^2 \mathbf{u})_\phi , \tag{3.95}$$

(6) Λ component:

$$\frac{Dv}{Dt} = - \frac{vw}{r} - \left(2\Omega + \frac{u}{r \cos \Lambda} \right) u \sin \Lambda$$

$$- \frac{1}{r} \left(\frac{\partial \Phi}{\partial \Lambda} + \frac{1}{\rho} \frac{\partial p}{\partial \Lambda} \right) + \frac{1}{\rho} (A \nabla^2 \mathbf{u})_\Lambda , \tag{3.96}$$

(7) r component:

$$\frac{Dw}{Dt} = \frac{v^2}{r} + \left(2\Omega + \frac{u}{r \cos \Lambda} \right) u \cos \Lambda$$

$$- \left(\frac{\partial \Phi}{\partial r} + \frac{1}{\rho} \frac{\partial p}{\partial r} \right) + \frac{1}{\rho} (A \nabla^2 \mathbf{u})_r . \tag{3.97}$$

Because the depth of the ocean is such a small fraction of the radius of the earth (0.08%), a useful approximation is to assume that r in these equations is simply R_e, and that the radial derivatives can be replaced by $\partial/\partial z$.

The full spherical coordinate system must be used, rather than the beta plane approximation, whenever the north–south scale of the motion is large enough for variations in the numerical value of β to be important; another instance is when the azimuthal connectedness of the motion plays a role, as is the case for the Antarctic Circumpolar Current.

3.12   Fluctuations, Reynolds Stresses, and Eddy Coefficients

The eddy coefficients for momentum, salinity, and heat (this last to be discussed in Chapter 4) have been introduced without a completely satisfactory justification, if indeed one exists. In this section, we will briefly develop an improved explanation of fluctuations and diffusion in fluids in terms of long- and short-time-scale processes and Reynolds stresses, i.e., correlations between quasi-random variations of the fluid fields such as $\mathbf{u}$, p, and T, due to the nonlinear terms in the equation of motion.

From the previous discussions, it is clear that for lower frequency variability in the sea, the higher frequency motions, random or not, are of little interest except to the extent that they contribute to the evolution of the longer term dynamics. This suggests that some sort of filtering operation could be imposed that would low-pass the equations of interest and suppress the higher frequency motion. However, simple linear filtering is not a correct operation because of the nonlinear character of the terms. Instead, the concept of a two-time-scale equation of motion has been developed by a number of workers, including Reynolds, Prandtl, and others. In this, a certain time scale, t_a, is selected as the lower limit for the dynamical processes of interest, and the fluid equations are averaged over t_a. The evolution is then allowed to proceed on longer time scales and is captured in certain terms that appear in the averaged equations, which will in turn be related to the eddy coefficients. The success of this procedure depends in part on the existence of a so-called *spectral gap* in the function that describes the distribution of energy among the various frequencies of the motion, i.e., the *energy frequency spectrum*. Examples of such spectra will be presented in Chapter 5, but for now, Fig. 3.18 may serve to illustrate the concepts. This figure shows three schematic spectra, one of which is termed *red*, i.e., has increasing amounts of energy as the frequency is lowered (the wavelength increased) down to some frequency f_{low}; the second shows a peak at f_{peak} and characterizes a *narrowbanded process*. The third spectrum illustrates the concept of the spec-

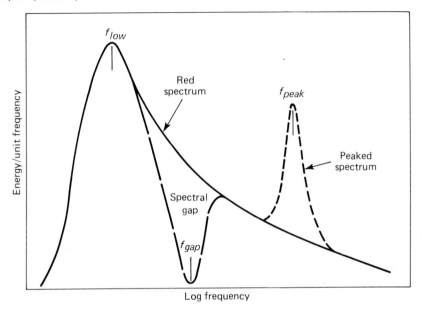

Fig. 3.18   Three types of energy spectra for oceanic motions. (a) Solid line: typical "red" spectrum showing monotonic increase toward lower frequencies up to the spectral maximum at f_{low}. (b) Dashed line: spectrum showing a gap at f_{gap}, which divides slow from fast motions. (c) Dotted line: spectrum showing a peak response at f_{peak}, characterizing narrowband processes such as tides and swell.

tral gap and is one for which an averaging over time scales shorter than $1/f_a$ would not seriously distort the description of the dynamics at lower frequencies. If such a gap can be shown to exist even approximately, than the procedure to be derived below would quite possibly be a valid one.

Consider a series of measurements of the generic fluid quantity, $\gamma(\mathbf{x},t)$, made at some point, $\mathbf{x}$, as a function of time. It is altogether possible, even likely, that another set of measurements made at the same place at another time would yield a quite different *time series*, because in geophysics (as with other branches of nonlaboratory science), randomness is a highly pervasive characteristic. The question then arises as to how to define the statistical moments of γ, i.e., values such as the mean, variance, and higher-order functions of the distribution. Now in statistical mechanics, the idea of *ensemble averaging* is a central concept, wherein an experiment of interest is repeated a large number of times, N, under presumably identical conditions, during which the macroscopic initial and boundary values of γ are maintained essentially constant. The ensemble average of γ may then be obtained by the operation

$$\langle \gamma \rangle = \frac{1}{N} \sum_{i=1}^{N} \gamma_i \, , \tag{3.98}$$

where the *bra* ($\langle$) and *ket* ($\rangle$) notation denotes an ensemble average.

While such repeated experiments may be practicable in the laboratory, they are next to impossible in geophysics (and totally impossible in astrophysics), because one cannot properly prepare the initial state of the natural system prior to each set of observations. Even in a system as controllable as one in the laboratory, it is often found that the detailed development of the fluid field under study is not reproducible from one run to the next, in spite of carefully prepared initial conditions; it is likely that only the values of the mean and standard deviation of the field are repeatable, while the finer-grained aspects of the motion are not. In fluids, this is often due to the existence of small-scale instabilities that grow unpredictably out of random fluctuations as the system evolves. Nevertheless, carefully defined ensemble averaging is the preferred method for arriving at the best estimates of the field.

An alternative method of averaging is to average over time, expecting that, if the series of observations is continued for a sufficiently long time, the system will ultimately pass through all of its *microstates,* with each state occurring as frequently as its statistics dictates. The mathematical process analogous to Eq. 3.98 is

$$\bar{\gamma} = \frac{1}{2t_a} \int_{-t_a}^{t_a} \gamma(t) \, dt \, , \tag{3.99}$$

where we have denoted the temporal average by an *overbar,* and the averaging interval by $2t_a$. If the random geophysical process is *statistically stationary*, i.e., the statistical moments of its distribution (mean, variance, kurtosis, etc.) are found to be the same when the experiment is repeated many times over, then the time average is expected to yield the same estimates of γ as the ensemble average, and in the sense of numerical equality, one has

$$\bar{\gamma} \leftrightarrow \langle \gamma \rangle \, . \tag{3.100}$$

The assumption of equivalence of temporal and ensemble averaging is termed the *ergodic hypothesis* of statistical mechanics.

If, however, the temporal mean is not statistically stationary but evolves in time, it is possible that the averaging interval may not have been extended sufficiently long to include all of the system microstates. On the other hand, it is often the case that the longer-time evolution of the system may constitute the very process of interest, and that the short-time averaging is sufficient to establish the mean and variance of the uninteresting processes. For example, one might wish to study the longer-period variations of oceanic current systems and not be at all interested in the tides, in which event a time series that encompasses a largish number of tidal periods but which is shorter than the characteristic time for change of the large-scale system (which may be months) could be used to average out the tides, while not doing serious damage to the measurement of low frequency variation. Nevertheless, even though there is a clear spectral gap between the semidiurnal/diurnal tides and the motions of interest (see Table 3.1), which presumably would allow one to separate out the low frequency signals from the high frequency ones, there are several things potentially troublesome about this procedure. One is that the tidal motions, through nonlinear effects, can modify the longer term dynamics. Secondly, the tides themselves have frequency components at fortnightly and longer periods and these components will appear directly in the average values. Thirdly, to thoroughly filter out the tidal components takes a *record length* of many tens of tidal cycles, during which the slower process itself may well be changing, i.e., the spectral gap is too narrow to allow the averaging to be done properly. Lastly, for many problems, no spectral gap may exist.

While there is no general solution to this problem except for intelligent consideration of alternatives, a reasonably useful procedure is to assume the possibility of dividing the motions into slow and fast processes, and to study the consequences for the fluid equations. To do this, we write the generic fluid quantity as the sum of a mean part, $\Gamma(\mathbf{x}, t_0)$, which is denoted by a capital letter and assumed to vary on longer time scales, t_0, and a fluctuating part, $\gamma'(\mathbf{x}, t)$, which is denoted by lower-case primed quantities and characterized by higher frequency variation. This latter variation may be either quasi-random or orderly. We will also assume that the temporal and ensemble averages are equivalent in some ergodic sense, so that we may interchange them at will, with the symbolic equivalences being represented by

$$\bar{\gamma} \leftrightarrow \Gamma \leftrightarrow \langle \gamma \rangle \ . \tag{3.101}$$

For notational reasons, the bra-ket average is the most convenient. Thus the velocity field, for example, will be decomposed into slowly varying and rapidly varying portions:

$$\mathbf{u}(\mathbf{x}, t) = \mathbf{U}(\mathbf{x}, t_0) + \mathbf{u}'(\mathbf{x}, t) \ , \tag{3.102}$$

where we have made explicit recognition of the low frequency variation of $\mathbf{U}$ via the use of t_0 for the slow time. By definition, the mean of $\mathbf{u}(\mathbf{x}, t)$ over time $2t_a$ is

$$\overline{\mathbf{u}} \equiv \mathbf{U} \equiv \langle \mathbf{u} \rangle , \tag{3.103}$$

while the mean of the fluctuations over the same interval is assumed to vanish:

$$\overline{\mathbf{u}'} = \mathbf{0} = \langle \mathbf{u}' \rangle . \tag{3.104}$$

Similar decompositions and notations can be applied to the other dependent variables in the dynamic and thermodynamic equations.

Our objective now is to derive relationships between the mean and fluctuating quantities on one hand, and the eddy viscosity and diffusivity tensors on the other. During the course of that program we will find that this will impose more precise statistical interpretations on the field variables, as well as improved insight into the meaning of $\mathbf{A}$ and κ. Additionally, we shall assume that the fluid is incompressible, in which case the continuity equation reduces to Eq. 3.69. Then upon substitution of Eq. 3.102 into the incompressibility condition, we obtain

$$\nabla \cdot (\mathbf{U} + \mathbf{u}') = 0. \tag{3.105}$$

Next, we perform our carefully chosen averaging operation on Eq. 3.105, which, being linear, commutes with other linear operators such as differentiation, integration, and addition, so that we obtain from that relationship

$$\langle \nabla \cdot (\mathbf{U} + \mathbf{u}') \rangle = \nabla \cdot \langle \mathbf{U} \rangle + \nabla \cdot \langle \mathbf{u}' \rangle = 0 . \tag{3.106}$$

Since $\langle \mathbf{u}' \rangle = 0$, so do each of the partial derivatives making up $\nabla \cdot \langle \mathbf{u}' \rangle$, and hence the divergences must vanish separately:

$$\nabla \cdot \mathbf{U} = 0 = \nabla \cdot \langle \mathbf{u}' \rangle . \tag{3.107}$$

This suggests that both the mean and the fluctuating components of velocity separately behave as incompressible fluids, perhaps, in retrospect, not a surprising result.

We now look at the effects of the field decomposition on the momentum equation, which we take to include only the more rigorously derived molecular viscosity, so that $A_v = A_h = 0$; however, we might expect that this parameterization will reappear during the manipulations. For ease of understanding we treat only the x component of that equation in the form of Eq. 3.84:

$$\left[\frac{\partial}{\partial t} + (\mathbf{U} + \mathbf{u}') \cdot \nabla\right] (U + u') = f(V + v') - e(W + w')$$

$$-\frac{\partial \Phi}{\partial x} - \frac{1}{\rho} \frac{\partial}{\partial x} (P + p') + \nu_m \nabla^2 (U + u') . \tag{3.108}$$

Here the horizontal and vertical Coriolis frequencies are, respectively,

$$f = 2\Omega \sin \Lambda \tag{3.109}$$

and

$$e = 2\Omega \cos \Lambda . \tag{3.110}$$

We next average Eq. 3.108, using the properties expressed by Eqs. 3.103, 3.104, and 3.106, a calculation that yields

$$\frac{\partial U}{\partial t_0} + U \frac{\partial U}{\partial x} + V \frac{\partial U}{\partial y} + W \frac{\partial U}{\partial z} - fV + eW$$

$$+ \frac{\partial \Phi}{\partial x} + \frac{1}{\rho} \frac{\partial P}{\partial x} - \nu_m \nabla^2 U$$

$$= -\frac{\partial}{\partial x} \langle u'u' \rangle - \frac{\partial}{\partial y} \langle v'u' \rangle - \frac{\partial}{\partial z} \langle w'u' \rangle . \tag{3.111}$$

The left-hand side of Eq. 3.111 is written entirely in terms of the slowly varying "mean" velocity, U, and pressure, P, and their temporal changes during characteristic times, t_0. The convective derivative must now be interpreted to be differentiation following this slowly varying motion. The right-hand side, which contains the only surviving terms from the nonlinear advective derivative that involve the fluctuating velocity, is in general nonzero because of the averages of quadratic quantities. These describe momentum fluxes carried by the fluctuations, where in Eq. 3.111 the horizontal (u') component can transport u', v', and w' components of momentum. These terms appear as body forces to the upper-case quantities that tend to reduce the acceleration from other forces, i.e., are frictionlike (unless negative, which happens on occasion), and can be considered to be elements of the divergence of the Reynolds stress tensor, τ_R. The slowly varying flow obeys an equation exactly like the primitive equation (Eq. 3.84) except for the stress divergence on the right-hand side in place of the eddy viscosity. For example, the z–x component of that tensor, which has the form

$$\tau_{zx} = -\rho \langle w'u' \rangle , \qquad (3.112)$$

represents the average flux of horizontal momentum carried across a surface, z = constant, by the fluctuations, in analogy to the arguments leading to Eq. 3.53. The Reynolds stress has a similar matrix representation:

$$\tau_R = -\rho \begin{bmatrix} \langle u'u' \rangle & \langle u'v' \rangle & \langle u'w' \rangle \\ \langle v'u' \rangle & \langle v'v' \rangle & \langle v'w' \rangle \\ \langle w'u' \rangle & \langle w'v' \rangle & \langle w'w' \rangle \end{bmatrix} . \qquad (3.113)$$

In a statistical sense the elements are to be thought of as the cross-correlation functions of the time-varying velocity components.

The next step in the development is to assume that the stresses are themselves proportional to the macroscopic derivatives of the mean velocity, and that under the influence of those gradients, fluid parcels execute random walks with an effective "mean free path" termed the *mixing length,* $\mathbf{l} = (l_x, l_y, l_z)$. In moving one mixing length, the parcel is assumed to pick up on the average a velocity increment, $\mathbf{u}'$, from the mean flow, which is thereby attenuated somewhat. Such down-gradient fluxes are the simplest assumptions applicable to diffusion processes and work well in kinetic theory of gases; but in this case of macroscopic fluctuations, the assumptions are more questionable, since there is no method yet known for calculating the coefficients involved. Nevertheless, we will take for the relationships the expressions:

$$\tau_{xx} = 2A_h \frac{\partial U}{\partial x} ,$$

$$\tau_{yy} = 2A_h \frac{\partial V}{\partial y} ,$$

$$\tau_{zz} = 2A_v \frac{\partial W}{\partial z} ,$$

$$\tau_{xy} = \tau_{yx} = A_h \left(\frac{\partial V}{\partial x} + \frac{\partial U}{\partial y} \right) ,$$

$$\tau_{xz} = \tau_{zx} = A_v \frac{\partial U}{\partial z} + A_h \frac{\partial W}{\partial x} ,$$

and

$$\tau_{yz} = \tau_{zy} = A_v \frac{\partial V}{\partial z} + A_h \frac{\partial W}{\partial y} . \tag{3.114}$$

A_v and A_h are the same quantities as discussed in Section 3.9, and they can be shown to be proportional to the product of the mixing length, the fluctuation velocity, and the density, viz:

$$A_h = \rho \langle l_x u' \rangle = \rho \langle l_y v' \rangle . \tag{3.115a}$$

and

$$A_v = \rho \langle l_z w' \rangle . \tag{3.115b}$$

Thus the diffusion tensor, $\mathbf{K} = \mathbf{A}/\rho$, is seen to be a measure of how rapidly the fluctuating velocity components transport momentum across one mixing length, on the average. The slow-scale vector equation of motion is then

$$\frac{\partial \mathbf{U}}{\partial t_0} + \mathbf{U} \cdot \nabla \mathbf{U} + 2\Omega \times \mathbf{U} + \frac{1}{\rho} \nabla P + \nabla \Phi = \frac{1}{\rho} \nabla \cdot \mathbf{A} \cdot \nabla \mathbf{U} , \tag{3.116}$$

which has a form identical to Eq. 3.83, except that the capitalized, averaged variables appear (which change over the longer times, t_0) rather than the instantaneous field variables. Thus the eddy viscosity is readily seen to be the result of fluctuations, but more importantly, the momentum equation in the form of Eqs. 3.83 or 3.116 must now be interpreted as one that has been averaged over the high frequency fluctuations, and which is thereby limited to describing the slower evolution of the ocean. What is meant by "high frequency" and "slow evolution" clearly depends on the processes of interest, and thus the numerical values to be used must also depend on them. The efficacy of the algorithm will depend, at the minimum, on the skill of the user, but under the proper circumstances, it appears to work satisfactorily.

For the remainder of this book, we will revert to the notation of Eq. 3.83 and will not explicitly refer to the averaging of that equation; however, the appearance of eddy coefficients in an equation will carry with it the implication that the equation must be considered to be the same type as Eq. 3.116, and that the choice of values for the elements of $\mathbf{A}$ and $\mathbf{K}$ must be made with the averaging process in mind.

3.13 Boundary Conditions in Fluid Dynamics

Solutions to partial differential equations are completed by imposing on them *initial conditions* specified over all space, and *boundary conditions* specified over all time. While most initial conditions are peculiar to the problem at hand, most boundary conditions are of a rather general nature. We shall discuss several of the latter.

Conditions At a Rigid, Impermeable Boundary

Here there can be no fluid flow across the boundary and no motion of that surface. If $\hat{n}$ is a unit normal to the boundary, the conditions on the velocity and the advective and diffusive fluxes are:

$$\mathbf{u} \cdot \hat{n} = 0 , \tag{3.117a}$$

$$\mathbf{J} \cdot \hat{n} = 0 , \tag{3.117b}$$

and

$$\mathbf{J}_d \cdot \hat{n} = 0 . \tag{3.117c}$$

Interfacial Boundary

A deformable boundary such as that between air and water, or between two different but immiscible fluids is such that no fluid particles cross the boundary as long as it maintains its integrity, i.e., no spray or mixing occurs. Thus, if $\xi(x,y,t)$ is the equation of the surface elevation above the equilibrium level, z_0, then the vertical coordinate, z, of a particle on the free surface is

$$z = z_0 + \xi(x,y,t) . \tag{3.118}$$

The requirement that a fluid particle on the surface will remain there as it moves along is equivalent to

$$\frac{D}{Dt}(z - z_0 - \xi) = 0$$

$$= \left(\frac{\partial}{\partial t} + u \frac{\partial}{\partial x} + v \frac{\partial}{\partial y} + w \frac{\partial}{\partial z} \right)(z - \xi) . \tag{3.119}$$

Since the coordinate z is independent of x, y, and t, we may differentiate ξ and solve for w to obtain

$$w = \frac{\partial \xi}{\partial t} + u \frac{\partial \xi}{\partial x} + v \frac{\partial \xi}{\partial y} \qquad (3.120)$$

as the *kinematic boundary condition* on the vertical velocity of a free surface. This boundary condition is clearly nonlinear, involving products of height and velocity. However, if the fluid velocities on the surface or the slopes, $\partial \xi / \partial x$ and $\partial \xi / \partial y$, respectively, are small, the terms involving the horizontal currents can be neglected compared with $\partial \xi / \partial t$; then the kinematic condition reduces to

$$w \simeq \frac{\partial \xi}{\partial t} . \qquad (3.121)$$

This form is useful in small-amplitude linear problems.

Beyond the kinematic condition, it is also clear that pressure differences across a free surface cannot be supported without disrupting that surface, unless a normal surface stress exists to balance the difference; intermolecular forces exert such stresses via a *surface tension, τ_s*, which resists the pressure upon distension. The value of τ_s for clean seawater is approximately 0.079 N m^{-1} (79 poise). Surface stress is found experimentally to be proportional to the *curvature* of the distorted free surface, or inversely proportional to the radius of curvature, R; in two dimensions, it is the sum of the curvatures, $1/R_x + 1/R_y$, that is the operative quantity. Then the pressure difference that can be supported is

$$p - p_a = -\tau_s \left(\frac{1}{R_x} + \frac{1}{R_y} \right) , \qquad (3.122)$$

where p_a is the pressure above the free surface (i.e., atmospheric pressure). The radii of curvature are counted positive when the center of curvature is above the surface, or outside the liquid (see Fig. 3.19). For the case shown, $R_x < 0$, and the interior pressure is thus greater than the exterior, as it must be under the influence of the surface tension. Now the curvatures are approximately the second derivatives of the surface elevation, so that the pressure just below the surface is

$$p = p_a - \tau_s \left(\frac{\partial^2 \xi}{\partial x^2} + \frac{\partial^2 \xi}{\partial y^2} \right) . \qquad (3.123)$$

This equation is termed the *dynamic boundary condition*, and its imposition

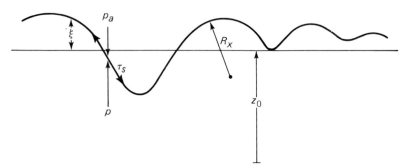

Fig. 3.19 Small-scale forces at air–water interface, showing pressure difference, $p - p_a$, balanced by surface tension, τ_s. The radius of curvature is R_x, which is negative if its center is beneath the surface of the liquid. Capillary waves have restoring forces due to surface tension.

leads to equations for capillary waves, for example. It is also required as a condition between layers of two dissimilar liquids, in which event τ_s clearly will have a numerical value different from the one cited above. Oils and other surfactants can cause considerable variation in τ_s, with even a thin film of oil on a surface reducing τ_s to less than one-half of its clean surface value.

Boundary Conditions for Viscous Fluids

A viscous fluid requires additional balances of forces at boundaries, the most obvious of which is that the tangential (as well as the normal) component of fluid velocity must match that of the boundary:

$$\mathbf{u} \times \hat{n} = \mathbf{u}_b \ . \tag{3.124}$$

If the boundary is fixed, this condition requires that the interior flow velocity go to zero at the boundary or, if moving, to move in unison with it. This transition takes place through a relatively thin layer such as the surface Ekman layer or the benthic boundary layer on the seafloor, whose thicknesses are counted in tens of meters. In such a region, a large velocity shear $(\partial u / \partial z)$ normal to the surface develops that may often render the boundary layer unstable, much as internal shear does (see Fig. 3.14). In the shear layer, small disturbances may grow rapidly via *shear flow instability,* leading to a turbulent boundary layer that transports fluid properties in directions normal to the flow via eddy fluxes at rates much greater than laminar flow allows. It is this process that the tensors **A** and κ attempt to describe. Other factors such as density gradients and the presence of contaminants can also influence the rate of growth of the turbulent layer. Shear flow instability in a two-layer fluid is discussed in Chapter 5. These examples of the use of the boundary conditions will be given during the process of solving for oceanic motions such as waves and currents. Additionally, the field variables for thermal, acoustical, electromagnetic, and optical fields have their own sets of boundary conditions, and these, along with their governing equations, will be developed further along the way.

For now we leave the mechanical portions of the physics of the sea and turn to thermodynamics and energy fluxes in order to understand the effects of heat and radiation in the ocean. We will return to hydrodynamics at the end of Chapter 4, where the coupled dynamical and thermodynamical equations will be developed.

Bibliography

Books

Batchelor, G. K., *An Introduction to Fluid Dynamics*, Cambridge University Press, Cambridge, England (1967).
Defant, A., *Physical Oceanography,* Vols. I and II, Pergamon Press, New York, N.Y. (1961).
Gill, A. E., *Atmosphere-Ocean Dynamics*, Academic Press, New York, N.Y. (1982).
Lamb, H., *Hydrodynamics*, 6th ed., Dover Publications, New York, N.Y. (1945).
Landau, L. D., and E. M. Lifshitz, *Fluid Mechanics,* Pergamon Press, Oxford, England (1959).
Neumann, G., and W. J. Pierson, Jr., *Principles of Physical Oceanography,* Prentice-Hall, Inc., Englewood Cliffs, N.J. (1966).
Prandtl, L., *Essentials of Fluid Dynamics,* Hafner Publishing Co., New York, N.Y. (1952).
von Arx, W. S., *An Introduction to Physical Oceanography,* Addison-Wesley Publishing Co., Reading, Mass. (1962).
Yih, C.-S., *Fluid Mechanics,* McGraw–Hill Book Co., New York, N.Y. (1969).

Journal Articles and Reports

Haxby, W. F., *1982-83 Lamont-Doherty Geological Observatory Yearbook 12*, Columbia University, Palisades, N.Y. (1983).
Marsh, J. G., A. C. Brenner, B. D. Beckley, and T. V. Martin, "Global Mean Sea Surface Based Upon the Seasat Altimeter Data," *J. Geophys. Res.,* Vol. 91, p. 3501 (1986).
Oceanographic Atlas of the North Atlantic Ocean, Section I, U.S. Naval Oceanographic Office, Washington, D.C. (1969).
Schwiderski, E. W., *Global Ocean Tides, Part II,* U.S. Naval Surface Weapons Center, NSWC TR-79-414, Dahlgren, Va. (1979).
Thorpe, S. A., "Experiments on the Instability of Stratified Shear Flows: Miscible Fluids," *J. Fluid Mech.* Vol. 46, p. 299 (1979).

Chapter Four

Thermodynamics and Energy Relations

*"The heat's on, the pressure's great,
The heat's on, we're in a state."*

Rock song, 1985

4.1 Introduction

The unusual thermodynamic properties of water that were mentioned in the previous chapters deserve a fuller explanation than the purely phenomenological one that has been given to date. A complete exposition would take us into the physics of the liquid state, a subject that is less well developed than the parallel topics of the solid or gaseous states. However, much can be explained through a better understanding of the molecular character of the main constituents of seawater: H_2O, Na^+, Cl^-, and Mg^{++}. Other ions are present as well, as can be seen in Table 4.1, but their contributions are fractionally small compared with the major species.

TABLE 4.1
Major Constituents of Seawater
(Salinity 35 psu)

Positive ions (g kg^{-1})		Negative ions (g kg^{-1})	
Na^+	10.752	Cl^-	19.345
Mg^{++}	1.295	Br^-	0.066
K^+	0.390	F^-	0.0013
Ca^{++}	0.416	SO_4^{--}	2.701
Sr^{++}	0.013	HCO_3^-	0.145
		$B(OH)_3$	0.027

H_2O: 965 ppt	Dissolved Materials: 35 ppt

In addition to the major constituents cited, seawater contains nearly all known natural elements, some of which are present in exceedingly minute amounts, to be sure. However, it is the properties of pure water on one hand and the dissolved salts on the other that account for the dominant characteristics of the medium.

4.2   Molecular Structure of H_2O

Water is a strongly polar molecule, with the larger oxygen atom bonding the smaller hydrogen atoms at a subtended angle of 104° 31′ and at a distance of 0.096 nm (0.96 Å). Due to the unshared electrons on the oxygen atom, the electronic orbitals produce a four-pronged molecule that is negative on the side containing the unshared negative charges, and positive on the side with the H atoms. Figure 4.1a shows the configuration for a single molecule. The resultant dipole moment (charge × separation), which is of order 1.84 debyes (1 debye = 3.3357×10^{-30} coulomb-meters), enables the water molecule to bond with its neighbors with intermediate strength bonds. This cohesive force is termed *hydrogen bonding* and is illustrated in Fig. 4.1b. Water also possesses a strong ability to dissolve salts that are held together

(a)   (b)

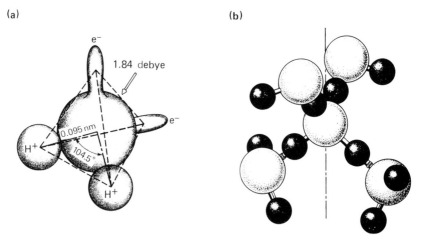

Fig. 4.1 (a) Schematic of atomic configuration and electronic orbitals for water molecule, with oxygen atom at center and hydrogen atoms at angle of 105°. Dipolar character comes from protons at H + positions and unshared electrons at e − locations; the direction of the dipole moment is along the symmetry axis. The atoms and the electronic orbitals have tetrahedral symmetry. (b) A fully bonded structure of water molecules, with hydrogen bonds represented by dark, disk-like rods, and OH bonds by lighter, smooth rods. This structure represents the "ice-like" microstate of water, and is about 10% less dense than the close-packed structure of unbounded molecules.

by ionic bonds, e.g., NaCl and the other ionic constituents of seawater. It is this polar character of the hydrogen bond that gives water its unusual thermodynamic properties such as elevated boiling and freezing points, large heat capacity, high latent heat of vaporization, and large dielectric constant.

Water enters into important oceanic processes in all three of its phases — gaseous, liquid, and solid. As was described in Chapter 2, the latent heat

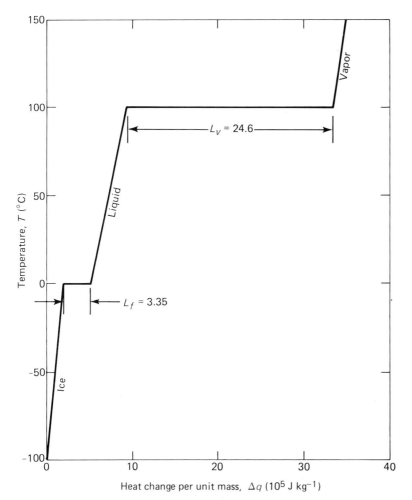

Fig. 4.2 Heats of liquification, L_f, and vaporization, L_v, and specific heat at constant pressure, C_p (as represented by the reciprocal slopes of the lines), for ice, water, and vapor, shown as a function of temperature. Even though seawater does not reach 100°C, it nevertheless goes through a transition to the vapor phase under the influence of warming.

of vaporization and condensation are the major means by which the sea communicates with the atmosphere. In the vaporous phase, each water molecule is free to move between collisions, and exerts a pressure proportional to its kinetic temperature; it is a triatomic gas and reflects that in its number of degrees of freedom and its specific heat. At 100°C and 1 atm pressure, it undergoes a first-order phase transition to the liquid state, releasing its heat of condensation without change of temperature in the process (Fig. 4.2). In the liquid state, a condition intermediate between gas and solid, its anomalous physical properties are more apparent. According to current theories of the liquid state, fluctuations in the medium lead to two types of microstates: (1) free water molecules not unlike those in the gaseous state, undergoing only weak interactions with their neighbors, and (2) a kind of transitory partial solid consisting of clusters of hydrogen-bonded water molecules with microproperties akin to those of ice, that persist for times of order 10^{-10} s. The binding energy and coordination of molecules in this microstate are greater than for the gas-like state. The number of clusters decreases with increasing temperature until at 100°C all are in the gaseous microstate. At the other end, the ice-like state is achieved at 0°C for pure water (although for

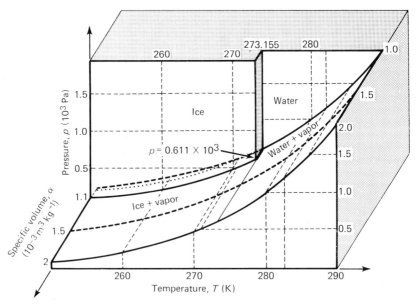

Fig. 4.3 Phase diagram for fresh water showing pressure–volume–temperature surface in the vicinity of the triple point, which is actually a line at $T = 273.155 + 0.0098$ K and $p = 0.611 \times 10^3$ Pa. Ice, liquid, and vapor can all coexist between approximately 1.0×10^{-3} and 1.09×10^{-3} m³ kg⁻¹. Addition of salt alters this surface. [From Fleagle, R. G., and J. A. Businger, *An Introduction to Atmospheric Physics* (1980).]

saline solutions, the freezing point is depressed below and the vapor temperature is elevated above the pure water values, as will be discussed later).

The crystaline structure of ice is hexagonal, with 24 molecules in the crystal lattice occupying approximately the same volume as 27 free molecules in the liquid state; therefore ice is less dense than water, with a density of 0.917×10^3 kg m^{-3} at $0°C$ and 1 atmosphere pressure. While this structure is relatively open, sea salts are not readily accommodated in the interstitials and are found instead as pockets of unfrozen brines within the sea ice. Upon crystallization, pure water releases its heat of fusion of 3.35×10^5 J kg^{-1}, or 80 cal g^{-1}. Figure 4.2 summarizes these phase changes in a temperature-heat change plot valid for sea level conditions. Figure 4.3 presents a three-dimensional phase surface showing the pressure–specific volume–temperature relationship for pure water in all three states.

4.3 Effects of Temperature, Salinity, and Pressure

The addition of salt to water alters its thermodynamic properties in various ways, some of which have important qualitative as well as quantitative effects. Consider first the density of pure water as a function of temperature (Fig. 4.4). At a temperature of $T_\rho = 3.98°C$, it is at its maximum density of $\rho = 1.000 \times 10^3$ kg m^{-3}; above and below that temperature, the density is lower. The addition of salt also increases the density, so that at normal oceanic salinities of some 35 psu, the densities are near 1.024×10^3 kg m^{-3}. The salt also depresses both the temperature of maximum density, T_ρ, and the freezing point temperature, T_f, as shown in Figs. 4.5 to 4.7; these two temperatures coincide at $-1.33°C$ and a salinity of 24.695. It is a fact of considerable profundity that fresh water is lighter at its freezing point than at slightly warmer temperatures; this property prevents lakes from freezing from the bottom up. As a lake cools from above, the water becomes more dense and begins to sink to intermediate depths; it is replaced by rising warmer subsurface water, until at approximately 4°C it is maximally dense. As this condition is approached, the cooled surface water can sink to the bottom. In this fashion, the entire water column of the lake turns over and becomes isothermal at 4°C before further cooling of the surface can take place. Only then may the surface temperature approach freezing and the formation of ice begin. This peculiar property of water allows life in fresh water bodies to survive at depth unless the entire water mass freezes, a fact that obviously has had important impacts on aquatic species during their evolutionary history.

The freezing point of water is also lowered by pressure at a rate of -7.53×10^{-3} °C bar^{-1} pressure increase. Figure 4.7 shows the depression of both

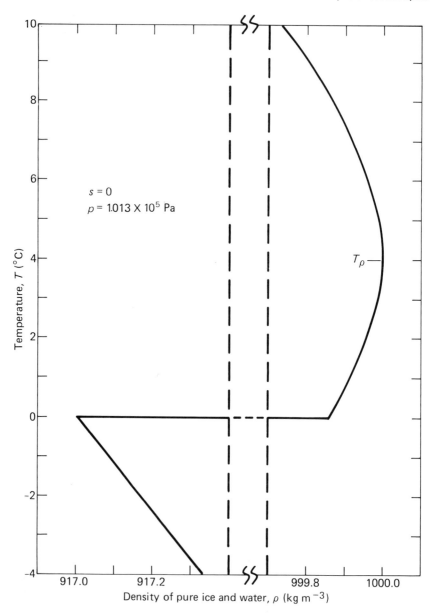

Fig. 4.4 Density of pure water and ice as a function of temperature. Ice is minimally dense at $T = 0°C$, while fresh water is maximally dense at $T = 3.98°C$. Salt depresses temperature of maximum density as well as increasing density. [Adapted from Gross, M. G., *Oceanography* (1982).]

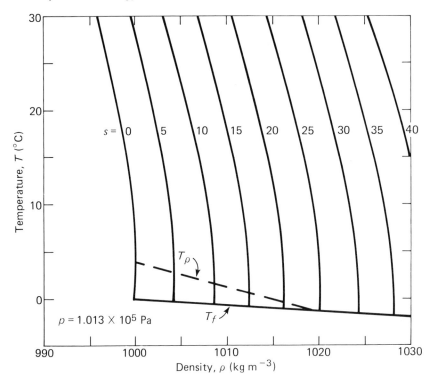

Fig. 4.5 Density of saline water as function of temperature and salinity. With increasing salinity, both freezing point, T_f, and temperature of maximum density, T_ρ, are lowered until at $T = -1.33°C$ and $s = 24.7$ psu, the two temperatures coalesce.

T_f and T_ρ with pressure. An equation relating the depression of the freezing point to salinity and pressure is:

$$T_f(s,p) = as + bs^{3/2} + cs^2 + dp , \qquad (4.1)$$

where $a = -0.0575$, $b = 1.710523 \times 10^{-3}$, $c = -2.154996 \times 10^{-4}$, and $d = -0.00753$, and where T is in degrees Celsius, s is in psu, and p is in bars.

From this phenomenological explanation, it is clear that the density of sea-water is a complicated function of the state variables, and that an equation of state is needed for an orderly and more complete description of the system. Therefore, we will proceed to the somewhat more abstract and powerful apparatus of thermodynamics as given by that subject's First and Second Laws applied to the fluid in question. We will find that there is a large element of empiricism present in the treatment because of the complications

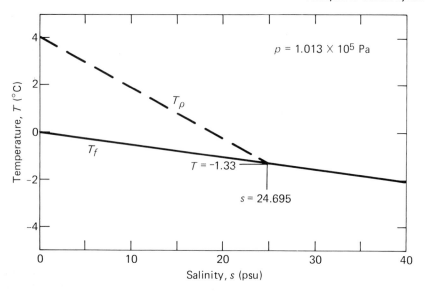

Fig. 4.6 Temperature of maximum density and freezing point as a function of salinity, at 1 atm.

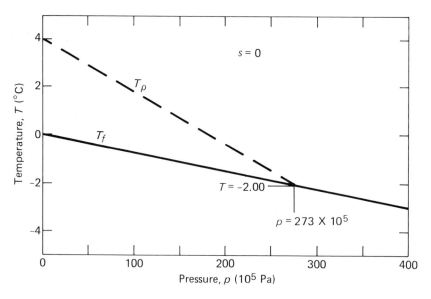

Fig. 4.7 Temperature of maximum density and freezing point as a function of pressure, at zero salinity.

of the medium and the need to derive very accurate values of densities for certain work in oceanography. However, this subject is not alone in that regard, for real systems invariably require a retreat to experimental data for most values of material properties.

4.4 Thermodynamics of Seawater

The task at hand is to use the basic laws of thermodynamics to obtain an equation of state of the form

$$\rho = \rho(s, T, p) , \qquad (4.2)$$

as well as the other equations needed to round out the dynamical statements of the last chapter, so that there are as many equations as unknowns. We will find that this task involves evaluating a number of thermomechanical coefficients, e.g., specific heat and compressibility, as functions of the *state variables* such as ρ, s, T, and p, and then using these quantities in relationships derived from the basic laws.

First Law of Thermodynamics

The First Law is a statement of conservation of certain kinds (not all) of energy, in particular, the internal energy of the thermodynamic system under study. As it is usually given, it relates small changes of the generalized store of *energy per unit mass, de,* to other small changes in *heat content per unit mass, dq,* and the *differential of work done on the system, dw.* The internal energy is a *state variable* in that the store of energy in the system, when in equilibrium in a given state, always has the same value no matter how the state was reached. This means that if s, T, and p are the independent state variables chosen to describe seawater, then the internal energy depends only on these values and not on the history of the processes that brought it to that state. Mathematically, this is expressed by saying that *de* is a *perfect differential.* However, q and w are not state variables and hence are not perfect differentials.

The First Law is written as

$$de = dq + dw \qquad \text{J kg}^{-1} . \qquad (4.3)$$

For example, if the system is confined under constant pressure, a decrease in the specific volume, i.e., the volume per unit mass, $d\alpha$, would mean that some external agent had done work *on the system,* and thus $dw = -p \, d\alpha$, i.e., the external agent had compressed it. Here $\alpha = 1/\rho$ is the specific volume of sea water, measured in cubic meters per kilogram. Or, if several chemical substances are added to a sample of water, each of which has a chemical potential, μ_i, and each in an amount of dn_i moles, the internal energy would be changed by $\Sigma \mu_i dn_i$; this could be due to the addition of salts, say, upon which energy is absorbed during the dissolution of the salts. Another source of change comes about when evaporation occurs, in which event the heat of vaporization would lead to a loss of heat, $L_v dn$, where L_v is the heat of vaporization per mole (see Eq. 2.9), and dn is the number of moles evaporated. Another source that can change the internal energy of the water is the induction of an electric dipole moment per unit volume, $d\mathbf{P}$, by an applied electric field, $\mathbf{E}$, which might come from an electromagnetic wave, for example. Both the polar water molecule and the dissociated salt ions can contribute to the electric polarization, as will be discussed in Chapter 8. Other processes could also occur that change the internal energy and which must be entered into the energy budget, so that a more general conservation equation can be written:

$$de = dq - p \, d\alpha + \sum_i \mu_i \, dn_i + L_v \, dn + \alpha \mathbf{E} \cdot d\mathbf{P} + \dots . \quad (4.4)$$

It is worth noting that each of the terms in this expression (with the exception of dq) is the product of an *intensive variable* giving the intensity of a quantity—pressure, chemical potential, etc.—and an *extensive variable,* such as α or n_i, giving the amount of substance present. We will presently see that the application of the Second Law will remove the lack of symmetry for dq.

In this formulation we have left out other types of energy, such as kinetic and potential energy of the fluid flow, since these are mechanical forms. While in principle they can be converted to thermodynamical forms through friction and other dissipative forces, in practice in the sea these changes are small compared to other energy inputs. We shall return to the mechanical energies at the end of the chapter, when rate equations for the various types of energy will be derived and the energy conversions will be discussed.

The chemical potential may be written as if there were only one chemical species affecting the internal energy; but we have seen that there are at least a dozen. However, by considering the mass fraction of these constituents as a fraction of the total salinity, $s,$ and further considering μ as a combined chemical potential of all of the salts relative to the potential of pure water, the chemical potential of the combined system can be expressed as

$$\mu \, ds = \sum_i \mu_i \, dn_i . \quad (4.5)$$

This is possible because the proportions of the various salts in the sea vary only slightly and can therefore be represented by a single quantity, the mass fraction or salinity, *s*.

The internal energy is one of several *thermodynamic potentials* of use in arriving at measurable thermomechanical coefficients. For fluids, a more convenient one is the *enthalpy per unit mass, h,* defined via

$$h = e + p\alpha , \tag{4.6a}$$

whose differential is

$$dh = de + p\,d\alpha + \alpha\,dp . \tag{4.6b}$$

If Eq. 4.4 is substituted into this enthalpy expression, the term in $p\,d\alpha$ immediately cancels, leaving the differential $\alpha\,dp$ representing pressure–volume changes. For fluid systems (as contrasted with gases), the desired coefficients can be determined much more readily at constant pressure than at constant density; hence the preference for enthalpy as a state variable.

Second Law of Thermodynamics

In Eq. 4.4, the heat differential, *dq,* was not expressed as the product of an extensive and an intensive variable, as were the other quantities; nor is *dq* a perfect differential. These deficiencies can be rectified by the introduction of the Second Law, which states that the change in *entropy per unit mass, dη,* when an increment of heat is added, is greater than or equal to *dq/T*, where *T* is the absolute temperature in kelvins. Thus, the Second Law of Thermodynamics is expressed as an inequality, viz:

$$d\eta \geq dq/T . \tag{4.7}$$

Now entropy is a measure of the *disorder* possessed by the system and is a state variable/perfect differential of the extensive type that is paired with the absolute temperature as an intensive variable. In fluid dynamics, as we shall see, entropy is proportional to the amplitude of the motion, among other quantities. Strictly speaking, the equality in Eq. 4.7 holds only for reversible processes, which, in going through a thermodynamic cycle that returns the state variables to their initial condition, also return entropy to its initial value. In the presence of rapid variations, however, the Second Law asserts that the entropy will *increase* during such a cycle. Calculations with Eq. 4.7 as an equality rather than an inequality will place a lower bound on the entropy, and quantities derived from it will be bounded estimates. In practice, the circumstances are somewhat unusual wherein the changes in the ocean are so rapid as to render the equation unusable.

By substituting the entropy expression (Eq. 4.7) and the equivalent chemical potential (Eq. 4.5) into the First Law, the combined form of the two laws is obtained:

$$de = T \, d\eta - p \, d\alpha + \mu \, ds + L_v \, dn + \alpha \mathbf{E} \cdot d\mathbf{P} + \dots \, . \qquad (4.8a)$$

The same procedure for the enthalpy expression yields

$$dh = T \, d\eta + \alpha \, dp + \mu \, ds + L_v \, dn + \alpha \mathbf{E} \cdot d\mathbf{P} + \dots \, . \qquad (4.8b)$$

This latter equation will form the basis for the discussion of the thermodynamics of the ocean to follow. In order to apply it to the time-dependent situations encountered in hydrodynamics, time derivatives are required, and these must be considered to be convective derivatives because the thermodynamic fields are both time- and space-dependent. Thus, we must write

$$\frac{Dh}{Dt} = T \, \frac{D\eta}{Dt} + \alpha \, \frac{Dp}{Dt} + \mu \, \frac{Ds}{Dt} + L_v \, \frac{Dn}{Dt} + \alpha \mathbf{E} \cdot \frac{D\mathbf{P}}{Dt} \, , \qquad (4.9)$$

and agree that we shall restrict ourselves to time variations that are slow enough to maintain the validity of the entropy equality.

A word about the propriety of using time derivatives in Eq. 4.9 is in order, since thermodynamics ordinarily deals with equilibrium states and, indeed, might more appropriately be called thermostatics. For a fluid in a nonequilibrium state, the time changes associated with its motion are to be thought of as occurring during a succession of closely related states, in each of which the departure from equilibrium is small. State variables such as density and internal energy clearly have definite values at any time during this process, and are independent of the existence of equilibrium. Additionally, other quantities that depend functionally on ρ and e, such as η and T, may be defined in nonequilibrium, time-dependent situations through equations relating them to ρ and e. Thus time-dependent thermodynamics can be derived from classical thermodynamics by considering the fluid states to be a succession of near-equilibrium states. In situations such as very high frequency sound waves, or shock conditions, the fundamental premises must be reexamined for correctness, but for most problems in oceanography, the near-equilibrium case may be assumed.

4.5 Additional Thermodynamic Equations

Equation 4.9 is an important statement that enlarges our repertoire of dynamical relationships, but which introduces more new dependent variables (h, η, μ, and T) than it contributes equations. We must therefore derive additional thermodynamic relationships from our previous work, in order to obtain as many final equations as unknowns. This task will take us into the complexities and abstractness of partial derivatives of thermodynamic state variables, a thorough understanding of which requires a certain amount of experience with thermodynamical functions. Since our objectives here are to study ocean physics, we shall only outline the derivation of the more important relationships between partial derivatives and refer the reader to the bibliography at the end of the chapter for more detail.

Consider the enthalpy per unit mass, h, to be a function of the variables η, p, and s, to coincide with Eq. 4.8:

$$h = h(\eta,p,s) \qquad \text{J kg}^{-1} . \tag{4.10}$$

We will regard the other functions (L_v and E) appearing in the equation for the enthalpy as being fully specified and contributing in a deterministic way to the variables in Eq. 4.10. Then small changes in h can be written as

$$dh = \left(\frac{\partial h}{\partial \eta}\right)_{ps} d\eta + \left(\frac{\partial h}{\partial p}\right)_{\eta s} dp + \left(\frac{\partial h}{\partial s}\right)_{\eta p} ds , \tag{4.11}$$

where the subscripts indicate which variables are held constant in the partial differentiations. By comparison with Eq. 4.8b, one immediately obtains expressions for the temperature, specific volume, and chemical potential as partial derivatives:

$$T = \left(\frac{\partial h}{\partial \eta}\right)_{ps} , \tag{4.12a}$$

$$\alpha = \left(\frac{\partial h}{\partial p}\right)_{\eta s} , \tag{4.12b}$$

and

$$\mu = \left(\frac{\partial h}{\partial s}\right)_{\eta p} . \tag{4.12c}$$

In this equation, T is measured in kelvins, α in cubic meters per kilogram, and μ in joules per kilogram. These quantities can themselves be further regarded as variables depending on η, p, and s, with small changes written as

$$d\alpha = \left(\frac{\partial \alpha}{\partial \eta}\right)_{ps} d\eta + \left(\frac{\partial \alpha}{\partial p}\right)_{\eta s} dp + \left(\frac{\partial \alpha}{\partial s}\right)_{\eta p} ds , \qquad (4.13)$$

$$dT = \left(\frac{\partial T}{\partial \eta}\right)_{ps} d\eta + \left(\frac{\partial T}{\partial p}\right)_{\eta s} dp + \left(\frac{\partial T}{\partial s}\right)_{\eta p} ds , \qquad (4.14)$$

and

$$d\mu = \left(\frac{\partial \mu}{\partial \eta}\right)_{ps} d\eta + \left(\frac{\partial \mu}{\partial p}\right)_{\eta s} dp + \left(\frac{\partial \mu}{\partial s}\right)_{\eta p} ds . \qquad (4.15)$$

It is clear that the nine partial derivatives in Eqs. 4.13 to 4.15 are second order in h, and many are recognizable as being related to the various thermomechanical coefficients mentioned previously. We want to write these equations in terms of *measurable coefficients* that may be used as necessary in the equation of state. Furthermore, our interests in changes in μ are slight, since variations in the enthalpy of the ocean due to changes in dissolved salt content are very small compared to those, for example, from radiation and pressure. Hence we will neglect Eq. 4.15 and consider μ as given when it appears elsewhere; this approximation can easily be removed, if necessary.

The common thermomechanical coefficients for seawater are modifications of the usual ones valid for $s = 0$, plus four new ones arising from the salinity derivatives. They are:

Specific Heats

Specific heat at constant volume and salinity, $C_{\alpha s}$:

$$C_{\alpha s} = \left(\frac{\partial q}{\partial T}\right)_{\alpha s} = T \left(\frac{\partial \eta}{\partial T}\right)_{\alpha s} . \qquad (4.16)$$

Specific heat at constant pressure and salinity, C_{ps}:

$$C_{ps} = \left(\frac{\partial q}{\partial T}\right)_{ps} = T \left(\frac{\partial \eta}{\partial T}\right)_{ps} . \qquad (4.17)$$

Specific heat at constant volume and pressure, $C_{\alpha p}$:

$$C_{\alpha p} = \left(\frac{\partial q}{\partial T}\right)_{\alpha p} = T\left(\frac{\partial \eta}{\partial T}\right)_{\alpha p}. \tag{4.18}$$

This last specific heat, while definable mathematically, may be operationally very difficult, if not impossible, to measure. We will have no further use for it here.

Volumetric Coefficients

Isobaric/isohaline thermal expansion coefficient, a_T:

$$a_T = \frac{1}{\alpha}\left(\frac{\partial \alpha}{\partial T}\right)_{ps}. \tag{4.19}$$

Isothermal/isohaline compressibility coefficient, a_p (reciprocal bulk modulus):

$$a_p = -\frac{1}{\alpha}\left(\frac{\partial \alpha}{\partial p}\right)_{Ts}. \tag{4.20}$$

Isobaric/isothermal saline contraction coefficient, a_s:

$$a_s = -\frac{1}{\alpha}\left(\frac{\partial \alpha}{\partial s}\right)_{Tp}. \tag{4.21}$$

Halinometric Coefficients

Pycnohalinity coefficient, π_α:

$$\pi_\alpha = \frac{1}{s}\left(\frac{\partial s}{\partial \alpha}\right)_{\eta p}. \tag{4.22}$$

Thermohalinity coefficient, π_T:

$$\pi_T = \frac{1}{s}\left(\frac{\partial s}{\partial T}\right)_{\eta p}. \tag{4.23}$$

Clearly $s\pi_\alpha$ and αa_s are related much as are the speed of sound and the bulk modulus, i.e., as isentropic and isothermal changes, respectively.

Miscellaneous

Speed of sound, c:

$$c^2 = \left(\frac{\partial p}{\partial \rho}\right)_{\eta s} = -\alpha^2 \left(\frac{\partial p}{\partial \alpha}\right)_{\eta s}. \tag{4.24}$$

Ratio of specific heats, γ_h:

$$\gamma_h = C_{ps}/C_{\alpha s}. \tag{4.25}$$

To arrive at the partial derivatives in Eqs. 4.13 and 4.14 in terms of measurable quantities, we first compare the coefficient of dp in Eq. 4.13 with Eq. 4.24, and obtain

$$\left(\frac{\partial p}{\partial \alpha}\right)_{\eta s} = -\rho^2 c^2 = -\frac{c^2}{\alpha^2}. \tag{4.26}$$

(Note that ρc is the acoustic impedance of seawater, as will be shown in Chapter 7.) Next consider the coefficient of $d\eta$ in Eq. 4.14, and compare it with Eq. 4.17. From this we obtain

$$\left(\frac{\partial T}{\partial \eta}\right)_{ps} = \frac{T}{C_{ps}}. \tag{4.27}$$

Similarly, a comparison of the coefficients of ds with the halinometric coefficients gives

$$\left(\frac{\partial \alpha}{\partial s}\right)_{\eta p} = \frac{1}{s\pi_\alpha} \tag{4.28}$$

and

$$\left(\frac{\partial T}{\partial s}\right)_{\eta p} = \frac{1}{s\pi_T}. \tag{4.29}$$

The remaining coefficients are more difficult to derive. Consider first the specific heat, $C_{\alpha s}$; from Eqs. 4.6 and 4.8 we obtain the well-known thermodynamic relationship,

$$C_{\alpha s} = T\left(\frac{\partial \eta}{\partial T}\right)_{\alpha s} = \left(\frac{\partial h}{\partial T}\right)_{\alpha s} - \alpha \left(\frac{\partial p}{\partial T}\right)_{\alpha s}. \tag{4.30}$$

Now when $d\alpha = 0 = ds$, Eq. 4.13 yields

$$d\alpha = 0 = \left(\frac{\partial\alpha}{\partial\eta}\right)_{ps} d\eta - \frac{\alpha^2}{c^2} dp \qquad (4.31)$$

$$\equiv Y\, d\eta - X\, dp\,,$$

while for the same case, Eq. 4.14 becomes

$$dT = \frac{T}{C_{ps}} d\eta + \left(\frac{\partial T}{\partial p}\right)_{\eta s} dp$$

$$\equiv Z\, d\eta + Y\, dp\,, \qquad (4.32)$$

where for convenience in writing, the coefficients have been abbreviated X, Y, and Z. The parameter Y is, from the equality of the cross-derivatives of h,

$$Y = \left(\frac{\partial^2 h}{\partial\eta\,\partial p}\right) = \left(\frac{\partial\alpha}{\partial\eta}\right)_{ps} = \left(\frac{\partial T}{\partial p}\right)_{\eta s} = \left(\frac{\partial^2 h}{\partial p\,\partial\eta}\right). \qquad (4.33)$$

Substituting Eq. 4.31 in Eq. 4.32 and solving for $d\eta/dT$, we obtain

$$\left(\frac{d\eta}{dT}\right)_{\alpha s} \equiv \left(\frac{\partial\eta}{\partial T}\right)_{\alpha s}$$

$$= \frac{C_{\alpha s}}{T} = 1/(1 + Y^2/XZ)Z. \qquad (4.34)$$

Restoring the values for X and Z and solving for Y^2 gives

$$Y^2 = \frac{(\gamma_h - 1)}{\gamma_h C_{\alpha s}} \frac{\alpha^2 T}{c^2}. \qquad (4.35)$$

Another expression for Y may be obtained using Eqs. 4.33, 4.17, and 4.19:

$$Y = \frac{\alpha}{c^2 a_T} (\gamma_h - 1). \qquad (4.36)$$

Equating Eq. 4.35 to the square of Eq. 4.36 results in a useful thermodynamic identity:

$$\gamma_h (\gamma_h - 1) C_{\alpha s} = c^2 a_T^2 T .$$ (4.37a)

An alternative form of this relationship that involves only experimentally observed or calculable quantities on the right-hand side may be obtained from the definitions of the coefficients c^2, a_T, and $C_{\alpha s}$:

$$C_{\alpha s} = C_{ps} + T \left(\frac{\partial \alpha}{\partial T} \right)_{ps}^2 \left(\frac{\partial p}{\partial \alpha} \right)_{Ts}$$

$$= C_{ps} - T a_T^2 / \rho a_p .$$ (4.37b)

The specific heat at constant pressure and the required derivatives may be obtained from an equation of state, the most recent internationally accepted version of which is given ahead. Numerical values for that quantity are near 3994 J (kg °C)$^{-1}$ at $T = 20°C$ and $s = 35$ psu; the specific heat at constant volume is at most 2% smaller, with the value of γ_h at those environmental conditions being approximately 1.014.

We are now in a position to write Eqs. 4.13 and 4.14 in terms of the measurable coefficients above; at this point of recasting, we shall also generalize them for time-varying problems and immediately obtain two additional equations that contribute to our goal. In addition, we shall write one, the temperature equation, in terms of convective heating rates, Dq/Dt, rather than the entropy derivative, $D\eta/Dt$, but will retain the latter in the equation for specific volume. These useful equations give time derivatives of α and T in terms of time variations of other state variables that can be calculated readily:

$$\frac{D\alpha}{Dt} = \frac{\alpha a_T T}{C_{ps}} \frac{D\eta}{Dt} - \frac{\alpha^2}{c^2} \frac{Dp}{Dt} + \frac{1}{s\pi_\alpha} \frac{Ds}{Dt}$$ (4.38)

and

$$\frac{DT}{Dt} = \frac{1}{C_{ps}} \frac{Dq}{Dt} + \frac{\alpha a_T T}{C_{ps}} \frac{Dp}{Dt} + \frac{1}{s\pi_T} \frac{Ds}{Dt} .$$ (4.39)

Equation 4.38 will be used in Chapter 7 in the derivation of the acoustic wave equation, while Eq. 4.39 is used immediately below.

4.6 Heat Conduction Equation

An independent equation for heat conduction may be derived from considerations similar to those in Section 3.10. Allowing again for an anisotropic thermal conductivity, κ_q, appearing in the heat conduction term, we may write

$$\frac{Dq}{Dt} = \frac{\partial q}{\partial t} + \mathbf{u} \cdot \nabla q$$

$$= \frac{1}{\rho} \left[\nabla \cdot \kappa_q \cdot \nabla T - \frac{1}{c_e} \frac{\partial \langle S \rangle}{\partial t} - \nabla \cdot \langle \mathbf{S} \rangle + Q(\epsilon,\mu,L_v) \right], \quad (4.40)$$

where the first term in the q derivative represents the rate of change of heat content and the second, heat convection. The right-hand side includes heat conduction and diffusion, and the heating rate per unit mass due to time changes and the convergence of the vector radiance, $\langle \mathbf{S} \rangle$ (see Eqs. 8.175, 8.177, 9.8, and 9.64); here c_e is the speed of light in seawater (see Eq. 8.19 and Table 9.2). Included in $Q(\epsilon,\mu,L_v)$ are the heating rate per unit volume, the viscous dissipation rate, ϵ, chemical reactions, μ, and other sources or sinks not accounted for elsewhere. Values of the scalar κ_q for molecular conduction in the sea are of order 0.596 W $(m^2 \, K \, m^{-1})^{-1}$; the corresponding eddy conductivities are orders of magnitude larger (see Appendix Three).

For processes in a source-free, incompressible, constant-salinity ocean, the left-hand side can be written in terms of the temperature as well, using Eqs. 4.37a, 4.38, and 4.39:

$$C_{\alpha s} \frac{DT}{Dt} = \frac{Dq}{Dt} = \alpha \nabla \cdot \kappa_q \cdot \nabla T. \quad (4.41)$$

This relationship, evaluated for constant, isotropic heat conductivity and no current, becomes the usual heat conduction equation:

$$\frac{\partial T}{\partial t} = \frac{\kappa_q}{C_{\alpha s} \rho} \nabla^2 T = K_q \nabla^2 T, \quad (4.42)$$

where $K_q \equiv \kappa_q / \rho C_{\alpha s}$ is the molecular thermal diffusivity, which has a value in the ocean of approximately $1.49 \times 10^{-7} \, m^2 \, s^{-1}$. In the general case, the more complete expressions (Eqs. 4.39 and 4.40) must be retained.

4.7 Specific Volume and Salinity Equations

We have already derived equations for density and salinity changes (Eqs. 3.68 and 3.81, respectively). Recasting the first into a form giving the convective derivative for specific volume, we obtain

$$\frac{D\alpha}{Dt} = \alpha \nabla \cdot \mathbf{u}. \quad (4.43)$$

The equivalent form for salinity is slightly more complicated, and makes use of a relationship for the quantity $\rho D\gamma/Dt$ that also invokes the continuity equation. Adding ρ times the convective derivative of a property, γ, to γ times the continuity equation, we obtain

$$\rho \frac{D\gamma}{Dt} \equiv \rho \left(\frac{\partial \gamma}{\partial t} + \mathbf{u} \cdot \nabla \gamma \right) + \gamma \left(\frac{\partial \rho}{\partial t} + \nabla \cdot \rho \mathbf{u} \right)$$

$$\equiv \frac{\partial (\rho \gamma)}{\partial t} + \nabla \cdot (\rho \gamma \mathbf{u}) , \tag{4.44}$$

which may be regarded as an identity. Applying this to Eq. 3.81 yields

$$\frac{Ds}{Dt} = \alpha \nabla \cdot \kappa_s \cdot \nabla (\rho s) + \alpha S , \tag{4.45}$$

where S represents source and sink terms for salinity as given by Eq. 3.81. Equations 4.40, 4.43, and 4.45 are in forms suitable for use in the enthalpy equation as given by Eq. 4.9, and in the specific volume and temperature rate equations (Eqs. 4.38 and 4.39, respectively).

4.8 Equation of State

We have now assembled all of the equations required to complete our hydrodynamic and thermodynamic system, except for an equation of state of the form

$$\rho = \rho(s, T, p) \tag{4.46a}$$

or

$$\alpha = \alpha(s, T, p) . \tag{4.46b}$$

An equation of state of sufficient accuracy to be used in the computation of the oceanic density field (which is needed for the so-called *dynamic method* of determining large-scale oceanic currents; see Chapter 6) is a very complicated one indeed. An internationally agreed upon equation of state (UNESCO, 1981) fits the available density measurements with an accuracy of order 3.5×10^{-6} over the normally encountered range of oceanic pressures, temperatures, and salinities. This equation is of the form

$$\rho(s,T,p) = 1/\alpha(s,T,p) = \frac{\rho(s,T,0)}{1 + \left\langle \frac{1}{\alpha}\left(\frac{\partial\alpha}{\partial p}\right)\right\rangle p} , \qquad (4.47)$$

where each quantity on the right-hand side, except pressure, is expressed as a polynomial series in s and T, expanded about values for zero salinity and a pressure of 1 bar. The *secant* or *mean compressibility*, $\langle -(1/\alpha)(\partial\alpha/\partial p)\rangle$, is defined as:

$$-\frac{\alpha(s,T,p) - \alpha(s,T,0)}{\alpha(s,T,0)p} \equiv \frac{1}{K_T(s,T,p)} , \qquad (4.48)$$

where $K_T(s,T,p)$ is the secant or mean bulk modulus, whose reciprocal is the mean of a_p (Eq. 4.20). It is a function of s, T, and p, and its variation at $s = 35$ psu is shown in Fig. 4.8 as a function of temperature and pressure. The compressibility decreases as temperature, pressure, and salinity increase, which is understandable in terms of the molecular character of matter: As higher pressure squeezes the molecules closer together, or as higher concentrations of ions interleave between the H_2O molecules, still further reductions in volume are increasingly resisted by the crowding.

The relationship of Eq. 4.47 is an improvement on the simplest possible linear equation of state, which is an expansion about zero values for T, p, and s of the type given by $\alpha \simeq \alpha_0(1 + a_TT - a_pp - a_ss)$. We may assume a slightly more complicated form for improved accuracy, viz: $\alpha \simeq \alpha_{00}(1 + a_TT)(1 - a_ss)(1 - \langle a_p\rangle p)$. Here the a_i's are given by Eqs. 4.19 to 4.21; $\rho(s,T,0) = 1/\alpha_{00}(1 + a_TT)(1 - a_ss)$; and α_{00} is the specific volume of pure water. The general form of Eq. 4.47 follows immediately from this latter approximation if we incorporate higher-order terms in T, s, and p.

We shall not give the complete development of the so-called International Equation of State for Seawater, EOS 80, in the interests of brevity; the reader may refer to Millero et al. (1980), UNESCO (1981), or Fofonoff (1985) for a more comprehensive specification.

The equation is most clearly displayed in four parts; the first gives the specific volume in the same form as Eq. 4.47:

$$\alpha(s,T,p) = \alpha(s,T,0)[1 - p/K_T(s,T,p)] . \qquad (4.49a)$$

The second part gives the density at surface pressure (indicated by 0):

$$\rho(s,T,0) = 1/\alpha(s,T,0) = A + Bs + Cs^{3/2} + Ds^2 . \qquad (4.49b)$$

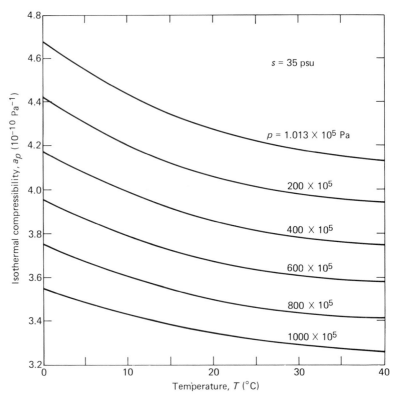

Fig. 4.8 Isothermal compressibility of seawater as a function of temperature and pressure. [Adapted from Knauss, J., *Introduction to Physical Oceanography* (1978)].

The third gives the secant bulk modulus, K_T:

$$K_T(s,T,p) = E + Fs + Gs^{3/2}$$

$$+ (H + Is + Js^{3/2})p$$

$$+ (M + Ns)p^2 . \tag{4.49c}$$

The fourth part consists of polynomials A, B, ..., N up to fifth degree in the temperature; these are listed in Table 4.2. The temperature is specified in degrees Celcius; the pressure is in bars, or 10^5 Pa; the salinity is in psu; the density is in kg m^{-3}; and the specific volume is in m^3 kg^{-1}. This equation is sufficiently accurate to determine the density field from observed values of salinity, temperature, and pressure to within a standard error of approximately 0.009 kg m^{-3} over the entire oceanic pressure range. Since variations in the composition of dissolved salts can lead to differences in the density of natural seawater of order 0.05 kg m^{-3}, the equation appears as accurate as necessary when salinities are determined by electrical conductivities, as they are. For this last reason, the relationship between salinity and conductivity is an important one to specify. It is given as a part of the UNESCO literature, with its inverse, i.e., conductivity as a function of salinity, being shown via Eqs. 8.33 to 8.36.

TABLE 4.2
Coefficients for the International
Equation of State for Seawater, EOS 80

	A	B	C
T^0	$+999.842594$	$+8.24493 \times 10^{-1}$	-5.72466×10^{-3}
T^1	$+6.793952 \times 10^{-2}$	-4.0899×10^{-3}	$+1.0227 \times 10^{-4}$
T^2	-9.095290×10^{-3}	$+7.6438 \times 10^{-5}$	-1.6546×10^{-6}
T^3	$+1.001685 \times 10^{-4}$	-8.2467×10^{-7}	
T^4	-1.120083×10^{-6}	$+5.3875 \times 10^{-9}$	
T^5	$+6.536332 \times 10^{-9}$		

	D	E	F
T^0	$+4.8314 \times 10^{-4}$	19652.21	$+54.6746$
T^1		$+148.4206$	-0.603459
T^2		-2.327105	$+1.09987 \times 10^{-2}$
T^3		$+1.360477 \times 10^{-2}$	-6.1670×10^{-5}
T^4		-5.155288×10^{-5}	

	G	H	I
T^0	$+7.944 \times 10^{-2}$	$+3.239908$	$+2.2838 \times 10^{-3}$
T^1	$+1.6483 \times 10^{-2}$	$+1.43713 \times 10^{-3}$	-1.0981×10^{-5}
T^2	-5.3009×10^{-4}	$+1.16092 \times 10^{-4}$	-1.6078×10^{-6}
T^3		-5.77905×10^{-7}	

	J	M	N
T^0	$+1.91075 \times 10^{-4}$	$+8.50935 \times 10^{-5}$	-9.9348×10^{-7}
T^1		-6.12293×10^{-6}	$+2.0816 \times 10^{-8}$
T^2		$+5.2787 \times 10^{-8}$	$+9.1697 \times 10^{-10}$

[Adapted from Fofonoff, N. P., *J. Geophys. Res.* (1985).]

Plots of density versus the three independent thermodynamic variables are difficult to construct, but one particular projection of the equation of state has proven to be very useful for a diagnostic technique called *water type analysis*. This is the form shown in Figs. 4.9 and 4.10, i.e., the loci of constant

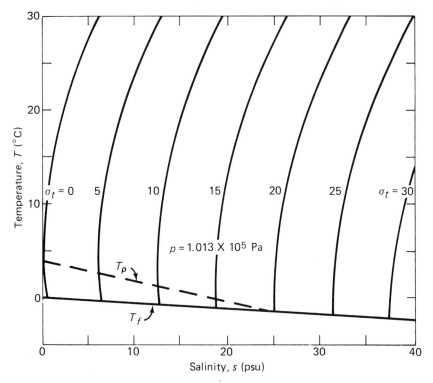

Fig. 4.9 Temperature–salinity–density diagram for seawater over range of normal variations of *T* and *s*. This is a cross-plot of Fig. 4.5. [Adapted from Dietrich, G., *Oceanography, An Introductory View* (1968).]

density on a plot giving temperature versus salinity. Figure 4.9 shows values of constant density, or *isopycnals*, in σ_t units, where σ_t is defined as

$$\sigma_t = \rho - \rho_{00} , \tag{4.50}$$

and where $\rho_{00} = 1000 \text{ kg m}^{-3}$ at 4°C is the density of fresh water. Thus σ_t represents the departure of density from the fresh water value.

The graphs of Figs. 4.9 and 4.10 are known as *T–s diagrams*, and a homogeneous water type is represented by a single point on the diagrams. A sequence of water samples measured at varying depths will show the kind of variation given in Fig. 4.10, with depth given as a parameter along the line; the variations of water type properties with depth are usually found to lie along a curve that is more or less characteristic of the area of the ocean being sampled. The characteristic *T–s* relationship, along with analyses of dissolved oxygen and silica, for example, allow the identification of the ori-

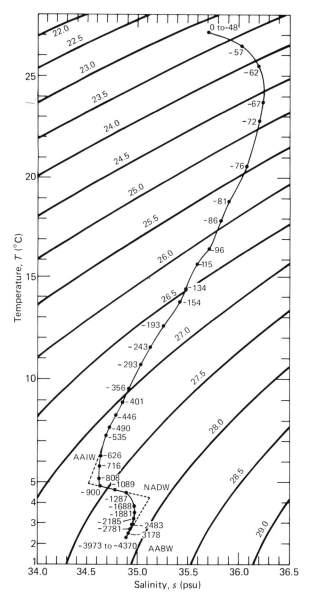

Fig. 4.10 Temperature–salinity–density diagram for an oceanographic station in the tropical Atlantic at 5°N, 25°W. The parameter is density in constant σ_t units, or isopycnals. The S-shaped curve represents the variation of water type properties as the depth increases from $z = 0$ m at the upper end to -4370 m at the bottom. Various water types are AAIW, NADW, and AABW (see Fig. 2.22 for legend). The straight line between AAIW and NADW indicates the mixing of these two types has occurred at depths between 900 to 1300 m. Similarly, mixing of NADW and AABW water has occurred over the range from 2000 m to the bottom.

gin of the water mass with reasonable certitude. The *T–s* curve also possesses the property that two overlying water masses that are initially homogeneous and thus represented by two distinct points in the *T–s* plot will, upon vertical diffusive mixing, appear on the diagram as points connecting an approximately straight line separating the two initial values. The example shown in Fig. 4.10 gives evidence of mixing between Antarctic Intermediate Water, North Atlantic Deep Water, and Antarctic Bottom Water (see Fig. 2.22). Thus the *T–s* diagram is a thermodynamic space that is useful in the analysis of mixing processes, in analogy to *p–T* or *p–α* plots in ordinary thermodynamics.

4.9 Combined Hydrodynamics and Thermodynamics

We have arrived at a point where a summary of the coupled hydrodynamic and thermodynamic equations is in order. The physical laws that underlie these are: Newton's equations (3); conservation laws for mass, salinity, and heat flow (3); the first and second laws of thermodynamics (2); an equation of state (1); and an equation for internal energy *or* enthalpy (1). This last, as summarized by Eq. 4.10, is implicitly carried by the various thermomechanical coefficients. In terms of rate equations, the coupled systems are:

1–3. Momentum:

$$\frac{D\mathbf{u}}{Dt} = -2\boldsymbol{\Omega} \times \mathbf{u} - \nabla\Phi - \alpha\nabla p + \alpha\nabla\cdot\mathbf{A}\cdot\nabla\mathbf{u} . \quad (4.51\text{--}53)$$

4. Continuity:

$$\frac{D\alpha}{Dt} = \alpha\nabla\cdot\mathbf{u} . \quad (4.54)$$

5. First and Second Laws:

$$\frac{De}{Dt} = T\frac{D\eta}{Dt} - p\frac{D\alpha}{Dt} + \mu\frac{Ds}{Dt} + L . \quad (4.55)$$

6. Specific volume:

$$\frac{D\alpha}{Dt} = \frac{\alpha a_T T}{C_{ps}}\frac{D\eta}{Dt} - \frac{\alpha^2}{c^2}\frac{Dp}{Dt} + \frac{1}{s\pi_\alpha}\frac{Ds}{Dt} . \quad (4.56)$$

7. Temperature:

$$\frac{DT}{Dt} = \frac{1}{C_{ps}} \frac{Dq}{Dt} + \frac{\alpha a_T T}{C_{ps}} \frac{Dp}{Dt} + \frac{1}{s\pi_T} \frac{Ds}{Dt} . \tag{4.57}$$

8. Heat Conduction:

$$T\frac{D\eta}{Dt} = \frac{Dq}{Dt}$$

$$= \alpha(\nabla \cdot \kappa_q \cdot \nabla T - \frac{1}{c_e} \frac{\partial \langle S \rangle}{\partial t} - \nabla \cdot \langle \mathbf{S} \rangle + Q) . \tag{4.58}$$

9. Salinity:

$$\frac{Ds}{Dt} = \alpha[\nabla \cdot \kappa_s \cdot \nabla (\rho s) + S] . \tag{4.59}$$

10. State:

$$\rho(s,T,p) = \frac{\rho(s,T,0)}{1 - p/K_T(s,T,p)} . \tag{4.60}$$

Here L, Q, and S (this last representing the quantities on the right side of Eq. 3.81) are sources or sinks of heats of transformation, other heat sources, and salinity, respectively. The 10 dependent variables that these equations can in principle yield are u, v, w, ρ, s, T, p, e, η, and q. An eleventh equation for the chemical potential could have been developed, if needed. However, a few variables (η and q in particular) can be eliminated quickly in favor of the remainder. Also, the contribution of chemical potential to heating and temperature changes is generally negligible. Additionally, except for acoustic waves, $\nabla \cdot \mathbf{u} = 0 = D\alpha/Dt$. The incompressibility assumption is valid when the current and phase speeds of motions are both small compared to c, and the vertical scale of the motion small compared to the scale height of 225 km. Except for acoustic waves, these requirements are exceedingly well satisfied in the ocean.

One other exception is the adiabatic compression and heating of water that occurs in the deep ocean because of the overlying hydrostatic pressure; here, compressibility must carefully be taken into account to derive *in situ* pressures and densities. These considerations lead to the concepts of *potential temperature*, θ, and *potential density*, ρ_θ, which are simply the temperature and density that a sample of seawater initially at some depth, z, would take on if it were lifted isentropically to a reference level, z_r. The potential temperature is defined as

$$\theta(s_0, T_0, p_0; p_r) = T_0 + \int_{p_0}^{p_r} \Gamma[s, \theta(s_0, T_0, p_0; p_r), p]\, dp\,, \qquad (4.61)$$

where $\Gamma(s, T, p)$ is the local *adiabatic lapse rate*, i.e., the change in temperature per unit change in pressure occurring during an isentropic vertical displacement. The extended arguments in θ indicate that it is the temperature an element of seawater would have if moved adiabatically and with no change of salinity from a region of initial salinity, temperature, and pressure of s_0, T_0, and p_0, respectively, to another level where the reference pressure is p_r; this level may be greater or less than p_0. From thermodynamic arguments similar to the ones just presented, one may derive, from Eqs. 4.19, 4.33, and 4.37, the relationship

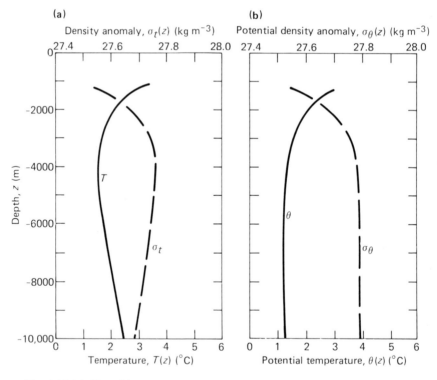

Fig. 4.11 (a) Plot of *in situ* temperature, T, and density anomaly, σ_t, versus depth; (b) same data re-plotted in terms of potential temperature, θ, and potential density anomaly, σ_θ. Below approximately 4000 m, the temperature increase and density decrease are due to adiabatic compression with depth; when θ and σ_θ are used as thermodynamic variables, the deep water compressive effects are removed.

$$\Gamma(s,T,p) \equiv \left(\frac{\partial T}{\partial p}\right)_{\eta s} = \frac{T}{C_{ps}}\left(\frac{\partial \alpha}{\partial T}\right)_{\eta s} \qquad °C\ bar^{-1}\ . \qquad (4.62)$$

In the deep ocean, the lapse rate is generally negative, so that the potential temperature is less than the *in situ* temperature. The equation for ρ_θ has a form similar to Eq. 4.61.

The UNESCO equation of state also has, as adjuncts, precise relationships for salinity, bulk modulus, density anomaly, specific heat, adiabatic lapse rate, and algorithms for calculating the so-called *buoyancy frequency*, which will be discussed in Chapter 5.

It is often the case that deep vertical sections, when plotted in terms of θ and ρ_θ, do not show changes with z in the deep ocean, indicating that the observed variations in the *in situ* values of T and ρ are due to the adiabatic compression alone. Under these conditions, θ and ρ_θ are termed *conserved quantities*. Figure 4.11 is a vertical section given in terms of both quantities. Figure 4.12 illustrates a use of potential temperature in specifying the volume distribution of essentially all of the waters of the world oceans among

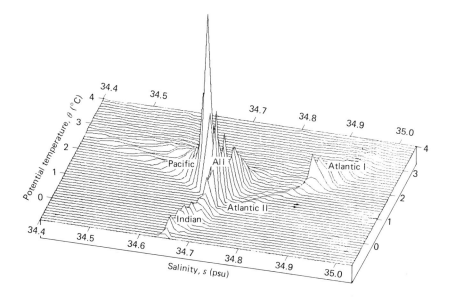

Fig. 4.12 Simulated potential temperature–salinity distribution for water masses of the three major world oceans. Vertical coordinate is proportional to oceanic volume having values of θ and s. Pacific bottom water is most abundant; Indian Ocean varies slightly in salinity; Atlantic has bimodal distribution. Most of the water in the ocean is cold—below $\theta = 4°C$. Total volume is about $1.32 \times 10^9\ km^3$. [From Worthington, L. V., in *Evolution of Physical Oceanography* (1981).]

extant values of θ and s. Since most of the water in the ocean is deep, the potential temperatures are low. The three major oceans have distinctive thermodynamic properties, as Fig. 4.12 shows.

4.10   Energy Flow and Energy Equations

As in ordinary mechanics, the momentum equations (Eqs. 4.51 to 4.53) can be converted to a rate equation for mechanical energy by taking their dot product with $\mathbf{u}$, thereby giving

$$\rho \, \frac{D(\tfrac{1}{2}\mathbf{u}^2)}{Dt} = -\rho \mathbf{u} \cdot \nabla \Phi \; - \; \mathbf{u} \cdot \nabla p \; + \; \nabla \cdot \mathbf{A} \cdot \nabla (\tfrac{1}{2} \, \mathbf{u}^2) \; - \; \rho \epsilon \; , \qquad (4.63)$$

where we have abbreviated the fluid *kinetic energy per unit volume* as

$$\tfrac{1}{2} \, \rho \mathbf{u}^2 \; = \; \tfrac{1}{2} \, \rho \mathbf{u} \cdot \mathbf{u} \; . \qquad (4.64)$$

In this equation, a rotational term does not appear, since $\mathbf{u}$ is everywhere perpendicular to the Coriolis force, which therefore does no work. The quantity ϵ is called the *dissipation rate* and is given by

$$\epsilon \; = \; \nu_m \left[\left(\frac{\partial \mathbf{u}}{\partial x} \right)^2 + \left(\frac{\partial \mathbf{u}}{\partial y} \right)^2 + \left(\frac{\partial \mathbf{u}}{\partial z} \right)^2 \right] , \qquad (4.65)$$

where

$$\left(\frac{\partial \mathbf{u}}{\partial x} \right)^2 = \left(\frac{\partial \mathbf{u}}{\partial x} \right) \cdot \left(\frac{\partial \mathbf{u}}{\partial x} \right) , \; \text{etc.} \qquad (4.66)$$

We can reconfigure Eq. 4.63 somewhat in order to aid in its interpretation. First, we use Eq. 4.44 as applied to $\tfrac{1}{2} \, \mathbf{u}^2$; then we manipulate the pressure gradient term to obtain from it

$$-\mathbf{u} \cdot \nabla p \; = \; - \nabla \cdot (p\mathbf{u}) \; + \; p \nabla \cdot \mathbf{u} \; . \qquad (4.67)$$

These result in a recast version of Eq. 4.63 for the mechanical energy:

$$\frac{\partial}{\partial t} \, (\tfrac{1}{2} \, \rho \mathbf{u}^2) \; + \; \nabla \cdot [(\tfrac{1}{2} \, \rho \mathbf{u}^2 + p)\mathbf{u} \; - \; \mathbf{A} \cdot \nabla (\tfrac{1}{2} \, \mathbf{u}^2)]$$

$$= \; - \rho \mathbf{u} \cdot \nabla \Phi \; + \; p \nabla \cdot \mathbf{u} \; - \; \rho \epsilon \; . \qquad (4.68)$$

A somewhat more specialized form of this can be obtained for cases where the potential energy is due to wavelike elevations, ξ. Using Eq. 3.120 for the kinematic boundary condition and assuming that the potential gradient has only a z component, we obtain

$$\rho \mathbf{u} \cdot \nabla \Phi = -\rho \mathbf{u} \cdot \mathbf{g} = \rho w g = \rho g \frac{D\xi}{Dt}$$

$$= \rho g \left(\frac{\partial \xi}{\partial t} + u \frac{\partial \xi}{\partial x} + v \frac{\partial \xi}{\partial y} \right) . \tag{4.69}$$

We can then rewrite Eq. 4.68 with terms in the amplitude, ξ, transformed to the left-hand side:

$$\frac{\partial}{\partial t} (\tfrac{1}{2} \rho \mathbf{u}^2 + \rho g \xi) + \nabla \cdot [(\tfrac{1}{2} \rho \mathbf{u}^2 + \rho g \xi + p) \mathbf{u} - \mathbf{A} \cdot \nabla (\tfrac{1}{2} \mathbf{u}^2)]$$

$$= p \nabla \cdot \mathbf{u} - \rho \epsilon . \tag{4.70}$$

This form of the energy equation is useful for wave propagation problems.

Returning to the more general case Eq. 4.68, define now a *mechanical energy flux vector*, $\mathbf{F}_u$, having dimensions of watts per square meter, by

$$\mathbf{F}_u = (\tfrac{1}{2} \rho \mathbf{u}^2 + p) \mathbf{u} - \mathbf{A} \cdot \nabla (\tfrac{1}{2} \mathbf{u}^2) . \tag{4.71}$$

Then Eq. 4.68 becomes

$$\frac{\partial}{\partial t} (\tfrac{1}{2} \rho \mathbf{u}^2) + \nabla \cdot \mathbf{F}_u = -\rho \mathbf{u} \cdot \nabla \Phi + p \nabla \cdot \mathbf{u} - \rho \epsilon . \tag{4.72}$$

This equation can be interpreted with the use of the parallelepiped of Fig. 4.13. First, the kinetic energy per unit volume within the parallelepiped is changing at a rate given by the time derivative on the left-hand side; the other term on the left is the divergence of the flux of mechanical energy *across* the faces of the small volume, $\nabla \cdot \mathbf{F}_u$. On the right-hand side are (1) the rate of working by or against gravitational forces when fluid crosses equipotential surfaces, $-\rho \mathbf{u} \cdot \nabla \Phi = -\rho w g$ (the latter form being valid when the earth's gravity is the only source of potential energy); (2) the rate of energy release or uptake by fluid expansion or compression, $p \nabla \cdot \mathbf{u} = (p/\alpha) D\alpha/Dt;$ and (3) the rate of loss of energy *within* the parallelepiped due to viscous forces. This latter term needs interpretation, an attempt at which will be given below.

Let us return to the energy flux vector, $\mathbf{F}_u$: Equation 4.71 shows it to be composed of the net flux rate of kinetic energy per unit volume, $\tfrac{1}{2} \rho \mathbf{u}^2 \mathbf{u}$, across all six faces; the net energy flux due to pressure forces, $p\mathbf{u}$; and the

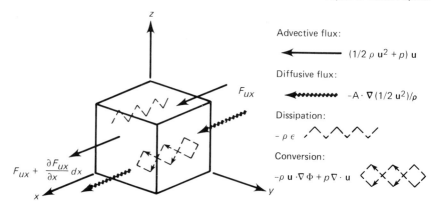

Fig. 4.13 Advective and diffusive fluxes of mechanical energy through an elemental volume of water, plus dissipation and conversion within the volume, as given by Eq. 4.72. [After Gill, A. E., *Atmosphere–Ocean Dynamics* (1982).]

diffusive flux of kinetic energy across the surfaces caused by eddy and molecular viscosities. As with all geophysical fluid eddy processes, this rate of energy diffusion in the vertical is much smaller than the rate in the horizontal.

The energy dissipation rate, as written, includes only molecular viscosity. The only true frictional dissipation in the system is due to the molecular forces and occurs at very small lengths, near $l_\nu = (\nu_m{}^3/\epsilon)^{1/4}$, which is of millimeter order. This dissipated mechanical energy is transformed into heat and, at short wavelengths, into potential energy of vertical mixing; these processes contribute to the term $Q(\epsilon,\mu,L_\nu)$ in Eq. 4.40. Now the rich variety of flows in fluid dynamics makes it a practical impossibility to deal numerically at the same time with scales ranging from the dimensions of an ocean basin (say 10^4 km) to the scale of the *dissipation length, l_ν,* a span of the order of 10^{10}. At any given scale of study, the uses of an eddy viscosity, **A,** and its duals for heat and salinity diffusion, κ_q and κ_s, respectively, represent an attempt to parameterize the so-called *sub-grid eddy processes,* as well as to remove energy from the model system, either analytically or numerically, in a moderately realistic way. As we have seen, numerical values of A_ν/ρ are of order 10^2 to 10^4 times molecular, while those for A_h/ρ may be 10^{10} to 10^{11} times greater (of the order of very viscous, heavy syrup). Nevertheless, it is only at the very small scales that conversion of mechanical energy to heat energy actually occurs; however, this heating rate, compared with heating from sunlight or cooling by evaporation, is essentially negligible. For example, a hydraulic flow that falls through a vertical drop of roughly 400 m and is completely randomized increases its temperature by only approximately 1°C.

It is a somewhat amusing fact that the initial energy inputs to the sea from wind stress and sunlight, as described in Chapters 2 and 9, and which occur at millimeter and submillimeter scales, are transformed by a wide variety of processes into global-scale flows, but then are ultimately dissipated by radiative and resistive processes again taking place at these small scales.

The budget of kinetic energy flux discussed in the context of Fig. 4.13 can be expanded to entire basin scales by considering the ocean to be composited from many such elemental volumes. By integrating Eq. 4.72 over an ocean basin, for example, an equation for the rate of change of total mechanical energy is obtained. Let us define the total kinetic energy, K, to be

$$K = \iiint_V \tfrac{1}{2}\, \rho \mathbf{u}^2 \; d^3\mathbf{x} \; . \tag{4.73}$$

Then applying the divergence theorem and the rule for differentiating under the integral to Eq. 4.73, we obtain

$$\frac{dK}{dt} = -\iint_S \mathbf{F}_u \cdot \hat{n}\, d^2\mathbf{x} - \iiint_V \rho\epsilon \; d^3\mathbf{x}$$

$$+ \iiint_V (-\rho\mathbf{u}\cdot\nabla\Phi + p\nabla\cdot\mathbf{u})\, d^3\mathbf{x} \; , \tag{4.74}$$

where (as before) $\hat{n}$ is the unit outward normal to the boundary surface, i.e., bottom plus free surface. On the right-hand side are the expressions for the fluxes of energy into the bottom and across the air–sea interface, and the dissipation of kinetic energy within the volume. The final volume integral gives the rate at which the geopotential and the compressive forces do work; these will be shown below to represent energy transformations from mechanical to other types.

There remain two other forms of energy to include in an overall energy rate equation: potential and internal. The geopotential, Φ, is the potential energy per unit mass, since its negative gradient gives the combined gravitational and centrifugal forces per unit mass. Its astronomical tidal part is time- and space-dependent (cf. Chapters 3 and 5), while the terrigeneous (earth-derived) portion depends only on z. An equation for its rate of change may be derived from the relationship for the convective derivative. The rate of doing work per unit volume, $\rho D\Phi/Dt$, is just the rate at which the geopotential does work on the fluid flow, $-\rho\mathbf{u}\cdot\mathbf{f}_g = \rho\mathbf{u}\cdot\nabla\Phi$. From Eqs. 3.50 and 4.44 we obtain

$$\rho \, \frac{D\Phi}{Dt} = \frac{\partial(\rho\Phi)}{\partial t} + \nabla \cdot (\rho\Phi\mathbf{u}) = \rho\mathbf{u} \cdot \nabla\Phi \, . \qquad (4.75)$$

This represents a budget equation for potential energy.

The rate equation for internal energy per unit mass, e, may be derived from Eqs. 4.44, 4.54, 4.55, 4.58, and 4.59. Collectively, these become

$$\rho \, \frac{De}{Dt} = \frac{\partial(\rho e)}{\partial t} + \nabla \cdot (\rho e\mathbf{u})$$

$$= \rho \left[T \, \frac{D\eta}{Dt} - p \, \frac{D\alpha}{Dt} + \mu \, \frac{Ds}{Dt} + L \right]$$

$$= \nabla \cdot \kappa_q \cdot \nabla T - \frac{1}{c_e} \frac{\partial\langle S\rangle}{\partial t} - \nabla \cdot \langle \mathbf{S}\rangle + Q - p\nabla \cdot \mathbf{u}$$

$$+ \mu\nabla \cdot \kappa_s \cdot \nabla(\rho s) + \mu S + L \, , \qquad (4.76)$$

or, rewriting this relationship slightly,

$$\frac{\partial(\rho e)}{\partial t} + \frac{1}{c_e} \frac{\partial\langle S\rangle}{\partial t} + \nabla \cdot [\rho e\mathbf{u} + \langle \mathbf{S}\rangle - \kappa_q \cdot \nabla T - \mu\kappa_s \cdot \nabla(\rho s)]$$

$$= \frac{\partial(\rho e)}{\partial t} + \frac{1}{c_e} \frac{\partial\langle S\rangle}{\partial t} + \nabla \cdot \mathbf{F}_e$$

$$= Q + \mu S + L - p\nabla \cdot \mathbf{u} \, . \qquad (4.77)$$

The quantity $\mathbf{F}_e$ may be called the *internal energy flux vector,* and is given by

$$\mathbf{F}_e = \rho e\mathbf{u} + \langle \mathbf{S}\rangle - \kappa_q \cdot \nabla T - \mu\kappa_s \cdot \nabla(\rho s) \, , \qquad (4.78)$$

which shows that internal energy is transported by conduction, radiation, and eddy and molecular diffusive heat transfer caused by both temperature and salinity gradients. The salinity contribution is undoubtedly small but could be appreciable under limited circumstances such as freezing or thawing of seawater. On the right-hand side are volumetric sources of heat, Q, chemical energy released by salinity changes, μS, heats of phase transformation, L, and heat released or absorbed by volumetric changes, $-p\nabla \cdot \mathbf{u}$.

4.11 Total Energy Equations

Collecting together the relationships for kinetic, potential, and internal energies, Eqs. 4.72, 4.75, and 4.77, respectively, we obtain an important equation for the rate of change of total energy per unit volume:

$$\frac{\partial}{\partial t} \left[\rho(\tfrac{1}{2} \mathbf{u}^2 + \Phi + e + \langle S \rangle / c_e) \right] + \nabla \cdot \mathbf{F}_t$$

$$= -\rho\epsilon + \mu S + L + Q, \tag{4.79}$$

where the total energy flux vector, $\mathbf{F}_t$, is

$$\mathbf{F}_t = \left[\rho(\tfrac{1}{2} \mathbf{u}^2 + \Phi + e) + p \right] \mathbf{u}$$

$$+ \langle S \rangle - \mathbf{A} \cdot \nabla(\tfrac{1}{2} \mathbf{u}^2) - \kappa_q \cdot \nabla T - \mu \kappa_s \cdot \nabla(\rho s). \tag{4.80}$$

This flux vector is composed (in order of appearance) of advective fluxes of kinetic, potential, and internal energies; the flux rate of pressure work; the vector radiance; and diffusive fluxes of kinetic energy, heat, and chemical energy, the last three transported by gradients. The source functions in the right-hand side of Eq. 4.79 have a certain element of arbitrariness about them, in that all can be considered to be volumetric heating rates and included in the definition of $Q(\epsilon,\mu,L)$. However, we have kept explicit the viscous dissipation, salinity, and heats of transformation, with remaining sources, if any, incorporated into Q. It is worth noting that the function Q in Eq. 4.76, which has ϵ and L shown implicitly, has had just those portions cancelled during the summation for the total energy. Care should be taken not to count these energies twice and not to omit them. Other cancellations that take place in Eq. 4.79 upon summation include the geopotential term, $\rho \mathbf{u} \cdot \nabla \Phi$ in Eq. 4.75, which has been cancelled by its negative in Eq. 4.72, as well as the compressibility term, $p \nabla \cdot \mathbf{u}$, appearing in that same equation, which has been cancelled by the identical term with opposite sign in the relation for internal energy, Eq. 4.77. These terms, which represent conversions of energy from one kind to another, do not appear in the total energy equation, but nevertheless represent important processes occurring within the ocean.

A total energy equation for a macroscopic ocean volume can be derived by integrating the rate equation, Eq. 4.80, over the volume of interest. Define the total internal energy within the volume, V, by I:

$$I = \iiint_V \rho e \, d^3\mathbf{x} \, , \tag{4.81}$$

the total potential energy, P, by

$$P = \iiint_V \rho \Phi \, d^3\mathbf{x} \, , \tag{4.82}$$

and the total energy, E, by

$$E = K + P + I \, . \tag{4.83}$$

Then Eq. 4.79 becomes, after integration over the volume V,

$$\frac{dE}{dt} + \iint_S \mathbf{F}_t \cdot \hat{n} \, d^2\mathbf{x} = \iiint_V (-\rho e + \mu S + Q + L) \, d^3\mathbf{x} \, , \tag{4.84}$$

which is a statement of a large-scale energy flux conservation and conversion, with interpretations analogous to those given for Eq. 4.74. Thus the total system energy varies because of fluxes out over the bounding surface, plus dissipation and energy transformations within the volume.

While Eq. 4.84 gives the total system energy, it does not mean that all of this energy is available in any actual process. An interpretation of the integrals evaluated over the entire ocean is that K is the energy that could be obtained by bringing the entire ocean current system to rest; P, the energy released by bringing it into coincidence with equipotential surfaces; and I, the energy released by making the salinity uniform and by lowering the temperature to absolute zero. For example, Gill has estimated that the mean available potential energy in a typical subtropical gyre is of the order of 10^5 J m^{-2} of surface area, or a total for the North Atlantic gyre of order 10^{18} J; this energy is stored in elevations or depressions of fluid away from equipotential surfaces because of geostrophic flow.

The subject of total energy in the sea is perhaps of less interest than the quantities of energy that actually change during some process being studied, since the processes required to release K, P, and I are thermodynamically impossible. We will apply some of these energy concepts in Chapters 5 and 6.

Bibliography

Books

Batchelor, G. K., *An Introduction to Fluid Dynamics,* Cambridge University Press, Cambridge, England (1974).

Dietrich, G., *Oceanography, an Introductory View,* John Wiley & Sons, Inc., New York, N.Y. (1968).

Eckart, C., "Equations of Motion of Sea-Water," in *The Sea,* Vol. 1, M. N. Hill, Ed., Interscience Publishers, New York, N.Y. (1962).

Eckart, C., *Hydrodynamics of Oceans and Atmospheres,* Pergamon Press, Inc., New York, N.Y. (1960).

Fofonoff, N. P., "Physical Properties of Sea Water," in *The Sea,* Vol. 1, M. N. Hill, Ed., Interscience Publishers, New York, N.Y. (1962).

Gill, A. E., *Atmosphere-Ocean Dynamics,* Academic Press, Inc., New York, N.Y. (1982).

Gross, M. G., *Oceanography,* 3rd ed., Prentice-Hall, Inc., Englewood Cliffs, N.J. (1982).

Hasted, J. B., *Aqueous Dielectrics,* Chapman and Hall, Ltd., London, England (1973).

Knauss, J., *Introduction to Physical Oceanography,* Prentice-Hall, Englewood Cliffs, N.J. (1978).

Leyendekkers, J. V., *Thermodynamics of Sea Water, Parts I and II,* Marcel Dekker, New York, N.Y., and Basel, Switzerland (1976).

Mamayev, O. I., *Temperature-Salinity Analysis of World Ocean Waters,* Elsevier Scientific, Amsterdam (1975).

Morse, P. M. *Thermal Physics,* W. A. Benjamin, New York, N. Y. (1964).

Sommerfeld, A., *Thermodynamics and Statistical Mechanics,* Academic Press, Inc., New York, N.Y. (1956).

Warren, B. A. and C. Wunsch, Eds., *Evolution of Physical Oceanography,* MIT Press, Cambridge, Mass. (1981).

Weyl, P. K., *Oceanography: An Introduction to the Marine Environment,* John Wiley & Sons, New York, N.Y. (1970).

Worthington, L. V., "The Water Masses of the World Ocean: Some Results of a Fine-Scale Census," in *Evolution of Physical Oceanography,* B. A. Warren and C. Wunsch, Eds., MIT Press, Cambridge, Mass. (1981).

Journal Articles and Reports

Fofonoff, N. P., "Physical Properties of Seawater: A New Salinity Scale and Equation of State for Seawater," *J. Geophys. Res.,* Vol. 90, p. 3332 (1985).

Millero, F. J., C.-T. Chen, A. Bradshaw, and K. Schleicher, "A New High Pressure Equation of State for Seawater," *Deep Sea Res.,* Vol. 27, p. 255 (1980).

UNESCO, "The Practical Salinity Scale 1978 and the International Equation of State of Seawater 1980. Tenth Report of the Joint Panel on Oceanographic Tables and Standards," UNESCO Technical Papers in Marine Science No. 36, UNESCO, Paris, France (1981).

Chapter Five

Geophysical Fluid Dynamics I: Waves and Tides

Of waves, we really can't fathom
How they get as big as we hav'em.
 To gargantuan size
 They occasionally rise
And our very best 'matics can't math'em.

5.1 Introduction

The temporal and spatial variability of ocean and atmosphere governed by the principles developed in the previous chapters is known as *geophysical fluid dynamics,* with the factors that distinguish the motions from ordinary hydrodynamical ones being strong rotation, vertical stratification, and global scale. In addition, both of these geophysical fluids are multicomponent ones, with water vapor in the case of the atmosphere and dissolved salts in the case of the ocean adding interesting and significant complications.

In this chapter and the next, we will develop some of the more important features of geophysical flows in the ocean, including surface and internal wave motion, boundary layer dynamics, and wind-driven currents. During the course of events, we will discuss a number of major oceanic phenomena such as wind-wave generation, the Ekman layers, upwelling and downwelling, western boundary currents such as the Gulf Stream, and Rossby waves. Along the way, a number of calculational conveniences of fluid dynamics will be developed, i.e., potential and stream functions, vorticity theorems, and instability analyses.

5.2 Quasi-Steady Motions

Even in the steady state, the effects of Coriolis force and stratification are dominant in geophysical fluids. We will now derive two equations that characterize the horizontal and vertical forces at work in the case of steady currents in a stratified ocean, the *geostrophic equation* and the *Brunt-Väisälä frequency relationship.*

In the steady state, time derivatives are zero, by definition. Additionally, we confine our attention for now to regions where the spatial variations of current are negligible, and will also neglect tidal forcing as well. Under these conditions, the horizontal component of the momentum equation, as was discussed in Section 2.8, takes on the simple form

$$(-2\Omega \times \mathbf{u})_h - \frac{1}{\rho} \nabla_h p = 0 , \tag{5.1}$$

which can be decomposed into x and y components using Eqs. 3.84 and 3.85:

$$2\Omega v \sin \Lambda = \frac{1}{\rho} \frac{\partial p}{\partial x} , \tag{5.2}$$

and

$$-2\Omega u \sin \Lambda = \frac{1}{\rho} \frac{\partial p}{\partial y} , \tag{5.3}$$

respectively, or, with a slight rearrangement,

$$\frac{\partial p}{\partial x} = \rho f v \tag{5.4}$$

and

$$\frac{\partial p}{\partial y} = -\rho f u . \tag{5.5}$$

These are the equations of *geostrophic balance,* which for the Northern Hemisphere state that the flow of a northward steady current, v, is subject to an eastward Coriolis force (Fig. 3.6) and is balanced by a westward horizontal pressure gradient; additionally, a steady eastward current, u (southward Coriolis force), is balanced by a northward pressure gradient. The pressure gradients are developed during the ocean's adjustment to equilibrium by way of an inclination or tilting of its constant density surfaces away from the equipotential surfaces, which sets up a *hydraulic head* (or elevation above the equipotential) of approximately the right amount necessary to balance the current against the Coriolis force. This is termed geostrophic flow, and will be explained more fully ahead. In the Southern Hemisphere, the Coriolis force and the sense of the gradients are reversed, but the dynamic balances remain the same.

The vertical component of the momentum equation in the steady state yields the simple *hydrostatic equation*:

$$0 = -\frac{\partial \Phi}{\partial z} \hat{k} - \frac{1}{\rho} \frac{\partial p}{\partial z} \hat{k}$$

$$= g - \frac{1}{\rho} \frac{\partial p}{\partial z} \hat{k} , \tag{5.6}$$

where **g** is now taken to be the effective gravity as given by the sum of Eqs. 3.24 and 3.46. Since $g = -g\hat{k}$, we have

$$\frac{\partial p}{\partial z} = -\rho(z)\, g . \tag{5.7}$$

This is immediately integrable if $\rho(z)$ is known from the surface ($z = 0$) down to the depth $z = -h$:

$$p(-h) = p_a - \int_{-h}^{0} \rho(z')\, g \, dz' , \tag{5.8}$$

where p_a is the atmospheric pressure, as before. Thus the absolute pressure at depth $-h$ can be calculated, knowing the vertical density distribution. To a first approximation, the density, which varies roughly between 1028 and 1050 kg m^{-3} under typical deeper oceanic conditions, can be assumed constant at, say, $\rho_c = 1036$ kg m^{-3}. Then the pressure is simply

$$p(-h) = p_a + \rho_c g h . \tag{5.9}$$

Since $g = 9.80$ m s^{-2},

$$p(-h) - p_a = 1036 \times 9.80 \, h$$

$$\simeq 10153 \, h \, , \tag{5.10}$$

where h is in meters and $p - p_a$ is in pascals. Because 10^5 Pa = 1 bar, this gives rise to the rule of thumb that the oceanic pressure increases at approximately 1 decibar per meter of depth. Oceanic pressures are therefore often quoted in dbar, which are numerically within 1 to 2% of the depth in meters.

It should be noted that the approximation of constant density is not sufficient for calculating the horizontal pressure gradients needed in Eqs. 5.4 or 5.5. There the full equation of state is required, with *in situ* values of temperature, salinity, and pressure being used in the calculation, as was discussed in Chapter 4.

In order to obtain a more tractable description of the variations of pressure and density, let us separate each of these quantities into vertically varying, steady state components, $p_0(z)$ and $\rho_0(z)$, respectively, plus fluctuating, time-dependent perturbation components, $p'(\mathbf{x},t)$ and $\rho'(\mathbf{x},t)$, respectively. The perturbation components are assumed to be small variations about the background equilibrium values of pressure and density. Thus we may write:

$$p(\mathbf{x},t) = p_0(z) + p'(\mathbf{x},t) \tag{5.11}$$

and

$$\rho(\mathbf{x},t) = \rho_0(z) + \rho'(\mathbf{x},t) \, . \tag{5.12}$$

Now an important and useful variant of the momentum equations may be derived by substituting these into Eqs. 4.51 to 4.53. Neglecting the tidal potential, Φ_t, the terms in pressure and gravity become

$$-\nabla[p_0(z) + p'(\mathbf{x},t)] + [\rho_0(z) + \rho'(\mathbf{x},t)]\mathbf{g} \tag{5.13}$$

$$= -\frac{\partial p_0}{\partial z} \, \hat{k} - \rho_0 g \hat{k} - \nabla p' - \rho' g \hat{k} \, , \tag{5.14}$$

which, because of the hydrostatic equation (Eq. 5.7), becomes a relationship with only the primed quantities appearing in the pressure and gravity terms of the momentum equation:

$$\rho \frac{D\mathbf{u}}{Dt} = -2\rho \mathbf{\Omega} \times \mathbf{u} - \nabla p' + \rho' \mathbf{g} + \nabla \cdot \mathbf{A} \cdot \nabla \mathbf{u} \, . \tag{5.15}$$

Note that the total density, $\rho_0 + \rho'$, has been retained in the acceleration and Coriolis terms. The quantity $\rho' \mathbf{g}/\rho$ is called *reduced gravity,* $\mathbf{g}'$, and is typically $10^{-3}\ g$ in the ocean because of the small size of density perturbations:

$$\frac{\rho'}{\rho}\mathbf{g} = \frac{\rho - \rho_0}{\rho}\mathbf{g} = \mathbf{g}' \, . \tag{5.16}$$

Its negative, $-\mathbf{g}'$, is sometimes called the *buoyancy force*, since a parcel of water of density near $\rho_0 + \rho'$ is buoyed up by its surroundings, so that the effective gravitational force acting on it is reduced to $\mathbf{g}'$. Reduced gravity is of importance in *baroclinic flows,* where the velocity and density vary rapidly in the vertical, especially near the surface layers; such motions are typically of much lower frequency and propagate more slowly than the *barotropic* flows, which are essentially uniform throughout the water column.

A further simplification to the momentum equations that is of considerable utility is called the *Boussinesq approximation,* in which the actual density multiplying the inertial term $D\mathbf{u}/Dt$ is replaced by a constant density, ρ_c. This retains the reduced gravity in the buoyancy forces but does not eliminate acoustic waves and compressibility and, as a bonus, simplifies certain analyses. It is especially useful in the shallow-water hydrostatic approximation. We will use the Boussinesq approximation for much (but not all) work ahead.

Returning to the hydrostatic equation, the unperturbed pressure and density may be substituted into Eq. 5.7 to obtain

$$\frac{dp_0}{dz} = -\rho_0 g = \frac{\partial p_0}{\partial \rho_0}\frac{d\rho_0}{dz} = c_0^2\,\frac{d\rho_0}{dz} \, , \tag{5.17}$$

where c_0 is the speed of sound. Equation 5.17 may be immediately integrated from the surface, where $\rho_0 = \rho_s$, to depth z:

$$\rho_0(z) = \rho_s \exp\left[-\int_z^0 \frac{g\, dz'}{c_0^2(z')}\right] \, . \tag{5.18}$$

For a constant speed of sound, c_0, this gives

$$\rho_0(z) = \rho_s \exp(-gz/c_0^2) = \rho_s \exp(-z/H_s) \, , \tag{5.19}$$

where z is negative below the surface of the sea. If c varies in the vertical, it is natural then to define a *local scale height,* $H_s(z)$, which is the height over which the density increases by e:

$$H_s(z) = c^2(z)/g .$$ (5.20)

Since H_s is approximately 225 km in the ocean, many problems having small amplitudes of motion allow the constant density approximation. Nevertheless, Eq. 5.15, in which we have not made the Boussinesq approximation, has greater generality and will be used in a number of problems requiring it.

5.3 Buoyancy and Stability

The buoyancy forces clearly establish the vertical stability of a water column, for if a water mass is introduced into a region where it is heavier than its surroundings, it will sink; if it is lighter, it will rise. As the sinking or rising motions carry the mass above or below its equilibrium depth where it is neutrally buoyant, oscillations in the vertical may result, called, appropriately enough, *buoyancy oscillations.* Consider Fig. 5.1, in which a water parcel

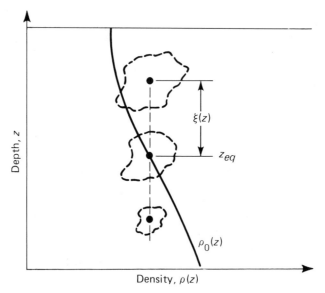

Fig. 5.1 Buoyancy oscillations in a stratified water column. The parcel of water is considered to be enclosed in a kind of flaccid balloon, and oscillates about equilibrium position, z_{eq}, if expansion and contraction due to pressure changes do not overcome buoyancy forces.

is initially at the equilibrium depth, z_{eq}, and conceptually assumed to be enclosed in a kind of flaccid balloon that prevents interchange of properties with its surroundings. It now receives an upward impulse and undergoes a vertical upward displacement, $\xi(z)$, from its equilibrium position. When it arrives at $z_{eq} + \xi$, the parcel will have expanded slightly due to decompression and will be a bit more buoyant as a result. If, however, the background density, $\rho_0(z)$, has decreased sufficiently rapidly over the height ξ, the parcel will find itself heavier than its surroundings, will halt, and will reverse its direction of motion. As it sinks, its kinetic energy carries it past its equilibrium position and it will move to a depth of approximately $z_{eq} - \xi$, where it is now slightly compressed. Here, if the background water column is sufficiently dense, the parcel will be forced upward once again. Assuming the profile $d\rho_0/dz$ permits it over the entire region of excursion, the parcel will oscillate about its equilibrium depth with a characteristic radian frequency, N, variously called the *buoyancy frequency,* the *Brunt–Väisälä frequency*, or the *stability frequency.* If the water column is only weakly stratified, however, the parcel might expand or contract sufficiently under pressure change to find itself increasingly lighter than its surroundings upon ascent, or increasingly heavier upon descent, and further rise or sink accordingly. In the moist atmosphere, the buoyant rising motion, which is accompanied by release of the latent heat of condensation of the water vapor and attendant warming of the rising parcel of air, often becomes completely unstable at some point, with the air rising to high altitudes. This process is termed *convective instability* and is a major source of the vertical uplifting of clouds discussed in Chapter 2. For the sea, analogous sinking motions exist as a result of surface cooling or salinity increases.

In the ocean, the background temperature, salinity, and compressibility collectively determine the vertical density profile. In its upper reaches (i.e., in the mixed layer and seasonal thermocline) the temperature and salinity effects are most pronounced and variable, while in the deep regions, the ocean usually ranges between neutrally stratified to very slightly stable. In this region, density and temperature changes are almost entirely due to adiabatic compression of seawater by the pressure of the overlying fluid (Fig. 4.11).

In order to derive a stability condition for vertical motions, consider again the vertical momentum and continuity equations in the near-steady-state case, allowing only the possibility of small vertical velocities. Under these conditions, we may neglect advective, Coriolis, and viscous forces and again expand pressure and density in terms of small perturbations:

$$\rho \frac{Dw}{Dt} = -\frac{\partial}{\partial z}(p_0 + p') - (\rho_0 + \rho')g \tag{5.21}$$

$$= -\frac{\partial p'}{\partial z} - \rho' g , \tag{5.22}$$

while from the continuity equation, one obtains

$$\frac{D\rho}{Dt} = -\rho \nabla \cdot \mathbf{u} \simeq -\rho_0 \nabla \cdot \mathbf{u} . \tag{5.23}$$

Since the vertical velocity is assumed small, we may neglect the product of ρ' and $\nabla \cdot \mathbf{u}$ in Eq. 5.23. For isentropic, isohaline vertical displacements of a fluid column, we obtain from Eq. 4.38,

$$\frac{D\rho}{Dt} = \frac{1}{c^2} \frac{Dp}{Dt} = \frac{1}{c^2} \frac{D}{Dt} (\rho g \xi) \simeq \frac{\rho_0 g w}{c^2} . \tag{5.24}$$

Here we have used the hydrostatic approximation, so that the perturbation pressure is proportional to the actual density times the vertical excursion, ξ:

$$p' = \rho g \xi . \tag{5.25}$$

Equation 5.24 states that for buoyancy oscillations, the temporal changes in the density are caused simply by vertical advection of the background profile, ρ_0, by the vertical velocity, w. Then the velocity divergence may be written as

$$\nabla \cdot \mathbf{u} \simeq -\frac{g w}{c^2} = -\frac{1}{\rho} \frac{D\rho}{Dt}$$

$$= -\frac{1}{\rho} \left(\frac{\partial \rho}{\partial t} + w \frac{\partial \rho}{\partial z} \right) . \tag{5.26}$$

The vertical component of the momentum equation (Eq. 4.53) becomes

$$\rho \frac{Dw}{Dt} = -\frac{\partial p}{\partial z} - \rho g . \tag{5.27}$$

We next totally differentiate Eq. 5.27 with respect to time and obtain a second-order differential equation for w:

$$\rho \frac{D^2 w}{Dt^2} = -\frac{\partial}{\partial z}\frac{Dp}{Dt} - g\frac{D\rho}{Dt}$$

$$\simeq -g\frac{\partial \rho}{\partial z}w - \frac{g^2}{c^2}\rho w , \tag{5.28}$$

where we have used Eq. 5.24 for the terms on the right. This may be rearranged to obtain

$$\frac{D^2 w}{Dt^2} = -g\left(\frac{1}{\rho}\frac{\partial \rho}{\partial z} + \frac{g}{c^2}\right)w$$

$$= N^2 w . \tag{5.29}$$

Here the radian buoyancy frequency, N, is defined as

$$N \equiv \left[-g\left(\frac{1}{\rho}\frac{\partial \rho}{\partial z} + \frac{g}{c^2}\right)\right]^{\frac{1}{2}} \quad \text{rad s}^{-1} . \tag{5.30}$$

The solutions to Eq. 5.29 are clearly either oscillatory or exponential, according to whether N is real, corresponding to buoyancy oscillations, or imaginary, corresponding to rising/sinking motions. Thus the buoyancy frequency measures the stability of the water column against small vertical perturbations. A stable water column is one that is sharply enough stratified so that $(1/\rho)(\partial \rho/\partial z)$, which is normally negative, will overcome the compressive change, g/c^2, so that $N^2 > 0$.

An alternative expression for the buoyancy parameter, N^2, may be derived from purely thermodynamic arguments by writing a generalized small density change, $d\rho$, in terms of changes in pressure, temperature, and salinity, and comparing it with an isentropic change $(d\rho)_\eta$. The result is (see Gill, loc. cit.),

$$N^2 = g\left[a_T \frac{dT}{dz} + \frac{ga_T^2 T}{C_{ps}} - a_s \frac{ds}{dz}\right], \tag{5.31}$$

which is useful when the temperature and salinity profiles are known separately.

Examples of actual vertical profiles of salinity, temperature, density anomaly, and buoyancy frequency are shown in Fig. 5.2. This case is for a tropical mediterranean (inland) sea with very little wind-mixed layer present, and

(a)   (b)

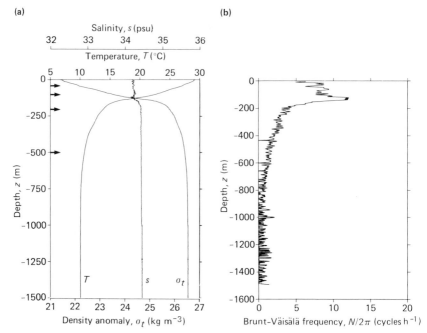

Fig. 5.2   (a) Vertical profiles of salinity, temperature, and density anomaly for a tropical, landlocked sea under low wind conditions. (b) Profile of buoyancy frequency calculated from data of (a). Density is controlled mainly by temperature, which is
characteristic of most nonpolar waters. [Adapted from Apel, J. R., *et al., J. Phys. Oceanography* (1985).]

nearly isohaline conditions existing at depths below approximately 200 m.
The maximum Brunt–Väisälä frequency occurs where the temperature gradient is largest (near 150 m depth) and has a value in excess of 10 cycles h^{-1}
(0.01745 rad s^{-1}). Another example, which represents more typical temperate Atlantic conditions, is shown in Fig. 5.3. Here N shows a secondary maximum near a depth of 750 dbar, probably due to the presence of a differing
water mass. The calculated quantity $d\theta/dp$ on the graph is the potential temperature gradient, and its constancy beneath some 2000 dbar, despite
decreasing N, indicates that the buoyancy frequency variations below that
depth are due to adiabatic compressibility alone (i.e., g/c^2).

5.4 Surface Waves

Surface waves are the most visible manifestation of time-dependent ocean
dynamics, and their study has occupied natural scientists for centuries. Much

has been learned of the linear properties of gravity and capillary waves, and some theoretical advances have been made in understanding their weakly nonlinear features. However, it is fair to say that in spite of much attention, not only are strongly nonlinear features such as cusping and breaking not well modeled, but the very processes by which the wind—by far the most ubiquitous source of wave energy—actually generates surface waves are not altogether clear. In another arena, an extremely long wavelength type of gravity wave, the *tsunami* or *seismic sea wave,* occurs with sensible amplitude so rarely that the available observational data base is insufficient to test theories of tsunami generation and propagation. The deep sea tide, another well-known kind of oceanic motion, is a forced, shallow-water, linear wave that has recently yielded in some degree to the computational power of numerical analysis, and reasonable scientific predictive capability now exists for open ocean tides, based on theoretical principles and observational data.

This is not to underestimate the predictive skill in surface waves and tsunami accumulated by oceanographers; indeed, operational forecasts and warnings for both types have been implemented, and produce acceptable alerts in normal situations. However, cases of severe storm waves are still misforecast with enough frequency to dilute the confidence of those who must rely upon such predictions. In addition to a lack of timely data, it is fair to say that there are still first-order unsolved research problems in nonlinear surface waves, and their study remains an area of active interest.

The wind is thought to generate surface waves through a sequence of processes that assume varying proportions with time. At very low wind speeds, small capillary waves develop on the surface, and are thought to be a time- and space-growing viscous instability. As the instability causes the capillaries to grow, the wavelengths and amplitudes increase so as to extend them into the gravity range (of order centimeters), whereupon new processes apparently take over to continue the growth. The first is direct forcing via wind pressure exerted on the steep parts of the wave slope; the second (and probably the less important) is shear stress exerted on the surface by (1) turbulent eddies in the wind, and (2) tangential surface stress from air flow over the small-scale, irregular wavelets on the surface. The pressure work done by the wind derives largely from those wind Fourier components whose wavelengths are near the water wavelengths and which are in phase with the wave slope. Figure 5.4 schematically illustrates these processes along the length of one gravity wave, and gives an indication of the streamlines of air flow in a coordinate system moving with the wave phase speed, c_p.

As the waves continue to increase in length and height, yet another process comes into play: nonlinear, finite amplitude wave/wave interactions, in which the quasi-random gravity waves, in scattering off one another, produce both longer and shorter waves. The scattering process can be thought

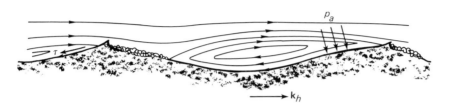

Fig 5.4 Schematic of weakly breaking surface waves showing the normal pressure force near the crest of the wave and the tangential wind-stress force. The streamlines of air flow over the crests show regions of closed circulation in a coordinate system traveling at the wave phase speed. [Adapted from Banner, M. L., and W. K. Melville, *J. Fluid Mech.* (1976).]

of as being caused by the interactions of four waves, two of which intersect to form an interference pattern that is considered to be a third, virtual wave. This virtual wave then scatters a fourth, real wave toward longer wavelengths. While weak, this process is thought to be largely responsible for the characteristic spectral shape and increasing wavelengths of water waves as time goes on. The scattering toward longer wavelengths results in a continual lengthening of the dominant waves as time progresses, while scattering toward shorter lengths generates waves that are lost among the spectral components already existing at small wavelengths.

On the short wavelength side, the wave energy also spreads itself among the various frequencies (or wavelengths) present, through long-wave/short-wave interactions, so that in near-equilibrium, the wave energy distribution reaches a saturation form that is more or less global in its level and functional form, somewhat as the molecules in a gas reach a Maxwellian distribution when in equilibrium. Examples of this spectrum will be given ahead.

In addition to the wind speed and direction, the generation of waves clearly depends on the length of time that the wind has been blowing, i.e., the *duration*, and the distance that the observation point is located offshore, that is, the *fetch*. Both the duration and fetch that are required to allow a "fully developed sea" to be formed depend on wind velocity (Fig. 5.5). For example, in order for a wind of 10 m s^{-1} to generate a fully developed, saturated wave spectrum, it must blow for nearly 18 h over an oceanic expanse of perhaps 320 km. Under those conditions, the so-called *significant wave height*, $\xi_{1/3}$, will be approximately 2 m, and the *wave period*, T_w, will be about 7.5 s. However, there is an entire spectrum of waves that actually make up the "significant" wave. Figure 5.5 also provides estimates of typical wave periods, lengths, and heights. While waves of the height of this example may also be generated by stronger winds under conditions of shorter duration and less fetch, they will not have equilibrated into the global spectrum illustrated in Fig. 5.6.

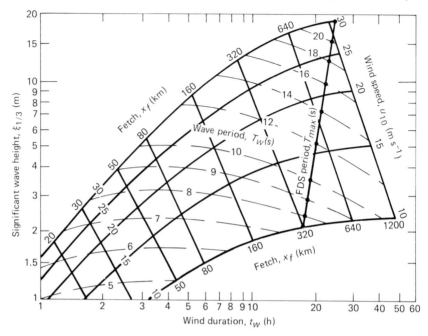

Fig. 5.5   Cumulative sea state diagram showing significant wave height, $\xi_{1/3}$, as a function of wind duration, fetch, and speed. A fully developed sea (FDS) is considered as having arisen from conditions shown along the near vertical line labeled "FDS period, T_{max}." [Adapted from Van Dorn, W. G., *Oceanography and Seamanship* (1974).]

This figure shows the distribution of squared wave heights, or wave energy, among various frequencies under fully developed equilibrium conditions.

As the waves propagate away from the immediate region of generation, or as the wind speed is reduced, they become *swell,* which is actually the far-radiation field of the surface wave source. Since longer waves travel faster than shorter ones (i.e., they are dispersive; see Section 5.6), the wavelengths and periods of swell gradually increase with the time and distance from the source; their amplitudes are also reduced due to spreading and friction, so that swell waves are usually linear, coherent, small-amplitude gravity waves generally having periods longer than several seconds.

Having briefly described how wind waves are generated and propagate, we will now deduce some of their linear properties from the hydrodynamic equations. The surface wave problem has many characteristics that allow the exercise of the theory developed to date, but does not necessitate most of the thermodynamics. In addition, real wind waves are almost always non-linear, but we will not consider that important complication here.

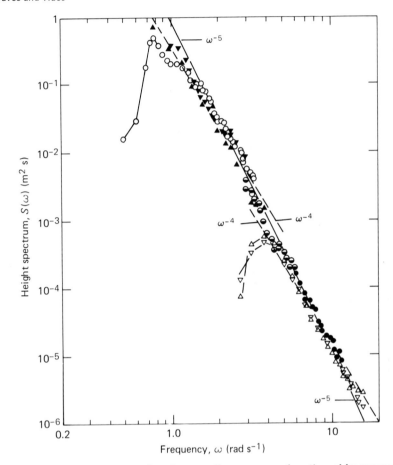

Fig. 5.6 Height spectrum of surface gravity waves as a function of frequency, for the "equilibrium range" beyond the spectral peak; the shape of the peak is shown for only three cases. Slope of the solid straight line is − 5 (cf. Eq. 5.109). However, more recent work suggests that the slope of the high-frequency region may actually be − 4 if analyzed differently (dashed lines). [Adapted from Phillips, O. M., *The Dynamics of the Upper Ocean* (1977).]

5.5 Linear Capillary and Gravity Waves

Surface waves have frequencies, ω, much greater than N or f, and to a first approximation may be considered as linear, so that $\mathbf{u} \cdot \nabla \mathbf{u} \approx 0$. In addition to making the linear assumption, we may neglect the Coriolis, buoyancy, and viscous terms in the momentum equation, although dissipation becomes increasingly important at higher frequencies. Under these conditions, it can be demonstrated (and we will do so after deriving expressions for the velocity) that the fluid motion will be *irrotational*, or will possess no vorticity. Fluid vorticity, ζ, is defined as the curl of the velocity field:

$$\zeta \equiv \nabla \times \mathbf{u} . \tag{5.32}$$

A vector field that is irrotational can be derived from a scalar potential, which for the fluid velocity is named appropriately enough the *velocity potential,* φ (not to be confused with the potential energy per unit mass, Φ, or the coordinate, ϕ). Thus we may write for the velocity,

$$\mathbf{u} = \nabla \varphi . \tag{5.33}$$

Additionally, the incompressibility condition, $\nabla \cdot \mathbf{u} = 0$, implies that the velocity potential satisfies Laplace's equation,

$$\nabla^2 \varphi = 0 . \tag{5.34}$$

A large body of mathematical methods is available to arrive at solutions to Eq. 5.34.

Consider now a uniform, frictionless ocean of depth H, whose surface is subject to a small wavelike perturbation, $\xi(x,y,t)$, as shown in Fig. 5.7. In the Boussinesq approximation, the total pressure in the fluid, p, is the hydrostatic pressure, $p_0 = p_a - \rho g z$, plus the perturbation pressure, $p' = \rho g \xi$:

$$p = p_a + \rho g (-z + \xi) . \tag{5.35}$$

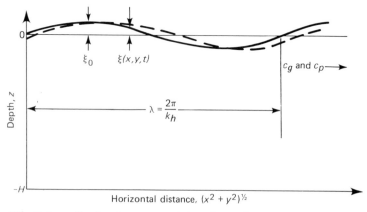

Fig. 5.7 Schematic showing nomenclature for surface waves.

Under the conditions assumed above, the equations of motion reduce to

$$\rho \frac{\partial \mathbf{u}}{\partial t} = -\nabla p' \tag{5.36}$$

and

$$\nabla \cdot \mathbf{u} = 0 , \tag{5.37}$$

since the perturbation density, ρ', equals 0, by assumption. Operating on Eq. 5.36 with the divergence operator, $\nabla \cdot$, one immediately obtains Laplace's equation for the perturbation pressure, p':

$$-\rho \frac{\partial}{\partial t} (\nabla \cdot \mathbf{u}) = \nabla \cdot \nabla p' = \nabla^2 p' = 0 . \tag{5.38}$$

Thus the perturbation pressure is a type of velocity potential in this instance.

Because the problem is linear, we may assume a traveling sinusoidal wave as a solution, since any arbitrary linear disturbance may be composed out of such basis functions via Fourier analysis. Now p' is proportional to ξ, so we may try as a solution

$$\xi(\mathbf{x},t) = \xi_0 \cos (\mathbf{k}_h \cdot \mathbf{x} - \omega t) , \tag{5.39}$$

where the *horizontal wave vector,* $\mathbf{k}_h$, is

$$\mathbf{k}_h = k\hat{\imath} + l\hat{\jmath} , \tag{5.40}$$

and the *radian frequency* is ω. The wave moves with a *phase speed,* c_p, given by

$$c_p = \omega / k_h , \tag{5.41}$$

where the scalar *horizontal wave number* is

$$k_h = (k^2 + l^2)^{\frac{1}{2}} . \tag{5.42}$$

We have not yet specified a z dependence, for this must come out of the analysis; nor have we applied the boundary conditions. Laplace's equation (using Eqs. 5.35 and 5.39) becomes

$$-k_h^2 p' + \frac{\partial^2 p'}{\partial z^2} = 0 . \tag{5.43}$$

The boundary conditions are those given by Eqs. 3.117a, 3.121, and 3.123, transformed to conditions on p'. Thus at the surface, $z = \xi$, and at the bottom, $z = -H$, we impose the requirements:

$$w(z) = \partial \xi / \partial t \qquad \qquad \text{at } z = \xi, \qquad (5.44a)$$

$$p(z) = p_0(z) + p'(z) = p_a \qquad \text{at } z = \xi, \qquad (5.44b)$$

$$w(z) = 0 \qquad \qquad \text{at } z = -H, \qquad (5.44c)$$

and

$$\partial p'(z)/\partial z = 0 \qquad \qquad \text{at } z = -H. \qquad (5.44d)$$

The free surface boundary conditions can be approximately satisfied by applying them at $z = 0$ rather than $z = \xi$, because of the small amplitude assumption. Hence Eqs. 5.44a and 5.44b will be evaluated at $z = 0$ when imposing the boundary conditions.

Now the time variation of all the time-dependent quantities must be harmonic, since $p \sim \xi \sim \cos \omega t$. The harmonic solution that satisfies the bottom boundary condition is

$$p' = p_0' \cosh k_h (z + H) \cos (\mathbf{k}_h \cdot \mathbf{x} - \omega t) \qquad (5.45)$$

and

$$w = w_0 \sinh k_h (z + H) \sin (\mathbf{k}_h \cdot \mathbf{x} - \omega t). \qquad (5.46)$$

The horizontal velocity components, similarly obtained from Eq. 5.36, are:

$$u = u_0 \cosh k_h (z + H) \cos (\mathbf{k}_h \cdot \mathbf{x} - \omega t) \qquad (5.47)$$

and

$$v = v_0 \cosh k_h (z + H) \cos (\mathbf{k}_h \cdot \mathbf{x} - \omega t). \qquad (5.48)$$

It is readily established that these solutions meet the surface and bottom boundary conditions. One additional condition must be met, however, if we are to incorporate capillary waves—continuity of pressure across the interface, from which Eq. 3.123 was obtained:

$$p = p_a - \tau_s \left(\frac{\partial^2 \xi}{\partial x^2} + \frac{\partial^2 \xi}{\partial y^2} \right). \qquad (5.49)$$

Here τ_s is the surface tension and p_a the atmospheric pressure. To proceed further, we need one of the several forms of the *Bernoulli equation* to describe the surface dynamics. By writing the velocity in terms of $\nabla \varphi$, our version of the momentum equation can be rearranged to give:

$$\nabla \left(\frac{\partial \varphi}{\partial t} + \tfrac{1}{2} \, \mathbf{u}^2 + \frac{p}{\rho} + gz \right) \equiv \nabla \chi = 0 , \qquad (5.50)$$

where χ is simply an abbreviation for the quantity in the parentheses. Integrating this along a streamline, ψ, just below the free surface, along which p and ρ are constant, we obtain

$$\int_\psi \nabla \chi \cdot d\mathbf{s} = \chi = \text{const.} \qquad (5.51)$$

This means that the sum

$$\frac{\partial \varphi}{\partial t} + \tfrac{1}{2} \, \mathbf{u}^2 + \frac{p'}{\rho} + g\xi = \frac{p_c}{\rho} = \text{const.} , \qquad (5.52)$$

is a constant, p_c/ρ, along the streamline. Neglecting $\tfrac{1}{2} \, \mathbf{u}^2$ as being of second order, we may combine Eqs. 5.44a, 5.49, and 5.52 and differentiate the result with respect to time to get

$$\frac{\partial^2 \varphi}{\partial t^2} + \frac{1}{\rho} \left[-\tau_s \left(\frac{\partial^2}{\partial x^2} + \frac{\partial^2}{\partial y^2} \right) + \rho g \right] \frac{\partial \varphi}{\partial z} \Bigg|_{z=0} = 0 , \qquad (5.53)$$

assuming that φ also behaves harmonically in time. This is an additional equation that the velocity potential must obey at the surface. Now φ will have the same functional form as p', so that Eq. 5.53 will result in the condition:

$$\cos (\mathbf{k}_h \cdot \mathbf{x} - \omega t)[\omega^2 \cosh k_h (z + H)$$
$$- k_h (g + k_h^2 \tau_s /\rho) \sinh k_h (z + H)] \Bigg|_{z=0} = 0 , \qquad (5.54)$$

which can be satisfied for all values x and t only if the quantity in the square brackets vanishes. Thus we obtain

$$\omega = \omega(\mathbf{k}) = \pm [(gk_h + k_h^3 \, \tau_s /\rho) \tanh k_h H]^{1/2} . \qquad (5.55)$$

This important relation is called the *dispersion equation* for small-amplitude gravity/capillary waves on the ocean surface, and is a fundamental result of the linear theory. We will return to it ahead in a more extensive discussion of propagation characteristics.

Next the amplitudes may be evaluated in terms of ξ_0. At $z = 0$, Eqs. 5.44a and 5.44b hold, so that

$$w(0) = \left.\frac{\partial \xi}{\partial t}\right|_{z=0} = \omega\xi_0 \sin\left(\mathbf{k}_h \cdot \mathbf{x} - \omega t\right)\Big|_{z=0} ,$$

or

$$w_0 \sinh k_h H = \omega^2 \xi_0 / \omega . \tag{5.56}$$

Using the dispersion relationship (Eq. 5.55) we finally obtain

$$w_0 = g\xi_0 \frac{k_h}{\omega} \frac{(1 + B)}{\cosh k_h H} , \tag{5.57}$$

where

$$B \equiv k_h^2 \tau_s / \rho g \tag{5.58}$$

is an abbreviation for the surface tension term.

From Eq. 5.36, one obtains for the remaining amplitudes:

$$p_0' = \rho g\xi_0 \frac{(1 + B)}{\cosh k_h H} , \tag{5.59}$$

$$u_0 = g\xi_0 \frac{k}{\omega} \frac{(1 + B)}{\cosh k_h H} , \tag{5.60}$$

and

$$v_0 = g\xi_0 \frac{l}{\omega} \frac{(1 + B)}{\cosh k_h H} . \tag{5.61}$$

It is worth noting that ω/k and ω/l are the projections of the phase speed, ω/k_h, on the x and y axes, respectively, and because $k, l \leq k_h$, these projected speeds are larger than the actual phase speed. This is due to the same effect as a wave encountering a bulkhead at a shallow angle. The speeds at which a constant phase point runs along the bulkhead, ω/k or ω/l, are larger than the forward phase speed, ω/k_h, one or the other of them becoming infinite when the crests are exactly parallel to the bulkhead. This is illustrated in Fig. 5.8. Also note that the fluid velocities u_0 and v_0 are *inversely* proportional to the quantities ω/k and ω/l. This is a consequence of the parallel polarization (**u** parallel to $\mathbf{k}_h$) of the gravity waves.

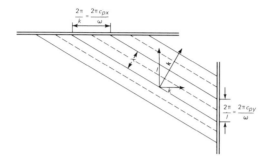

Fig. 5.8 Propagation of surface waves at an angle to vertical walls. Projected speeds of progression of constant phase points along the walls are greater than the actual wave speed, ω/k_h.

It is probably worth noting that while the direction of the phase *velocity,* $\mathbf{c}_p$, is that of the wave vector, $\mathbf{k}_h$, it is not true that its decomposition into x,y components of the form

$$\omega k/(k^2 + l^2)$$

and

$$\omega l/(k^2 + l^2)$$

yields the propagation speed of the points of constant phase in the x and y directions. By considering in Fig. 5.8 the limiting case of a wave train incident nearly normally on a bulkhead oriented along the x axis, it is clear that $k \approx 0$ and $k_h \approx l$, and that the speed of progression of a point of constant phase along the x axis, c_{px}, is nearly infinite, while that of a point along the y axis, c_{py}, is slightly *greater* than ω/k_h. Neither component is as given by the expressions just above, which would predict $c_{px} \approx 0$ and $c_{py} < \omega/k_h$. A more nearly correct quantity with which to describe directional phase propagation is the *slowness,* **s,** where

$$\mathbf{s} = \frac{\mathbf{k}}{\omega} = \frac{k\hat{i} + l\hat{j}}{\omega} . \tag{5.62}$$

The reciprocals of the components of the slowness are the propagation speeds of constant phase points in the x,y directions, and its absolute magnitude is the reciprocal of the scalar phase speed:

$$s_x = \frac{k}{\omega} = \frac{1}{c_{px}} , \tag{5.63}$$

$$s_y = \frac{l}{\omega} = \frac{1}{c_{py}} , \tag{5.64}$$

and

$$|s| = \left(\frac{k^2 + l^2}{\omega^2} \right)^{1/2} = \frac{1}{c_p} . \tag{5.65}$$

In the phase arguments of oscillating functions, the slowness appears as

$$(\mathbf{k}_h \cdot \mathbf{x} - \omega t) = \omega(\mathbf{s} \cdot \mathbf{x} - t) . \tag{5.66}$$

5.6 Dispersion Characteristics of Surface Waves

The propagation characteristics of surface waves possess many interesting features. We will first briefly discuss some of the general properties of dispersive waves using a one-dimensional *dispersion diagram*, or a plot of ω versus k_h. Such a plot for water waves is shown in Fig. 5.9, where the dispersion characteristic suggests that low frequency waves have small wave numbers, so that both ω and k_h tend toward zero together. At higher frequencies, the (ω, k_h) plot has an inflection point where the slope begins to increase once again; this will be shown to be due to capillary waves. The phase speed, c_p, is given by the slope of the chord, ω/k_h, in Fig. 5.9, while the group speed, c_g, is $\partial\omega/\partial k_h$, or is the slope of the tangent line passing through ω and k_h, at least in one dimension. In two dimensions, the group velocity, $\mathbf{c}_g$, which is also the velocity of propagation of energy, is a quantity given by the wave vector gradient of the dispersion relation:

$$\mathbf{c}_g \equiv \nabla_\mathbf{k} \omega(\mathbf{k}) = \frac{\partial\omega}{\partial k} \hat{i} + \frac{\partial\omega}{\partial l} \hat{j} , \tag{5.67}$$

where $\omega(\mathbf{k})$ comes from Eq. 5.55. Carrying out the required manipulations yields

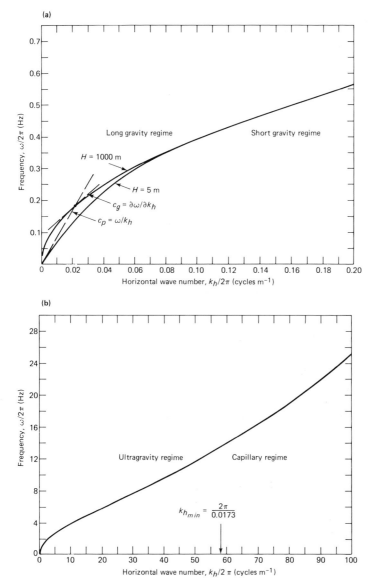

Fig. 5.9 Form of dispersion equation for surface gravity–capillary waves. (a) Low frequency regime, with upper end of the range corresponding to a wavelength of 5 m, for two values of water depth, H. The slope of the chord through any point, ω/k_h, gives the phase speed, while the slope of the tangent gives the group speed. The speed of sound, c, is the theoretical limit for long length gravity waves, but other factors such as shallow water and Coriolis effects actually limit the speeds. (b) High frequency regime, with an upper range of 1 cm. Phase speed minimum occurs at $\lambda = 0.0173$ m.

$$c_p^2 = (g/k_h + k_h \tau_s/\rho) \tanh k_h H \tag{5.68}$$

and

$$\mathbf{c}_g = (c_p/2) [2k_h H \operatorname{csch} 2k_h H + (1 + 3B)/(1 + B)]\hat{h}, \tag{5.69}$$

where $\hat{h}$ is a horizontal unit vector in the direction of propagation, i.e., that of $\mathbf{k}_h$.

5.7   Surface Gravity Waves

We now divide the discussion according to regions of low and high wave number. At low wave numbers or long wavelengths, the capillary terms of Eqs. 5.55 through 5.58 are negligible and the restoring forces are purely gravitational; then the dispersion equation reduces to that for *surface gravity waves:*

$$\omega = \pm [gk_h \tanh k_h H]^{1/2} . \tag{5.70}$$

Two sub-cases can be distinguished: deep and shallow water.

Deep Water

Here $k_h H \gg 1$ and $\tanh k_h H \simeq 1$, so the deep water gravity wave phase speed reduces to

$$c_p \simeq (g/k_h)^{1/2}, \qquad k_h H \gg 1, \tag{5.71}$$

while the deep water group speed becomes

$$c_g \simeq c_p/2 . \tag{5.72}$$

Thus surface gravity waves are *dispersive*, with the speeds depending inversely on the square root of the wave number, or directly on the square root of the wavelength; therefore, longer gravity waves travel faster than shorter ones. In addition, the group speed of a packet of deep-water surface gravity waves is one-half the phase speed, so that within a traveling packet, wave crests first appear toward the rear of the group, propagate through it, and then disappear near the front. The energy in the packet moves approximately with the speed of the centroid, which is c_g. In addition, an initially localized packet consisting of many Fourier components spreads out in space and time as it propagates, so that at a distant observation point, the long wavelength components, which travel the fastest, also appear earliest. Often the presence of a distant storm at sea is first announced by the appearance of long wavelength, rapidly propagating swell coming from the approximate direction of the storm.

Shallow Water

The deep water dispersion relation predicts infinite velocities at $k_h = 0$, but, in fact, the speeds of gravity waves are theoretically limited by the neglected speed of sound, c. Long before that speed is approached, however, shallow water limitations come into play. Under those conditions, $k_h H \ll 1$, $\tanh k_h H \simeq k_h H$, and the shallow water phase and group speeds become

$$c_p = c_g \simeq (gH)^{1/2}, \qquad k_h H \ll 1 . \qquad (5.73)$$

Thus the phase and group speeds of long length, shallow water gravity waves are the same and are given the square root of gH; they are dispersionless as well, and travel at speeds set only by the depth of water. The dependence on H means that shallow water waves will undergo refraction when encountering varying depth water. Figure 5.10 shows the phase speed of long wave-

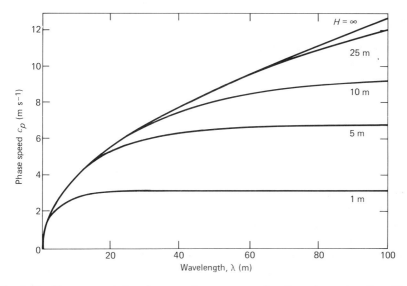

Fig. 5.10 Phase speed of surface gravity waves as a function of wavelength, with water depth as a parameter. [Adapted from Neumann, G., and W. J. Pierson, Jr., *Principles of Physical Oceanography* (1966).]

length gravity waves as a function of wavelength, with water depth as a parameter. The rule of thumb for gravity waves is that they begin to feel the effects of shoaling when the water depth is approximately one-half wavelength, and they are slowed significantly when it reaches one-quarter wavelength.

Typical bottom topographies and their refractive effects are illustrated in Figs. 5.11a to 5.11c. Figure 5.11a shows how constant phase fronts and the associated orthogonal ray paths bend when encountering an offshore tapering ridge, which acts as a focusing lens for gravity waves approaching parallel to the axis of the ridge. This has the effect of concentrating wave energy on the shoreline near the center of the ridge. Such topography exists immediately offshore of the harbor at Hilo, Hawaii, for example, where bathymetric focusing of seismic sea waves has caused much damage to the waterfront. Figure 5.11b demonstrates the opposite topography, that of a submarine canyon; here wave energy tends to be diverged from the canyon axis and instead to be focused near the edges. Something like this geometry exists in the vicinity of the Hudson Canyon offshore of New York City, with the consequence that storm wave energy tends to be concentrated on the shore of western Long Island or northern New Jersey, depending on the wind direction. Figure 5.11c demonstrates refraction around a circular island with a resultant crossing wave field on the downwave side of the island. Seas in such a region tend to arrive from a range of directions and to be somewhat confused.

In calculating refraction effects, the general relationship of Eq. 5.68 must be used for the phase speed. Ray paths in water of slowly varying depth may be computed from the *eikonal equation*, which is developed in Chapter 7 in the context of acoustic propagation, or by other methods involving conservation of *wave action* (see Section 8.9).

5.8 Surface Capillary Waves

At the other end of the frequency spectrum from long gravity waves lie surface capillary waves, whose restoring force is surface tension. From the general relationship (Eq. 5.68) it is clear that the phase velocity has a minimum value, which can be found by setting $\partial c_p / \partial k_h = 0$. The minimum phase speed in seawater, $c_{p_{min}}$, is found to be 0.23 m s^{-1} and occurs at a wavelength, λ_{min}, of 0.0173 m:

$$c_{p_{min}} = (4g\tau_s/\rho)^{1/4} = 0.234 \quad \text{m s}^{-1} . \tag{5.74}$$

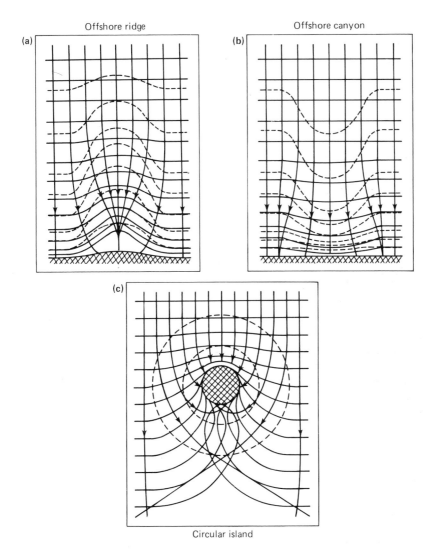

Offshore ridge

Offshore canyon

(a)

(b)

(c)

Circular island

Fig. 5.11 Plan views of wave refraction around various subsurface bathymetries, with waves incident from top: (a) tapering offshore ridge; (b) offshore canyon; (c) circular island. Approximate contours of constant depth are shown as dotted lines; constant phase fronts and rays constitute an orthogonal grid. [Adapted from *Techniques of Forecasting Wind Waves and Swell* (1951).]

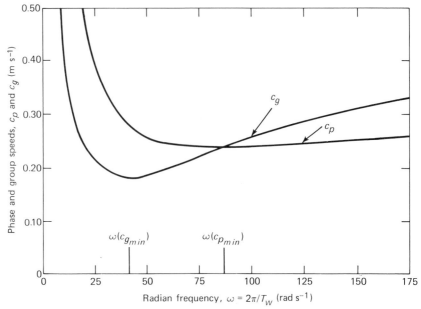

Fig. 5.12   Phase and group speeds of mixed gravity–capillary waves, showing speed minima in the centimeter wavelength region. [Adapted from Neumann, G., and W. J. Pierson, Jr., *Principles of Physical Oceanography* (1966).]

Similarly, the group speed (Eq. 5.69) has a minimum of 0.178 m s^{-1} at λ_{min} = 0.0438 m. Plots of the variation of c_p and c_g in the vicinity of the minima are presented in Fig. 5.12. For wavelengths near λ_{min}, the waves are mixed gravity–capillary ones, while for higher frequencies, they are pure capillary waves. While the frequency at the minimum, $\omega_{min}/2\pi$, is 13.4 Hz, capillary waves may exist at upward of 100 Hz or beyond before molecular viscous damping reduces them to very small amplitudes and lifetimes. Because of the short wavelengths involved, $k_h H \gg 1$ and thus oceanic capillary waves are nearly always deep water waves; a possible exception is seen at the beach, where capillary waves on the thin sheets of water receding down the sand following a breaker demonstrate some shallow depth character. Capillary waves are *anomalously dispersive*, with short waves traveling faster than longer ones; in a wave packet, the shorter, more highly damped capillaries are found at the front. Packets of *parasitic capillary waves* are readily generated by breaking or steep gravity waves and are trapped near the regions of generation by their slow speeds. Such short waves have wavelengths close to those of the electromagnetic waves used by many radars, and they may serve as *Bragg scatterers* for radar energy incident on the sea surface; this interaction is described more fully in Chapter 8.

5.9   Surface Wave Vorticity

To demonstrate that the fluid motion is indeed irrotational and vortex-free, we calculate the velocity derivatives using the definition of vorticity. From Eq. 5.32,

$$\zeta = \begin{vmatrix} \hat{\imath} & \hat{\jmath} & \hat{k} \\ \dfrac{\partial}{\partial x} & \dfrac{\partial}{\partial y} & \dfrac{\partial}{\partial z} \\ u & v & w \end{vmatrix} . \tag{5.75}$$

Evaluating the derivatives in the expansion, one obtains from Eqs. 5.46 and 5.48,

$$\zeta_x = \frac{\partial w}{\partial y} - \frac{\partial v}{\partial z}$$

$$= l w_0 \sinh k_h (z + H) \cos (\mathbf{k}_h \cdot \mathbf{x} - \omega t)$$

$$- k_h v_0 \sinh k_h (z + H) \cos (\mathbf{k}_h \cdot \mathbf{x} - \omega t) ,$$

or

$$\frac{l k_h}{\rho \omega} - \frac{k_h l}{\rho \omega} = 0 . \tag{5.76}$$

Similar calculations show that $\zeta_y = \zeta_z = 0$, so that the solution for linear gravity waves has zero vorticity, as asserted initially. Since we have earlier assumed that the fluid was also incompressible, we have established conditions necessary for the existence of a *stream function*, ψ, whose derivatives give the velocity components.

5.10   The Stream Function

Plots of streamlines of fluid flow have an intuitive appeal to the viewer, but their interpretation in time-varying problems must be made carefully. A streamline is a line in the fluid whose tangent is everywhere instantaneously parallel to $\mathbf{u}$. A family of streamlines at any time represents solutions to the set of equations

$$\frac{dx}{u(\mathbf{x},t)} = \frac{dy}{v(\mathbf{x},t)} = \frac{dz}{w(\mathbf{x},t)} . \tag{5.77}$$

In any plane (say the x,z plane), the slope of the streamline, dz/dx, is given by

$$\frac{dz}{dx} = \frac{w}{u} ,$$

(5.78)

and is therefore proportional to the ratio of velocity components; e.g., a streamline can be obtained by integrating Eq. 5.78 in the x-z plane.

The *path* of a fluid parcel does not in general coincide with a streamline except in the case of steady flow. The formulation of hydrodynamics that gives the motion in terms of the paths of fluid particles is termed *Lagrangian*, and is akin to ordinary particle dynamics; it is useful in oceanography in many circumstances, both observational and theoretical. The more usual formulation, however, is *Eulerian*, wherein the velocity, pressure, etc., are considered to be fluid fields specified in space and time; it is this formulation that we have been dealing with so far in this book. From the Eulerian perspective, the stream function is a useful concept and the streamlines helpful in visualizing the flow.

If further, the flow is incompressible (or if compressible, then steady), the continuity equation takes on the form, $\nabla \cdot \mathbf{u} = 0$. Additionally, if it is two-dimensional, or the coordinate axes can be rotated to make it so, the velocity divergence has only two derivatives. For the surface wave problem, the directions of interest are chosen to be x and z, so that the velocity field may be written as

$$\mathbf{u} = (u, 0, w,) .$$

(5.79)

The continuity equation is then

$$\frac{\partial u}{\partial x} + \frac{\partial w}{\partial z} = 0 .$$

(5.80)

From these equations, it is clear that $(u\,dz - w\,dx)$ is the exact differential of a scalar stream function, $d\psi$, and that

$$u = \frac{\partial \psi}{\partial z}$$

(5.81)

and

$$w = - \frac{\partial \psi}{\partial x} .$$

(5.82)

The *Langrangian stream function*, $\psi = \psi(x,z,t)$, is defined by the line integral

$$\psi(x,z,t) = \psi_a + \int (u \, dz - w \, dx) , \qquad (5.83)$$

where ψ_a is a function independent of x and z, and the line integral is taken along a curve starting at a reference point, 0, to the field point, (x,z). There is no flow across streamlines, so that the stream function is a *constant* along a given streamline, and families of such lines may be labeled by various constant values of ψ. It is this representation that we use here to display the flow characteristics of surface waves.

From Eqs. 5.46, 5.47, and 5.83, the stream function for surface waves in the x,z plane is found to be

$$\psi(x,z,t) = \psi_a + \psi_0 \sinh k(z + H) \cos (kx - \omega t) , \qquad (5.84)$$

where the stream function amplitude, ψ_0, is, from Eqs. 5.47, 5.56, and 5.60,

$$\psi_0 = 2g\xi_0/\omega \cosh kH , \qquad (5.85)$$

and where we have rotated the axes so that

$$k_h = k, \qquad l = 0 . \qquad (5.86)$$

In this instance, we have worked backward, deriving ψ from u and w, but often a more useful procedure is to solve Laplace's equation for ψ in the form

$$\frac{\partial^2 \psi}{\partial x^2} + \frac{\partial^2 \psi}{\partial z^2} = 0 , \qquad (5.87)$$

and then derive u and w by differentiation. The form of Eq. 5.87 follows from the incompressibility condition and the definition of u and w in terms of ψ.

Families of instantaneous streamlines for propagating surface waves are shown in Fig. 5.13a for the cases of deep, intermediate, and shallow water, as viewed in a coordinate system at rest. It is seen that the subsurface velocities are directed upward in the region of wave phase in front of the crest and are downward in the rear of the crest at 180° behind the upward-flowing region. Immediately at the crests and troughs, the velocities are completely horizontal. Figure 5.13b shows the Lagrangian representation of the same flow field, with particle orbits being generally elliptical and rotating clock-

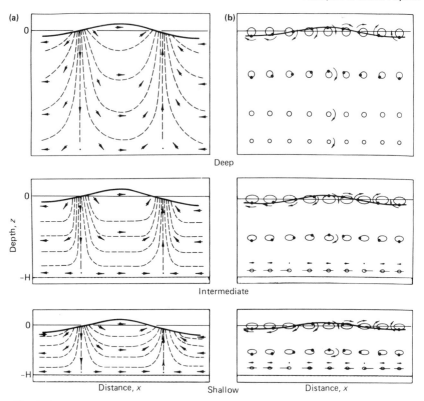

Fig. 5.13 (a) Instantaneous streamlines for surface gravity waves propagating in deep, intermediate, and shallow water. (b) Particle orbits for the same three conditions. The former is an Eulerian viewpoint, while the latter is Lagrangian. [Adapted from Kinsman, B., *Wind Waves* (1965).]

wise for the case illustrated. The minor axes of the ellipses decrease with depth approximately exponentially as $\exp(-k|z|)$, collapsing to straight lines at the bottom to meet the bottom boundary conditions. In the case of very deep water, the orbits are circular, but their radii fall off exponentially with the same vertical scale.

5.11   Energy Flow in Surface Waves

The flux of mechanical energy in a surface wave can be found using the equations developed in Chapter 4, especially Eq. 4.70. For the dissipationless, incompressible motion we are studying, the equation for conservation of mechanical energy becomes

$$\frac{\partial}{\partial t} \left(\tfrac{1}{2} \, \rho \mathbf{u}^2 + \rho g \xi \right) + \nabla \cdot [(\tfrac{1}{2} \, \rho \mathbf{u}^2 + \rho g \xi + p) \mathbf{u}] = 0 , \qquad (5.88)$$

where the del operator, ∇, is, for the (x,z) dimensions,

$$\nabla = \hat{i} \, \frac{\partial}{\partial x} + \hat{k} \, \frac{\partial}{\partial z} . \qquad (5.89)$$

Fluxes of energy and momentum per unit horizontal width of ocean may be derived by integrating Eq. 5.88 over depth, from the bottom to the free surface height, ξ. There are numbers of subtleties involved in such calculations, especially when more realistic cases of free surface and bottom boundary layers are considered, and the reader is referred to books such as that by Phillips for the correct formulation. As an example, we calculate the mean kinetic energy per unit surface area for infinitesimal, dissipationless waves:

$$\frac{dK}{dA} = \left\langle \int_{-H}^{\xi} \tfrac{1}{2} \, \rho \mathbf{u}^2 \, dz \right\rangle$$

$$\approx \int_{-H}^{0} \tfrac{1}{2} \, \rho \, \langle \mathbf{u}^2 \rangle \, dz , \qquad (5.90)$$

where the averaging process denoted by $\langle \ \rangle$ is now integration over a single wavelength, λ:

$$\langle \gamma \rangle = \frac{1}{\lambda} \int_{-\lambda/2}^{\lambda/2} \gamma(x) \, dx . \qquad (5.91)$$

The upper limit of integration in Eq. 5.90 is taken as $z = 0$, which is correct to second order in u. Using Eqs. 5.46, 5.47, 5.57, and 5.60, one obtains

$$\langle \mathbf{u}^2 \rangle = \langle u^2 + w^2 \rangle = \tfrac{1}{2} \, w_0^2 \cosh 2k(z + H) , \qquad (5.92)$$

which, upon integration over depth and substitution into Eq. 5.90, becomes

$$\frac{dK}{dA} = \tfrac{1}{4} \, \rho g \xi_0^2 \, (1 + k^2 \tau_s / \rho g) . \qquad (5.93)$$

A similar integration of the average potential energy per unit mass in Eq. 5.88, i.e., $\langle \rho g \xi \rangle$, yields an identical expression for dP/dA, which is a reflection of the fact that the average kinetic and potential energies of a linear harmonic oscillator are the same. Thus the total wave energy per unit area, $dE/dA = dK/dA + dP/dA$, is

$$\frac{dE}{dA} = \tfrac{1}{2}\, \rho g \xi_0^2 \, (1 + k^2 \tau_s / \rho g) \, . \tag{5.94}$$

For gravity waves, the second term in the parentheses is negligible and the energy becomes

$$\left(\frac{dE}{dA}\right)_{grav} \simeq \tfrac{1}{2}\, \rho g \xi_0^2 = \rho g \langle \xi^2 \rangle \, , \tag{5.95a}$$

indicating that the energy may be ascribed to the mean-square wave *amplitude*; gravity wave energy spectra are therefore often given in such terms. For capillary waves, however, the surface tension term dominates and

$$\left(\frac{dE}{dA}\right)_{cap} \simeq \tfrac{1}{2}\, \tau_s (k \xi_0)^2 = \tau_s \langle (\nabla \xi)^2 \rangle \, , \tag{5.95b}$$

indicating that for capillary waves, the energy arises from stretching of the surface against the restoring force of surface tension, and resides in the mean-square *wave slope*, $(\nabla \xi)^2$.

5.12 Statistical Descriptions of Surface Gravity Waves

The theory for small amplitude linear gravity waves just presented is a deterministic one, treating a single wave having parameters of amplitude, ξ_0; wave vector, $\mathbf{k}_h$; frequency, ω; and (without explicitly stating it) phase, ϑ. In actuality, there is a myriad of waves having varying parameters that make up a sea surface; to the extent that they are independent and interact only weakly, if at all, an instantaneous sea surface may be composed out of a summation of deterministic waves such as that given by Eq. 5.39:

$$\xi(\mathbf{x},t) = \sum_j \xi_{0j}(z) \cos (\mathbf{k}_{hj} \cdot \mathbf{x} - \omega_j t + \vartheta_j) \, . \tag{5.96}$$

Because of the randomness of the sea surface, the individual Fourier components making up the description cannot in general be followed during their motion (Laplace notwithstanding), but only their average properties determined. Thus statistical descriptions are required, as they are in so much of geophysics. There are several statistical functions that are commonly used for ocean gravity waves, of which we shall here be concerned with but three: the *probability distribution functions* (or frequency of occurrence) for (1) amplitude and (2) phase, these denoted by $p(\xi)$ and $p(\vartheta)$, respectively; and (3) the relative distribution of squared wave heights, or energy per unit frequency interval, denoted by $S(\omega)$. This latter is the same *wave energy spectrum* introduced in Section 5.4.

To a reasonable approximation, the amplitude distribution for surface gravity waves is found to be *Gaussian*, or *normal*, and the phase distribution is found to be *uniform*. By this we mean that the probability of observing at any given point a wave whose amplitude departs from mean sea level by an amount lying in the interval between ξ and $\xi + d\xi$, is

$$\text{Prob } (\xi, d\xi) \equiv p(\xi) \, d\xi$$

$$= \frac{1}{(2\pi)^{1/2} \, \sigma_\xi} \exp\left(-\frac{\xi^2}{2\sigma_\xi^2}\right) d\xi , \qquad (5.97)$$

where σ_ξ^2 is the variance of the surface about mean sea level, and where the range of ξ is

$$-\infty < \xi < \infty . \qquad (5.98)$$

Similarly, the probability of the phase falling between ϑ and $\vartheta + d\vartheta$ is more or less uniformly distributed between 0 and 2π; thus

$$\text{Prob } (\vartheta, d\vartheta) \equiv p(\vartheta) \, d\vartheta$$

$$= \frac{1}{2\pi} \, d\vartheta , \qquad (5.99)$$

where ϑ ranges over 360°, modulo 2π:

$$0 \leq \vartheta \leq 2\pi . \qquad (5.100)$$

It is usually assumed that the joint probability for the simultaneous occurrence of both random variables, ξ and ϑ, is the product of their independent probabilities:

$$p(\xi,\vartheta)\ d\xi\ d\vartheta\ =\ \frac{1}{(2\pi)^{3/2}\sigma_\xi}\ \exp\left(-\frac{\xi^2}{2\sigma_\xi^2}\right)\ d\xi\ d\vartheta\ . \qquad (5.101)$$

The probability distribution functions are both normalized to unity according to

$$\int_{-\infty}^{\infty} p(\xi)\ d\xi\ =\ 1 \qquad (5.102)$$

and

$$\int_{0}^{2\pi} p(\vartheta)\ d\vartheta\ =\ 1\ , \qquad (5.103)$$

which express the facts that all waves are certain (probability $=$ 1) to lie somewhere within the ranges given by Eqs. 5.98 and 5.100. A process that has the statistics indicated by Eq. 5.101 is termed a Gaussian random process. Furthermore, if σ_ξ does not change in time or only changes slowly, it is considered to be statistically stationary, in the sense that was described in Section 3.12.

The *energy frequency distribution* gives the division of wave energy among the various frequencies present in the wave field. Thus $S(\omega)\ d\omega$ gives the amount of wave height energy (as measured by squared amplitude) falling into the frequency interval between ω and $\omega + d\omega$. In electrical engineering, the quantity analogous to $S(\omega)$ is often termed the *power spectral density*, so named because it is proportional to the signal voltage squared; in oceanography it is not power but the square of a geophysical variable that is generally calculated from the measurements.

As is shown in works on generalized Fourier analysis, the frequency spectrum of a stationary random process may be computed via two integral operations: *autocorrelation* and *Fourier transformation*. The autocorrelation function or *autocovariance* of the wave amplitude, $C_\xi(t')$, is a measure of how well one portion of a record of wave heights, $\xi(t)$, is correlated with other parts removed from the time in question by an amount t', and is given by

$$C_\xi(t')\ =\ \lim_{T\to\infty} \frac{1}{2T} \int_{-T}^{T} \xi(t)\ \xi(t + t')\ dt\ . \qquad (5.104)$$

The time separation, t', is called the *lag time*. A wave height time series, $\xi(t)$, having periodic or quasiperiodic components will be positively correlated with itself whenever the lag time equals an integral number of periods of any one of the cyclic components, and negatively correlated when the lag is an odd number of half-periods. Thus the quadratic function, $C_\xi(t')$, will itself be approximately periodic or quasiperiodic, according to the periodicity of the time series. However, as the amplitude at any time t is multiplied by the amplitude further and further lagged in time, random variations in the periodicities and phases will cause increasing decorrelation. Thus the autocorrelation function, while oscillating with the periods of its components, also has an envelope that monotonically decays with increasing lag. The spectral content of $C_\xi(t')$ therefore reflects both the frequency content and the randomness of the original wave train.

Now the frequency distribution of any continuous, square-integrable function may be described by its Fourier transform, and it can readily be proven that the Fourier transform of the autocorrelation function gives the power spectral density of the original time series of measurements. In complex notation, the wave height power spectral density is then given by the transformation

$$C_\xi(\omega) = \int_{-\infty}^{\infty} C_\xi(t') e^{i\omega t'} dt' , \tag{5.105}$$

which is generally a complex function of frequency. However, in a statistically stationary wave field, it can be shown that $C_\xi(t')$ is an even function of t', and that $C_\xi(\omega)$ is therefore real. The wave height energy spectrum is then the real Fourier transform,

$$S(\omega) = 2 \int_0^{\infty} C_\xi(t') \cos \omega t' dt' . \tag{5.106a}$$

The inverse transformation gives an alternative expression for the autocorrelation function:

$$C_\xi(t') = \frac{1}{2\pi} \int_{-\infty}^{\infty} S(\omega) e^{-i\omega t'} d\omega . \tag{5.106b}$$

From its definition, the dimensions of $S(\omega)$ are seen to be m^2 $(\text{rad s}^{-1})^{-1}$, or m^2 Hz^{-1}, depending on whether ω or $f = \omega/2\pi$ is used.

$S(\omega)$ is generally normalized so that the sum of all of the energy present equals the mean-square amplitude, or variance, σ_ξ^2, appearing in Eq. 5.97. The variance is also the value of the autocovariance function at zero lag: $\sigma_\xi^2 = C_\xi(0)$. Since the probability distribution is Gaussian, the percentage of all wave heights lying between $\pm 1\sigma_\xi$ of mean sea level is approximately

68%. Another quantity called the *significant wave height,* $\xi_{1/3}$, is often used and is defined such that

$$(\tfrac{1}{4} \, \xi_{1/3})^2 \simeq \sigma_\xi^2 = C_\xi(0) = \frac{1}{2\pi} \int_{-\infty}^{\infty} S(\omega) \, d\omega . \qquad (5.107)$$

This quantity is so named because it corresponds to the average of the heights of the highest one-third of the waves in a random time series of $\xi(t)$ (see Fig. 5.5). It also roughly corresponds to the "average" wave height estimated via eye observations made by experienced personnel at sea. From the properties of the normal distribution, it can be shown that $\xi_{1/3}$ is approximately equal to 4 standard deviations of the height:

$$\xi_{1/3} \simeq 4\sigma_\xi . \qquad (5.108)$$

The prescriptions given by Eqs. 5.104 and 5.105 are abstractions whose realizations in practice require a number of important and somewhat subtle modifications, most of which may be found in the references. The subject of time series analysis (and its two-dimensional analogs for spatially periodic functions such as wave photographs or images) is highly developed and widely used in oceanography; the technique is not confined to surface gravity waves by any means, but is applied to many different types of measurements. The propriety of such procedures must be considered in the light of the issues of ergodicity and spectral separability discussed in Section 3.12.

Figure 5.14 illustrates several wave frequency spectra calculated from amplitude measurements made at fetches, x_f, ranging between 9.5 and 80 km during an offshore wind. While the high frequency tails of all of the spectra approach the kind of equilibrium distributions shown in Fig. 5.6, they nevertheless show departures from equilibrium as the spectral peak is approached. This phenomenon is termed *overshoot*, and represents a less than fully developed condition due to the limited fetch. From the deep water dispersion relation (Eq. 5.71), the lowest frequency peak at $\omega \simeq 1.5$ rad s^{-1} would have an associated wavelength of $\lambda = 2\pi/k \simeq 27$ m, and a phase speed of $c_p \simeq 6.5$ m s^{-1}.

The global nature of the equilibrium wave height spectrum is characterized by two general properties: a spectral peak at a radian frequency, ω_m, that depends on wind speed, u_{10}, and fetch; and a high frequency tail that falls off as ω^{-5} (cf. Fig. 5.6). A somewhat complicated formula that summarizes recent work in wind wave spectra is given by Eq. 5.109:

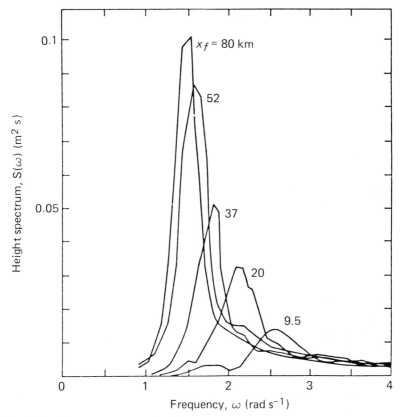

Fig. 5.14 Wind wave spectra for five increasing fetches, x_f. [Adapted from Hasselmann, K., et XV al., *Deutsch. Hydrogr. Z.* (1973).]

$$S(\omega) = \frac{\alpha(u_{10}, x_f)\, g^2}{(2\pi)^4} \exp\left[-\frac{5}{4}\left(\frac{\omega}{\omega_m}\right)^{-4}\right] \frac{F(\omega, u_{10})}{\omega^5} . \qquad (5.109)$$

This function is plotted schematically in Fig. 5.15. The major physics that it summarizes is fourfold: (1) waves having frequencies below ω_m are those traveling at phase speeds greater than the wind speed, and thus they have no significant net forcing causing them to grow; (2) waves in the high frequency region are dominated by breaking and cusplike crests, whose frequency spectra may be shown to behave as ω^{-5}; (3) wind speed and limited fetch, as incorporated into the dimensionless function $\alpha(u_{10}, x_f)$, modify the overall energy level of $S(\omega)$ and hence the value of $\xi_{1/3}$ without changing its shape; and (4) the function $F(\omega, u_{10})$ increases the peakedness of the spec-

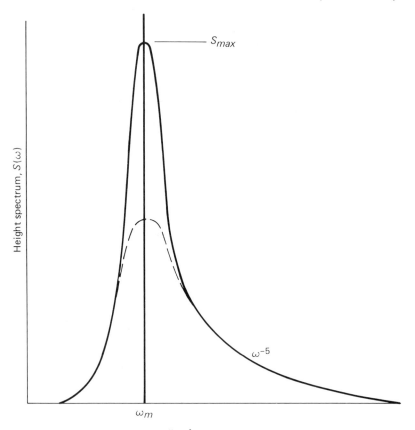

Fig. 5.15 Schematic form of JONSWAP semiempirical wind wave spectrum. The dotted line gives the Pierson–Moskowitz spectral shape, for which Fourier components are essentially independent; the solid line is more peaked due to nonlinear interactions. [Adapted from Hasselmann, K., et XV al., *Deutsch. Hydrogr. Z.* (1973).]

tral maximum and parameterizes the weak nonlinear wave/wave scattering interactions that cause the spectrum to be more narrow and cohesive than it would be in the case of linear, independent Fourier components. Figure 5.15 illustrates these characteristics. However, recent analyses indicate that the power law behavior of the high frequency range of the spectrum may actually be ω^{-4}, with the apparent dependence of ω^{-5} perhaps being an artifact of the methods used in analyzing the data. The reader is referred to Phillips (1985) for a discussion of this evolving issue.

It should be noted that long wavelength swell, as defined in the first part of Section 5.4, is not incorporated into Eq. 5.109, since it is not locally produced but rather is related to earlier sources of wind located elsewhere. At the other end of the spectrum, neither is this form of $S(\omega)$ correct for capillary waves, since both the production and damping of these short waves depends on quite different physics from that summarized by Eq. 5.109.

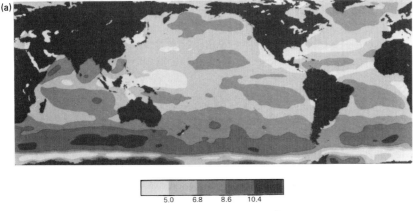

Surface wind speed, u_{10} (m s^{-1})

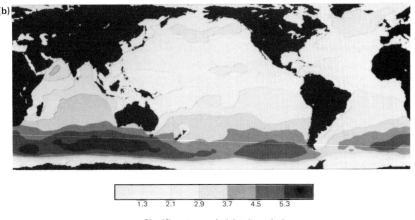

Significant wave height, $\xi_{1/3}$ (m)

Fig. 5.16 (a) Averaged surface wind speed from Seasat spacecraft. (b) Averaged significant wave height, including both wind waves and swell. High correlation exists between local winds and waves. [Adapted from Chelton, D. B., et al., *Nature* (1981).]

Figure 5.16 shows global maps of surface wind speed, u_{10}, and significant wave height, $\xi_{1/3}$, made with a satellite radar altimeter and averaged over approximately three months of the Austral winter of 1978. There is clear correlation between wind speeds and wave heights, especially in the area of the Antarctic Circumpolar Current, where certain regions show average values of $\xi_{1/3}$ in excess of 5 m. Such data go into making up a *wave climatology*, the average statistics of waves as a function of month and geographical location.

In nonstationary conditions, the evolution of the wave spectrum may be described by the *wave action equation*. Section 8.9 contains a development of this useful relationship, and its application to wave-current interactions.

The subject of surface waves is a highly evolved one and much literature exists on the subject, access to which may be obtained by consulting the references; clearly we have only scratched the surface of the topic. Instead we turn to discussions of two somewhat different types of very long gravity waves—tsunami and tides. But first we must derive the long wave, shallow water equations, whose use transcends surface gravity waves.

5.13 Long-Wave, Shallow Water Equations

It was apparent from Eq. 5.73 that when the water depth is shallow compared with a wavelength (this measure determined by the condition that $k_h H \ll 1$), both the phase and group speeds of gravity waves are given by $(gH)^{1/2}$. Although it is not immediately obvious, it may also be shown that the assumption of shallow water propagation is equivalent to the assumption that vertical accelerations are small, with the result that the pressure is purely hydrostatic; in this case, the mean and perturbation pressures are given by

$$\frac{\partial p_0}{\partial z} = -\rho_0 g \tag{5.110}$$

and

$$p' = \rho_0 g \xi , \tag{5.111}$$

which are relationships that apply at all points within the fluid. Wave motion under these conditions is termed *barotropic*, in that the properties of the fluid are essentially uniform from the top to the bottom of the water column. This is in contrast to *baroclinic modes*, in which density, velocity, etc., vary significantly in the vertical. The shallow-water barotropic approximation is widely applicable in the study of long wave oceanic motions, and we will derive the differential equations governing such dynamics.

We start with the Navier–Stokes equations (Eqs. 3.84 to 3.86) with **A** assumed constant, and substitute into them the hydrostatic equations immediately above. By writing for the total pressure gradient the expression

$$\nabla p = -\rho g \hat{k} + \rho g \nabla_h \xi , \tag{5.112}$$

where ∇_h is the horizontal gradient operator, we may recast the vector momentum equation as

$$\frac{\partial \mathbf{u}}{\partial t} + \mathbf{u} \cdot \nabla \mathbf{u} = -2\Omega \times \mathbf{u} - g \nabla_h \xi + K_h \nabla_h^2 \mathbf{u} + K_v \frac{\partial^2 \mathbf{u}}{\partial z^2} . \tag{5.113}$$

Thus to the extent that surfaces of elevation, ξ, have horizontal gradients (i.e., slope with respect to equipotential surfaces), gravity will play a role in establishing the motion.

To proceed further, we will require the services of the vertically integrated equation of continuity, written using the incompressible approximation. By integrating $\nabla \cdot \mathbf{u} = 0$ from the bottom, $z = -H$, to the equilibrium surface, $z = 0$, and assuming that u and v are uniform over z, we obtain

$$\int_{-H}^{0} \nabla \cdot \mathbf{u} \, dz = H \nabla_h \cdot \mathbf{u} + \int_{-H}^{0} \frac{\partial w}{\partial z} \, dz = 0$$

$$= H \nabla_h \cdot \mathbf{u} + \frac{\partial \xi}{\partial t} + \mathbf{u} \cdot \nabla_h \xi , \tag{5.114}$$

where the kinematic boundary condition has been used. We may next eliminate u and v from Eqs. 5.113 and 5.114 by taking the x and y derivatives of the x and y components of Eq. 5.113, then adding, and next invoking Eq. 5.114. In the β plane approximation of Eq. 3.41, these operations result in:

$$-\frac{1}{H} \frac{\partial^2 \xi}{\partial t^2} = f \left(\frac{\partial v}{\partial x} - \frac{\partial u}{\partial y} \right) - \beta u - g \nabla_h^2 \xi , \tag{5.115}$$

or, through a slight rearrangement,

$$\nabla_h^2 \xi - \frac{1}{c_h^2} \frac{\partial^2 \xi}{\partial t^2} = \frac{1}{g} (f\zeta - \beta u) . \tag{5.116}$$

Here the shallow water phase speed, c_h, is

$$c_h = (gH)^{1/2} , \qquad (5.117)$$

and the vertical component of fluid vorticity, ζ, is (see Eq. 5.75)

$$\zeta = \frac{\partial v}{\partial x} - \frac{\partial u}{\partial y} . \qquad (5.118)$$

Equation 5.116 is a two-dimensional wave equation for the vertical displacement, ξ, with the apparent source function on the right-hand side being due to fluid vorticity, ζ, and the beta-effect term, $\beta u/g$. This implies that variations in both fluid vorticity and planetary vorticity may propagate as shallow-water barotropic waves. We shall see in Chapter 6 that analogous baroclinic waves may also propagate in accordance with a similar equation, but with the displacement and speed interpreted somewhat differently.

In addition, other sources of motion may serve to excite long-wave shallow water waves, and their source functions will also appear on the right-hand side of Eq. 5.116. Two prominent ones are astronomical tides and seismic displacements of the sea floor.

5.14   Tsunami and Seismic Sea Waves

"Tsunami" is a Japanese word meaning "harbor wave," so named because of the amplification of water elevations that often occurs when such waves approach confined embayments (see Fig. 5.11a, for example). Tsunami are thought to be produced by sharp vertical motions of the sea floor during earthquakes or other seismic activity, although only a small fraction of quakes on the sea floor actually produce detectable tsunami. As was mentioned in Chapter 2, the Pacific Basin is especially active in terms of plate tectonics, and frequent small tsunami occur, with the result that seismic sea waves in that region have been the most thoroughly studied of any oceanic area.

During an earthquake on the sea floor, the vertical motion of the bottom can be considered to be distributed over an extended region, which then radiates waves outward. By the time the waves have propagated a few hundred kilometers away from the source, they appear to spread cylindrically; beyond the near-field of the source, the seismically excited motions propagate as shallow water waves whose speed is approximated by Eq. 5.117. In the Pacific, where a typical depth is $H \approx 5500$ m, the speeds are near 230 m s^{-1}, or some 15% of the speed of sound in water. The observed wave periods near shore are of the order of 15 to 100 min, with the shorter periods typically occurring closer to the source. The shorter period and speed then imply a wavelength of order 200 km, clearly a figure much greater than the water depth and one that thereby justifies the use of shallow water theory. The amplitudes of large tsunami in mid-ocean have been inferred from a few deep-sea bottom pressure gauges, and probably do not exceed a meter or so.

Because of their long wavelengths, small amplitudes, and consequent shallow slopes, tsunami at sea are essentially undetectable by ships; however, as a tsunami approaches shore, its wave energy, which is uniformly distributed from top to bottom, is compressed by the shoaling water and the amplitude increases accordingly.

Figure 5.17 illustrates four time series of water elevations at harbors along the northeast coast of the Japanese island of Hokkaido made on 12 August 1969, after an offshore earthquake. The tsunami signals are superimposed on the semidiurnal tide and have maximum amplitudes of perhaps 1.5 m, and were caused by an earthquake with a Richter magnitude of 7.8; the epicenter was located near the Japan Trench at 42° 42′N, 147° 37′E, approximately 170 km away from the region shown. These elevations represent a more or less typical local response to a noncatastrophic seismic sea wave, and probably occur a few times per year in the Pacific basin.

However, major earthquakes that are characterized by significant vertical motions may result in much larger amplitudes and in highly destructive wave action. The great Alaska earthquake of 27 March 1964 occurred with its epicenter just north of Prince William Sound, with a magnitude of 8.4 and an uplifted area of some 250,000 km^2. The resultant tsunami was one of the most damaging in the history of the North American continent. At least 122 casualties occurred, with water elevations in the local regions near the epicenter averaging 10 m, and with an extreme elevation of 20 m occurring on Kodiak Island. Cresent City, California, recorded a 4 m elevation, in considerable part due to harbor focusing and resonance effects, while various ports in Chile recorded elevations of 1 to 2 m. Figure 5.18 illustrates the calculated positions of this tsunami at successive times up to 1900 sec after the earthquake, while Fig. 5.19 demonstrates refractive effects on the wavefronts due to variable bathymetry in the Northeastern Pacific. Such travel time charts are obtained by integrating shallow water equations similar to Eq. 5.116, using the observed bathymetry of the sea floor.

The travel time of tsunami from Alaska to Hawaii is of order 5 h, and the time to propagate across the entire north–south extent of the Pacific is roughly 15 h. The *e*-folding decay time appears to be near 22 h, so that the background activity of a large tsunami may be detected for a few days after its time of origination.

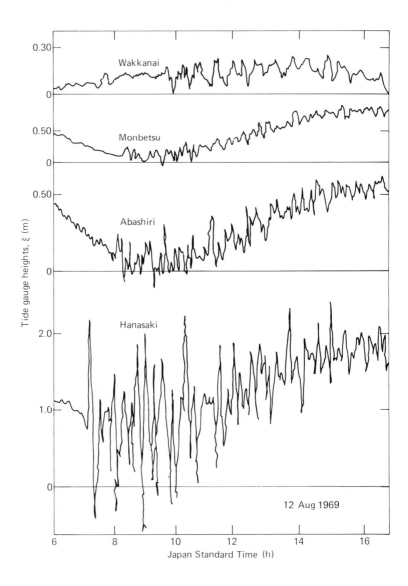

Fig. 5.17 Tidal and tsunami amplitudes at four stations on the northeast coast of Hokkaido, Japan, showing waveforms on 12 August 1969. [From Murty, T. S., *Seismic Sea Waves: Tsunamis* (1977). Originally due to Hatori, T. (1970).]

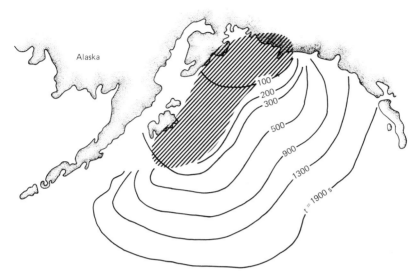

Fig. 5.18 Calculated positions of the leading distrubance of the tsunami result-ing from the Alaska earthquake of March 1964. Intervals shown are in seconds. The shaded area is the region of significant vertical ground motion. [Adapted from Murty, T. S., *Seismic Sea Waves: Tsunamis* (1977). Originally due to Hwang, L.-S. (1970).]

The rim of the Pacific basin frequently suffers tsunami damage. After a devastating wave in 1946, the International Pacific Tsunami Warning System was established under the auspices of UNESCO. The observational network consists of seismographs, tide gauges, and sundry other instrumentation for recording tsunami activity, linked via radio communications. Any earthquake on the sea floor exceeding a magnitude of 7.0 results in an alert being sent from the Warning Center near Honolulu to nations bordering the basin. Except for tsunami generated nearby, there is generally several hours' warning available for marine interests around the basin to take defensive action, evacuate low-lying coastal areas, and put ships to sea. The tsunami warning system is one of the very few successful oceanic forecast or warning services extant as of this writing (if only because of the difficulty in monitoring the broad expanse of the sea). Another such service is coastal and open ocean tide forecasts, to which we now turn our attention.

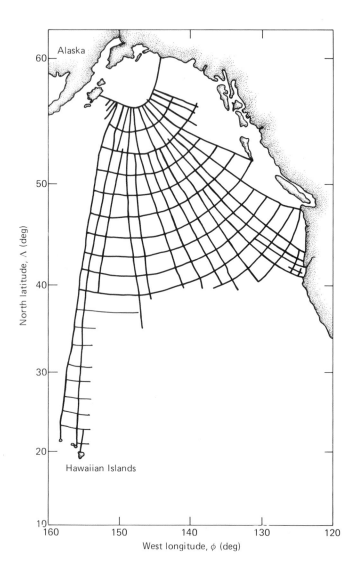

Fig. 5.19 Propagation/refraction diagram showing rays and phase fronts calculated for the Alaska tsunami of March 1964. [Adapted from Murty, T. S., *Seismic Sea Waves: Tsunamis* (1977). Originally due to Keulegan, G. H., and J. Harrison (1970).]

5.15   Deep Sea and Coastal Tides

A preliminary discussion of tidal forcing was given in Chapter 3, where the potential due to the astronomical tides, $\Phi_t(r,\theta,\phi,t)$, was left unspecified. Also, an incomplete explanation of the origins of the myriad of tidal constituents was given. We shall now outline a somewhat simplified derivation of the forced tidal equations and discuss the oceanic response, without going too deeply into a calculation that is, in its complete form, very complicated.

The sun and the moon each exert on the earth both gravitational forces and centrifugal forces due to rotation (not daily spin), with the tide-producing component defined as that part of those forces which does not produce motion of the earth as a whole, i.e., the residual left after subtracting the forces acting on the center of mass of the earth. In Figs. 3.11a and 3.11b, the force diagram and vectors give an indication of how these residual forces add up around the earth, and show that the force differentials are very nearly, but not exactly, symmetrical about the lines joining the centers of mass of the earth and the attracting bodies. The imaginary *equilibrium tidal elevation* developed by the net force would deform the shape of the combined solid and liquid earth into an ellipsoid of the general configuration shown in Fig. 5.20, with the bulge on the zenith side nearest the astronomical body being due to an excess of gravitational over centrifugal force, and the bulge on the opposite (or nadir) side arising from the larger centrifugal force compared with the inverse square law there. The equilibrium tidal theory also assumes that the oceans cover the entire earth, and that there is no time-dependence in the elevations other than that induced in the fluid by the motions of the sun and moon in their apparent paths.

While the actual *solid earth* tides appear to take on much of the character of an equilibrium tide because of the large speed of propagation of elastic waves in the solid earth, the speeds of propagation of the real oceanic tides are limited to the shallow water barotropic speed, c_h, which is near 220 m s^{-1} for averaged ocean depths. Now at mid-latitudes, the zenith point moves at an azimuthal speed of roughly 330 m s^{-1}, so that over most of the surface, the astronomical forcing function propagates around the earth much more rapidly than the oceanic tidal wave can follow.

The forcing function for the solar equilibrium tide has periods as seen on the rotating earth of $T_s = 24.0$ h for the daily spin, and $T_y = 365.242$ days for the mean tropical year. For the moon, the equivalent periods are $T_l = 24.841$ h for the lunar day (this being longer than the solar day because of the co-rotation of moon and earth, as was discussed in Chapter 2 with respect to the sidereal and solar days), and $T_m = 27.321$ days for the lunar sidereal month. Because of the nearly symmetric distribution of the equilibrium tide about the line joining the centers of mass (as shown in Fig. 5.20),

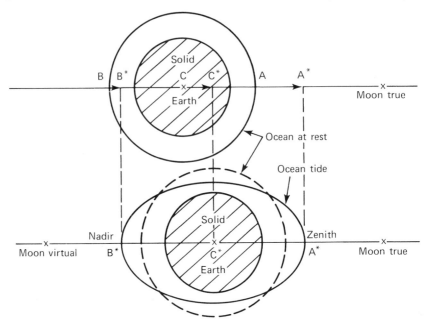

Fig. 5.20 Distribution of Newtonian equilibrium tidal amplitude under joint influences of gravitational and centrifugal forces. The mean displacement, C − C*, represents a center-of-mass translation. [Adapted from Schwiderski, E. W., *Rev. Geophys. and Space Phys.* (1980).]

these forces will have frequency components at twice the actual rotation rates as well as at those rates themselves.

We will now derive an expression for the tidal potential due to either one of these astronomical bodies in terms of the inverse square forces, Gm_j/R_j^2, and the centrifugal forces, $\Omega_j^2 R_j'$, where the distances are measured from the rotation axis through the center of mass of the body (see Fig. 5.21). Here $j = (s,m)$ is the index representing either sun or moon. The frequencies Ω_j give the rotation rates of the earth about the sun, and the moon about the earth: $\Omega_y = 2\pi/T_y$ for the mean annual rotation, and $\Omega_m = 2\pi/T_m$ for the mean monthly rotation. These forces are assumed to act on a unit mass at the position, P, located at (r,θ,ϕ) on the rotating earth, as shown in Fig. 5.21. We denote the distance between the centers of mass of the earth and of the astronomical body as R_j; the distance from the body's center of mass to the field point, P, as R; the perpendicular distance from the rotation axis to point P as R_j'; the distance from the center of the earth to P as r; and the angle between R_j and r as γ. Then the geometry of the problem gives:

$$R^2 = r^2 + R_j^2 - 2rR_j \cos \gamma \tag{5.119}$$

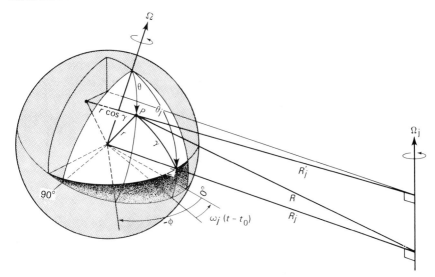

Fig. 5.21 Geometry for tidal force resolution on a unit mass located at point P with coordinates (r, θ, ϕ), when an astronomical object is above the zenith point, Z, with polar angles (θ_j, ϕ_j).

and

$$R_j' \simeq R_j - r \cos \gamma . \tag{5.120}$$

(We are implicitly considering the geometry of the sun–earth system, which is conceptually simpler than the moon–earth system, for which the system center of mass lies within the earth; however, the final results can be shown to hold for both geometries.) Now the scalar values of the two forces per unit mass are

$$|\mathbf{f}_{gj}| = \frac{-Gm_j}{r^2 + R_j^2 - 2rR_j \cos \gamma} \tag{5.121}$$

and

$$|\mathbf{f}_{cj}| = \Omega_j^2 R_j' = \Omega_j^2 (R_j - r \cos \gamma) . \tag{5.122}$$

At the earth's center, these two forces must balance in order for the planet to be in an equilibrium orbit. Equating Eqs. 5.121 and 5.122 with $r = 0$ gives a form of Kepler's third law: $\Omega_j^2 = Gm_j/R_j^3$. In one sense, the tides can be viewed as being due to (1) the earth's orbital position with its center of mass at the minimum of a fictitious potential energy well, $\Phi' = -Gm_j/R_j - \frac{1}{2}\Omega_j^2 R_j'^2$, and (2) the finite extent of the earth's distribution of fluid mass, which reaches to slightly higher energy levels on either side of the equilibrium point of the potential well and hence is subject to differential forces.

Let $\Phi_{0j} = -Gm_j/R_j$ be the mean geopotential at distance R_j; then the tidal potential, Φ_{tj}, corresponding to these forces is (from Eqs. 5.121 and 5.122)

$$\Phi_{tj}(r,\theta,\phi,t) = \Phi_{0j}\left\{1 - \frac{r}{R_j}\cos\gamma - \left[1 - \frac{2r}{R_j}\cos\gamma + \left(\frac{r}{R_j}\right)^2\right]^{-\frac{1}{2}}\right\}$$

$$\simeq \Phi_{0j}\left\{1 - \frac{1}{2}\left(\frac{r}{R_j}\right)^2(1 - 3\cos^2\gamma)\right\}. \tag{5.123}$$

Here we have expanded the inverse square root term to first order in powers of r/R_j, which is clearly much less than unity. In more detailed tidal models, higher order expansions are often used. The horizontal and vertical (radial) components of tidal force per unit mass are then $(-1/r)(\partial\Phi_{tj}/\partial\gamma)$ and $-(\partial\Phi_{tj}/\partial r)$, respectively. It is the horizontal component that is the tide-producing force; the vertical component is negligible, contributing a change in surface gravity of order 10^{-7} g.

As a measure of the relative effectiveness of the sun and moon in producing tides, we form the ratio $\Phi_{0m}R_s/\Phi_{0s}R_m$, which equals 2.182 when using the numerical values cited in Chapter 2. Thus the lunar theoretical equilibrium tidal force is more than twice as large as that of the sun, a fact that accounts for the moon's dominant role in tide production.

In order to relate the angle γ to θ and ϕ (the spherical coordinates of P), as well as to θ_j and ϕ_j (those of the zenith point, Z), the spherical triangle of Fig. 5.22 must be solved. It is worth noting that the azimuth angle ϕ_j is the hour angle for the apparent rotation rate of the earth with respect to the body, and $\theta_j = \theta_j(t)$ is the *co-declination* of the body. For the sun, the instantaneous declination varies at the annual rate over a range of $\pm 23.5°$, while for the moon, it varies at the lunar monthly rate over an additional range of $\pm 5.1°$. A short retreat to spherical trigonometry yields the required relationship:

$$\cos\gamma = \cos\theta\cos\theta_j + \sin\theta\sin\theta_j\cos\phi_j, \tag{5.124}$$

where $\phi_j = \omega_j(t - t_0) + \vartheta$ is the time-varying azimuth phase; we will discuss the frequencies ω_j below. Substituting this equation into Eq. 5.123, squaring, and grouping the trigonometric terms according to frequency, we obtain:

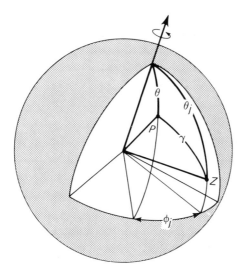

Fig. 5.22 Spherical triangle relating the angle γ to polar coordinates of the field point, P, and the zenith point, Z.

$$\Phi_{tj} = \Phi_{0j} \left(\frac{r}{2R_j}\right)^2 \left\{ (1 - 3\cos^2\theta_j)(1 - 3\cos^2\theta) \right.$$

$$+ 3\sin 2\theta_j \sin 2\theta \cos[\omega_j(t - t_0) + \vartheta]$$

$$\left. + 3\sin^2\theta_j \sin^2\theta \cos[2\omega_j(t - t_0) + 2\vartheta] \right\}. \qquad (5.125)$$

The frequencies ω_j are the rates at which the zenith point, Z, rotates with respect to the field point, P, and are the radian frequencies for the solar and lunar days, $\omega_s = 2\pi/T_s$ and $\omega_l = 2\pi/T_l$, respectively. These are not the only effective frequencies present, however, because of the slow modulations due to solar and lunar co-declinations, $\theta_j(t)$.

It should be observed from the time dependencies in Eq. 5.125 that there are three distinct groupings of frequencies, or "tidal species" present: semi-diurnal, diurnal, and long-period, denoted by $\nu = 2$, 1, and 0, respectively, and with each having a different latitude factor, $G_\nu(\theta)$, that shapes the equilibrium tide in the meridional coordinate. The semidiurnal and diurnal components come from $2\omega_j$ and ω_j, respectively. The long-period tides arise from variations in the lunar and solar co-declinations, θ_j, at semimonthly and semiannual periods, plus even longer-term variations caused by changes in orbit characteristics.

Rather than to attempt explicit designations of the forms of $\theta_j(t)$, which would retain only the solar and lunar ω_j as operative frequencies, it has proven advantageous in tidal calculations to expand the trigonometric functions involving $\theta_j(t)$ in Eq. 5.125 in Fourier series that use six basic frequencies: ω_l, corresponding to the lunar day; ω_m, to the lunar sidereal month; ω_y, to the tropical year; and three others arising from the orbital variation: $\omega_p = 2\pi/8.84$ years; $\omega_n = 2\pi/18.6$ years; and $\omega_{pl} = 2\pi/20,940$ years. Then the frequency of any given *tidal constituent, ω_i,* can be written in terms of positive and negative harmonics, n_d, of those six basic frequencies according to:

$$\omega_i = n_1\omega_l + n_2\omega_m + n_3\omega_y + n_4\omega_p + n_5\omega_n + n_6\omega_{pl} . \quad (5.126)$$

The integers n_d are termed *Doodson numbers,* and they model interference or intermodulation effects among forces acting at the six basic frequencies. These forces have given rise to the concept of fictitious astronomical bodies rotating at the apparently nonphysical periods listed in Table 3.1. Up to 400 distinct frequencies, ω_i, and equilibrium amplitudes, A_i, have been incorporated into detailed models used for long-term coastal tide forecasts. These 400 frequencies are also grouped into the same three species of semidiurnal, diurnal, and long-period rates. The 11 most important are listed in Table 5.1.

Numerical solutions for tidal amplitudes, ξ_i, assume that each of the tidal constituents is independent of the others, and use equilibrium tidal potential functions of the form

$$\Phi_{ti} = g\eta_i = gA_iG_\nu(\theta) \cos (\omega_i t + \nu\vartheta_i + \chi) , \quad (5.127)$$

where A_i is the equilibrium amplitude for the tide evaluated at $r = R_e$; $G_\nu(\theta)$ is the latitude factor for tidal species $\nu = 0, 1, 2$ (see Eq. 5.125); and frequency and phase, ω_i and ϑ_i, respectively, are those resulting from the intermodulation products in the Fourier expansions cited above. These terms and factors are also given in Table 5.1 for the 11 dominant constituents. The potentials are then used as forcing terms in the shallow-water hydrodynamic equations written in spherical coordinates, with departures from the equilibrium tide assumed to give the net forcing.

In the linearized, shallow-water approximation, the momentum equations and the vertically integrated continuity equation for each tidal constituent take on the forms (see Eqs. 3.95 and 3.96)

$$\frac{\partial u}{\partial t} - 2\Omega v \cos\theta = \frac{-g}{R_e \sin\theta} \frac{\partial}{\partial\phi} (\xi_i - \Phi_{ti}/g) + \frac{1}{\rho}(A\nabla^2\mathbf{u})_\phi , \quad (5.128)$$

TABLE 5.1

Characteristics of Principal Tidal Consitutents

Tidal species	Nomenclature	Latitude factor, $G_\nu(\theta)$	n_1	n_2	Doodson number n_3	n_4	Equilibrium amplitude, A_i (m)	Period, T_i (h)	Radian frequency, ω_i $(10^{-4}\ s^{-1})$
Semidiurnal	$\nu = 2$	$\sin^2 \theta$							
Principal lunar	M_2		2	0	0	0	0.242334	12.421	1.40519
Principal solar	S_2		2	2	-2	0	0.112841	12.000	1.45444
Lunar elliptic	N_2		2	-1	0	1	0.046398	12.658	1.37880
Lunisolar	K_2		2	2	0	0	0.030704	11.967	1.45842
Diurnal	$\nu = 1$	$\sin 2\theta$							
Lunisolar	K_1		1	1	0	0	0.141565	23.935	0.72921
Principal lunar	O_1		1	-1	0	0	0.100514	25.819	0.67598
Principal solar	P_1		1	1	-2	0	0.046843	24.066	0.72523
Elliptic lunar	Q_1		1	-2	0	1	0.019256	26.868	0.64959
Long period	$\nu = 0$	$\frac{1}{2}(1 - 3\cos^2 \theta)$							
Fortnightly	Mf		0	2	0	0	0.041742	327.86	0.053234
Monthly	Mm		0	1	0	-1	0.022026	661.31	0.026392
Semiannual	Ssa		0	0	2	0	0.019446	4383.05	0.003982

$$\frac{\partial v}{\partial t} + 2\Omega u \cos \theta = - \frac{g}{R_e} \frac{\partial}{\partial \theta} (\xi_i - \Phi_{ti}/g) + \frac{1}{\rho} (A \nabla^2 \mathbf{u})_\theta , \qquad (5.129)$$

and

$$\frac{\partial \xi_i}{\partial t} + \frac{1}{R_e \sin \theta} \left\{ \frac{\partial}{\partial \theta} [H(\theta,\phi) v \sin \theta] + \frac{\partial}{\partial \phi} [H(\theta,\phi) u] \right\} = 0 . \qquad (5.130)$$

Here we have allowed for the variable depth, H, of the ocean by using $H(\theta,\phi)$; the Φ_{ti} are given by Eq. 5.125; and the right-hand-most quantities are the eddy viscosity terms of Eqs. 3.91 and 3.92. These equations may be solved numerically with appropriate frictional bottom and free-surface boundary conditions, and solutions of the type shown in Fig. 3.12 may be obtained for each tidal constituent. Considerable care must be taken in arriving at the solutions, which are quite sensitive to the numerical parameters and to the bathymetry of the ocean. Observational data are used to constrain and calibrate the calculation, so that an overall internal precision of approximately ±0.10 m tide elevation error may be claimed for such models.

Tides in local geographical areas such as in regional seas and embayments may be solved using the simpler beta-plane or *f*-plane approximations. By this means, one may obtain analytical solutions to tides in channels, embayments, and boxes, as well as considerable insight into tidal dynamics. The publications cited at the end of this chapter contain examples of both approaches.

Figure 5.23 illustrates the analytical solution for a tidal *seiche*, or standing wave oscillation, in an open, rectangular embayment in the Northern Hemisphere. In Fig. 5.23a, it is assumed that at $t = 0$, high tide exists at the head of the bay (which is taken as north). As the tidal current begins to ebb toward the opening, the Coriolis force deflects the water particles to the right, which in turn elevates the water level on the western side of the bay (Fig. 5.23b). It will appear that the high tide initially at the bay head has moved from that location to the western margin, while the low tide has moved from the bay mouth to the eastern side of the bay. As the oceanic tide moves toward the south and the elevation at the mouth rises (Fig. 5.23c), the tide at the head has become low. To complete the cycle, the now-incoming tide hugs the eastern shore of the bay, again under the influence of the Coriolis force (Fig. 5.23d), and the surface is elevated there. It is seen that during this cycle the tidal wave elevation appears to rotate counterclockwise about a nodal point near the center of the embayment; this point is one of essentially zero change in tidal elevation, and is termed an *amphidromic point*. The currents also rotate counterclockwise about the amphidrome, and have

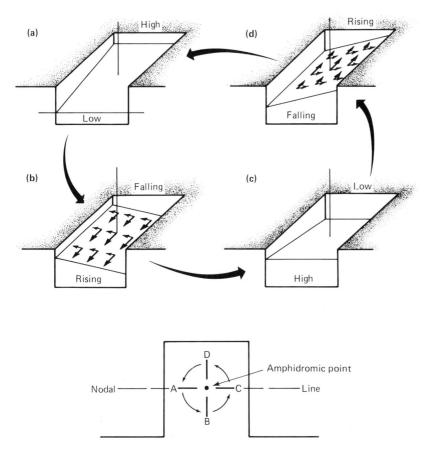

Fig. 5.23 Schematic of rise and fall of tides in an embayment in the Northern Hemisphere. Tidal elevations appear to rotate counterclockwise about the amphidromic point. [Adapted from von Arx, W. S., *An Introduction to Physical Oceanography* (1962).]

stronger along-axis components than cross-bay flow in the case of a narrow bay. A polar plot of tidal current speed versus direction describes an approximate ellipse termed the *tidal ellipse*.

To the extent that one of the natural tidal periods for an embayment coincides with the local half-pendulum day, the tidal amplitudes may become quite large. This is the case for Frobisher Bay on Baffin Island, Canada, under the influence of the semidiurnal tidal species, with the result that the tidal range there can be as large as 8 to 10 m (see Fig. 3.10a). The Bay of Fundy, located beween the U.S. and Canada, appears to be similarly resonant. Closer to the equator, the Gulf of Mexico at 30°N has a local Coriolis period near 24 h, so that Gulf Coast tides are mainly diurnal (see the tidal time series for Pensacola, Florida, in Fig. 3.10d).

The system of amphidromes and rotating tidal ellipses also exists in the open ocean, as the numerical solution illustrated in Fig. 3.12 shows for the M_2 lunar semidiurnal constituent. The orthogonal networks illustrated are composed of lines of constant tidal elevation, ξ_i, or *co-range lines,* and lines of constant phase, ϕ_i, measured in degrees with respect to the passage of the astronomical body over the Greenwich Meridian ($\phi = 0°$ in Fig. 5.21); these latter are *co-phase* lines. Each of the 11 (or 400) constituents of the tidal Fourier decomposition can be analyzed and charts analogous to Fig. 3.12 can be produced.

Along the coasts, the problem of predicting the tide for a particular harbor is accomplished by measuring the tidal amplitude as a time series of elevations of perhaps 29 days duration. This series is then approximated by a local expansion of the form $\Sigma \, \xi_i \sin (\omega_i t + \vartheta_i)$, and the amplitudes and

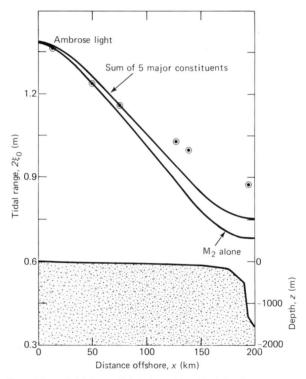

Fig. 5.24 Transition of tidal amplitude from New York harbor across the continental shelf to the shelf break. [Adapted from Schwiderski, E. W., *Rev. Geophys. and Space Phys.* (1980).]

phases for each consitituent evaluated numerically from the data. Residual elevations due to wind or atmospheric pressure forcing cannot be so fitted; these *meteorological tides* must be forecast, if at all, by other means.

The coastal tides are significantly modified from the deep sea tides by changes in depth and bottom friction, which lead to amplitude and phase changes of considerable magnitude in the transition zone from harbor to deep sea. Figure 5.24 illustrates the decay of tidal amplitude from Ambrose Lightship off New York City to the edge of the continental shelf, as well as its merger with the open-ocean tidal prediction of Fig. 3.12; the tidal *range*, or *double amplitude*, varies from approximately 1.5 m in New York Harbor to less than 0.7 m at the shelf break.

5.16 Internal Wave Dynamics

The surface waves discussed in Sections 5.4 to 5.12 are essentially *edge waves* that decay exponentially beneath the surface. Another possible wave mode of a geophysical fluid involves more nearly *body waves* within the interior of the fluid, and is termed an *internal wave*. Such waves are most familiar as oscillations visible in a two-layer fluid contained in a clear plastic box often sold in novelty stores. In the box, two immiscible and differently colored fluids fill the entire volume; when tilted or otherwise disturbed, a slow, large-amplitude wave is observed to propagate along the interface between the fluids. This is the internal wave, and while it has its maximum amplitude at the interface, its displacements are zero at the top and bottom. It owes its existence to the stratified density structure of the two fluids, with a very sharp density change occurring along the interface and with the properties that the smaller the density contrast, the lower the wave frequency, and the slower the propagation speed.

Similar modes exist in the geophysical fluids of the atmosphere and the ocean. Solar radiation is absorbed in the near-surface layers, resulting in warmer water and lower density in that region and leading to a stratified fluid. Upper ocean temperature and salinity gradients are relatively sharp under most conditions, and any excitation that disturbs the pycnocline will tend to propagate away from the region of generation as an internal wave. Such excitations include, in rough order of importance: (1) tidal flow of stratified water against bathymetric features such as islands, sea mounts, and continental shelf edges; (2) atmospheric forcing via barometric pressure changes and surface stress; (3) surface/internal wave interactions; and (4) any other processes producing pycnocline displacements.

Internal waves have frequencies that span the range from an upper limit of the maximum Brunt-Väisälä, or buoyancy frequency of the water column, down to the local Coriolis frequency, as will be demonstrated below. The frequencies generally fall in the range from a few tens of cycles per hour to one cycle per Coriolis period; the associated wavelengths range from a few hundred meters to many kilometers, and the phase speeds range from a fraction of a meter per second to perhaps 1 to 2 m s^{-1}. At the low frequency end are the *internal tides,* which have characteristic tidal periods and wavelengths of the order of hundreds of kilometers and which exhibit strong baroclinicity, i.e., vertically varying properties.

We earlier discussed (Section 5.3) the vertical motion of a flaccid balloon, which may be considered as the upper frequency limit of internal wave motion and one that essentially does not propagate away from the region of origin. That discussion, including the derivation of the Brunt–Väisälä frequency, forms the starting point for the theoretical development of the topic of internal waves. We shall neglect the contribution of the compressibility term to the buoyancy frequency in what follows, however, as it represents but a small correction to the internal wave motions. In addition, the approximation of a hydrostatic base state will continue to be made, a simplification that will hold for all but the shortest wavelength internal modes. Viscosity will also be neglected, since its inclusion is by now understood to bring about damping and attenuation, but in general, no fundamental new phenomena beyond boundary layer formation.

Our objective is to derive an equation that governs the motions of internal waves, as well as the associated dispersion relation. We will do this by successively eliminating u and v in favor of w and p' in the momentum equations, and then invoking the continuity equation to obtain our desired result. In a procedure that has by now become familiar, we write the linearized, small-perturbation momentum equations as

$$\frac{\partial u}{\partial t} - fv = -\frac{1}{\rho_0}\frac{\partial p'}{\partial x} \; , \tag{5.131}$$

$$\frac{\partial v}{\partial t} + fu = -\frac{1}{\rho_0}\frac{\partial p'}{\partial y} \; , \tag{5.132}$$

$$\frac{\partial w}{\partial t} = -\frac{1}{\rho_0}\frac{\partial p'}{\partial z} - \frac{\rho'}{\rho_0}g \; . \tag{5.133}$$

We eliminate u and v successively by taking the divergence of the horizontal components (Eqs. 5.131 and 5.132) and then invoking the incompressibility condition, $\nabla \cdot \mathbf{u} = 0$, which results in:

$$\frac{\partial^2 w}{\partial z\, \partial t} + f\left(\frac{\partial v}{\partial x} - \frac{\partial u}{\partial y}\right) = \frac{1}{\rho_0}\, \nabla_h^2 p' \ . \tag{5.134}$$

Next we take the curl of the same two momentum equations to obtain another relationship between the vertical vorticity, ζ, and the w velocity component:

$$\frac{\partial}{\partial t}\left(\frac{\partial v}{\partial x} - \frac{\partial u}{\partial y}\right) = \frac{\partial \zeta}{\partial t} = f\frac{\partial w}{\partial z} \ . \tag{5.135}$$

Taking the time-derivative of Eq. 5.134 and eliminating ζ yields an equation relating the horizontal divergence,

$$\nabla_h \cdot \mathbf{u} = \frac{\partial u}{\partial x} + \frac{\partial v}{\partial y} = -\frac{\partial w}{\partial z} \ , \tag{5.136}$$

to the horizontal Laplacian of the perturbation pressure:

$$\left(\frac{\partial^2}{\partial t^2} + f^2\right)\frac{\partial w}{\partial z} = \frac{1}{\rho_0}\frac{\partial}{\partial t}\,\nabla_h^2 p' \ . \tag{5.137}$$

Another equation relating w and p' is obtained from the continuity and vertical momentum equations, and it is one that does not involve rotation. From Eq. 5.133 and from

$$\frac{\partial \rho'}{\partial t} \simeq -w\frac{d\rho_0}{dz} \ , \tag{5.138}$$

(which is correct to first order), one obtains a relationship for the vertical motion that involves the buoyancy frequency, N, written for the incompressible case, $c = \infty$:

$$\frac{\partial^2 w}{\partial t^2} + N^2 w = -\frac{1}{\rho_0}\frac{\partial^2 p'}{\partial z\, \partial t} \ . \tag{5.139}$$

Here

$$N^2 = -\frac{g}{\rho_0}\frac{d\rho_0}{dz} \tag{5.140}$$

is the square of the Brunt–Väisälä frequency. It is now possible to eliminate p' between Eqs. 5.137 and 5.139 by taking the horizontal Laplacian of the latter and substituting the result into the former. Assuming that N^2 is constant, we then obtain an equation in w alone, which, upon making the Boussinesq approximation in the form of

$$\frac{1}{\rho_0} \frac{\partial}{\partial z} \left(\rho_0 \frac{\partial w}{\partial z} \right) \simeq \frac{\partial^2 w}{\partial z^2} , \tag{5.141}$$

becomes our desired equation for the vertical velocity:

$$\frac{\partial^2}{\partial t^2} \nabla^2 w + \left(N^2 \nabla_h^2 + f^2 \frac{\partial^2}{\partial z^2} \right) w = 0 . \tag{5.142}$$

The dispersion equation for internal waves in a rotating, stratified fluid may be simply derived from this equation by assuming harmonic space and time dependence, thus:

$$w = w_0 \exp[i(kx + ly + mz - \omega t)] . \tag{5.143}$$

Substitution of the plane wave solution into Eq. 5.142 yields the associated dispersion equation, which may be written in several equivalent forms:

$$\omega^2 = \frac{N^2(k^2 + l^2) + f^2 m^2}{k^2 + l^2 + m^2} , \tag{5.144}$$

$$\omega^2 = \frac{N^2 k_h^2 + f^2 m^2}{k^2} , \tag{5.145}$$

and

$$\omega^2 = N^2 \sin^2 \theta + f^2 \cos^2 \theta . \tag{5.146}$$

In Eq. 5.146, the colatitude in wavenumber space, θ, is the same as its equivalent in configuration space:

$$\theta = \sin^{-1} \left(\frac{k^2 + l^2}{k^2 + l^2 + m^2} \right) , \tag{5.147}$$

and is the angle made by the wave vector with respect to the z axis.

Equation 5.144 is the dispersion relation for internal gravity/inertial waves in an infinite medium, free of the effects of boundaries. However, most internal gravity/inertial waves are strongly governed by boundaries as well as by the intrinsic frequency–wave vector relationship just derived. This is the case of the two fluids in a box, so we will next solve the problem of such waves propagating in a constant depth ocean before discussing some of the consequences of our results to this point.

Instead of assuming a plane wave solution, in the finite boundary case we separate variables by taking the vertical velocity to be the product of a horizontally traveling plane wave and a vertical eigenfunction, $\hat{w}(z)$, which must meet the boundary conditions at the surface and the bottom. Thus

$$w(\mathbf{x},t) = \hat{w}(z) \, \exp[i(kx + ly - \omega t)] \, . \tag{5.148}$$

By substituting Eq. 5.148 into Eq. 5.142, one obtains the fundamental equation obeyed by $\hat{w}(z)$:

$$\frac{d^2\hat{w}}{dz^2} + k_h^2 \left[\frac{N^2(z) - \omega^2}{\omega^2 - f^2} \right] \hat{w} = 0 \, . \tag{5.149}$$

The quantity multiplying $\hat{w}$ is the square of the effective vertical wave number, m^2, which varies with z because of the buoyancy profile. It is clear that it vanishes when $\omega^2 = N^2$, a condition that is termed a *cutoff*, and it has a pole when $\omega^2 = f^2$, a condition called a *resonance* (see Fig. 6.55). For $\omega > N$ or $\omega < f$, the vertical wave number is imaginary, and the solutions are *evanescent;* between f and N, they are oscillatory, with forms determined by the functional behavior of $N(z)$. Thus the ocean constitutes a bandpass waveguide for internal waves, and allows propagation only between the two characteristic frequencies, f and N. Since $N \gg f$ in the ocean (usually by two or more orders of magnitude), internal waves are most nearly like the buoyancy oscillations of Section 5.3. It is only near f that they take on the characteristics of inertial oscillations (see Sections 3.5 and 6.3).

Although the solution of Eq. 5.149 is usually accomplished numerically, using observed profiles of N such as the one in Fig. 5.2, the case of $N = $ constant is solvable in terms of sinusoidal functions. In order to determine the allowed values of vertical wave number, the boundary conditions must be imposed. The bottom condition is simply satisfied by $\hat{w}(-H) = 0$, but the surface condition needs a bit of discussion. Although the surface is free, it is easy to show that all else being equal, the surface amplitude of an internal wave is reduced from that of a surface wave by a factor of the order of the effective gravity ratio, which is ordinarily of order 10^{-3}. However, typical subsurface amplitudes of internal waves are of order 10 m, but with recorded cases of very large, nonlinear waves having peak-to-trough ranges of 100 m. An internal wave having a 10 m amplitude at depth will be approximately 10 cm high at the surface, which implies that the upper boundary condition can be taken to be one of zero surface amplitude, to a good approximation. Indeed, the internal-wave-in-a-box has precisely zero surface and bottom excursions. Thus $\hat{w}(0) = 0$ will be taken as the upper condition, a requirement that quantizes the vertical wave number, m, into half-integer lengths:

$$m_n = n\pi/H, \qquad n = -1, -2, -3, \ldots . \qquad (5.150)$$

Here we have chosen negative integers, n, to distinguish these modes from much higher but similarly quantized acoustic modes to be studied in Chapter 7 (see Eq. 7.37).

The complete solutions may then be written in terms of summations over the normal modes,

$$w(\mathbf{x},t) = \sum_{n=-1}^{-\infty} w_{0n} \sin m_n (z + H)$$

$$\times \cos [(kx + ly - \omega_n t)] , \qquad (5.151)$$

where the eigenfrequencies are (from Eqs. 5.144 and 5.150)

$$\omega_n^2(\mathbf{k}) = \frac{N^2 k_h^2 H^2 + f^2 (n\pi)^2}{k_h^2 H^2 + (n\pi)^2} . \qquad (5.152)$$

From the solutions for $w(\mathbf{x},t)$ one may obtain the remaining dynamical quantities, u, v, and ρ', using Eq. 5.139 for the pressure, and combinations of Eqs. 5.131, 5.132, and 5.136 for the velocity components. For example, the pressure can be written as

$$p'(\mathbf{x},t) = \sum_n \frac{\rho_0 w_{0n}}{\omega_n m_n} (N^2 - \omega_n^2) \cos m_n (z + H)$$

$$\times \sin (kx + ly - \omega_n t) . \qquad (5.153)$$

Similar-appearing relationships can be derived for u and v.

Figure 5.25 illustrates the first three normalized eigenfunctions for the constant-N case, and Fig. 5.26 is a schematic of the dispersion relation for

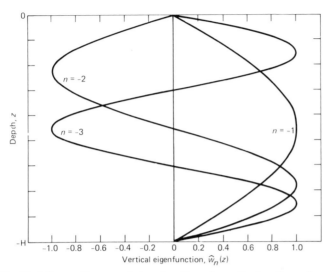

Fig. 5.25 First three orthonormal eigenfunctions for vertical motions, for a constant-N buoyancy profile.

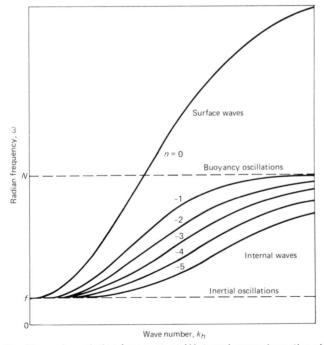

Fig. 5.26 Dispersion relation for constant-N internal waves (negative n's), plus that for surface gravity waves ($n = 0$). Internal modes are bounded above by N and below by f. [Adapted from Eckart, C., *Hydrodynamics of Oceans and Atmospheres* (1960).]

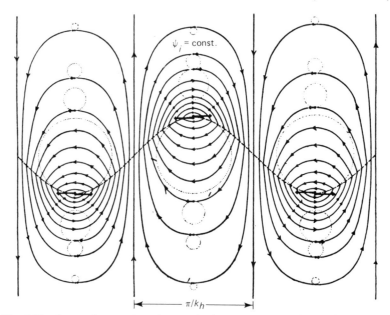

$\psi_i = \text{const.}$

π/k_h

Fig. 5.27 Streamlines and particle orbits for a propagating internal wave in a two-layer fluid. Three half-periods are shown for a wave traveling to right. [From Defant, A., *Physical Oceanography*, Vol. II (1961).]

those same modes. The bandpass character of the internal waves is clear from this diagram. In Fig. 5.27 is shown the vertical displacements and the streamlines for 1½ cycles of an interfacial wave in a two-layer fluid (see Section 5.17); the cellular nature of the wave currents is characteristic of internal waves in shallow water systems. In the region of phase where the near-surface internal currents are convergent, they can concentrate shorter surface waves whose group speeds are close to the current speed of the internal waves, with the result that lines of enhanced surface roughness can exist over coherent internal waves, making them visible to the eye and to remote sensing instruments. An image of such surface signatures made with an imaging radar is shown in Fig. 5.28.

A more realistic case of internal wave behavior is shown in Figs. 5.29 to 5.31, which show solutions of Eq. 5.149 for the profile shown in Fig. 5.2. Figure 5.29 shows the eigenfunctions. Here the peaking of the buoyancy frequency near 150 m depth weights the eigenfunctions so that the maximum of the first mode is at approximately 300 m. Solutions to the dispersion equation are illustrated in Fig. 5.30, where, because of the closeness of the observations to the equator, the influence of f is negligible. Figure 5.31 shows the phase and group speeds for the first two modes, illustrating that the long-

Fig. 5.28 Surface signatures of nonlinear internal waves on the continental shelf, made with a synthetic aperture imaging radar of 0.21 m wavelength. Internal currents modify the surface roughness, making waves visible to imaging devices. [From Gasparovic, R. F., et al., *Johns Hopkins APL Tech. Digest* (1985).]

est waves travel the fastest, and that the group speeds, as for surface waves, are significantly less than the phase speeds.

Figure 5.32 (Plate 2) shows a photograph of very large, nonlinear internal waves known as *solitons* in the Strait of Gibraltar, which are formed by tidal flow over the sill seen in Fig. 2.24. The photo was taken from the space shuttle in October 1984. Such internal solitons have distinctly nonlinear dynamics, and their current shear, $\partial u/\partial x$, modifies the surface wave spectrum; the theory for this process is discussed in Chapter 8.

In an actual oceanic environment, one does not generally see such coherent, modally decomposed waves as have been discussed to this point, but rather observes a quasi-random, continuously distributed spectrum of waves, as was described in the section on surface waves. The kinetic energy spec-

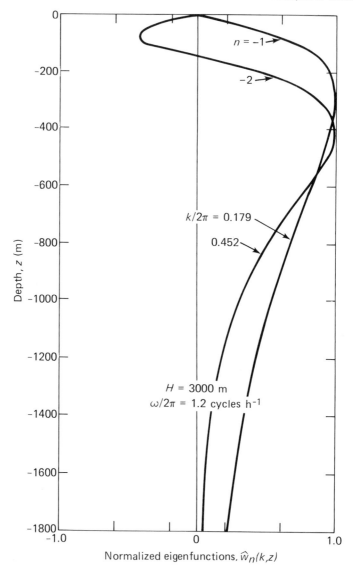

Fig. 5.29 First two vertical eigenfunctions for a variable-N water column, where density is as shown in Fig. 5.2. Compare with the constant-N case in Fig. 5.25.

trum of internal waves, $S(\omega)$, is specified by giving the distribution of the square of the velocity per unit frequency interval. The integral of $S(\omega)$ over all frequencies gives the total energy per unit mass in the wave motion. For

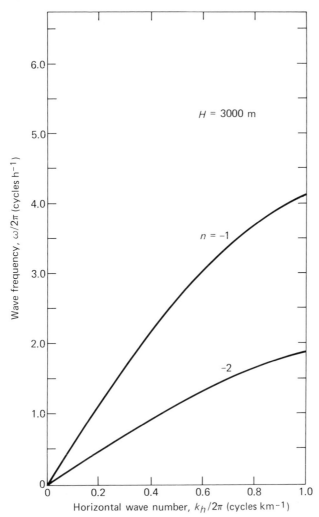

Fig. 5.30 Dispersion relation for first two modes as shown in Fig. 5.29.

internal waves it is found observationally that the energy spectrum, when measured at distances well removed from the immediate sources of waves, is approximately a global function, with nearly the same functional form found at most places in the ocean, much as is the equilibrium surface wave spectrum. The frequency spectrum is in the form of a power law, and the bandwidth over which the form holds is the one cited above for internal waves: $f \lesssim \omega \lesssim N$. The behavior of this *Garrett–Munk* spectrum is shown in Fig. 5.33 and is given by

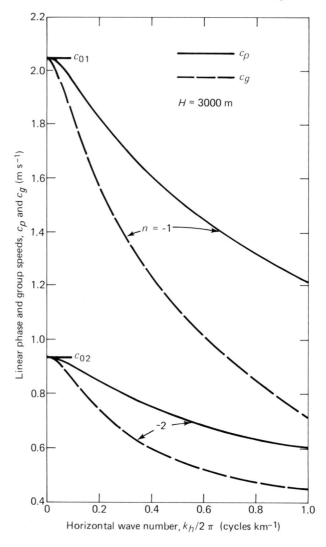

Fig. 5.31 Phase and group speeds for first two modes shown on Fig. 5.29.

$$S(\omega) = S_0 \left(\frac{\omega_0}{\omega} \right)^2. \qquad (5.154)$$

Thus the kinetic energy in the horizontal motions falls off as ω^{-2} over the approximately two decades separating f and N. It is thought that the global

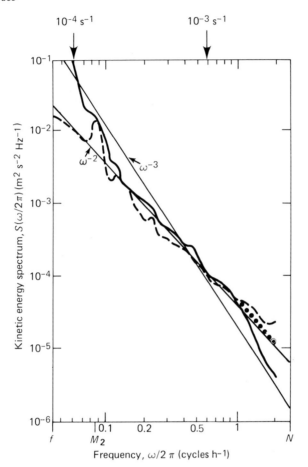

Fig. 5.33 Spectrum of kinetic energy of internal waves per unit frequency interval. Data come from beneath the seasonal thermocline at $z = -600$ m. The spectrum falls off approximately as ω^{-2} from f to N. [Adapted from Gill, A., *Atmosphere-Ocean Dynamics* (1982). Original due to Müller, Olbers, and Willebrand, *J. Geophys. Res.* (1978).]

nature of the spectrum is due to a statistical equilibration of energy among the modes according to an exponential distribution, much as occurs among the molecules of gas.

5.17 The Two-Layer Fluid

It is by now clear that a major effect of the stratification of a fluid is to introduce a new mode of motion, the internal wave. As a means of analytically modeling the effects of stratification and developing intuition in the subject, the *two-layer fluid model* has much to offer, for it allows a reasonable theoretical explication without inordinate simplification or complication. In the two-layer model, two immiscible fluids of differing parameters are superposed on one another, with the upper having constant density, ρ_1, constant mean speed, U_1, and equilibrium depth, h_1, while the lower has similarly defined quantities, ρ_2, U_2, and h_2. For this analysis, we locate the coordinate origin at the interface. The small oscillations of the free upper surface and the interface are described by displacements ξ_1 and ξ_2, which have a common wave number, k, and frequency, ω. It will be seen that this system has two normal modes of oscillation, one corresponding to the surface waves of Section 5.4 and the other to the interfacial wave of the novelty fluid-filled box mentioned in Section 5.16. The two-layer model is widely used in oceanography as a first-approximation analytical tool that offers much guidance to the veracity of more realistic and more complicated systems.

We will now derive the dynamical equations and dispersion relation for waves in the two-layer system. The notation used will be similar to that for surface and internal waves, with a potential function, φ, and perturbation pressure, p, (neglecting primes on these two quantities for now), and gravity, g, appearing in the equations as before. However, the sharp density discontinuity at the interface,

$$\Delta\rho = \rho_2 - \rho_1 , \qquad (5.155)$$

which occurs over an infinitely thin layer, implies that the buoyancy frequency is infinite at $z = 0$ and vanishes elsewhere. Thus the formulation leading to Eq. 5.149, for example, is not applicable to this case, and we must to some extent begin anew. Nevertheless, based on our earlier experience with waves, and the relationships illustrated in Fig. 5.34, we may expect that the amplitudes and velocity potentials will have the forms (see Eqs. 5.39, 5.44a, and 5.46):

$$\xi_1 = \xi_{01} \cos (kx - \omega t) + h_1 , \qquad (5.156)$$

$$\xi_2 = \xi_{02} \cos (kx - \omega t) , \qquad (5.157)$$

$$\varphi_1 = (Ae^{kz} + Be^{-kz}) \sin (kx - \omega t) , \qquad (5.158)$$

and

$$\varphi_2 = (De^{kz} + Ee^{-kz}) \sin (kx - \omega t) . \qquad (5.159)$$

Starting at the free surface at $z = \xi_1 + h_1$, we impose, respectively, the linearized Bernoulli equation for a moving medium (Eq. 5.52) and the kinematic boundary condition for the same case. Now the first-order convective derivative for a fluid layer flowing with constant horizontal velocity, **U**, is readily shown to be (neglecting nonlinear terms)

$$\frac{D}{Dt} = \frac{\partial}{\partial t} + \mathbf{U} \cdot \nabla_h \, , \tag{5.160}$$

which, when operating on an equation for a traveling plane wave of the type under consideration (e.g., Eq. 5.158) generates the negative of the *Doppler-shifted frequency*, ω_D:

$$\omega_D \equiv \omega - \mathbf{U} \cdot \mathbf{k}_h \, . \tag{5.161}$$

The Doppler frequency is called the *intrinsic frequency* in fluid dynamics and is usually denoted by $\sigma(\mathbf{k})$ in writings on that subject; the wave frequency in a moving medium, ω, is termed the *observed frequency* or *apparent frequency* and is often denoted by n by hydrodynamicists. It is ω_D that satisfies the dispersion relationship, $\omega(\mathbf{k})$, in a moving medium, so that such a relationship is written as $\omega_D = \omega_D(\mathbf{k}) = \omega - \mathbf{U} \cdot \mathbf{k}$, or (in fluid terminology), $\sigma(\mathbf{k}) = n - \mathbf{U} \cdot \mathbf{k}$.

In terms of the convective derivative applied to constant horizontal flow in the x direction, U_1, the Bernoulli equation for the upper surface becomes

$$p_1 = -\rho_1 \frac{D_1 \varphi_1}{Dt} - \rho_1 g(z - h_1) = 0 \qquad \text{at } z = h_1 + \xi_1 \, , \tag{5.162}$$

where we have chosen the reference level pressure at the free surface to be zero. The kinematic boundary condition at that surface is similarly written as

$$w = \frac{\partial \xi_1}{\partial t} + U_1 \frac{\partial \xi_1}{\partial x} = \frac{\partial \varphi_1}{\partial z} \qquad \text{at } z = h_1 + \xi_1 \, . \tag{5.163}$$

These two conditions, when applied to the upper surface, yield two relationships involving the amplitudes:

$$-\omega_{D1} (A e^{kh_1} + B e^{-kh_1}) + g \xi_{01} = 0 \tag{5.164}$$

and

$$\omega_{D1} \xi_{01} = k(A e^{kh_1} - B e^{-kh_1}) \, . \tag{5.165}$$

At the bottom, we require that the vertical velocity vanish, or that $\partial\varphi_2/\partial z = 0$ at $z = -h_2$, which leads to the condition

$$k(De^{-kh_2} - Ee^{kh_2}) = 0 \ . \tag{5.166}$$

The interface represents the most complicated case. First, the kinematic boundary condition must be imposed on each side of that surface, with the requirement that the amplitudes be the same. This yields, at $z = \xi_2$,

$$-\omega_{D1}\xi_{02} + k(A - B) = 0 \tag{5.167}$$

and

$$-\omega_{D2}\xi_{02} + k(D - E) = 0 \ . \tag{5.168}$$

Next, the Bernoulli equations for the interface at $z = 0$ must be developed. Because we wish to allow for the possibility of capillary wave development on the interface, the pressures across the interface may differ due to the surface tension. Recalling Eq. 3.123, we obtain

$$p_2 - p_1 = -\tau_s \frac{\partial^2 \xi}{\partial x^2} \qquad \text{at } z = \xi_2 \ , \tag{5.169}$$

where τ_s is now the relative surface tension of fluid 1 with respect to fluid 2. Finally, taking the pressures from the Bernoulli equation, which for layer 2 is of the form

$$p_2 = -\rho_2 \frac{D_2\varphi_2}{Dt} - \rho_2 gz + \rho_1 gh_1 \ , \tag{5.170}$$

we obtain

$$p_2 - p_1 = -\rho_2 \left(\frac{D_2\varphi_2}{Dt} + g\xi_2 \right) + \rho_1 \left(\frac{D_1\varphi_1}{Dt} + g\xi_2 \right) \tag{5.171}$$

$$= -\tau_s \partial^2\xi_2/\partial x^2 \ .$$

This yields another amplitude equation, after cancellation of common factors:

$$\rho_2\omega_{D2}(D + E) - \rho_1\omega_{D1}(A + B) - \Delta\rho g\xi_{02} = k^2\tau_s\xi_{02} \ . \tag{5.172}$$

In Eqs. 5.164 through 5.168 and 5.172, we have six equations in the unknowns ξ_{01}, ξ_{02}, A, B, D, and E, which can be solved by standard algebraic methods to obtain the amplitudes in terms of one, say ξ_{01}. As was the case for surface waves, the requirement for satisfying the interfacial Bernoulli equation leads to the dispersion equation for waves in the two-layer system. To spare the reader the tedium of the algebra, only the result is cited:

$$\rho_1 \omega_{D1}^4 + \rho_2 \omega_{D1}^2 \omega_{D2}^2 \coth kh_1 \coth kh_2$$

$$- \rho_2 gk (\omega_{D1}^2 \coth kh_1 + \omega_{D2}^2 \coth kh_2)$$

$$+ (\rho_2 - \rho_1) g^2 k^2 + k^3 \tau_s (gk - \omega_{D1}^2 \coth kh_1) = 0 . \quad (5.173)$$

This dispersion relation is biquartic in the upper-layer Doppler frequency and biquadratic in the lower-layer Doppler frequency. Its solutions for $\omega(k)$ will have four roots in general, some of which may be complex for a range of values of wave number. Such solutions indicate the potential for fluid instability, since they predict that the amplitude of a small oscillation will grow exponentially with time, until nonlinear effects set in. One simplified unstable case, the Kelvin–Helmholtz instability, will be examined in the next section.

In the limit of identical mean flows in both layers and zero density differ-ence, Eq. 5.173 is readily shown to describe surface gravity and capillary waves in a moving layer; this mathematical limit is most readily reached by setting $h_1 = 0$, in which event the equation reduces to

$$\rho_2 \omega_{D2}^2 (\omega_{D2}^2 \coth kh_2 - gk - k^3 \tau_s / \rho_2) = 0 , \quad (5.174)$$

or

$$\omega = kU_2 \pm [(gk + k^3 \tau_s / \rho_2) \tanh kh_2]^{1/2}$$

$$= kU_2 \pm \omega_D (k) , \quad (5.175)$$

which is just Eq. 5.55, the dispersion relation for surface waves as applied to a moving medium. This expression helps make clear the difference be-tween apparent and intrinsic frequencies, ω and ω_D, respectively.

In the general case, the solutions to Eq. 5.173 are much more complicated than the example just discussed. However, it is readily shown that there ex-ist two modes, corresponding to barotropic and baroclinic oscillations, whose character is indicated schematically in Fig. 5.34 for $\tau_s = 0$. The barotropic mode corresponds to the surface wave mode studied previously, with an ex-ponentially decreasing amplitude beneath the surface, and with both top and bottom layers moving in phase. The baroclinic mode shown in the lower part of Fig. 5.34 has its maximum vertical excursion at the interface and has a

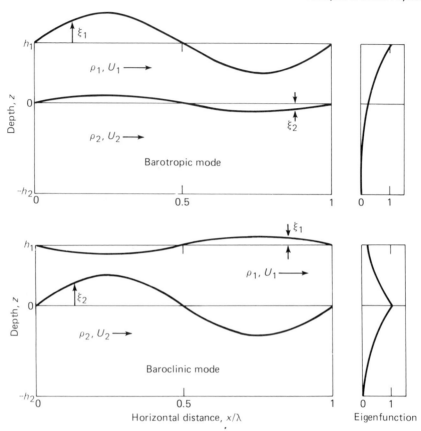

Fig. 5.34 Schematic of barotropic and baroclinic wave motions in a two-layer stratified flow. Surface displacement of the baroclinic mode is typically 10^{-3} times the internal displacement in the ocean. The vertical eigenfunctions are on the right.

greatly reduced amplitude at the upper surface. In it, the upper and lower surfaces move 180° out of phase in a kind of "breathing" oscillation. This motion corresponds to the internal wave of Section 5.16, with the density gradient concentrated in the infinitesimally thin layer at $z = 0$. The streamlines and particle orbits are as shown in Fig. 5.27.

To convince one's self of this assertion, we write the solutions for displacement and potential using the constants obtained from the solution to the amplitude equations above. The interfacial displacements can be shown to be

$$\xi_1 = \xi_{01} \cos (kx - \omega t) + h_1 \tag{5.176}$$

and

$$\xi_2 = \xi_{02} \cos (kx - \omega t) , \qquad (5.177)$$

where the amplitudes are related as

$$\xi_{02} = \xi_{01} \sinh kh_1 (\coth kh_1 - gk/\omega_{D1}^2) . \qquad (5.178)$$

The velocity potentials for the two layers are

$$\varphi_1 (x,z,t) = (g\xi_{01}/\omega_{D1}) \left[\cosh k(z - h_1) \right.$$

$$\left. + (\omega_{D1}^2/gk) \sinh k(z - h_1) \right] \sin (kx - \omega t) , \qquad (5.179)$$

and

$$\varphi_2(x,z,t) = (\xi_{02} \omega_{D2}/k) \cosh k(z + h_2)$$

$$\times \operatorname{csch} kh_2 \sin (kx - \omega t) . \qquad (5.180)$$

The asymmetries between the two layers in the forms of the velocity potentials and the dispersion relation are due to the imposition of a free-surface boundary condition at the upper surface, and the rigid bottom condition at $z = -h_2$. If we had imposed a *rigid lid* on the upper surface as well, parameters for both layers would have appeared symmetrically in all equations. The name "rigid lid" is something of a misnomer because free surface displacements are nevertheless required to give the necessary perturbation pressure gradients in the upper layer, via $\nabla p' = \rho g \nabla \xi$. The name is justified in that a rigid lid, if present, would provide identical pressure gradients.

Equations 5.173, 5.179, and 5.180 are most instructive in case of zero mean flow and zero viscosity (since we know the effects of these parameters), in the shallow layer approximation when $kh \ll 1$, and for nearly equal densities, $\rho_1, \rho_2 \simeq \rho$, and $\rho_2 - \rho_1 = \Delta\rho$. Then, in analogy to the discussion leading up to Eq. 5.73, we approximate the hyperbolic cotangents by $1/kh$, which leads to a biquadratic relationship for ω:

$$\omega^4 - \omega^2 k^2 g(h_1 + h_2) + k^4 g^2 h_1 h_2 \Delta\rho/\rho \simeq 0. \qquad (5.181)$$

It is recognized from Eq. 5.117 that the quantity $[g(h_1 + h_2)]^{1/2}$ is the shallow-water surface speed for long gravity waves, c_h, and that $g\Delta\rho/\rho = g'$ is the reduced gravity of Eq. 5.16. Solving for ω^2 and expanding the resultant square root for small values of the third term of Eq. 5.181, we obtain two roots, $\omega_\pm$:

$$\omega_{\pm}^2 \simeq k^2 c_h^2 \left[\tfrac{1}{2} \pm \left(\tfrac{1}{2} - g'h'/c_h^2 \right) \right] . \tag{5.182}$$

Here the quantity h' is the *effective depth* or harmonic mean of the two layer depths:

$$h' = \frac{h_1 h_2}{h_1 + h_2} , \tag{5.183}$$

which is smaller than either h_1 or h_2. The root ω_+^2 corresponds to the shallow water surface gravity wave, or barotropic mode of Fig. 5.34a:

$$\omega_+^2 \simeq k^2 g (h_1 + h_2) = k^2 g H . \tag{5.184}$$

The other root gives the dispersion relationship for a shallow water internal (or more correctly, interfacial) wave, or the baroclinic mode of Fig. 5.34b:

$$\omega_-^2 \simeq k^2 g' h' = k^2 \left(g \frac{\Delta\rho}{\rho} \right) \left(\frac{h_1 h_2}{h_1 + h_2} \right) . \tag{5.185}$$

Thus both shallow water modes are dispersionless, with phase speeds of

$$c_+ = \omega_+ / k = (gH)^{1/2} , \tag{5.186a}$$

and

$$c_- = \omega_- / k = (g'h')^{1/2} . \tag{5.186b}$$

Now a representative oceanic density contrast is $\Delta\rho/\rho \approx 10^{-3}$, and a typical summertime upper-layer depth is 30 m. On the continental shelf, where internal wavelengths are long compared with the depth, h_2 might be 40 m. Thus the two-layer speed for the two modes are $c_+ = c_h = 26$ m s^{-1}, and $c_- = c_{int} = 0.41$ m s^{-1}. (Actually, the total depth in this example is great enough so that only very long surface waves would propagate as shallow water waves.) Thus the limiting long internal wave baroclinic phase speed is very much less than the equivalent barotropic surface wave speed. In the ocean, internal waves are usually very slow and moderately long, with typical amplitudes of order 5 to 10 m.

Now the slowness of the baroclinic mode is not confined to internal waves alone. In Chapter 6, we will study much-longer-length mesoscale oscillations such as Rossby waves, and will discover that in general, a shallow stratified fluid supports both barotropic and baroclinic modes and that the differences in their speeds are very large indeed.

In addition to the description of internal waves, the two-layer model can be applied to the development of small waves such as capillary/ultragravity waves on the air–water interface. Here the density contrast is very large, and the distinction between densities must be maintained; however, the theoretical development goes over with very little change from that given above.

5.18   Kelvin–Helmholtz Instability

It has been mentioned several times that a fluid with sufficiently large vertical shear, $\partial u/\partial z$, will become unstable with respect to small wavelike perturbations that grow exponentially, at least for a limited time. An example of such unstable growth was shown in Fig. 3.14. A measure of the potential for instability of a sheared, stratified fluid is the dimensionless *Richardson number,* Ri, which is the square of the ratio of the buoyancy frequency, N, to the vertical shear (this latter quantity also having the dimensions of a frequency). Thus by definition,

$$\text{Ri} \equiv \frac{N^2}{(\partial u/\partial z)^2} . \tag{5.187}$$

It is generally a necessary (but not sufficient) condition for *shear flow instability* to occur for Ri to be less than ¼ somewhere in the flow; conversely, if Ri > ¼ everywhere in the flow, the system is stable. This means that in the ocean, a weakly stratified fluid column that also has a sufficiently strong baroclinic velocity structure may become unstable and oscillate with increasing amplitude, perhaps even becoming turbulent after sufficiently large growth has occurred.

Another type of instability that arises in a stratified fluid comes about when a denser fluid overlies a lighter one. In the absence of external restraining forces (and, in special cases, such do exist), this system will overturn, with the denser water sinking and replacing the lighter fluid at depth, as was discussed in Section 4.3; this instability is termed *gravitational* or *convective instability,* for obvious reasons.

The *Kelvin–Helmholtz instability* incorporates these two growth mechanisms in a version of the two-layer model, an analysis of which yields the unstable frequencies and growth rates with relative ease. The geometry considered is a minor variant of the two-layer system shown in Fig. 5.34; in fact, with some little difficulty, the dispersion relation of Eq. 5.173 could have been solved for its complex frequencies in the limit of very great layer depths. A complex frequency, which may be written as

$$\omega = \text{Re}(\omega) + i \,\text{Im}(\omega)$$

$$= \omega_r + i\omega_i \, , \qquad (5.188)$$

means that an oscillation in a fluid behaves according to

$$\gamma = \gamma_0 \, e^{\omega_i t} \cos \omega_r t \, , \qquad (5.189)$$

and hence is exponentially growing or damped, depending on whether ω_i is positive or negative, respectively. Thus complex roots with imaginary parts in the upper-half complex frequency plane correspond to unstable waves.

An example of a simple system that contains the essential physics of the instability is the two-layer shear-flow model of Section 5.17. However, there is a lack of symmetry between upper and lower layers evident in Eq. 5.173 that is due to the differing boundary conditions at the top and bottom. This asymmetry requires us to slightly reformulate the system of hydrodynamic equations (Eqs. 5.156 to 5.159) for the symmetric model of interest here. First, the free-surface, upper-boundary condition of Fig. 5.34 is replaced by the same condition that was applied to the lower boundary, i.e., zero displacement at $z = h_1$. Next both layer depths are allowed to become infinitely large, whereupon this system describes short or long waves in deep water.

Without going through the entire procedure of the previous section, it is sufficient to write the velocity potential in the form of a horizontally propagating sinusoid that exponentially decays away from the interface, viz:

$$\varphi_1 = \varphi_{01} e^{-kz} \sin (kx - \omega t) \, , \qquad z > 0 \qquad (5.190)$$

and

$$\varphi_2 = \varphi_{02} e^{kz} \sin (kx - \omega t) \, , \qquad z < 0 \, , \qquad (5.191)$$

and then impose the boundary conditions that the velocity potential should vanish at great heights and depths:

$$\varphi_1 \rightarrow 0 \qquad \text{as} \qquad z \rightarrow +\infty$$

and

$$\varphi_2 \rightarrow 0 \qquad \text{as} \qquad z \rightarrow -\infty \, . \qquad (5.192)$$

These now lead to an interfacial pressure difference given by

$$p_2 - p_1 = - \rho_2 (-\omega_{D2} \varphi_{02} + g\xi_0) + \rho_1 (-\omega_{D1} \varphi_{01} + g\xi_0)$$

$$= k^2 \, \tau_s \, \xi_0 \, , \qquad (5.193)$$

where the amplitude of the interface displacement is ξ_0. The linearized kinematic boundary conditions on the interface are, as before, given by:

$$\frac{\partial \varphi_1}{\partial z} = \frac{D_1 \xi}{Dt} = \frac{\partial \xi}{\partial t} + U_1 \frac{\partial \xi}{\partial x}, \tag{5.194}$$

and

$$\frac{\partial \varphi_2}{\partial z} = \frac{D_2 \xi}{Dt} = \frac{\partial \xi}{\partial t} + U_2 \frac{\partial \xi}{\partial x}. \tag{5.195}$$

Upon elimination of the three amplitude factors from Eqs. 5.193, 5.194, and 5.195, we obtain an implicit dispersion relation:

$$\rho_2 (\omega - U_2 k)^2 + \rho_1 (\omega - U_1 k)^2 + \Delta\rho g k = - k^3 \tau_s. \tag{5.196}$$

The solutions to this equation for the frequencies $\omega_\pm - k\tilde{U} = \omega(k)$ are:

$$\omega_\pm = k \left[\frac{\rho_1 U_1 + \rho_2 U_2}{\rho_1 + \rho_2} \right] \pm \left[\omega_0^2 - k^2 \rho_1 \rho_2 \left(\frac{U_1 - U_2}{\rho_1 + \rho_2} \right)^2 \right]^{1/2}$$

$$\equiv k\tilde{U} \pm \omega_1, \tag{5.197}$$

where the non-Doppler-shifted frequency for waves on the interface is ω_0:

$$\omega_0^2 \equiv \frac{1}{\rho_1 + \rho_2} [g(\rho_2 - \rho_1)k + k^3 \tau_s]. \tag{5.198}$$

It is clear from the earlier discussion on surface waves that Eq. 5.198 is the dispersion relation for interfacial gravity and capillary waves in deep water having a density $\rho_1 + \rho_2$ and no streaming velocity. The frequencies $\omega_\pm$, which are possibly complex, are Doppler-shifted from $\pm\omega_1$ by the streaming velocities, U_1 and U_2, to new values that are symmetrically distributed about a straight line in the ω–k plane whose slope is given by the density-weighted mean velocity, $\tilde{U}$:

$$\tilde{U} \equiv \frac{\rho_1 U_1 + \rho_2 U_2}{\rho_1 + \rho_2}. \tag{5.199}$$

The square root term in Eq. 5.197, which we have defined as $\pm\omega_1$, can be real, zero, or imaginary, according to whether the terms under the radical are positive, zero, or negative. Clearly the unstable case corresponds to imaginary values of $\pm\omega_1$. Equation 5.197 is the dispersion equation for an infinitely deep, streaming, two-layer system having surface tension on the interface, and is completely symmetric in the parameters describing the layers, in contrast to the example of the previous section. It describes two distinct physical growth mechanisms: gravitational and shear-flow instabilities.

Gravitational Instability

In the absence of streaming velocities ($U_1 = U_2 = 0$), the frequencies $\omega_\pm$ are purely imaginary conjugate pairs if the condition of a negative discriminant under the square root is met. This condition may occur for small wave numbers (or long wavelengths) that co-exist with a state of sufficiently large density difference:

$$k_c^2 \leq \frac{g(\rho_2 - \rho_1)}{\tau_s} . \tag{5.200}$$

Equation 5.200 indicates the presence of a gravitational instability when the upper-layer density is greater than that of the lower layer ($\rho_1 > \rho_2$) by an amount that provides a pressure difference sufficient to overcome the surface tension. If the fluid is unbounded (as our present example implicitly assumes) there are always small enough wave numbers for an instability to occur, with the result that the denser fluid sinks and overturns. On the other hand, if the range of wave numbers is limited to sufficiently high values, the fluid may instead oscillate but not become unstable, since ω_0 remains real. An example of such a seemingly unlikely condition is water contained in a small, hollow, vertical cylinder having an open lower end, wherein the surface tension between fluid and container holds the water in place against gravity at the opening in a downwardly curved meniscus, even when oscillating. Nevertheless, in the ocean, the presence of a heavier fluid overlying a light one inevitably results in a gravitational instability and overturning.

Shear Flow Instability

From Eq. 5.197, if the shear is nonzero but small enough, stable oscillations may still be possible for low wave numbers, provided that

$$\omega_0^2 > k^2 \rho_1 \rho_2 [(U_1 - U_2)/(\rho_1 + \rho_2)]^2 . \tag{5.201}$$

In this case, the dispersion equation for waves in the fluid has two real roots termed the *slow branch* and the *fast branch*, where "slow" and "fast" are measured with respect to the weighted streaming velocity, $\tilde{U}$. It can be shown that this *stability boundary* is set by conditions of synchronous wave propagation between the upper and lower layers. Figure 5.35 shows the dispersion

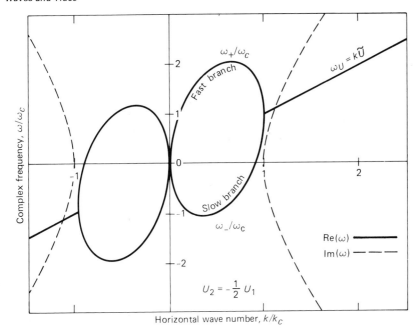

Fig. 5.35 Dispersion diagram for Kelvin–Helmholtz instability in a two-layer flow. Complex frequency roots exist for real wave numbers greater than the critical value, k_c, indicating a temporally growing instability. Inclusion of surface tension would reduce complex frequencies to zero for a large enough wave number. The system is stable for low wave numbers if the upper layer density is less than the lower layer density.

diagram, where in the region of small wave numbers, the two branches denoted by $\omega_\pm$ form an ellipse that is symmetrically distributed about a straight line whose slope is given by Eq. 5.199, and which has the equation

$$\omega_{\bar{U}} = k\tilde{U} . \tag{5.202}$$

The displaced frequencies about this linear streaming term are $\pm\omega_1$, which are, from Eq. 5.197,

$$\pm\omega_1 = \pm k\left[c_0^2 - \rho_1\rho_2\left(\frac{U_1 - U_2}{\rho_1 + \rho_2}\right)^2\right]^{1/2} , \tag{5.203}$$

where $c_0 = \omega_0/k$, as obtained from Eq. 5.198. These frequencies are real for small wave numbers but clearly become imaginary if the wave number is large enough. Instability sets in when

$$\rho_1 \rho_2 \left(\frac{U_1 - U_2}{\rho_1 + \rho_2} \right)^2 \geq c_{0_{min}}^2 \tag{5.204}$$

and occurs at the value of k for which c_0 is a minimum. For example, at the air–water interface, shear flow instability begins when

$$\rho_a \rho_w \left(\frac{\Delta U}{\rho_w} \right)^2 \geq c_{0_{min}}^2 = (0.23 \text{ m s}^{-1})^2 , \tag{5.205}$$

or at a velocity difference between air and water, ΔU, given by

$$\Delta U \geq (\rho_w / \rho_a)^{1/2} c_{0_{min}} \approx 6.7 \text{ m s}^{-1} . \tag{5.206}$$

Thus flow instability can be expected for wind speeds in excess of perhaps 7 m s^{-1}, as measured very close to the sea surface.

If surface tension is zero there is a lower critical wave number given by

$$k_c = [g(\rho_2 - \rho_1)(\rho_2 + \rho_1)] / [\rho_1 \rho_2 (U_1 - U_2)^2] . \tag{5.207}$$

The associated critical frequencies, $\omega_{c\pm}$, are given by

$$\omega_{c\pm} \equiv k_{c\pm} \tilde{U} . \tag{5.208}$$

Thus there is instability for large enough wave numbers for any value of velocity, $\tilde{U}$.

Figure 5.35 shows the dispersion relation for this two-stream, density-stratified fluid, with frequencies measured in terms of ω_c and wave numbers in units of k_c; the figure is evaluated for the case of zero surface tension and oppositely streaming layers having $U_2 = -\frac{1}{2} U_1$. The diagram indicates that for long wavelengths, the system is stable against growing waves, provided that the density contrast allows it to be intrinsically stable $(\rho_2 > \rho_1)$. However, for sufficiently high wave numbers, as given by Eq. 5.207, an instability sets in, with frequencies occurring in complex conjugate pairs, $\omega = \omega_r \pm i\omega_i$. The real part of the frequency in this unstable regime above $k = k_c$ is linear and single-valued and is given by

$$\omega_r = k\tilde{U} , \tag{5.209}$$

while the imaginary parts are given by Eq. 5.203:

$$\omega_i = \pm \omega_1 , \tag{5.210}$$

and are symmetric about the k axis.

When surface tension is included, it is found that at even higher wave numbers, the imaginary parts of ω reach conjugate maxima and minima and then decrease until they become zero once again for waves in the capillary region. This is due to the stabilizing effect of surface tension, and occurs when ω_i = 0 once again. Thus the two-layer, density-stratified, streaming fluid exhibits a band of instabilities bounded on the high end by viscous effects, and on the low end by conditions that are related to synchronous wave propagation in the two fluids.

In the ocean or at the air–sea interface, analogous instabilities occur in various guises. For example, small-scale unstable internal waves in the upper layers of the sea are observed to take the form of roll vortices under the influence of vertically sheared velocities. They may be made visible by dyeing layers of limited vertical extent and then observing the evolution of the waves, which have an appearance similar to those in Figs. 3.14b and 3.14c. The process of *spindrift*, the local effect of strong wind gusts producing a spray of drops during surface wave growth, is thought to be due to the Kelvin–Helmholtz instability. Still other instances exist in the literature.

The Kelvin–Helmholtz example has many characteristics of other, more complicated instabilities, and its analysis thus serves as a prototype for the study of these more difficult cases. Two of considerable interest in large-scale geophysical fluid dynamics are the barotropic and the baroclinic instabilities, which are thought to play important roles in the meandering of mesoscale current systems, and which involve the release and/or exchange of both kinetic energy of flow and potential energy stored in the geostrophic setup of the current systems. We will discuss limited aspects of the baroclinic instability in Chapter 6.

Bibliography

Books
Batchelor, G. K., *An Introduction to Fluid Dynamics,* Cambridge University Press, Cambridge, England (1967).

Defant, A., *Physical Oceanography,* Vols. I and II, Pergamon Press, Oxford, England (1961).

Drazin, P. G., and W. H. Reid, *Hydrodynamic Stability*, Cambridge University Press, Cambridge, England (1981).

Eckart, C., *Hydrodynamics of Oceans and Atmospheres,* Pergamon Press, Oxford, England (1960).

Gill, A. K., *Atmosphere-Ocean Dynamics,* Academic Press, New York, N.Y. (1982).

Hendershott, M. C., and W. H. Munk, "Tides," in *Ann. Rev. Fluid. Mech.,* Vol. 2, p. 205, Annual Reviews, Inc., Palo Alto, Calif. (1970).

Kinsman, B., *Wind Waves,* Dover Publications, Inc., New York, N.Y. (1965).

Lamb, H., *Hydrodynamics,* 6th ed., Dover Publications Inc., New York, N.Y. (1945).

Landau, L. D., and E. M. Lifshitz, *Fluid Mechanics,* Pergamon Press, Oxford, England (1959).

LeBlond, P. H., and L. A. Mysak, *Waves in the Ocean,* Elsevier Scientific Publishing Co., Amsterdam, Netherlands (1978).

Lighthill, J., *Waves in Fluids,* Cambridge University Press, Cambridge, England (1979).

Murty, T. S., *Seismic Sea Waves: Tsunamis,* Fisheries and Marine Service, Ottowa, Canada (1977).

Neumann, G., and W. J. Pierson, Jr., *Principles of Physical Oceanography,* Prentice-Hall, Inc., Englewood Cliffs, N.J. (1966).

Phillips, O. M., *The Dynamics of the Upper Ocean,* 2nd ed., Cambridge University Press, Cambridge, England (1977).

Techniques for Forecasting Wind Waves and Swell, U.S. Navy Hydrographic Office Pub. No. 604, U.S. Government Printing Office, Washington, D.C. (1951).

Turner, J. S., *Buoyancy Effects in Fluids,* Cambridge University Press, Cambridge, England (1973).

Van Dorn, W. G., *Oceanography and Seamanship,* Dodd, Mead & Co., New York, N.Y. (1974).

von Arx, W. S., *An Introduction to Physical Oceanography,* Addison-Wesley Publishing Co., Inc., Reading, Mass. (1962).

Yih, C.-S., *Fluid Mechanics,* McGraw–Hill Book Co., New York, N.Y. (1969).

Journal Articles and Reports

Apel, J. R., J. R. Holbrook, A. K. Liu, and J. J. Tsai, "The Sulu Sea Internal Soliton Experiment," *J. Phys. Oceanogr.,* Vol. 15, p. 1625 (1985).

Banner, M. L., and W. K. Melville, "On the Separation of Air Flow Over Water Waves," *J. Fluid Mech.,* Vol. 77, p. 825 (1976).

Chelton, D. B., K. J. Hussey, and M. E. Parke, "Global Satellite Measurements of Water Vapor, Wind Speed and Wave Height," *Nature,* Vol. 294, p. 529 (1981).

Gasparovic, R. F., J. R. Apel, A. Brandt, and E. S. Kasischke, "Synthetic Aperture Radar Imaging of Internal Waves," *Johns Hopkins APL Tech. Digest,* Vol. 6, p. 338 (1985).

Hasselmann, K., et XV al., "Measurements of Wind–Wave Growth and Swell Decay During the Joint North Sea Wave Project (JONSWAP)," *Deutsch. Hydrogr. Z.,* Suppl. A (1973).

Müeller, P., D. J. Olbers, and J. Willebrand, "The IWEX Spectrum," *J. Geophys. Res.* Vol. 83, p. 479 (1978).

Phillips, O. M., "Spectral and Statistical Properties of the Equilibrium Range in Wind-Generated Gravity Waves," *J. Fluid Mech.,* Vol. 156, p. 505 (1985).

Schwiderski, E. W., "On Charting Global Ocean Tides," *Rev. Geophys. Space Phys.,* Vol. 18, p. 243 (1980).

Chapter Six

Geophysical Fluid Dynamics II: Currents and Circulation

Into this Universe, and why *not knowing,*
Nor whence, *like Water willy-nilly flowing;*
And out of it, as Wind along the Waste,
I know not whither, *willy-nilly blowing.*

Rubáiyát of Omar Khayyám
Quatrain XXIX

6.1 Introduction

The division of oceanic dynamics into (1) waves and tides and (2) currents and circulation is partly a matter of convenience and partly a reflection of the qualitatively different character of the motions with regard to both spatial dimensions and temporal scales. Waves tend to be reasonably periodic, or at least moderately well described in terms of oscillating basis functions. Larger-scale currents tend to be more nearly aperiodic (with a few important exceptions) and highly complex, with relatively few analytic solutions available to describe them. What are available, however, are solutions for certain *processes* such as Ekman pumping, planetary waves, etc., whose dynamics underlie the basin-scale motions. In addition, numerical solutions to many important problems on the mesoscale do exist, as well as solutions for the general oceanic circulation; with few exceptions, the latter tend not to resolve the individual eddies and fronts, however, because of a lack of both physical understanding and computer capacity.

In many ways, the problems of the mesoscale and planetary scale circulations are very much like their atmospheric counterparts. It has been mentioned earlier that a considerable portion of our understanding of dynamics of the atmosphere has made its way to ocean problems, with the point of view and even the vocabulary revealing that heritage.

In this chapter, we will develop the basics of the more important processes that go into making up the mesoscale and basin-scale circulation, undertaken along the lines in Chapter 2, i.e., proceeding from cause to effect. As our discussion shifts to basin-wide scales, it will by necessity become more qualitative and descriptive, for we will have arrived at the point where the frontiers of the subject are visible, and where the topics are rather highly specialized. As before, we begin with the action of the wind on the sea surface, this time not so much concerned with the wave-making processes, but with the integrated response of the ocean via wind-driven currents.

6.2 Surface Wind Stress and Ekman Layer Dynamics

The action of the wind on the ocean's surface layers was discussed in general terms in Chapter 2, where it was pointed out that the wind-driven circulation was more the result of a chain of indirect processes working on the entire bulk of the fluid than it was the direct effects of the stress acting on the surface layers. This is not to say that the direct response to the wind is inconsequential, but that the large-scale dynamics of the main oceanic gyres must be understood in terms of several subtle and pervasive forces, i.e., Ekman boundary layer transport and downwelling/upwelling, Sverdrup interior transport, conservation of potential vorticity, frictional western boundary currents, geostrophic balance, and quasi-geostrophic meandering. In the next several sections, these processes will be discussed and an attempt will be made to synthesize their consequences into a picture of the general wind-driven oceanic circulation.

Surface Stress

The stress exerted by the wind on the surface of the sea was given earlier by Eq. 2.10 in terms of the wind velocity, $\mathbf{u}_w$, atmospheric density, ρ_a, and the drag coefficient, c_d:

$$\tau = \rho_a c_d \mathbf{u}_w |\mathbf{u}_w| . \quad N m^{-2} . \tag{6.1}$$

This stress is a force per unit area of the surface, and it couples downward into the water column through frictional forces (i.e., eddy and molecular viscosities) so that the *horizontal momentum* is transferred *vertically*. Thus the surface stress put into the water by the wind is transformed by friction into a body stress throughout a surface boundary layer known as the *Ekman layer,* with the induced subsurface horizontal flow being called *Ekman*

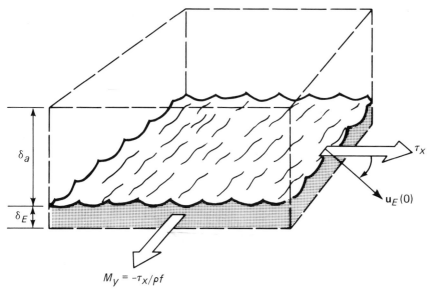

$$M_y = -\tau_x / \rho f$$

Fig. 6.1 Three-dimensional box encompassing atmospheric and oceanic surface boundary layers. The wind direction is eastward; the surface Ekman drift is at 45° to the right of the wind; the depth-integrated transport, M_y, is southward (see Eq. 6.22b). This process occurs in the regions of westerly winds.

flow. Figure 6.1 is a schematic diagram of a box containing the atmospheric and oceanic boundary layers, and illustrates the wind stress, surface velocity, and mass transport directions, the first and last via large arrows.

The flow will be shown below to cause a convergence of near-surface water in certain regions of the ocean such as the middle of the subtropical gyres, and a divergence in other regions such as in the equatorial currents and upwelling areas near the west coasts of continents. In the convergence regions, the surface influx induces slow but important downwelling flow as a result of the accumulation of mass; there is then a divergent bottom flow outward from the region, which takes place in another thin Ekman layer, or *benthic boundary layer,* along the bottom. In the equatorial or coastal surface divergence regions, flow of opposite sign occurs, with upwelling taking place to replenish the surface water transported away, and with a convergent or benthic layer (or other deep flow) carrying water to the upwelling region so as to satisfy the continuity equation (see Fig. 6.11).

The Benthic Ekman Layer

We will now derive the equations describing the Ekman flows and boundary layer characteristics, but for didactic reasons, will first do so for the bottom boundary layer, which is somewhat simpler than the surface one. This does not offend logic so much if it is realized that the atmospheric boundary layer is a consequence of the same physics as the benthic boundary layer and is the atmospheric analogue to it, albeit with quite different numerical values for layer thickness, velocities, etc. It is this atmospheric layer that provides the stresses that drive the slightly more complicated oceanic surface layer. We will use the linearized, incompressible momentum equations with perturbation pressure, p', and constant values of the vertical eddy viscosity, A_v. Thus we write (see Eqs. 3.84, 3.85, and 3.57)

$$\frac{\partial u}{\partial t} - fv = -\frac{1}{\rho}\frac{\partial p'}{\partial x} + \frac{1}{\rho}\frac{\partial \tau_x}{\partial z} \tag{6.2}$$

and

$$\frac{\partial v}{\partial t} + fu = -\frac{1}{\rho}\frac{\partial p'}{\partial y} + \frac{1}{\rho}\frac{\partial \tau_y}{\partial z} , \tag{6.3}$$

where the stress derivatives (which use simplified notation indicating the stress direction only) are related to the eddy viscosity terms via $\partial \tau_x/\partial z = A_v \partial^2 u/\partial z^2$ and $\partial \tau_y/\partial z = A_v \partial^2 v/\partial z^2$. We neglect the horizontal components of the frictional term since those velocity derivatives are small compared with the vertical one in the Ekman layer, at least at points well removed from western boundary currents. Next the velocities are separated into a pressure-driven component, $\mathbf{u}_p$, and the Ekman component, $\mathbf{u}_B$; since the equations are linear, this separation is possible. The pressure-driven portion is identical to the geostrophic component under steady-state conditions, for which $\partial \mathbf{u}/\partial t = 0$. Thus we write

$$\mathbf{u} = \mathbf{u}_p + \mathbf{u}_B . \tag{6.4}$$

For the pressure gradient equations, we have

$$\frac{\partial u_p}{\partial t} - fv_p = -\frac{1}{\rho}\frac{\partial p'}{\partial x} \tag{6.5a}$$

and

$$\frac{\partial v_p}{\partial t} + fu_p = -\frac{1}{\rho}\frac{\partial p'}{\partial y} , \tag{6.5b}$$

while for the Ekman component, there results

$$\frac{\partial u_B}{\partial t} - f v_B = \frac{A_v}{\rho} \frac{\partial^2 u_B}{\partial z^2} \qquad (6.6a)$$

and

$$\frac{\partial v_B}{\partial t} + f u_B = \frac{A_v}{\rho} \frac{\partial^2 v_B}{\partial z^2} . \qquad (6.6b)$$

From the form of Eqs. 6.5, it is clear that in the steady state, $u_p = u_g = -(1/\rho f)\partial p'/\partial y$. The boundary conditions for the benthic layer require that both normal and tangential components of velocity vanish at the bottom, $z = -H$, while at heights well above the bottom, the total velocity is required to merge smoothly into the geostrophic flow, which is termed the *free stream condition*. To effect this simply, we realign the x and y axes so that x is in the direction of u_g. Then mathematically, the free-stream condition is given by

$$\left.\begin{array}{l} u = u_g \\ v_B = 0 \\ w_B = 0 \end{array}\right\} \quad \text{at } z + H \gg \delta_B . \qquad (6.7)$$

Here u_g is the steady geostrophic flow speed well above the bottom. The distance δ_B is termed the *benthic Ekman layer thickness,* and its dependencies will be derived next.

Ekman Layer Thickness

In the steady state, Eqs. 6.6 represent a balance between Coriolis force and eddy friction, analogous to the way in which the steady state versions of Eq. 6.5 reflect the geostrophic balance between Coriolis force and pressure gradient. Equations 6.6 can be manipulated by differentiation to eliminate the coupled variables and to obtain fourth-order differential equations in u_B and v_B; we will deal only the u_B equation, since the other component has a quite similar description:

$$\frac{d^4 u_B}{dz^4} + \frac{\rho^2 f^2}{A_v^2} u_B = 0 . \qquad (6.8)$$

Solutions for u_B and v_B meeting the boundary conditions are:

$$u_B(z) = -u_g \exp[-(z + H)/\delta_B] \cos[(z + H)/\delta_B] \qquad (6.9)$$

and

$$v_B(z) = u_g \exp[-(z + H)/\delta_B] \sin[(z + H)/\delta_B]. \qquad (6.10)$$

In terms of the total velocity (Eq. 6.4) the solutions are

$$u(z) = u_g\{1 - \exp[-(z + H)/\delta_B] \cos[(z + H)/\delta_B] \qquad (6.11)$$

and

$$v(z) = u_g \exp[-(z + H)/\delta_B] \sin[(z + H)/\delta_B], \qquad (6.12)$$

where the Ekman layer thickness, δ_B, is

$$\delta_B = \left(\frac{2K_v}{f}\right)^{1/2}. \qquad (6.13)$$

Thus the velocity exponentially approaches the geostrophic velocity while oscillating slightly about that value in the vertical coordinate. The range of depths directly affected by friction is a few times δ_B, and this thickness increases with larger values of eddy diffusivity, but decreases with larger values of the Coriolis parameter. Thus the Ekman layer can be expected to be thin in polar regions and thicker in the tropics. For nominal values of vertical diffusion coefficients, the thickness of the benthic boundary layer is 10 to 20 m. In the atmosphere, the equivalent thickness is near 1 km.

Figure 6.2 is a plot of the solutions above, shown as a function of height above the sea floor in units of the Ekman thickness; the companion graph, Fig. 6.3, is a polar plot of the Ekman velocity, with vertical height above the bottom as the parameter. The velocity at several times δ_B is the geostrophic velocity, $\mathbf{u}_g$, but as the bottom is approached, it becomes increasingly deflected away from $\mathbf{u}_g$ toward the left (in the Northern Hemisphere) until, at the bottom, the velocity vector has rotated through 45° and has shrunk to zero.

The physics of the Ekman layer reflects not only the local development of a turbulent boundary layer at the sea floor, but the very important process of dissipation of the input wind stress momentum. To see this, recall that within the bulk of the water column *above* the benthic Ekman layer, the x directed flow represents a balance between the y component of pressure and the same component of Coriolis force. As u_B decreases in the Ekman layer due to eddy viscosity, the velocity-dependent Coriolis force must

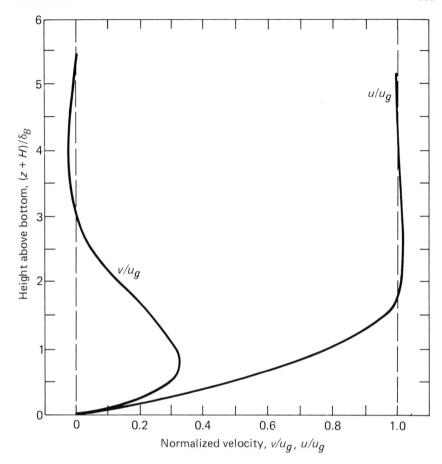

Fig. 6.2 Velocity profiles through the benthic Ekman layer for the Northern Hemisphere. Speeds are in units of near-bottom geostrophic velocity, u_g; height above sea floor is in units of Ekman layer thickness, δ_B. Depth-integrated water transport through the layer, M_y, is to the north. [Adapted from Pedlosky, J., *Geophysical Fluid Dynamics* (1979)].

weaken there also; but in the presence of a now-unbalanced pressure gradient, a y velocity must develop and produce cross-isobar flow, i.e., v_B, which is directed along y from high to low pressure. Thus the pressure force does work on the fluid through the agent of the viscous boundary layer, and energy is required to maintain the geostrophic flow in the presence of such friction. Clearly this work must be coming from the wind stress at the surface, and makes its way through the water column to the benthic layer.

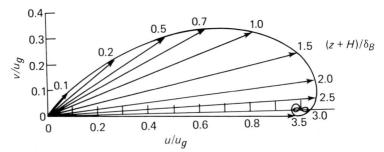

Fig. 6.3 Variation of near-bottom water velocity in form of a hodograph, or polar plot of u and v components, taken from Fig. 6.2. The parameter labeling velocity vectors is the height above the bottom, $(z + H)/\delta_B$. Northern Hemisphere velocity measurements made during a vertical descent would reveal that the velocity vector rotates to the left and is reduced from the interior geostrophic flow, as the bottom is approached. [Adapted from Pedlosky, J., *Geophysical Fluid Dynamics* (1979).]

Spin-Down Time

Now work is being done on the fluid within the Ekman layer by the pressure gradient at the rate required to maintain the layer against the dissipative forces within it. Without this energy supply, the frictional boundary layer forces would slowly reduce the geostropic flow velocity via viscous dissipation. From Eq. 4.63, the rate at which the north–south horizontal pressure gradient does work, W, throughout the entire Ekman layer is

$$\frac{d^2 W}{dt\, dA} = -\int_{-H}^{0} \mathbf{u} \cdot \nabla p \, dz$$

$$= -\int_{-H}^{0} v(z)\, \frac{\partial p}{\partial y}\, dz$$

$$\approx \tfrac{1}{2}\, \rho u_g^2 f\, \delta_B \ . \tag{6.14}$$

This quantity must also be the rate at which energy is being supplied to each unit area of the layer. The interior geostrophic water column of height H_g lying above the benthic boundary layer has a kinetic energy per unit area of approximately (see Eq. 5.90)

$$\frac{dK}{dA} = \int_{-H_g}^{0} \tfrac{1}{2}\, \rho \mathbf{u}_g^2 \, dz$$

$$\approx \tfrac{1}{2}\, \rho u_g^2\, H_g \tag{6.15}$$

when in geostrophic balance. Unless the energy were being transferred more or less continuously, the time, t_s, required for the geostrophic flow to decay would be of order of the reciprocal of

$$\frac{1}{t_s} \approx \frac{d^2 W}{dt\, dA} \,\bigg/\, \frac{dK}{dA} = \frac{f\delta_B}{H_g} . \tag{6.16}$$

This time measures the natural decay time of geostrophic flow in the presence of friction, and is termed the *spin-down* (or *spin-up*) *time*. From Eq. 6.16, it is independent of the velocity and depends only on the eddy diffusivity, the geostrophic column thickness, and the Coriolis frequency. The flow will be approximately steady and in geostrophic balance if the decay time is long compared with the inertial period, or if

$$ft_s = \frac{H_g}{\delta_B} = \left(\frac{H_g^2 f}{2K_v}\right)^{1/2} \gg 1 . \tag{6.17}$$

Another way of stating this is that we require the *Ekman number,* Ek, to be small, where that number is the reciprocal of the square of Eq. 6.17:

$$\text{Ek} \equiv \frac{2K_v}{fH_g^2} = \left(\frac{1}{ft_s}\right)^2 . \tag{6.18}$$

The smallness of the Ekman number is, from Eq. 6.16, equivalent to the neglect of the time derivative, or to the assumption of near-steady-state conditions.

From this discussion, it follows that the wind stress puts energy into the water column, which is then transformed by methods to be discussed ahead into essentially geostrophic flow; that energy is then dissipated in the benthic boundary layer through viscosity. (While this description is approximately correct, the actual situation is much more complicated. The reader is referred to Pedlosky [1979] or Gill [1982] for more extended discussions.)

Mass Transport

The *horizontal mass transport* per unit width of the Ekman layer is the vertical integral of the mass flux, $\rho\mathbf{u}_B$, and is expressed in units of kilograms per meter-second; its components are given by

$$(M_{xB}, M_{yB}) = \int_{-H}^{0} (\rho u_B, \rho v_B)\, dz . \tag{6.19}$$

This vector bottom flux can be related to the surface stress by vertically integrating Eqs. 6.2 and 6.3 and noting that the pressure gradients must vanish at the top and bottom of the sea. The result is

$$-\frac{\partial M_{xB}}{\partial t} + fM_{yB} = \tau_x \tag{6.20a}$$

and

$$-\frac{\partial M_{yB}}{\partial t} - fM_{xB} = \tau_y , \tag{6.20b}$$

which, in the steady state, gives the *benthic Ekman transport* per unit width in terms of the surface stress:

$$-M_{xB} = \tau_y/f \tag{6.21a}$$

and

$$M_{yB} = \tau_x/f . \tag{6.21b}$$

This vector is directed to the left of the wind at the bottom, so that an *eastward* wind stress gives a *northward* bottom transport, while a *northward* stress drives a *westward benthic transport*. In either event, the net, integrated transport in the benthic boundary layer is to the *left* of and at 90° to the wind, a fact of fundamental importance in ocean (and atmospheric) dynamics. This effect is due to the frictionally induced cross-isobar flow described above. As an estimate of the numerical values of transport, a nominal average value of stress is $\tau \approx 0.1$ N m^{-2}, in which event the transport in mid-latitudes is $M_B \approx 10^3$ kg (m s)$^{-1}$, a relatively small value of transport per unit width. Near the equator, the theory suggests that larger values of transport exist. Nevertheless, when integrated over an entire ocean basin, this represents a significant flow.

The Surface Ekman Layer

At the ocean surface, a similar development can be carried out and analogous solutions obtained, with the velocity now being labeled u_E; the boundary layer scale, δ_E, may be near 20 to 40 m. The boundary conditions are again such that the flow must merge with the geostrophic flow at depth; however, we no longer take the u component of $\mathbf{u}_E$ to be aligned with the x axis, but instead orient it at 45° to the right of the wind. The solutions for the surface layer case are illustrated in Fig. 6.4. As before, we note that the sur-

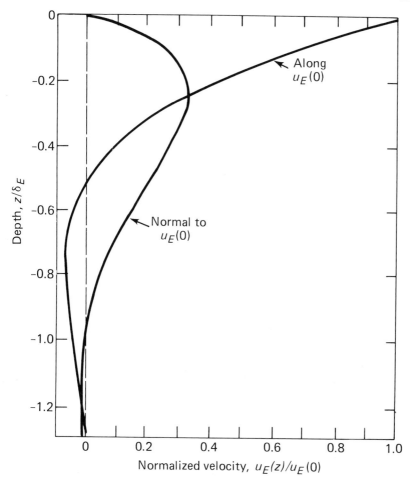

Fig. 6.4 Drift velocity profiles through the near-surface Ekman layer, resolved into components along the surface drift velocity, $\mathbf{u}_E(0)$, and at right angles to it. This is superimposed on the deeper geostrophic velocity. The presence of stratification alters this behavior.

face flow, $u_E(0)$, is at 45° to the right of the wind, and that when measured with respect to the geostrophic velocity at the bottom of the boundary layer, $\mathbf{u}_g(-5\delta_E)$, the flow veers to the right, as is illustrated in Fig. 6.5. Figure 6.5 is a hodograph of the surface Ekman layer velocity vector as it rotates and diminishes with depth (cf. Fig. 2.19).

A schematic diagram summarizing the velocity distribution with depth is given in Fig. 6.6, which illustrates the surface and benthic boundary layers superimposed on the interior, geostrophic velocity. The total flow is the vec-

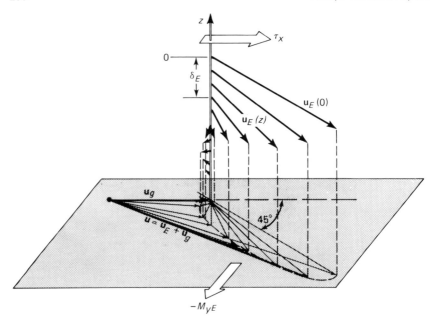

Fig. 6.5 Hodograph for the oceanic surface boundary layer for the Northern Hemisphere. Wind stress is to the east, as in Fig. 6.1. The surface Ekman velocity, $u_E(0)$, is at 45° to the right of the wind, and continues to rotate and shrink with depth. At several Ekman depths, δ_E, the velocity merges with the geostrophic velocity, u_g, here assumed to have the same direction as the surface wind stress. (It is not necessary that τ and u_g have exactly the same direction.) Velocity measurements made during a vertical descent would show the total subsurface velocity, u, rotating counterclockwise and attenuating with depth.

tor sum of u_E and u_g, and the summed velocities actually veer mainly counterclockwise to the *left* when observed during a descent through the entire water column.

The mass transports in the surface Ekman layer are opposite in sign to those in the benthic layer, with an *eastward* wind stress being associated with a *southward* transport, and a *northward* stress giving an *eastward* transport; Fig. 2.20 illustrates these transports for the Northern Hemisphere via wide arrows. For either component, the integrated surface flows are in a direction 90° to the right of the wind in the Northern Hemisphere, and have the opposite direction to the benthic flows. This flow also satisfies the continuity equation, which requires that the net inflow be equal to the net outflow. The transport equations for this case are

$$M_{xE} = \tau_y / f \qquad (6.22a)$$

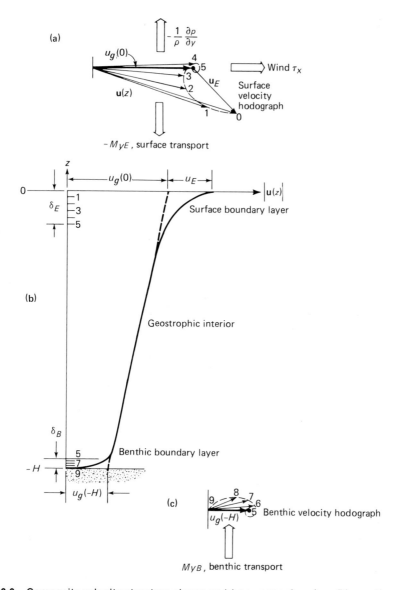

Fig. 6.6 Composite velocity structure shown as (a) an upper plan view, (b) a vertical profile, and (c) a lower plan view. When surface and benthic Ekman velocities are superimposed on the geostrophic interior flow, a complex variation in velocity magnitude and direction results. The geostrophic interior flow may itself vary in speed and direction due to other causes. The case shown assumes the geostrophic velocity to have the same direction throughout, which is that of the surface wind, but to be mildly baroclinic. In descending from the surface to the bottom, the total velocity from this profile would rotate counterclockwise almost continuously in the Northern Hemisphere.

and

$$-M_{yE} = \tau_x / f \,, \tag{6.22b}$$

and yield a transport that is to the right of the wind.

For either the surface or benthic layers, the vector equation for the total mass transport per unit area of the boundary layer can be written as

$$\mathbf{M}_{E,B} = (+,-) \, \tau \times \hat{k}/f \,, \tag{6.23}$$

where $\hat{k}$ is the unit vector in the vertical direction. Additionally, it can be shown that the mass transport in the surface layer is the same as in the benthic layer, and depends only on the wind stress and not on the details of the turbulent mixing.

The physical reason for the Ekman spirals is approximately as follows: At any depth, the stress that has been transmitted downward from above imparts a velocity to the layer immediately below, which, under the influence of the Coriolis force, would like to move in a direction 90° to the right; however, the effect of eddy viscosity is to reduce the speed and to partially resist this right-angle movement, with the net result that the velocity vector is continuously attenuated as it rotates with depth. This is described by the Ekman spiral. The requirement for its congruence to the geostrophic interior velocity and the origins of the latter will be explained ahead.

Something like the Ekman layers are actually observed in the ocean and the atmosphere, although at any given instant the boundary layer profiles may depart dramatically from the simple description given here. However, when averaged over several days, an Ekman-like surface profile is generally obtained, but with the surface "drift" velocity very roughly near 30° rather than 45°, although a wide range of angles is observed. The average depth of the surface layer varies from some 10 to 100 m, which is, of course, approximately the depth of the mixed layer, so that the mixed and the Ekman layers generally coincide. However, they are not synonymous, since one is defined in terms of temperature distribution and the other in terms of velocity shear.

6.3 Inertial Oscillations

Sudden bursts of wind put energy and momentum into the ocean surface layers and excite what are called *inertial oscillations*. These are the rotary motions in the inertial circle described in Chapter 3, and can frequently be seen in current meter records. They are properly considered a mainly vertically propagating wave that is near the lower frequency limit for internal waves, as given by Eq. 5.146, with $\theta \approx 0$. Presumably the integrated result of these impulsive forces (as well as the steadier wind stress input) on the large-scale circulation is ultimately to bring about the condition of geostrophic flow. To model inertial motions in an approximate fashion, we include in the equations of motion (1) a constant horizontal pressure gradient directed toward the north that represents geostrophic balance, and (2) a scale-dependent eddy viscosity such as might derive from a Fourier-analyzed friction term. For the latter, we write

$$\frac{1}{\rho} \nabla \cdot \mathbf{A} \cdot \nabla \mathbf{u} \rightarrow -K_h (k^2 + l^2)\mathbf{u} \equiv -\gamma(\mathbf{k})\,\mathbf{u}, \qquad (6.24)$$

where the damping term, $\gamma(\mathbf{k})$, is small for long-wavelength motions such as geostrophic flow, and larger for shorter-length inertial oscillations. Its reciprocal is the damping time, $t_d(\mathbf{k})$, for the motion.

With this approximation, the horizontal momentum equations are

$$\frac{\partial u}{\partial t} - fv = -\gamma u \qquad (6.25)$$

and

$$\frac{\partial v}{\partial t} + fu = -\gamma v - \frac{1}{\rho}\frac{\partial p}{\partial y}$$

$$= -\gamma v + fu_g, \qquad (6.26)$$

where the constant north–south pressure gradient is assumed to be geostrophically balanced:

$$fu_g = -\frac{1}{\rho}\frac{\partial p}{\partial y}. \qquad (6.27)$$

As initial conditions, we assume that a small parcel of water at (x_0, y_0) is given an eastward impulsive start with a velocity, u_0, by the wind. We solve this problem in four increasingly complicated approximations: (1) no pressure term or damping, (2) pressure but no damping, (3) no pressure, but with damping, and (4) with both pressure gradient and damping.

No Pressure or Damping

Under these conditions, the solutions to Eqs. 6.25 and 6.26 are readily found to be

$$x = x_0 + r_i \sin ft , \tag{6.28}$$

$$y = y_0 - r_i (1 - \cos ft) , \tag{6.29}$$

and

$$r_i = u_0/f , \tag{6.30}$$

where the radius, r_i, is that of the inertial circle (cf. Eq. 3.39). Squaring and adding these equations shows that the parcel motion is simply a horizontal oscillation in a constant radius inertial circle centered at $(x_0, y_0 - r_i)$, which is the case discussed in Chapter 3. The motion is shown in Fig. 6.7a, which illustrates clockwise rotation in a circle 2000 m in diameter.

Pressure Gradient but No Damping

The presence of the northward horizontal pressure gradient introduces an interesting new effect, the steady *drift* of the water parcel to the east at right angles to the gradient, while spiraling in inertial circles that are drawn out by the drift velocity. The solutions for the displacement in this case are

$$x = x_0 + u_g t + r_i' \sin ft \tag{6.31}$$

and

$$y = y_0 - r_i' (1 - \cos ft) . \tag{6.32}$$

Here the drift velocity is easily proven to be the geostrophic velocity, u_g, of Eq. 6.27, and the radius of oscillation is now proportional to the velocity difference between the impulsive and geostrophic speeds:

$$r_i' = (u_0 - u_g)/f . \tag{6.33}$$

Figures 6.7b to 6.7d show this motion for three values of the geostrophic velocity: small, medium, and large; the same value of the initial velocity, u_0, has been used in all three cases. It is seen from the figures that for small values of the geostrophic velocity, the motion is nearly circular but with a slow drift to the east added; for large values, the motion is nearly rectilinear with small inertial oscillations superimposed. The latter case describes impulsive, low speed wind events imposed on a large geostrophically balanced flow. At a critical intermediate value, when $u_0 = u_g/2$, the motion is cycloidal, while when $u_0 = u_g$ there are no inertial oscillations at all, but the wind puts energy into the system at exactly the value required to meet the geostrophic conditions.

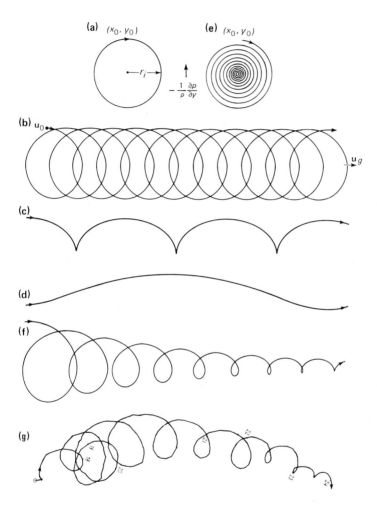

Fig. 6.7 Motion of a water parcel under the influences of Coriolis force, a horizontal pressure gradient arising from geostrophic balance, and damping due to eddy viscosity: (a) no pressure gradient or damping, in which case the motion is in an inertial circle; (b) (c) and (d) pressure gradient but no damping, with (b) small, (c) intermediate and (d) large values of the gradient and resultant drift; (e) no pressure gradient but with damping; (f) both pressure gradient and damping; (g) progressive vector diagram showing inertial oscillations from current meter measurements. [Fig. 6.7g adapted from Gustafson, T., and B. Kullenberg, *Sven. Hydrogr.–Biol. Komm. Skr., Hydrogr.* (1933).]

No Pressure, but With Damping

The solutions to Eqs. 6.25 and 6.26 are now

$$x = x_0 + r_i'' \left[e^{-\gamma t} \left(\sin ft - \frac{\gamma}{f} \cos ft \right) + \frac{\gamma}{f} \right] \qquad (6.34)$$

and

$$y = y_0 + r_i'' \left[e^{-\gamma t} \left(\cos ft + \frac{\gamma}{f} \sin ft \right) - 1 \right] , \qquad (6.35)$$

where the radius of the inertial circle is decreased by the damping to a value of

$$r_i'' = \frac{u_0 f}{f^2 + \gamma^2} . \qquad (6.36)$$

By forming the squares of $x - x_0 - r_i'' \gamma/f$ and $y - y_0 + r_i''$ and adding, one can show that this motion is an inward logarithmic spiral of the form

$$r(t) = \frac{u_0 e^{-\gamma t}}{(f^2 + \gamma^2)^{1/2}} , \qquad (6.37)$$

which asymptotically converges to the point $(x_0 + r_i'' \gamma/f, y_0 - r_i'')$ as time goes on. This motion is shown in Fig. 6.7e, and clearly models damped inertial oscillations.

Pressure Gradient and Damping

This is the case of a damped, drifting spiral, and the equations of motion must therefore accommodate a decreasing geostrophic drift velocity, since friction erodes both the drift and the inertial oscillations. By successively eliminating v from Eq. 6.25 and u from Eq. 6.26 and allowing for the possibility of a slowly varying u_g, the equations for the parcel velocity become

$$\frac{\partial^2 u}{\partial t^2} + 2\gamma \frac{\partial u}{\partial t} + (f^2 + \gamma^2) u = f^2 u_g , \qquad (6.38)$$

and

$$\frac{\partial^2 v}{\partial t^2} + 2\gamma \frac{\partial v}{\partial t} + (f^2 + \gamma^2) v = f \left(\gamma u_g + \frac{\partial u_g}{\partial t} \right) . \qquad (6.39)$$

These are recognized as the equations for a damped harmonic oscillator having an inhomogeneous forcing term. It can be readily verified that solutions to Eqs. 6.38 and 6.39 meeting the initial conditions $u = u_0$ and $v = 0$ at $t = 0$, are

$$u(t) = U_g \, e^{-\gamma_g t} + (u_0 - U_g) \, e^{-\gamma t} \cos ft \tag{6.40}$$

and

$$v(t) = -(u_0 - U_g) \, e^{-\gamma t} \sin ft . \tag{6.41}$$

Here the damped geostrophic velocity is given by

$$u_g = U_g \, e^{-\gamma_g t} . \tag{6.42}$$

The corresponding particle displacements are found by integration to be

$$x = x_0 + \frac{U_g}{\gamma_g} (1 - e^{-\gamma_g t})$$

$$+ r_i''' \left[\frac{\gamma}{f} + e^{-\gamma t} \left(\sin ft - \frac{\gamma}{f} \cos ft \right) \right] \tag{6.43}$$

and

$$y = y_0 - r_i''' \left[1 - e^{-\gamma t} \left(\cos ft + \frac{\gamma}{f} \sin ft \right) \right] . \tag{6.44}$$

In Eqs. 6.40 to 6.43, we have allowed for the fact that the damping rate, $\gamma(\mathbf{k})$, will have different numerical values depending on the scale of the motions (cf. Eq. 6.24), and we have denoted that rate for the geostrophic motion by γ_g. These equations describe the motion of a parcel of water in decreasing inertial circles, with a more slowly attenuated geostrophic drift to the east superimposed. The radius of oscillation is now found to be

$$r_i''' = \frac{(u_0 - U_g)f}{f^2 + \gamma^2} , \tag{6.45}$$

so that the circular motion depends on the departure of the initial velocity from the geostrophically balanced case, and is somewhat smaller than the undamped radius as given by Eq. 6.33. The displacement due to the geostrophic velocity is

$$x_g = (U_g/\gamma_g)(1 - e^{-\gamma_g t})$$

$$\simeq U_g t \qquad \text{for small } \gamma_g , \tag{6.46}$$

so that the linear drift speed appearing in Eq. 6.31 is recovered for lightly damped flow. Figure 6.7f illustrates this case, as derived from Eqs. 6.43 and 6.44 for a parcel at $\Lambda = 45°$ that receives an initial impulsive speed increment, u_0, of 0.3 m s^{-1} superimposed on an initial geostrophic flow of U_g = 0.04 m s^{-1}, and having damping times of $t_d = 1/\gamma = 2.89$ days and t_g = $1/\gamma_g = 28.9$ days. The corresponding eddy diffusion coefficient is approximately $K_h \simeq 1.29$ m^2 s^{-1}. The speed u_0 is what might be expected from a surface wind impulse of $u_w \simeq 12$ m s^{-1}, which induces a stress, τ_x, of 0.25 N m^{-2} (see Eq. 2.10). The resultant surface speed is approximately $u_0 = 2 \tau_x/\rho f \delta_E$ (see Eqs. 6.9, 6.19, and 6.22) for a surface Ekman depth, δ_E, of 16 m at mid-latitudes. The motion shown in Fig. 6.7f illustrates the slow reduction in the radius of inertia as eddy forces take their toll, and an asymptotic approach toward a geostrophic balance. However, in this case, the friction will also result in a small *ageostrophic* component of velocity in the direction of the gradient as well, just as occurs in the bottom Ekman layer, so that there will be a northward down-gradient flow in addition to the eastward, along-isobar drift in our example.

Something of this kind of motion is often seen in the ocean in current meter records. To illustrate the existence of decreasing, drifting spirals in observational studies, an analysis of measurements of the two horizontal components of current may be made using a technique called a *progressive vector diagram,* in which the time series of current vectors are laid end to end, with length indicating speed and angle indicating direction, producing a pseudo-trajectory for the water parcel. The diagram would also represent the actual trajectories of water parcels if the entire region of the ocean around the point of observation were partaking of the same motion at the same time. The form taken by damped inertial oscillations is shown in Fig. 6.7g, which is a five-day progressive vector diagram showing oscillations having a radius of less than 2 km. Figure 6.8 is a schematic illustration of the ageostrophic, cross-isobar flow that viscosity induces. Figure 6.9 is a photograph of an oil spill from a tanker that shows sinuous motions thought to be inertial oscillations, with drift superimposed.

The approximation used here of a small parcel of water moving in an externally established geostrophic pressure field is a useful didactic device, but is not overly rigorous. The input of wind energy into ocean currents must proceed through some analogous mechanism, however, with the long-term outcome of the forcing being large-scale geostrophic flow. As we have discussed earlier and will expand upon in the next section, this Ekman stress is not necessarily the direct source of large-scale oceanic currents, but only the first step in the establishment of the wind-driven oceanic circulation.

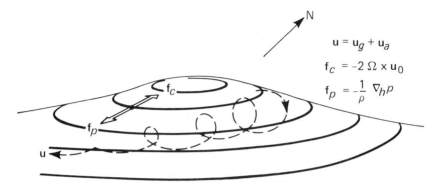

Fig. 6.8 Schematic of damped inertial oscillations drifting about an oceanic high pressure system with an ageostrophic component. [Adapted from von Arx, W. S., *An Introduction to Physical Oceanography* (1962).]

Fig. 6.9 Oil spill off Nantucket Island from a tanker. The sinuous shape of the oil distribution may be inertial oscillations superimposed on drift. The lighter striations are thought to be internal waves. [Figure courtesy of the National Oceanic and Atmospheric Administration and the National Aeronautics and Space Administration.]

Large-Scale Effects of Ekman Transport

On the scale of an oceanic basin, Ekman transport has very important effects. The prevailing surface wind stress patterns across the basin are set by the easterly trade winds in the tropics and the mid-latitude westerlies closer to the poles; from the directions of surface transport as given by Eq. 6.23, it is seen that both trades and westerlies drive net near-surface currents that converge on the region between them, which is the interior of the subtropical gyre (also called the subtropical convergence, for obvious reasons). Figures 2.20 and 2.21 illustrate this convergence for the North Atlantic gyre. Now in the bottom boundary layer, water is transported in just the opposite direction to the surface (see Fig. 6.6), which requires that the water in the interior, in order to join up these two flows, must first move downward and then rise near the perimeter, in order to satisfy continuity of flow. The schematic form of the resultant flows is shown in Fig. 6.10, which illustrates a large oceanic box having Ekman flows in the boundary layers, anticyclonic (clockwise in the Northern Hemisphere) geostrophic flows in the center, interior downwelling, and perimetric upwelling. This is the simplest model containing the basic current system of the gyre. Indeed, it is too greatly simplified, for it does not emphasize sufficiently two exceedingly important features of the real gyre—a major southward transport throughout much of the interior geostrophic region called *Sverdrup transport*, and the strong intensification of the current along the western boundary. We will return to these later, but will first compute the downwelling circulations.

The inward Ekman transport has two effects: (1) It causes the accumulation of a large amount of warm, near-surface water within the gyre, raising the sea height by a meter or so in the process; (2) more importantly, the slow downward flow mentioned above depresses the deeper water by several hundred meters. To calculate this downward speed, we integrate the continuity equation for the Ekman velocity over the entire depth of the layer, a thickness in excess of some $5\delta_E$, to obtain

$$
\int_{-5\delta_E}^{0} \nabla \cdot \mathbf{u}_E \, dz = 0
$$

$$
= \frac{\partial}{\partial x} \int u_E \, dz + \frac{\partial}{\partial y} \int v_E \, dz + \int \frac{\partial w_E}{\partial z} \, dz
$$

$$
= \frac{1}{\rho} \left[\frac{\partial M_{xE}}{\partial x} + \frac{\partial M_{yE}}{\partial y} \right] + w(0) - w(-5\delta_E) \, . \quad (6.47)
$$

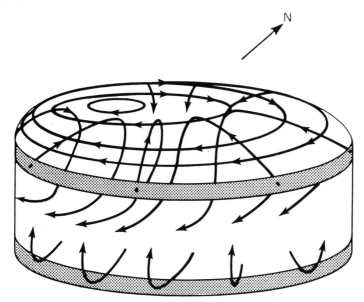

N

Fig. 6.10 Schematic of an oceanic "pillbox," showing surface and benthic layers, upwelling and downwelling, and anticyclonic horizontal circulation.

Now the vertical velocity at the surface is zero. We denote the vertical velocity at the bottom of the Ekman layer by w_E, and use Eqs. 6.22 as substitutes for the transports in the expression above. Then we arrive at an important relationship for the vertical velocity in terms of the z component of the *curl* of the surface wind stress:

$$w_E = \frac{1}{\rho f} \left(\frac{\partial \tau_y}{\partial x} - \frac{\partial \tau_x}{\partial y} \right)$$

$$= \frac{1}{\rho f} (\nabla \times \tau)_z . \tag{6.48}$$

From Eq. 6.48, if the curl is negative, the velocity describes *downwelling;* this is the case in the subtropical gyres, where the wind circulation has the opposite sign to the spin of the earth (which defines the sign of the vorticity and the curl; the right-hand rule applies here, as a mnemonic device). Thus the downward velocity at the bottom of the Ekman layer is the result of the clockwise, negative curl of the wind stress. On the other hand, if the curl is positive, Eq. 6.48 indicates that *upwelling* will occur; this is the case in the equatorial region, where the northeast and southeast trade winds both transport surface waters away from the equator, with an upward pumping

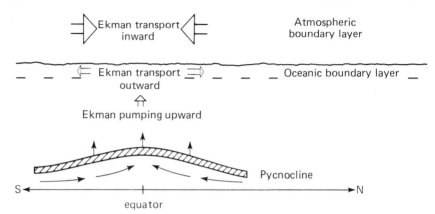

Fig. 6.11   Equatorial upwelling shown schematically in a north–south vertical section. Surface Ekman transport away from equator raises the thermocline and pycnocline to shallow depths in that region. [Adapted from Gill, A. E., *Atmosphere–Ocean Dynamics* (1982).]

of subsurface water termed *Ekman suction* (see Fig. 6.11) working to replace the displaced surface fluid. The colder upwelled fluid is the same as that shown as rising along the southern boundary of the box in Fig. 6.10. In the region close to the equator, one result is the thinning and cooling of the surface mixed layer, just as the opposite effect occurs in the center of the gyres, i.e., a warming and thickening of the upper layers. The latter will be discussed ahead in the section on properties of the subtropical gyre.

The vertical velocities are small and not at all directly measurable: $w_E \approx 10^{-6}$ m s^{-1} or approximately 30 m yr^{-1}, which is perhaps an order of magnitude larger than the vertical speeds resulting from the replacement of sunken polar waters (cf. Section 2.10).

6.4 Vorticity: Relative, Planetary, Absolute, and Potential

We have now arrived at the point where we are dealing with motions whose horizontal scales are much greater than the depth of the ocean, and it is useful to specialize the fluid equations to a shallow-water hydrostatic system that is vertically homogeneous. The more realistic case of vertically varying properties can be approximately modeled out of a sequence of such layers, known appropriately as multi-layered models; however, the two-layer model often can suffice to describe the salient features of both the barotropic and baroclinic modes of a system (see Section 5.17).

In order to proceed further with our understanding of large-scale ocean dynamics, we must digress slightly to develop more fully the concepts of *fluid* (or *relative*) *vorticity, planetary vorticity,* and their sum, termed *absolute vorticity,* as well as a closely related function known as *potential vorticity.* Relative vorticity, ζ, is a measure of the local angular velocity, ω, of the fluid, and is defined by

$$\zeta \equiv \nabla \times \mathbf{u} = 2\omega \tag{6.49}$$

$$= \left(\frac{\partial w}{\partial y} - \frac{\partial v}{\partial z} \right) \hat{i} + \left(\frac{\partial u}{\partial z} - \frac{\partial w}{\partial x} \right) \hat{j} + \left(\frac{\partial v}{\partial x} - \frac{\partial u}{\partial y} \right) \hat{k} . \tag{6.50}$$

Now the convective derivative can be rewritten in terms of the relative vorticity through an identity that states

$$\frac{D\mathbf{u}}{Dt} \equiv \frac{\partial \mathbf{u}}{\partial t} + \nabla \left(\tfrac{1}{2} \mathbf{u}^2 \right) + \zeta \times \mathbf{u} , \tag{6.51}$$

which can readily be proven by expanding the definition of the convective derivative (Eq. 3.8) into its constituent terms, applying it to one component of velocity (say u), and adding and subtracting the x derivatives of the other two components squared.

A scalar quantity called the *circulation,* Γ, is related to the fluid vorticity. The circulation is an integral kinematic property of the velocity field, and is defined as

$$\Gamma \equiv \oint \mathbf{u} \cdot d\mathbf{s} = \oint (u \, dx + v \, dy + w \, dz) . \tag{6.52}$$

Here the line integral is taken on a path around any arbitrary, open surface in space, and $d\mathbf{s}$ is an element of arc length along the path. From the definition of vorticity and from Stokes' theorem, Eq. 6.52 can be rewritten as

$$\Gamma = \oint \mathbf{u} \cdot d\mathbf{s} = \int\int \nabla \times \mathbf{u} \cdot d\mathbf{A} = \int\int \zeta \cdot d\mathbf{A}$$

$$= \int\int \zeta \cos \theta \, d^2\mathbf{x} , \tag{6.53}$$

where $d\mathbf{A}$ is an element of directed area with an outward normal, $\hat{n}$. Thus the circulation around a closed path, which has units of square meters per second, is a measure of the normal component of vorticity passing through the area encompassed by the path. The vorticity can therefore be interpreted

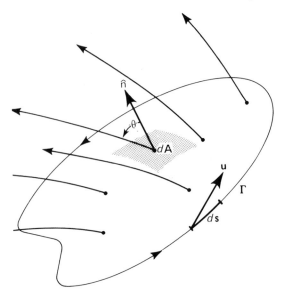

Fig. 6.12 Circulation, Γ, around an arbitrary circuit within the fluid in terms of the velocity along the boundary and the normal component of vorticity through the bounded area.

as the fluid circulation per unit area, or the flux density. This is shown in Fig. 6.12 for a general element within the fluid. Thus the distribution of the vertical component of vorticity of an oceanic current system, for example, could be measured by integrating the observed horizontal velocities around a succession of small closed paths in the fluid and dividing by the areas encompassed by the paths.

Figure 6.13 relates the net circulation, $d\Gamma$, around an infinitesimal closed horizontal path, to the vertical vorticity component, ζ_z within it. In Fig. 6.13, the spatially varying velocity is expanded in a series about the midpoint, and the sum of the velocity components along each leg is taken in a counterclockwise sense. Along the top and bottom x elements, the net contribution to the circulation is $(-\partial u/\partial y)\, dx\, dy$, while along the vertical elements, it is $(+\partial v/\partial x)\, dx\, dy$. For the total path, the circulation is therefore

$$d\Gamma = \left(\frac{\partial v}{\partial x} - \frac{\partial u}{\partial y}\right) dx\, dy$$

$$= \zeta\, dx\, dy \ . \tag{6.54}$$

Similar contributions arise from other paths; thus the circulation around a finite closed circuit can be composed from such infinitesimal elements.

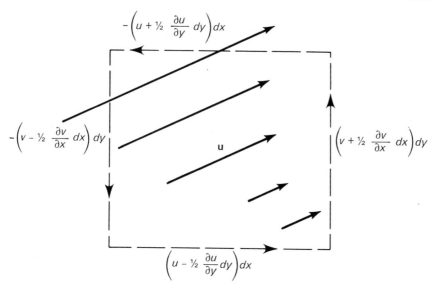

Fig. 6.13 Derivation of vorticity as the curl of a varying velocity field. Velocity components are expanded about the center of the element, and $\mathbf{u} \cdot d\mathbf{s}$ is summed around the circuit.

An important result from fluid dynamics known as Kelvin's circulation theorem relates the *time rate of change* of circulation around a material circuit (i.e., a closed curve that always encompasses the same fluid particles) to the convective acceleration of the fluid within it. For a small contribution to the circulation, $d\Gamma = \mathbf{u} \cdot d\mathbf{s}$, the convective rate of change is given by

$$\frac{D}{Dt}(d\Gamma) = \frac{D}{Dt}(u\,dx + v\,dy + w\,dz)$$

$$= \frac{Du}{Dt}\,dx + \frac{Dv}{Dt}\,dy + \frac{Dw}{Dt}\,dz + u\,du + v\,dv + w\,dw .$$

$$(6.55)$$

For any finite segment of the circuit having end points 1 and 2, the integral of Eq. 6.55 yields

$$\frac{D\Gamma_{12}}{Dt} = \frac{D}{Dt}\int_1^2 \mathbf{u} \cdot d\mathbf{s} = \int_1^2 \frac{D\mathbf{u}}{Dt} \cdot d\mathbf{s} + \frac{1}{2}\,\mathbf{u}^2 \Big|_1^2 . \qquad (6.56)$$

For a closed circuit, the right-hand term has identical values at the end points, a result that immediately gives:

$$\frac{D\Gamma}{Dt} = \oint \frac{D\mathbf{u}}{Dt} \cdot d\mathbf{s} \; . \tag{6.57}$$

The Navier–Stokes equations can next be used to substitute for $D\mathbf{u}/Dt$, and various results obtained. For the example of a homogeneous, inviscid fluid, the acceleration is derivable from a gradient of scalars via $D\mathbf{u}/Dt = -\nabla[(p/\rho) + \Phi]$, and therefore its integral around a closed curve vanishes. Then the substitution of this expression into Eq. 6.57 yields the important *Kelvin circulation theorem:*

$$\frac{D\Gamma}{Dt} = 0 \; , \quad \text{or} \quad \Gamma = \text{constant} \; . \tag{6.58}$$

Thus for a homogeneous, inviscid fluid, the circulation around any closed material circuit in the fluid remains constant as time goes on, and is a property of that parcel of fluid. An inviscid fluid that initially possesses no vorticity will remain vortex-free as time advances. In particular, when the motion of an inviscid fluid is generated from rest, it will necessarily remain irrotational motion in this idealized case. (For a stratified, rotating fluid, the results are more complicated and will not be given here. The result analogous to Eq. 6.58 is termed Ertel's theorem, a simplified form of which is given ahead by Eq. 6.72; see Pedlosky [1979].)

This theorem leads to the concept of *vortex tubes* frozen into the fluid. A vortex tube is a conceptual volume made up of all *vortex lines* passing through a particular material circuit, with the sides of the tube parallel to lines of vorticity, as Fig. 6.14 suggests. A vortex line or *vortex filament* is a material line of fluid particles that is everywhere in the direction of the vorticity vector. If, as time goes on, the circulation remains constant, then the vortex tube moves as a material volume in the fluid, preserving its mass and vortex flux, Γ. Now the divergence of the curl of any vector is identically zero, i.e.,

$$\nabla \cdot \boldsymbol{\zeta} = \nabla \cdot \nabla \times \mathbf{u} \equiv 0 \; , \tag{6.59}$$

so that the net vortex flux out of any volume is zero as well. From the divergence theorem, for any volume, V,

$$\iiint_V \nabla \cdot \boldsymbol{\zeta} \, d^3\mathbf{x} = \oiint_S \boldsymbol{\zeta} \cdot d\mathbf{A} = 0 \; , \tag{6.60}$$

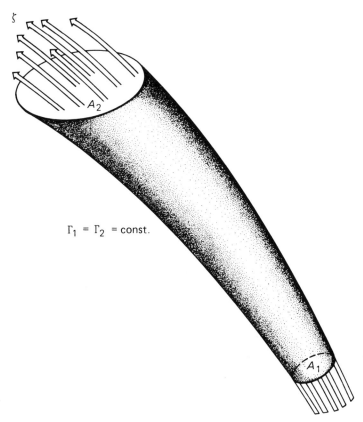

ζ

A_2

$\Gamma_1 = \Gamma_2 = \text{const.}$

A_1

Fig. 6.14 Conceptual view of a vortex tube made up of bundles of vortex filaments. In a homogeneous, inviscid fluid, the circulation, Γ, around a material path surrounding a vortex tube is preserved, and the vortex tube moves as a material volume.

where S is now the closed surface bounding V. If the volume is a vortex tube, the flux out of the sides is zero, by definition, so that the fluxes through the ends of the tube of areas A_1 and A_2 of Fig. 6.14 must therefore be equal. The circulation around any material circuit of a given vortex tube is equal to that for any other circuit, and the circulation then has a unique value for that tube. In this sense, the vorticity of a homogeneous, inviscid fluid is said to be frozen in the fluid.

Returning now to the discussion leading to Eq. 6.51: In the hydrostatic approximation, the perturbation pressure is purely due to pressure "head," or elevation, ξ, which causes both horizontal and vertical pressure gradients. As we have written several times,

$$\nabla p' = \rho g \nabla \xi , \tag{6.61}$$

where the density is assumed to be homogeneous in the vertical. In the absence of friction, the nonlinear momentum equation can be recast, using Eq. 6.51, into the form

$$\frac{\partial \mathbf{u}}{\partial t} + \nabla (\tfrac{1}{2} \mathbf{u}^2) + \zeta \times \mathbf{u} + 2\Omega \times \mathbf{u} = -g \nabla \xi . \tag{6.62}$$

By rearranging this slightly and writing it in terms of the horizontal components of the velocity and the vertical component of vorticity, ζ, we obtain

$$\frac{\partial u}{\partial t} - (f + \zeta)v = -\frac{\partial}{\partial x} [g\xi + \tfrac{1}{2}(u^2 + v^2)] \tag{6.63}$$

and

$$\frac{\partial v}{\partial t} + (f + \zeta)u = -\frac{\partial}{\partial y} [g\xi + \tfrac{1}{2}(u^2 + v^2)] . \tag{6.64}$$

Here the sum of the z components of the relative fluid vorticity and Coriolis frequency appear together in a combination called the *absolute vorticity,* i.e., the vorticity relative to a fixed frame of reference; this quantity is just twice the vertical component of the *total* fluid rotation rate seen by an observer viewing the earth from inertial space. Both ζ and f have the dimensions of reciprocal time, and because of its role in establishing the overall fluid angular momentum, it is natural to call f the *planetary vorticity* as well as the Coriolis frequency.

One of the advantages in dealing with the vorticity equation is that it does not contain the pressure, as we may see by operating with the vertical component of the curl on Eqs. 6.63 and 6.64, and subtracting one from the other; the result is

$$\frac{\partial}{\partial t} \left(\frac{\partial v}{\partial x} - \frac{\partial u}{\partial y} \right) + (f + \zeta) \left(\frac{\partial u}{\partial x} + \frac{\partial v}{\partial y} \right)$$

$$= - \left(u \frac{\partial}{\partial x} + v \frac{\partial}{\partial y} \right) \left(\frac{\partial v}{\partial x} - \frac{\partial u}{\partial y} \right) . \tag{6.65}$$

Bringing the right-hand side to the left and recalling the definition of the convective derivative (Eq. 6.51), we may write

$$\frac{D_h \zeta}{Dt} + (f + \zeta) \nabla_h \cdot \mathbf{u} = 0 , \tag{6.66}$$

which expresses the rate of change of relative vorticity in terms of the horizontal divergence of the velocity. One further modification can be made by replacing ζ with $f + \zeta$ in the convective derivative (assuming for now that f is constant), so that we obtain

$$\frac{1}{(f + \zeta)} \frac{D_h (f + \zeta)}{Dt} = - \nabla_h \cdot \mathbf{u} . \tag{6.67}$$

This result is termed the *vorticity equation,* and it relates the changes in absolute vorticity, $f + \zeta$, to the horizontal convergence, $- \nabla_h \cdot \mathbf{u}$. In words, Eq. 6.67 states that positive, cyclonic absolute vorticity is generated by horizontal *convergence of flow,* or that anticyclonic vorticity is generated by flow *divergence.* This is known as *vortex line stretching and shrinking,* and will be discussed further below, after the derivation of the vertically integrated continuity equation.

We next integrate the continuity equation for incompressible flow from the bottom at $z = - H$ to the elevated surface, ξ, taking advantage of the assumed barotropic, vertically constant character of the velocity; we also recall the surface kinematic boundary condition, Eq. 3.120:

$$\int_{-H}^{\xi} \left(\frac{\partial u}{\partial x} + \frac{\partial v}{\partial y} \right) dz = - \int_{-H}^{\xi} \frac{\partial w}{\partial z} dz = - w(\xi)$$

$$= - \frac{\partial \xi}{\partial t} - \mathbf{u} \cdot \nabla_h \xi . \tag{6.68}$$

By taking the constant water depth, H, inside the convective derivative, the continuity equation takes on a form symmetrical to the vorticity equation:

$$(H + \xi) \nabla_h \cdot \mathbf{u} = \left(\frac{\partial}{\partial t} + \mathbf{u} \cdot \nabla_h \right) (H + \xi) , \tag{6.69}$$

or

$$\frac{1}{(H + \xi)} \frac{D_h (H + \xi)}{Dt} = - \nabla_h \cdot \mathbf{u} , \tag{6.70}$$

where the horizontal convective derivative is

$$\frac{D_h}{Dt} \equiv \frac{\partial}{\partial t} + \mathbf{u} \cdot \nabla_h . \tag{6.71}$$

We next subtract Eq. 6.67 from Eq. 6.70 to eliminate the horizontal divergence and regroup terms to obtain the very important theorem of *conservation of potential vorticity*, Π, for homogeneous, shallow water:

$$\frac{D_h}{Dt} \left(\frac{f + \zeta}{H + \xi} \right) \equiv \frac{D_h \Pi}{Dt} = 0 , \tag{6.72}$$

or

$$\Pi = \frac{f + \zeta}{H + \xi} = \text{const.} \tag{6.73}$$

A more explicit rendition of this equation can be made by expanding upon the various expressions for the quantities that it contains, and allowing variations in water depth and Coriolis parameter:

$$\Pi = \frac{f_0 + \beta y + \dfrac{\partial v}{\partial x} - \dfrac{\partial u}{\partial y}}{H(x, y) + \xi(x, y)} . \tag{6.74}$$

The theorem asserts that the ratio of the quantities shown is a constant as the parcel moves about, in analogy to the constancy of the circulation, Γ, of Kelvin's theorem (Eq. 6.58). (We have not yet justified its application to the situation where the Coriolis frequency and depth are assumed variable, but that will be done below.)

Some discussion of Eqs. 6.72 and 6.73 is needed to explain their significance; this is assisted by Fig. 6.15, which shows four configurations of a time-varying fluid cylinder assumed to be rotating in a system having a constant planetary vorticity, f_0, in the Northern Hemisphere. In Figs. 3.15c and 3.15d, the fluid is assumed initially to be configured in the shallow, flat, shaded cylinder; as the water undergoes convergent flow, the vorticity equation (Eq. 6.67) states that positive vorticity is produced as vortex lines are stretched either upward or downward. For example, the extrusion could be due to a whirlpool, which stretches lines downward, or to a cyclone, which stretches them upward. This comes about because as the flow lines converge toward the center, they are deflected to the right by the Coriolis force, and the summation of these current filaments on the final, unshaded inner cylinder has the sense of cyclonic vorticity. In addition, the shallow water continuity relation (Eq. 6.70) states that horizontal convergence must be accompanied by increases in the overall fluid height, $H + \xi$, in order to con-

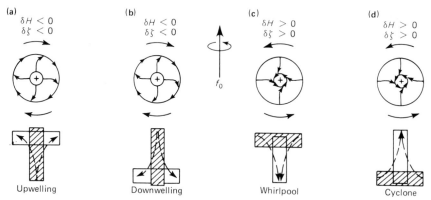

Fig. 6.15 Changes in potential vorticity from vortex line stretching and shrinking in the Northern Hemisphere. The initial state is shown hatched. (a) and (b) show shrinking during upwelling and downwelling, with loss of relative vorticity. (c) and (d) show vortex line stretching and increases in relative vorticity during vortex intensification. [Adapted from von Arx, W.S., *An Introduction to Physical Oceanography* (1962).]

serve fluid. This height increase is seen as going into a small change in surface elevation and a much larger increase in subsurface depression for the case of the whirlpool (Fig. 6.15c), and the converse for the cyclone (Fig. 6.15d). The combination of these two processes, as governed by Eq. 6.73, requires that a height increase be accompanied by vortex line stretching and compensating increases in relative vorticity, and that the ratio of the two quantities, $(f + \zeta)/(H + \xi)$, be constant as the cylinder moves about. All of this is a detailed but important way of saying that the total angular momentum of the parcel of fluid must stay constant in the absence of frictional viscous torques.

Figures 6.15a and 6.15b are two examples of the opposite case of horizontally divergent flow. In this case, the outward flowing water again is deflected to the right by the Coriolis force, but the summation around the new, larger, outer cylinder has the sense of negative, or anticyclonic, vorticity. The height must decrease, as must the relative vorticity, in order to maintain the potential vorticity constant. The two cases shown correspond to upwelling and downwelling, with the latter generally the case in the mid-ocean gyres.

The concept of potential vorticity conservation is of great importance in geophysical fluid motions, as two examples will show. The first is the process of *cyclogenesis,* or the generation of cyclonic vorticity in storms, which is shown in stylized form in Fig. 6.15d. As air flows horizontally inward and upward in a low pressure area due to ageostrophic forces (friction), its vortex lines stretch out, and it must increase its positive fluid vorticity in order to maintain the potential vorticity balance; thus it must spiral faster, becoming a lower pressure, higher velocity system in the process. The second example is the flow of ocean current from deep water up onto a shallower area; here the height and the vortex lines are compressed, requiring that the water column take on additional anticyclonic relative vorticity to conserve its potential vorticity. Such a case is shown in Fig. 6.15a. This occurs, for example, as the Gulf Stream flows north from Florida. Off Georgia it encounters a shoal area known as the Blake Plateau, at which it turns sharply to seaward, increasing its negative relative vorticity while decreasing its overall height. Many other examples can be cited, since the dynamics of large-scale motions on a rotating planet are governed to a major extent by this conservation theorem. The simplicity of Eq. 6.73 belies its power in the broad analysis of fluid motion on a rotating earth. We will return to it later in another, more local application of the equation in an example of western boundary currents and the Rossby radius of deformation.

From the discussions above, it is clear that a column of inviscid water will move so as to keep the ratio of its absolute vorticity to its total height a constant. Assuming for now that height changes do occur, the implications of Eq. 6.73 are that compensating changes in latitude and relative vorticity must follow. To systematize this notion, especially in the beta-plane formulation, we return to the nonlinear, frictionless shallow water equations, but now take note of the variation of f with latitude. Writing

$$\frac{\partial u}{\partial t} - (f_0 + \beta y + \zeta) v = - \frac{\partial}{\partial x} [g\xi + \tfrac{1}{2} (u^2 + v^2)] \qquad (6.75)$$

and

$$\frac{\partial v}{\partial t} + (f_0 + \beta y + \zeta) u = - \frac{\partial}{\partial y} [g\xi + \tfrac{1}{2} (u^2 + v^2)] , \qquad (6.76)$$

we take the curl of these relationships, as before, to obtain the equation for relative vorticity, which now includes another source of vorticity, βy, in addition to the horizontal divergence.

$$\frac{D_h \zeta}{Dt} = - (f_0 + \beta y + \zeta) \nabla_h \cdot \mathbf{u} - \beta v . \qquad (6.77)$$

However, this extra term may be transposed to the left-hand side and added to the constant, f_0, to give the total vorticity, $f_0 + \beta y + \zeta$, as the operand for the convective derivative. Thus the conservation theorem for potential vorticity (Eq. 6.73) remains valid for the case of variable planetary vorticity as well. The interpretation of this equation is that as a parcel of water moves

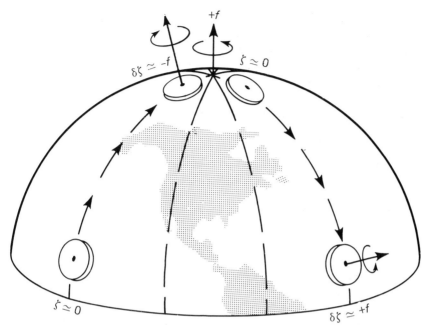

Fig. 6.16 Acquisition of relative vorticity, ζ, during meridional motions. A parcel of fluid near the equator having no initial absolute vorticity must acquire negative (clockwise) relative vorticity to offset the acquisition of positive planetary vorticity in going north. Similarly, a parcel near the pole having no relative but only planetary vorticity will exchange the latter for positive relative vorticity by the time it arrives near the equator.

north, say, it will acquire increased anticyclonic negative relative vorticity to offset the increase in the horizontal component of Coriolis force (or cyclonic, positive planetary vorticity). Conversely, a parcel that moves south will pick up cyclonic relative vorticity as it loses its planetary vorticity, keeping its absolute vorticity constant all the while. It is this vorticity balance that the beta effect attempts to model, at least over limited north-south excursions. Figure 6.16 illustrates how a parcel of water near the equator that has neither relative nor planetary vorticity will pick up anticyclonic relative vorticity as it moves poleward to offset the increasing cyclonic planetary vorticity to which it is subject. To an observer fixed in space, however, the parcel will have essentially zero absolute vorticity either at the equator or at the pole. It is easy to convince one's self of this by holding a pan of water with a floating toothpick in it, and turning while observing the toothpick; the water appears to rotate opposite to the observer's movement, but actually stays pointing in the same direction relative to the surroundings. Similarly, a parcel near the pole having no relative vorticity but only planetary vorticity, f, will acquire cyclonic relative spin as it moves equatorward to offset its loss of planetary vorticity. To the observer in inertial space, this parcel maintains f as its value all the while. The beta approximation describes this process for tangent plane motions.

6.5 Sverdrup Transport

To return to our study of the subtropical gyres: The accumulation of water in the center of the gyre and the downwelling velocity given by Eq. 6.48 bring about the situation shown in Fig. 6.15b, as well as that in Fig. 6.17. As the convergent flow deepens the central part of the gyre through downwelling, as well as elevating its surface slightly, the deeper water spreads out horizontally and shortens and compresses the vortex lines it brought with it, wanting to develop anticyclonic vorticity in the process. If the initial relative vorticity were ζ, the vorticity tendency upon sinking and spreading would be to drive it toward $\zeta - \delta\zeta$. However, here neither the depth, H, nor the surface setup, ξ, can change enough to offset the increased tendency toward anticyclonic fluid vorticity, and only one quantity can vary enough to provide the potential vorticity conservation required—the planetary vorticity, f, which must *decrease*. To lose planetary vorticity and regain its weakening relative vorticity, the water parcel must move equatorward. In Fig. 6.17, the parcel is shown as moving southward from where the planetary vorticity is $f_0 + \beta y$ to where it is $f_0 + \beta(y - \delta y)$, this expression derived from the beta plane approximation of Eq. 3.41. Thus we see that it must move equatorward in an *interior* flow in order to maintain its absolute vorticity a constant (barring depth changes) as it spreads and shrinks its vortex lines. For our Northern Hemisphere example, it must acquire a southward velocity, which we denote by $-v_I$. In the linear approximation (which is valid because this velocity is relatively small) the vorticity equation (Eq. 6.67) is

$$\frac{\partial \zeta}{\partial t} + f\nabla_h \cdot \mathbf{u} = -\beta v_I \,, \tag{6.78}$$

which, in the steady state, becomes

$$\beta v_I = f \frac{\partial w}{\partial z} \,, \tag{6.79}$$

where we have assumed no initial relative vorticity of the fluid. This is the equation that translates the downwelling divergence into vortex tube shrink-

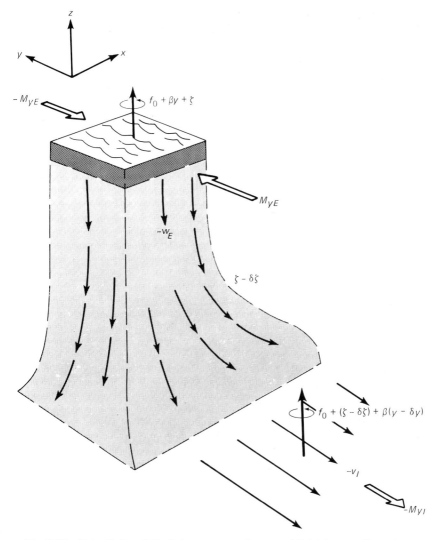

Fig. 6.17 Potential vorticity balance as an element of fluid downwells and moves equatorward. The surface convergence caused by wind stress curl brings about a subsurface divergence and vortex line compression. The tendency to lose relative vorticity, $\delta\zeta$, must be offset by the parcel moving south to exchange planetary for relative vorticity (cf. Fig. 6.16).

ing and a southward velocity, or the converse. This interior southward flow is illustrated as part of the main central flow in Fig. 2.21, and it occurs over a very large portion, but not all, of the subtropical gyre. We may obtain an estimate of the volume transport per unit width that it carries if we integrate Eq. 6.79 from the bottom of the surface Ekman layer to the top of the benthic layer (i.e., over the interior of the ocean). Under these conditions, the vertical velocity at the near-surface upper limit is $-w_E$, and that at the near-bottom limit is zero. This can be converted to a mass transport by multiplying by a depth-averaged density to obtain the relationship for the *interior Sverdrup transport*:

$$-M_{yl} = \frac{\rho f}{\beta} w_E = \frac{1}{\beta} (\nabla \times \tau)_z . \qquad (6.80)$$

Here we have used Eq. 6.48 for the wind stress curl. This equation is a simple yet important result; it states that in the interior of the gyre, in regions removed from the Ekman layers and the western boundary currents, the mass transport is established only by the horizontal derivatives of the surface wind stress, i.e., the curl. Of course, Ekman transport delivers the water to the interior, and potential vorticity conservation plus continuity determine where it goes, but it is the wind that is the causative agent. Figure 6.18 shows two graphs for the North Atlantic gyre that give an indication of the extent of this transport. The left graph shows contours of negative meridional (southward) transport existing over much of the gyre south of the Gulf Stream and north of approximately 20°N, indicating that the interior Sverdrup flow is a pervasive component of the gyre circulation. The graph on the right shows the measured dynamic topography of the same region (see "The Dynamic Method," in Section 6.6).

6.6 Western Boundary Currents and Geostrophic Flow

The southward Sverdrup flow must return to the north through some conduit in order to satisfy continuity, and it does so mainly by way of a boundary current along the western edge of the ocean basin. That a northward return flow should appear as a region of current localized along the edge of the continent is probably no surprise, but its intensity on the western boundary *is* surprising; there is no equivalent intensification on the eastern edge of the ocean, although there are weaker boundary currents there. The reasons for the western boundary currents are found in the requirement for an overall vorticity balance for the entire gyre, else it must either spin up or spin down under the influence of a nonzero basin-wide vorticity. (There exists an alternative and equivalent explanation in terms of the accumulation of slow Rossby waves on the western side of the oceans that will be given later.)

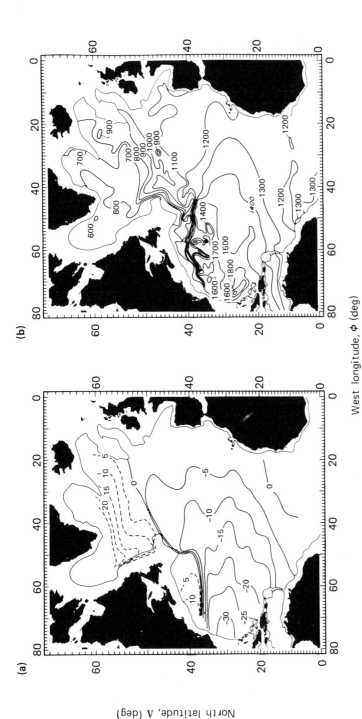

Fig. 6.18 (a) Contours of annual mean geostrophically balanced transport, $(M_{yI} - M_{yE})/\rho$, integrated along constant latitude lines from east to west, in units of 10^6 m^3 s^{-1}. Negative numbers indicate southward net transport. (b) Dynamic topography of the 100 dbar surface relative to the 1500 dbar surface, in units of millimeters. [From Leetmaa, A., and A. F. Bunker, *J. Mar. Res.* (1978).]

Before developing the vorticity argument further, it is worthwhile to look at the morphology of the gyre, with some special attention paid to the western edge. Figure 2.21 schematically shows the main pattern of the near-surface circulation, with (1) the eastward-directed current flowing under the influence of the westerlies; (2) the southward Sverdrup current; (3) the westward current in the region of the trades; and (4) the northward, concentrated return flow along the edge of the continent. Every major ocean basin has such a flow, but with considerable variation brought about by local geography and wind stress curl.

The vertical profiles of properties across the gyre are also of interest. Figure 6.19 shows schematically an idealized density section through the North Atlantic gyre that exhibits the deepening of the warm, light water in the interior brought about by the inward Ekman flows and, to some degree, by subsequent solar warming. Also apparent is the asymmetry in the vertical

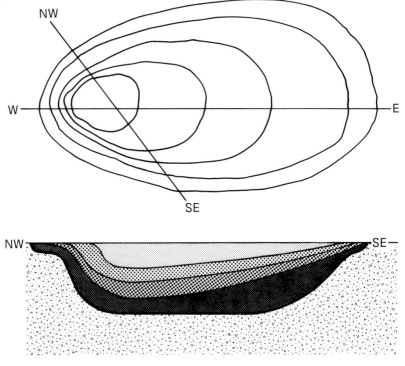

Fig. 6.19 Schematic density section of the North Atlantic gyre along the northwest-southeast direction. The deepening of the warmer, lighter water just inside the Gulf Stream is apparent. [Adapted from von Arx, W. S., *An Introduction to Physical Oceanography* (1962).]

profile of the lower density water, with the sharper gradients and deeper thermocline found on the west and north sides of the gyre, and more gradual gradients and shallow thermoclines to the south and east. Actual profiles of temperature and salinity are illustrated in Fig. 6.20, with the sections taken along the vertical lines on Fig. 6.21 marked as "slope" and "central." The slope thermocline is shallow, with its maximum at perhaps 100 m depth,

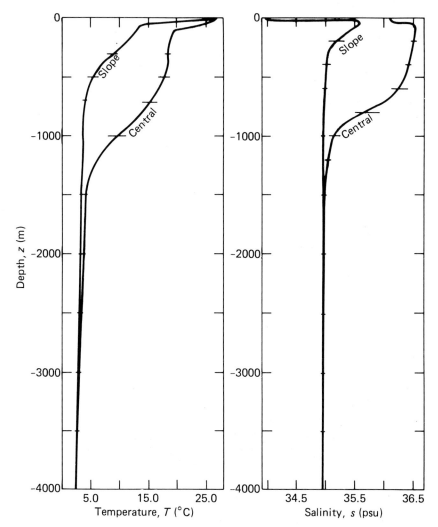

Fig. 6.20 Temperature and salinity profiles of the North Atlantic gyre taken on the continental slope and in the gyre center. Isotherm deepening in the center is due to downwelling. [Adapted from Stommel, H., *The Gulf Stream* (1966).]

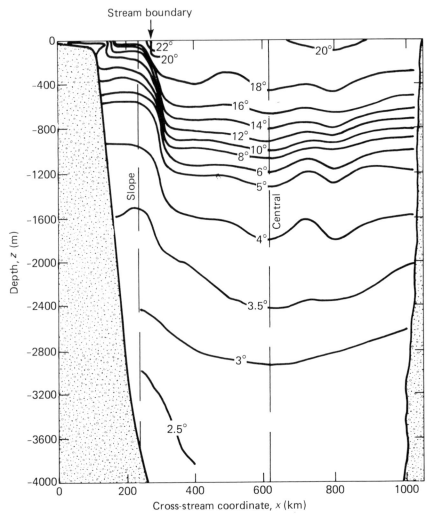

Fig. 6.21 Temperature section from the northeast of the U.S. to Bermuda. The steep slope of the isotherms indicates the northwest wall of the Gulf Stream. The stream boundary is indicated by the small arrrow. Vertical dashed lines indicate the approximate positions of the profiles in Fig. 6.20. [Adapted from Stommel, H., *The Gulf Stream* (1966).]

while the central section shows the main thermocline to be at approximately 800 m. The two sections are separated by a very sharp dip in the isotherm structure visible at the arrow marked "stream boundary" in Fig. 6.21. At the so-called *northwest wall* of the Gulf Stream, the discontinuities in properties are very large; it is, in fact, the oceanic analogue of an atmospheric frontal system. An operational definition of the edge of the Gulf Stream is the (x,y) locus of the 15°C isotherm at a depth of 200 m. Note that the slope of the isotherms persists down to at least 4000 m.

An examination of the baroclinic velocity structure that goes along with the temperature and density sections lends further credence to the discontinuous nature of the western boundary current. Figure 6.22 shows a schematic vertical east–west section through the Gulf Stream, with *isotachs* (contours of constant velocity) plotted. The core of the current is found near $x = 140$ km and has a speed in excess of 1.4 m s^{-1}. The boundary is found some 50 to 60 km farther west. There is a level of zero velocity whose depth varies with cross-stream distance, as given by the isotach labeled 0; there is also evidence of oppositely directed return currents both inshore and offshore of the main current system. A so-called *level of no motion* is assumed to

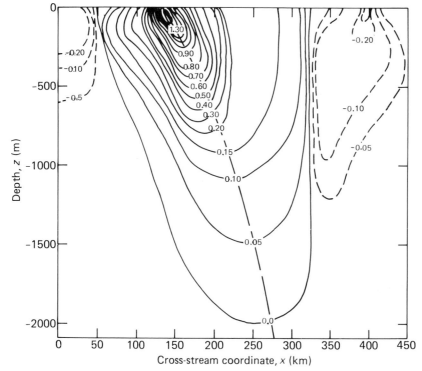

Fig. 6.22 Schematic of contours of constant northward velocity, or isotachs, through a western boundary current. Northward flow exists where contours are labeled with positive speeds, while southward flows are negative. Units are meters per second. [Adapted from Neumann, G., and W. J. Pierson, Jr., *Principles of Physical Oceanography* (1966).]

exist at a depth of approximately 2000 m, i.e., at a depth where the density field no longer slopes. It is worth noting that this section was obtained by the "dynamic method," wherein the slopes of density surfaces are converted to speeds through the use of the geostrophic and hydrostatic equations; this will be discussed below in more detail. The dynamic method remains a mainstay in the measurement of strong geostrophic currents, for it is still impossible to moor direct-measurement devices in the Gulf Stream when the instruments are at depths less than some 500 m, because of excessive drag on the mooring array from the current.

Figure 6.23 is a more schematic vertical section through a western boundary current. The geostrophic current, $\mathbf{u}_g$, is directed into the paper. *Isopycnals* (constant density surfaces) are denoted by dashed lines and parametrically labeled by ρ, and it is seen that these slope steeply across the front. A shorter dashed line joins the points where the isopycnals are ap-

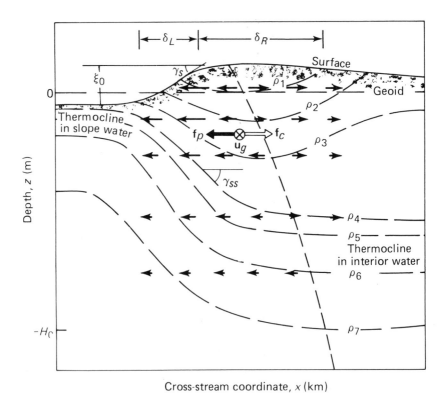

Cross-stream coordinate, x (km)

Fig. 6.23 Schematic of constant density surfaces and pressure gradient forces through a western boundary current in geostrophic balance. The flow is into the paper to the left of the zero-slope line, and out of the paper to the right of the line. [Adapted from Stommel, H., *The Gulf Stream* (1966).]

proximately level, that is, in coincidence with *isobars* (constant pressure surfaces); the latter are very nearly horizontal in the ocean, because pressure is mainly proportional to depth. The slope of the isopycnals with respect to the isobars provides the horizontal pressure gradient necessary to balance the Coriolis force, which is directed toward the right in Fig. 6.23. The isopycnals make an angle, γ_{ss}, of importance to the determination of the subsurface current. The maximum subsurface slope is greatly exaggerated in the diagram, and is approximately 5×10^{-3} for typical Gulf Stream conditions. Surface slopes are typically 10^{-5} in the strongest current regions. The pressure gradient forces are distributed more or less as shown by the short, dark arrows, whose lengths are proportional to the gradient; the force is seen to vanish along the curved dashed line where the constant density surfaces are level. Table 6.1 lists typical numerical values for some of the parameters in Figure 6.23.

TABLE 6.1
Typical Numerical Values For Gulf Stream Parameters

ξ_0	≈ 1 m, elevation across western boundary
δ_L	≈ 50 km, scale of front
δ_R	≈ 50 km, transverse scale of the southeastern portion of the Gulf Stream
H_{tc}	≈ -100 m, slope water thermocline depth
	≈ -800 m, interior water thermocline depth
H_0	≈ -2000 m, level of no motion
u_{max}	≈ 2 m s^{-1}, speed of velocity maximum
γ_s	$\approx 10^{-5}$, surface slopes of isopycnals
γ_{ss}	$\approx 5 \times 10^{-3}$, subsurface slopes of isopycnals
$M_y(\delta_L + \delta_R)/\rho$	$\approx 60 \times 10^6$ m^3 s^{-1}, volume transport of Gulf Stream

The geostrophic current in Fig. 6.23 is directed into the plane of the paper and has approximately the distribution shown in Fig. 6.22, including the sloping of its axis down and to the right. Its assumed level of no motion is taken as the depth where the density shows no slope, although, in general, it is not possible to define such a level uniquely. In fact, it is one of the unsolved problems of physical oceanography as to how to determine the *total* geostrophic current, both baroclinic and barotropic. Only the baroclinic component may be determined by the method of isopycnal slopes, by definition, which leaves an unknown barotropic component, uniform from top to bottom, to be measured by other methods. Satellite altimetry may provide such determinations for the surface velocity. The method uses measurements of the *surface setup*—or surface elevation, ξ, and *surface slope*, γ_s, measured with respect to the equipotential geoidal surface—to determine the surface geostrophic current. Both barotropic and baroclinic components are included in the observation, so that the assumption of a level of no motion no longer must be made. However, the subsurface flows are left undetermined by this method and must be found by other means, such as the dynamic method.

Geostrophic Currents

Having established the structure of a western boundary current, we will now outline the method by which observations of the density field and the surface slopes can lead to the velocity field. To do this, we recall the geostrophic equations, as well as the hydrostatic equation for the perturbation pressure given in terms of the surface elevation, ξ. These are:

$$-fv_g = -\frac{1}{\rho}\frac{\partial p'}{\partial x} = -g\frac{\partial \xi}{\partial x}\,, \tag{6.81}$$

$$fu_g = -\frac{1}{\rho}\frac{\partial p'}{\partial y} = -g\frac{\partial \xi}{\partial y}\,, \tag{6.82}$$

and

$$p' = \rho g \xi\,. \tag{6.83}$$

These equations are diagnostic point statements that hold anywhere within a geostrophically balanced flow; however, they are degenerate in that they do not allow the solution for a current system to be obtained, but only tell what the relationships between arbitrary currents and pressure gradients must be. Similar equations hold in meteorology, where they are known as the "thermal wind equations." At the sea surface, they take on forms that, when written in terms of surface slopes, $\tan \gamma_s$ (see Fig. 6.23), become

$$u_{gs} = \frac{-g}{f}\tan \gamma_{sy} = -\frac{g}{f}\frac{\partial \xi}{\partial y}\bigg|_s \tag{6.84}$$

and

$$v_{gs} = \frac{g}{f}\tan \gamma_{sx} = -\frac{g}{f}\frac{\partial \xi}{\partial x}\bigg|_s\,. \tag{6.85}$$

Thus the cross-stream slopes with respect to the geoid are measures of the surface geostrophic velocity components at right angles to the section. To the extent that they are truly geostrophic, these components are parallel to the isobars. For the Gulf Stream, the maximum surface slopes are of order 10^{-5}, but in other regions of geostrophic flow such as the interior of the subtropical gyres, they are easily two orders of magnitude smaller. Such slopes are very small and difficult to determine.

The Dynamic Method

At depths other than $z = 0$, measurements of the slopes of the density field via a sequence of vertical profiles made across a section can be converted to velocity determinations using the *dynamic method*. To derive these relationships, we differentiate the hydrostatic equation with respect to cross-stream direction, and the geostrophic equations with respect to depth; from this we obtain the *vertical shear* of the geostrophic current in terms of the horizontal slope of the density field:

$$\frac{\partial u_g}{\partial z} = -\frac{g}{f\rho}\frac{\partial \rho}{\partial y} \tag{6.86}$$

and

$$\frac{\partial v_g}{\partial z} = \frac{g}{f\rho}\frac{\partial \rho}{\partial x}. \tag{6.87}$$

(In Eqs. 6.86 and 6.87, we have neglected a term in the vertical density gradient that is proportional to N^2, which can be shown to be about 100 times smaller than the horizontal density derivative, except perhaps for the equatorial regions.) For example, by integrating Eq. 6.87 in the vertical from any desired depth, z', down to the level of no motion, $-H_0$, the vertical profile of velocity can be calculated from the density observations.

$$\int_{-H_0}^{z'} \frac{\partial v_g}{\partial z}\, dz = v_g(z') - v_g(-H_0) = \frac{g}{f}\int_{-H_0}^{z'} \frac{1}{\rho}\frac{\partial \rho}{\partial x}\, dz. \tag{6.88}$$

This can be evaluated via a series of density values, $\rho(z)$, computed from the equation of state (Eqs. 4.49), using salinity and temperature observations made as a function of depth. These are usually taken along a lengthy oceanographic *section*; each section is developed from a series of vertical *stations* at which the $s(z)$ and $T(z)$ measurements are made with a C–T–D, or a conductivity–temperature–depth instrument. Conductivity is converted to salinity via electrochemical relationships (e.g., the inverses of Eqs. 8.33 to 8.36; see the UNESCO reference in Chapter 4). The velocity section of Fig. 6.22 was obtained by this method, and the isotachs contoured by joining points of constant speed. This neglects the current, $v_g(-H_0)$, in Eq. 6.88, which is the integration "constant." H_0 is chosen so that the velocity at that depth is assumed to be zero; hence its name as the "level of no motion."

Gradient Flow

The method just presented suffices for computing rectilinear geostrophic currents. However, when the current system curves appreciably in the horizontal plane, more complicated expressions must be used that take into account the centripetal accelerations that are developed. Instead of the geostrophic equation, a generalization termed the gradient flow equation is used that includes the centripetal terms.

When flow lines are significantly curved, the centrifugal force exerted on the water parcel enters into the balance of forces and modifies the dynamics. In tightly curved currents, the magnitude of the centrifugal term may be comparable to (but generally less than) the Coriolis and pressure gradient terms. This effect can be introduced into the equations of motion most conveniently by transforming to a local coordinate system in which the fluid velocity is locally oriented parallel to, and the centrifugal term is normal to, an isobar. Figure 6.24 shows such a local system in which n is the normal direction and s is the tangential direction. The balance of forces for flow around either a high and low pressure system are:

$$\frac{\partial u}{\partial t} = -\frac{1}{\rho}\frac{\partial p}{\partial s} = 0 \tag{6.89}$$

and

$$\frac{u^2}{R} + fu = -\frac{1}{\rho}\frac{\partial p}{\partial n}, \tag{6.90}$$

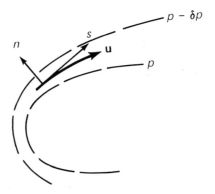

Fig. 6.24 Local coordinate system for derivation of the gradient flow equations.

where R is the local radius of curvature, and is positive for cyclonic and negative for anticyclonic flow. For motion purely parallel to an isobar, $\partial p/\partial s = 0$, and $p(s) = $ const. By solving Eq. 6.90 for u, we obtain the equation for *gradient flow:*

$$u = -\frac{fR}{2} \pm \left[\left(\frac{fR}{2}\right)^2 - \frac{R}{\rho}\frac{\partial p}{\partial n}\right]^{1/2}, \qquad (6.91)$$

where the geostrophic speed in this coordinate system is given by

$$u_g = -\frac{1}{\rho f}\frac{\partial p}{\partial n}. \qquad (6.92)$$

We substitute this into Eq. 6.91, and obtain a modified form of the gradient flow relationship involving u_g:

$$u = -\frac{fR}{2}\left[1 \pm \left(1 + \frac{4u_g}{fR}\right)^{1/2}\right]. \qquad (6.93)$$

The ratio of geostrophic to centripetal velocities, u_g/fR, is termed the *Rossby number,* Ro, and its smallness is a measure of the goodness of the geostrophic approximation. In general, Ro is defined as the ratio of centrifugal forces to Coriolis forces. For example, if L is a characteristic scale of the fluid motion, and U a characteristic speed, then the Rossby number is

$$\text{Ro} \equiv \frac{U^2/L}{f_0 U} = \frac{U}{f_0 L}. \qquad (6.94)$$

The balance of forces in gradient flow for Northern Hemisphere low and high pressure systems is shown in Fig. 6.25. While there are four possible solutions to Eq. 6.93, depending on combinations of the signs of $\partial p/\partial n$ and R as well as the $\pm$ sign, only the two solutions shown are both physically and mathematically sensible in geophysical fluids. Thus flow around an oceanic high-pressure feature represents a balance between the outwardly directed centrifugal and pressure gradient forces, and the inwardly directed Coriolis force. Around an oceanic low, the balance is between outward centrifugal and Coriolis forces, and the inward pressure gradient. In the Southern Hemisphere, the sense of the velocities is reversed about high and lows, but the balance of forces remains the same. In tightly curved current meanders, for example, the gradient flow equation must be used in the dynamic method in order to arrive at more nearly correct estimates of the velocity.

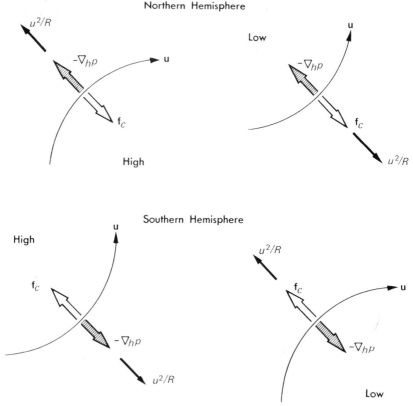

Fig. 6.25 Balance of forces in gradient flow around oceanic high and low pressure systems, for Northern and Southern Hemispheres. [From von Arx, W. S., *An Introduction to Physical Oceanography* [1962].]

6.7 Vorticity Balance in a Western Boundary Current

By the time that the wind-driven ocean currents have reached a steady state under the applied wind stress, the density and pressure fields must have adjusted themselves to a condition of dynamic balance. The winds for the North Atlantic, when averaged over time, have the general form of the streamlines shown in Fig. 6.26; they circulate clockwise in a very large, anticyclonic high pressure system having negative relative vorticity. Based on intuition, we would expect that the steady-state circulation in the gyre would have the same general form as the wind, that is, a more or less symmetric circulation, as shown by the solid lines in Fig. 6.26. We shall see that this intuition is incorrect, however, when the *variation* in the Coriolis force is introduced.

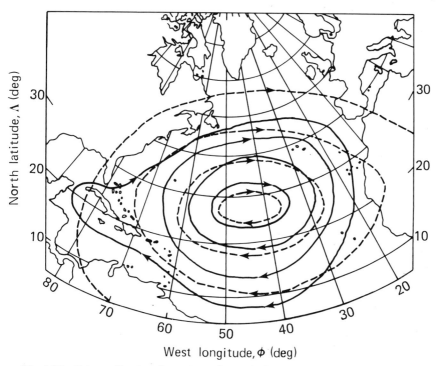

Fig. 6.26 Schematic of surface streamlines of wind stress (dashed lines) and current (solid lines), for a symmetrical ocean basin with a constant rotation rate. [From Stommel, H., *The Gulf Stream* (1966).]

The simplest model of the subtropical gyre demonstrating the very important phenonemon of western intensification of the current is due to Stommel. The gyre is assumed to be contained in a rectangular basin having horizontal boundaries at $(0, y_n)$ and vertical boundaries at $(x_e, 0)$, where x_e and y_n are the coordinates of the eastern and northern edges, respectively (see Fig. 6.27). The coordinate origin is at the southwest corner of the basin. In order to model the effects of the baroclinic structure in the actual ocean, the calculation uses an effective depth, H, of 200 m (see Eq. 5.183), which is much less than the actual depth. This value is chosen to make the long-wavelength baroclinic phase speed of Eq. 5.186b have the correct value. The north–south variation of zonal winds from westerlies to trade winds is described by a cosine distribution of stress, so that

$$\tau_x(y) = -\tau_0 \cos(\pi y / y_n) . \tag{6.95}$$

Thus along $y = 0$, the trade wind stress is directed toward the west and is

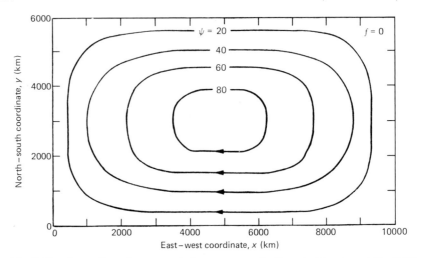

Fig. 6.27 Streamlines for a nonrotating ocean under the wind forcing of Fig. 6.26. Numerical values of ψ with units of 10^{-4} m^2 s^{-1} label the stream function. An ocean with constant rotation has streamlines that differ only slightly from this case. [From Stommel, H., *The Gulf Stream* (1966).]

negative, while at $y = y_n$, the stress due to the westerlies is toward the east. The basin is assumed to be subject to either a fixed or a variable Coriolis force. In the beta-plane approximation, the variable Coriolis frequency is

$$f = f_0 + \beta y \; ; \qquad (6.96)$$

however, the problem will be solved in three differing approximations: $f = 0$, $f = f_0$, and $f = f_0 + \beta y$, to illustrate the effects of rotation. Eddy viscosity is included as a frictional body force of the same type as was used in the problem of inertial oscillations, i.e., $-\gamma u$ and $-\gamma v$ (see Eq. 6.24). The linearized shallow water momentum and continuity equations are used, with the free surface at ξ and the effective depth at $-H$; also, it is assumed that u and v are independent of depth. In the steady state, those equations may then be written as

$$-fv = -g \frac{\partial \xi}{\partial x} - \gamma u - \frac{\tau_0}{\rho H} \cos (\pi y/y_n) \, , \qquad (6.97)$$

$$fu = -g \frac{\partial \xi}{\partial y} - \gamma v \, , \qquad (6.98)$$

and

$$\nabla_h \cdot \mathbf{u} = 0 \, . \qquad (6.99)$$

Thus we regard the wind as essentially a body force distributed over depth H, the magnitude of whose surface stress is τ_0. This parameterization is a highly simplified way of introducing the effects of direct wind forcing on subsurface layers.

The vorticity equation for this case can be obtained, as before, by cross-differentiation and subtraction to eliminate the pressure term, followed by the use of the continuity equation in the outcome. The result is the shallow water vorticity equation with a frictional term:

$$\beta v + \frac{\tau_0 \pi}{\rho H y_n} \sin (\pi y/y_n) + \gamma \left(\frac{\partial v}{\partial x} - \frac{\partial u}{\partial y} \right)$$

$$= \beta v + \frac{1}{\rho H} (\nabla \times \tau)_z + \gamma \zeta = 0 . \qquad (6.100)$$

A horizontal stream function, $\psi = \psi(x,y)$, is useful in this problem, with the velocity components being related to it via $u = \partial \psi / \partial y$ and $v = -\partial \psi / \partial x$. Upon substitution into the vorticity equations, an inhomogeneous partial differential equation for ψ is obtained, viz:

$$\nabla_h^2 \psi + \frac{\beta}{\gamma} \frac{\partial \psi}{\partial x} = \tilde{\omega} \sin (\pi y/y_n) , \qquad (6.101)$$

where the vorticity constant, $\tilde{\omega}$, may be found from

$$\tilde{\omega} = \tau_0 \pi / \rho H y_n \gamma , \qquad (6.102)$$

and measures the ratio of wind stress to damping force. The boundary conditions are that the shoreline of the ocean should be a streamline of the flow having zero value.

The complete solution to Eq. 6.101 is expected to be the sum of the homogeneous general solution and a particular solution proportional to the inhomogeneous term. The physics of the problem suggests trying a normal mode decomposition into sinusoids in the meridional coordinate, and hyperbolic functions in the zonal coordinate, with a separation of variables attempted:

$$\psi (x,y) = X(x) Y(y) + \psi_0 \sin (\pi y/y_n) , \qquad (6.103)$$

where the trial separation functions $X(x)$ and $Y(y)$ are chosen to be

$$X(x) = \sum_j [p_j \exp(K_{pj}x) + q_j \exp(K_{qj}x)] \qquad (6.104)$$

and

$$Y(y) = \sum_j (c_j \sin k_j y + d_j \cos k_j y) . \qquad (6.105)$$

The right-hand term in Eq. 6.103 is the particular solution to the inhomogeneous equation. The constants K_{pj} and K_{qj} are reciprocal lengths for the x dimension that are set by the combinations

$$\left.\begin{array}{c} K_{pj} \\ K_{qj} \end{array}\right\} = -\frac{\beta}{2\gamma} \pm \left[\left(\frac{\beta}{2\gamma}\right)^2 + k_j^2\right]^{\frac{1}{2}}, \qquad (6.106)$$

where the k_j's are quantized meridional wave numbers.

When the boundary conditions are imposed, it is found that the solution for the gravest (lowest) mode, $k_1 = \pi/y_n$, may be written as

$$\psi(x,y) = (\tilde{\omega}/k_1^2) (\sin k_1 y) (pe^{Ax} + qe^{Bx} - 1), \qquad (6.107)$$

where the subscripts, j, have been dropped. The constants p and q turn out to be:

$$p = \frac{1 - e^{Bx_e}}{e^{Ax_e} - e^{Bx_e}} \qquad (6.108a)$$

and

$$q = 1 - p . \qquad (6.108b)$$

It is clear that these solutions are not normal mode oscillations of a box (the problem is time-independent), but something else; they are probably best visualized with the aid of numerical solutions. Stommel used the following numerical values to apply the solution to the North Atlantic:

$$x_e = 10^4 \text{ km}$$

$$y_n = 2\pi \times 10^3 \text{ km}$$

$$H = 0.2 \text{ km}$$

$$\tau_0 = 0.1 \text{ N m}^{-2} .$$

The friction coefficient, γ, is an adjustable parameter of the problem, with a value of 10^{-6} s^{-1} giving best agreement with the observed velocities. This corresponds to a damping time, $1/\gamma$, of 11.6 days, a not unreasonable value.

Figures 6.27 to 6.29 show numerical results for the cases of $f = 0$ and

$f = f_0$. Without rotation ($f = 0$), the streamlines do indeed mimic the wind stress and a symmetric flow results, as Fig. 6.27 shows. The surface heights, ξ, are given in Fig. 6.28 and exhibit *setup* of some 0.08 m in the northeast and southwest sectors, and a similarly sized *setdown* in the opposite quadrants. This is caused by the wind stress accumulating water along the coastlines where it acts onshore, and depleting it where it acts offshore. Next, the basin is allowed to rotate at a constant value of f_0 appropriate to the middle latitudes, with the result obtained for surface elevations shown in Fig. 6.29; the flowlines (not shown) are barely distinguishable from the $f = 0$ case. Although a basin-wide anticyclonic gyre with a broad scale setup of perhaps 1.50 m has evolved, there is importantly no evidence of western boundary current development or other major asymmetry visible in the model.

The source of this failure of the theory using a constant rotation rate is to be found in the need for an overall vorticity balance in the gyre that was mentioned at the beginning of this section. The variation of Coriolis force in the meridional direction, as described by the beta effect, introduces a significant source of negative vorticity (of relative magnitude -1.0, say, when measured in terms of the wind-induced vorticity, $\tau_0 \pi/\rho H u$) along the western boundary, and a positive source ($+1.0$) along the eastern boundary (cf. Fig. 6.16). Next, an attempt to arrive at an overall vorticity balance for the basin will be made, first by assuming symmetrical and then asymmetrical currents; the total vorticity along each boundary will be normalized to that of the wind stress. Table 6.2 summarizes the three sources of vorticity in

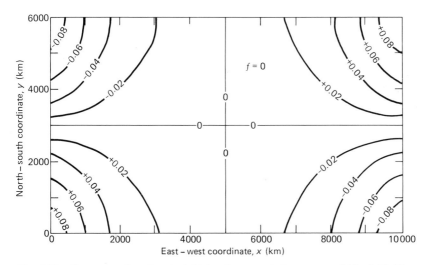

Fig. 6.28 Contours of surface height for the nonrotating ocean of Fig. 6.27. Elevations are in meters. [From Stommel, H., *The Gulf Stream* (1966).]

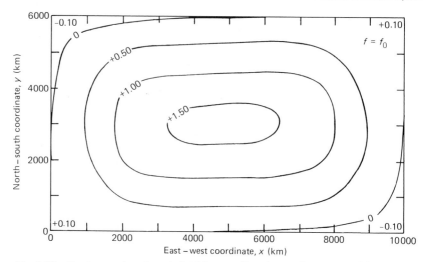

Fig. 6.29 Contours of surface height for a uniformly rotating ocean with $f_0 = 0.25 \times 10^{-4}$ s^{-1}. Streamlines are similar to those of Fig. 6.27, but a symmetrical surface setup in excess of 1.50 m is established that counterbalances the Coriolis force. [From Stommel, H., *The Gulf Stream* (1966).]

the symmetrical problem: wind stress curl, eddy viscosity, and the beta effect, as well as their relative magnitudes.

TABLE 6.2

Relative Vorticities for Symmetrical Circulation

Source	Western Current	Eastern Current
Wind Stress	− 1.0	− 1.0
Eddy Viscosity	+ 0.1	+ 0.1
Variation in f	− 1.0	+ 1.0
Total	**− 1.9**	**+ 0.1**

From Table 6.2, it is seen that there is a lack of balance in the total vorticity between the eastern and the western boundaries caused by the fact that the wind puts negative vorticity into the ocean along both boundaries that cannot be offset by equal current strengths there. If it existed, this system would by necessity undergo accelerations and its vorticity distribution would change toward one that is in closer balance.

If we now let the current become asymmetrical, as is actually observed, and again attempt to do a vorticity balance, we will find that there is perhaps 9 times the amount of negative vorticity in the west as exists in the east because of the concentration of current there, but that there is also about 100 times as much frictionally induced cyclonic vorticity in the west as well, because of the narrowness of the current and the large shear there. Table 6.3 summarizes this distribution, and indicates that while there is very little adjustment to the vorticity balance required in the area of the eastern boundary current, in the west there is a large loss (-9.0) in relative vorticity associated with the beta effect acting on the large currents; these, in turn, are accompanied by a hundredfold increase in cyclonic vorticity in the west, due to the eddy viscosity ($+10$).

TABLE 6.3

Relative Vorticities for Asymmetrical
Circulation

Source	Western Current	Eastern Current
Wind Stress	-1.0	-1.0
Eddy Viscosity	$+10.0$	$+0.1$
Variation in f	-9.0	$+0.9$
Total	**0.0**	**0.0**

Thus an overall balance is possible if friction (i.e., eddy viscosity) can produce the required vorticity. It can do so if the velocity shear is large enough, which requires that the current be both strong and narrow. Under these conditions the current sheds cyclonic eddies along its northwest boundary in copious amounts as it traverses its circuit. Thus western intensification can be understood in terms of an overall balance in the vortex production in the gyre. The alternative view in terms of Rossby waves accumulating along the western boundary will be given in Section 6.10.

This analysis can be extended to cyclonic and anticyclonic wind systems in both hemispheres with the same result: western boundary currents with large frictional losses along their inshore edges are dynamic requirements in the global ocean, and every ocean has such a system, as a glance at Fig. 2.16 tells.

To further quantify this view, when the Stommel model is used with $f = f_0 + \beta y$, it produces the results shown in Figs. 6.30 and 6.31, which are to be compared with the more realistic oceanic streamline configuration shown

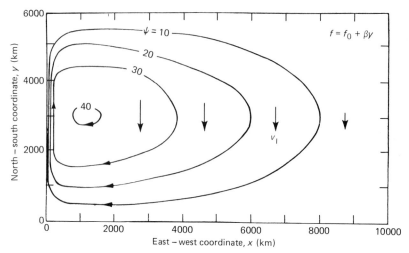

Fig. 6.30 Streamlines for a rotating ocean with a linearly varying Coriolis parameter. A strong western boundary current has developed in the process of reaching an overall vorticity balance. [From Stommel, H., *The Gulf Stream* (1966).]

in Fig. 6.32. Now the streamlines are greatly concentrated on the western side, and the contours of surface elevation are at a maximum in the model "Sargasso Sea." Thus this model contains the rudiments of a real oceanic gyre and clearly warrants further development and refinement.

A slightly more realistic model of gyre circulation for the Pacific by Munk

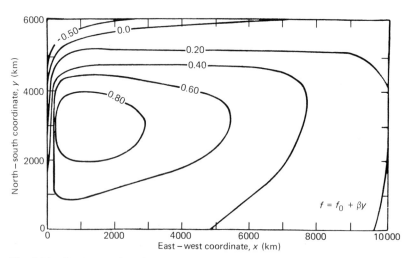

Fig. 6.31 Contours of surface height for the flow shown in Fig. 6.30. Units are in meters above the equipotential surface. [From Stommel, H., *The Gulf Stream* (1966).]

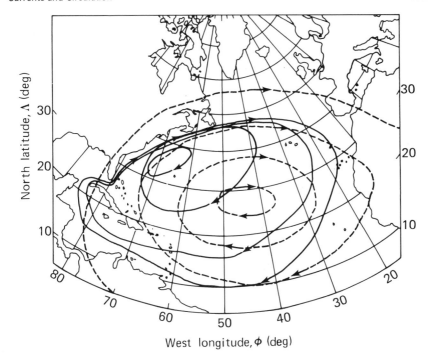

North latitude, Λ (deg)

West longitude, Φ (deg)

Fig. 6.32 Schematic of surface streamlines of wind stress (dashed lines) and current (solid lines) for an asymmetric ocean basin with variable Coriolis parameter. [From Stommel, H., *The Gulf Stream* (1966).]

makes use of observed wind stress and also incorporates an eddy viscosity term, but continues to neglect nonlinearities. Again the basin is taken as rectangular, and the equations used are the shallow water approximations. The steady-state momentum equations are

$$-\rho f v = -\frac{\partial p}{\partial x} + A_h \nabla_h^2 u + \frac{\partial}{\partial z}\left(A_v \frac{\partial u}{\partial z}\right) \qquad (6.109)$$

and

$$\rho f u = -\frac{\partial p}{\partial y} + A_h \nabla_h^2 v + \frac{\partial}{\partial z}\left(A_v \frac{\partial v}{\partial z}\right). \qquad (6.110)$$

These equations represent a geostrophic flow that is frictionally eroded by horizontal eddy viscosity, but replenished by wind momentum acting at the surface. They are next integrated over depth and the vertical shear stress term in A_v evaluated in terms of the surface wind stress, τ_x and τ_y. The integrated pressure terms are eliminated by cross-differentiation, and a fourth-order linear equation for the *transport stream function,* ψ_M, is obtained. The cross-differentiation also introduces the curl of the wind stress, as before. The equation for the stream function so obtained is:

$$\left[A_h \left(\frac{\partial^4}{\partial x^4} + 2\frac{\partial^4}{\partial x^2 \, \partial y^2} + \frac{\partial^4}{\partial y^4}\right) - \beta \frac{\partial}{\partial x}\right]\psi_M = (\nabla \times \tau)_z . \quad (6.111)$$

In Eq. 6.111, the transport stream function is related to the velocity stream function in the same fashion as are their parent quantities; thus

$$M_x = \int_{-H}^{\xi} \rho u \, dz = -\frac{\partial \psi_M}{\partial y} = -\int_{-H}^{\xi} \rho \frac{\partial \psi}{\partial y} dz \quad (6.112)$$

and

$$M_y = \int_{-H}^{\xi} \rho v \, dz = \frac{\partial \psi_M}{\partial x} = \int_{-H}^{\xi} \rho \frac{\partial \psi}{\partial x} \, dz . \quad (6.113)$$

The partial derivatives of the transport stream function (with units of kilograms per meter-second), represent the mass transport rates per unit horizontal distance, and their division by ρ represent the *volume* transport rates per meter. With the sign convention given, $\psi_M(x,y)$ can be interpreted as the vertically and zonally integrated northward transport rate between the western boundary at $x = 0$ and some interior point, x. The steady-state solution of Eq. 6.111 can most readily be appreciated by presenting the formulas for the surface elevation, $\xi(x,y)$, and the transport function, $\psi_M(x,y)$, for the region near the western boundary, and by studying the graphical representation of the solution. Munk's solutions for the western sector are:

$$\psi_M(x,y) = \psi_0 x_e \, X(x)\frac{1}{\beta} \frac{\partial \tau_x}{\partial y} , \quad (6.114)$$

where the surface elevation, $\xi_w(x,y)$ is related to $X(x)$ by

$$\xi_w(x,y) = \xi_0(y) \, X(x)$$

$$= \xi_0(y)\left[1 - \frac{2}{\sqrt{3}} \, e^{-kx/2} \cos\left(\frac{\sqrt{3}}{2}kx - \frac{\pi}{6}\right)\right] , \quad (6.115)$$

and where $\xi_0(y)$ is the meridional variation in the surface elevation along the western boundary, which must be specified through other sources. The separation function in the square brackets, $X(x)$, gives the east–west variation of both transport stream function and surface elevation. The important reciprocal length scale, k, gives the characteristic dimension of the boundary current on the left-hand, inshore side:

$$k \equiv \left(\frac{\rho \beta}{A_h} \right)^{1/3}. \tag{6.116}$$

Clearly the spatial scale, $1/k$, is set by the horizontal eddy viscosity, A_h, and this is an adjustable parameter of the theory. The best agreement with the observed dimensions (see Table 6.1, for example) is obtained for $A_h / \rho = 5 \times 10^3$ m^2 s^{-1}, which indicates the presence of a very sizable frictional dissipation. In this event, $1/k \simeq 60$ km, a reasonable estimate of the scale of the Kuroshio. This dissipation is to be identified with the eddy and ring generation along the intense portions of the boundary currents and is also a global property of such systems.

Figure 6.33 shows Munk's solution and gives plots of the separation function, $X(x)$, for the transport, and its normalized derivative,

$$\frac{1}{k} \frac{dX}{dx} = \frac{X'}{k} \simeq \frac{2}{\sqrt{3}} e^{-kx/2} \sin \left(\frac{\sqrt{3}}{2} kx \right), \tag{6.117}$$

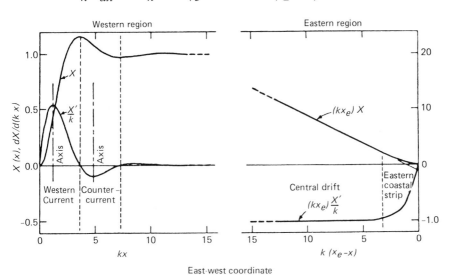

Fig. 6.33 Plot of zonal variation in the transport stream function, as given by $X(x)$ (which is also proportional to surface elevation), and north–south transport velocity, proportional to $dX/d(kx)$. The left-hand side is for the western boundary region, and the right-hand side is for the eastern boundary current. [From Munk, W. H., *J. Meteorol.* (1950).]

which is proportional to the north–south transport velocity. The solution shown is for both the western and eastern boundary currents, with a change of scale made on the vertical axis. The western current distribution has a characteristic dimension of order $1/k$ and a full width of perhaps 4 times that. Immediately to its east, the existence of a southward countercurrent is suggested by the negative portion of X', which amounts to perhaps 20% of the strength of the main, northward flow. Still further to the east, the velocity oscillates weakly and becomes essentially constant in the region of the central drift, which is where the Sverdrup flow occurs (note the portion termed "central drift" on the right-hand side of Fig. 6.33). A small eastern boundary current brings the velocity there to zero, to complete the flow.

Figure 6.34 shows the solution for streamlines and transport function, with the wind stress and its north–south derivative plotted on the left. In the background is a stippled-in Pacific coast line shown for purposes of orientation. With only the observed annual wind stress as the driving force, the model reproduces the major Pacific gyres in a surprisingly realistic way. The main

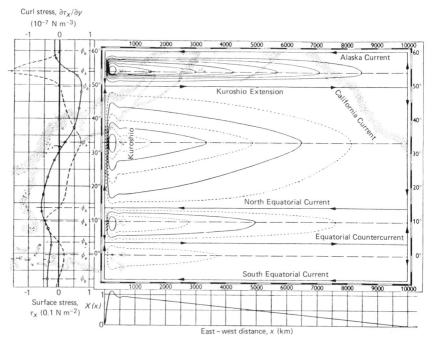

Fig. 6.34 Transport streamlines for a rectangular ocean basin approximating the North and Central Pacific. The annual mean zonal wind stress, τ_x, and its curl, $(\nabla \times \tau)_z$, are plotted on the left. At the bottom is the transport separation function, $X(x)$. Volume transport between adjacent lines is $10^7 \text{ m}^3 \text{ s}^{-1}$. [From Munk, W. H., *J. Meteorol.* (1950).]

subtropical circulation is accompanied by the subpolar gyre to the north and by the more complex equatorial current systems to the south. In all of the gyres, there is a small but significant *return circulation* just to the east of the boundary current, in the vicinity of the inflection points of the streamlines. Such return circulation is also found in observations and in models of the Gulf Stream. There is considerable recirculating flow inferred from the available data, as Fig. 6.35 indicates for the deep circulation; this figure shows the empirical stream function for deep water having potential temperature, θ, less than 4°C, and illustrates the recirculation region southeast of the mouth of the St. Lawrence River.

While the Pacific Ocean model is successful in many respects, it underpredicts the total Kuroshio transport by perhaps a factor of 2, and somewhat overpredicts the scale of the northwest front due to the value of the friction coefficient used. Nevertheless, it appears to contain the essential physics of the steady-state problem, with the possible exception of the neglected

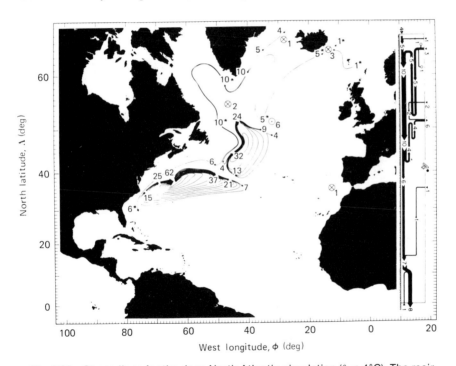

Fig. 6.35 Streamlines for the deep North Atlantic circulation ($\theta < 4°C$). The recirculation region southeast of St. Lawrence carries of order 60×10^6 m^3 s^{-1} offshore of the main Gulf Stream. Water descends from the lower thermocline in regions marked with $\otimes$, and ascends where indicated by $\odot$. [From Worthington, L. V., *On the North Atlantic Circulation* (1976).]

nonlinear terms. Further improvements have been made on this basic calculation and more realistic results have been obtained by others, some of which will be discussed ahead briefly.

The same type of calculation can be applied to an entire ocean, using the winds as input, with a schematized streamline configuration for a basin having the general topology of the Pacific being shown in Fig. 6.36. With the exception of the Antarctic Circumpolar Current, which is connected around the world, this gyre configuration is essentially correct in its broad outline. From such considerations, one can build up a mathematical description of

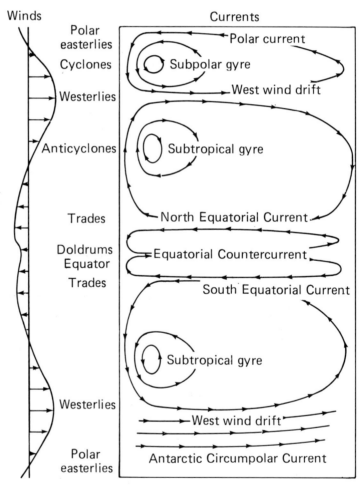

Fig. 6.36 Schematic configuration of current gyres in the Pacific as driven by wind stress curl.

the general oceanic circulation that produces a reasonable picture. It possesses a major fault, however, in that it lacks the subsurface baroclinic variability that actually exists in the ocean, due to its use of simple, barotropic, vertically integrated equations.

6.8 The Rossby Radius and the Western Boundary Current

While eddy friction seems to establish the cross-stream dimension on the western side of a boundary current (see Eq. 6.116), that on the eastern side appears to be more nearly set by conservation of potential vorticity. The horizontal scale that characterizes that side (and, in fact, which establishes dimensions of the boundary of *any* oceanic flow that is strongly influenced by rotation), is the *Rossby radius of deformation*, or more simply, the Rossby radius. There are actually two types of Rossby radii, one for barotropic flow and one for baroclinic flow; since the latter generally involves a number of internal modes such as were discussed in Section 5.16, there will be a Rossby radius for each mode. We will now derive a simple equation for the Gulf Stream velocity profile on the Sargasso Sea side that produces this parameter.

Consider a cross-stream section in a two-layer configuration much like the one in Fig. 6.23, shown schematically in Fig. 6.37. The upper layer is assumed to be of variable height, $\xi(x)$, while the lower layer is quiescent and very deep. We apply Eqs. 6.81 and 6.82 to the wedge of water of constant density, ρ_1, located between the surface and the second underlying layer of

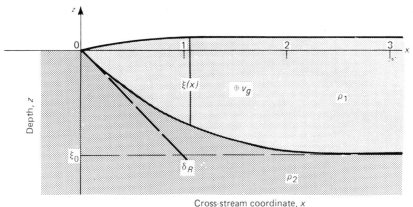

Fig. 6.37 Schematic cross section of a western boundary current flowing toward the north. The Rossby radius, δ_R, establishes the scale of the eastern velocity gradient.

constant density, ρ_2. We use the geostrophic and potential vorticity equations in this geometry, taking into account the fact that the vertical restoring force along the interface is actually the reduced or effective gravity, g', where, in this case,

$$g' = \frac{\rho_2 - \rho_1}{\frac{1}{2}(\rho_2 + \rho_1)} g . \tag{6.118}$$

Such a configuration is termed a *reduced gravity model*, and is essentially the simplest baroclinic model known that is also capable of yielding reasonable physics (see Section 5.17).

The geostrophic equation for this surface density slab is

$$-fv_g = -\frac{1}{\rho}\frac{\partial p'}{\partial x} = -g'\frac{\partial \xi}{\partial x} , \tag{6.119}$$

while the conservation theorem for potential vorticity reads

$$\Pi = \frac{f + \partial v_g/\partial x}{\xi(x)} = \frac{f}{\xi_0} , \tag{6.120}$$

where $\xi(x)$ is the thickness of the isopycnal upper layer at any point along the x coordinate, and ξ_0 is its thickness well to the east, where the density surface is level. Solving for the velocity shear in Eq. 6.120, we obtain

$$\frac{\partial v_g}{\partial x} = \frac{f}{\xi_0}[\xi(x) - \xi_0] . \tag{6.121}$$

Now we differentiate Eq. 6.119 with respect to x and substitute the resultant second derivative of isopycnal height into Eq. 6.121:

$$\frac{\partial^2 \xi}{\partial x^2} = \frac{f^2}{g'\xi_0}(\xi - \xi_0) \equiv \frac{1}{\delta_R^2}(\xi - \xi_0) . \tag{6.122}$$

A solution to this differential equation that satisfies the boundary condition of $v_g = 0$ at $x = +\infty$ is:

$$\xi(x) = \xi_0[1 - \exp(-x/\delta_R)] . \tag{6.123}$$

The geostrophic speed profile is

$$v_g(x) = v_0 \exp(-x/\delta_R) , \tag{6.124}$$

where the parameter v_0 is an integration constant. We term the e folding distance, δ_R, the *baroclinic Rossby radius of deformation*, since it sets the horizontal scale for the Gulf Stream. Its definition is given by

$$\delta_R \equiv (1/f)(g'\xi_0)^{1/2} .\qquad (6.125)$$

Thus it can be seen that δ_R is the ratio of the speed of a shallow water internal gravity mode, $(g'\xi_0)^{1/2}$, in a layer of thickness, ξ_0, and reduced gravity, g', to the Coriolis frequency. To calibrate the integration constants and other parameters of the equations, we invoke the known total transport of the Gulf Stream, and use it with other reasonable numerical values to evaluate the Rossby radius. Now the total volume transport, V, is computed from

$$V = \int_0^\infty v_g(x)\,\xi(x)\,dx = \tfrac{1}{2}\,\frac{g'}{f}\,\xi_0^2 .\qquad (6.126)$$

If we use numbers characteristic of the Gulf Stream off Cape Hatteras, then with $\xi_0 = 800$ m, $g' = 2 \times 10^{-3}\,g$, $v_0 = 4.0$ m s^{-1}, and $V = 64$ Sv, we find that $\delta_R = 40$ km. Figure 6.38 illustrates the exponentially decaying cross-stream speed profile and compares it with one obtained from the dynamic method. Although the model is very simple, the agreement with observed quantities is reasonable, suggesting that the interior region of the Gulf Stream has fairly uniform potential vorticity. However, this method must fail on the outer, western wall, since there is considerable production of vor-

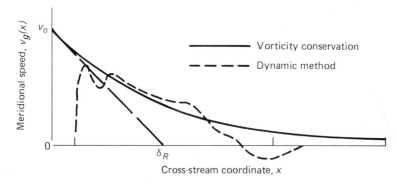

Fig. 6.38 Solution for the velocity profile across the Gulf Stream compared with a profile obtained from the dynamic method. Comparison suggests that potential vorticity is conserved on southeast side of the current. [Adapted from Stommel, H., *The Gulf Stream* (1966).]

ticity there from eddy viscosity, as was discussed in the previous section. The calculation is thus terminated before $x = 0$.

The value of the Rossby radius found here is fairly typical for mid-latitudes. However, as the equator is approached, f tends to zero and the radius as defined becomes very large. It appears, nevertheless, that geostrophic balances hold to within 1 to 2° of the equator, and perhaps may even bridge the singularity there in some fashion. Elsewhere in the ocean, the Rossby radius is the dimension over which large-scale dynamics (either barotropic or baroclinic) come into play, and it approximately divides the mesoscale motions from sub-mesoscale ones.

6.9 Meanderings and Other Mesoscale Motions

The major current systems of the ocean are actually highly variable, and as improved observational techniques have been developed, the extent of this variability has become more acutely appreciated. Since the Gulf Stream is probably the best studied system in this regard (as well as in others) we will continue to center the discussion on it; the concepts, however, are applicable to almost all other energetic current systems.

It is something of an axiom that in geophysical fluids, the variance in the motion of the system (where variance is used in the technical, statistical sense), is of the order of the square of the mean. The diffusive mass transport by eddies, for example, may exceed that of the steady mean current in some situations. As a case of well-documented meandering motion, the variations in the position of the northwest wall of the Gulf Stream over an eight month period are shown in Fig. 6.39. The system shows propagating, amplifying waves that are apparently driven to instablity by the shear-flow process of baroclinic instability. Over some interval downstream of Cape Hatteras, the meandering results in detachments of entire rings, or loops of Gulf Stream water, to both the north and south, which then live independent existences for many months.

A composite of the thermal fronts of a major portion of the entire western boundary current is shown in Fig. 6.40, and includes (1) the Gulf of Mexico Loop Current, which is a highly variable anticyclonic system that occasionally detaches from the main flow; (2) the Florida Current, which is a relatively stable jet whose near-surface flow occasionally exceeds 3 m s^{-1}, and (3) the Gulf Stream, which is mildly unstable off Georgia and which becomes strongly unstable after leaving the continental land mass at Cape Hatteras. This latter region is the one in which the data of Fig. 6.39 were taken (which meanders are reproduced in Fig. 6.40). Only where the current is essentially pinned against the land boundary is it relatively stable.

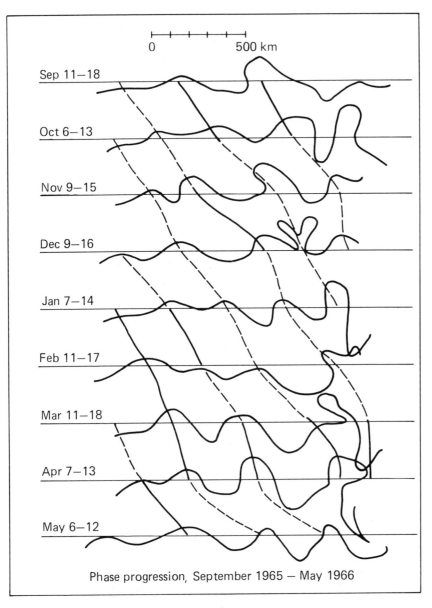

Sep 11–18

Oct 6–13

Nov 9–15

Dec 9–16

Jan 7–14

Feb 11–17

Mar 11–18

Apr 7–13

May 6–12

0   500 km

Phase progression, September 1965 – May 1966

Fig. 6.39 Time and space evolution of meanders in the northwest wall of the Gulf Stream over eight months. Lines define locus of 15°C isotherm at −200 m depth (see Fig. 6.21). Dashed and solid lines show phase propagation of major oscillations [From Hansen, D. V., *Deep-Sea Res.* (1970).]

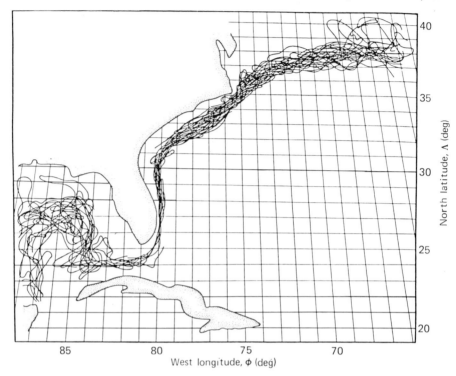

Fig. 6.40   Composite view of the boundary of the Gulf Stream system built out of combined ship and satellite infrared observations. Variability shown occured over nine months. [From Maul, G., et al., *J. Geophys. Res.* (1978).]

Such wide-scale views as Fig. 6.40 have been greatly aided by the availability of satellite images made with sensors that detect the heat radiation given off by warm bodies, using infrared wavelengths near 10 to 12 μm, where the earth's atmosphere is relatively transparent (see Section 8.10). For example, Fig. 6.41 is an image of the Gulf Stream east of Cape Hatteras, made with a satellite sensor sensitive to thermal infrared heat radiation, which shows spatially growing waves and detached eddies both to the north and south of the Gulf Stream. The eddies are termed *rings* when they enclose nonlocal water. Those to the north have warm cores, are anticyclonic, and generally contain large segments of Sargasso Sea water that have been pinched off during an extreme meander. Those to the south enclose continental slope water, are cyclonic, and are cooler than their surroundings, at least for several weeks after their detachment. While the rings tend to lose their surface thermal contrast with respect to their surroundings as time goes on, their gradient flow

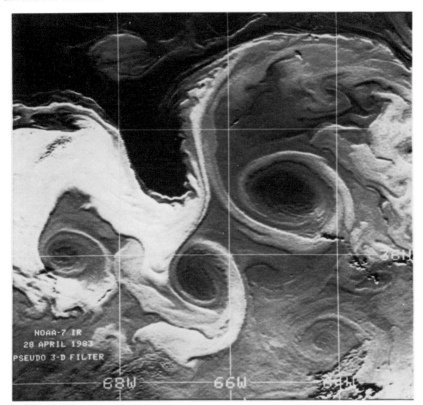

Fig. 6.41 Image of the Gulf Stream region off the U.S. east coast, made from a NOAA polar orbiting satellite with a thermal infrared imaging device. Visible are the Gulf Stream, warm and cold-core rings and eddies, cooler slope water, and warmer Sargasso Sea water. Lighter shades are warmer, and the image has been enhanced to give the appearance of elevations and depressions. [From Hawkins, J., M. Lybanon et XIII al., *EOS,* (1985). Image courtesy of the United States Navy.]

velocity structure persists for months, and the rings and eddies may have lifetimes of order 1 year.

Such satellite data, while very useful, must have subsurface information appended to them to be of value in quantitative studies. An example of subsurface variability is given in Fig. 6.42a, which shows the positions of the 15°C isothermal surface at several depths in the Gulf Stream region; the data cover a four month interval in 1975. Nine cyclonic rings to the south and three anticyclonic rings to the north are also visible. Figure 6.42b shows a vertical temperature section across the stream and through two of the southern, cyclonic cold-core rings. Within the rings, colder water maintains the shallow levels it had when entrapped near the continental slope by the mean-

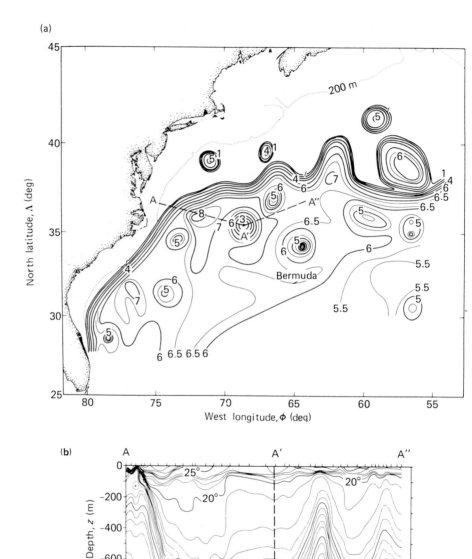

Fig. 6.42 (a) Positions of the 15°C isotherm at depths between 100 and 700 m, showing the Gulf Stream, nine cyclonic, and three anticyclonic eddies. Data were taken by subsurface drifting floats and other means. (b) Subsurface profile along a line segment A–A′–A″ in Fig. 6.42a, showing the Gulf Stream boundary at left, and cold-core rings at 650 and 850 km. [From Richardson, P., et al., *J. Geophys. Res.* (1978)].

dering Gulf Stream. Conversely, the rings on the north side are warm, anticyclonic, and have the deep thermocline structure of the Sargasso Sea, from whence they came.

The process of detachment of a ring is shown in Fig. 6.43, where a cold core ring is seen to result from the separation of a large meander, which usually takes several days to occur. Schematics of the interior structures of warm and cold core eddies are shown in Fig. 6.44, where the four characteristics of temperature, salinity, density, and sign of vorticity are used to differentiate between them. Within such an eddy, the inclination (baroclinicity) of the constant density surfaces with respect to the constant pressure surfaces is quite clear. Warm core rings typically have surface elevations of order 1 m

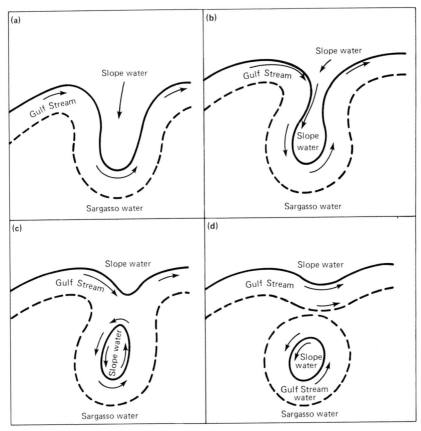

Fig. 6.43 Diagram of separation of cold-core, cyclonic ring from the Gulf Stream starting with (a) meander motion, (b) extreme nonlinear meander, (c) detachment of ring and rejoining of main current system, and (d) independent ring. [From Parker, C. E., *Deep-Sea Res.* (1971).]

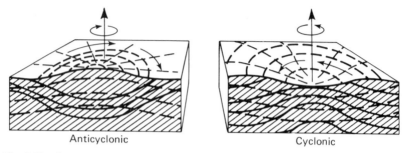

Anticyclonic   Cyclonic

Fig. 6.44 Cross sections of anticyclonic and cyclonic eddies in the Northern Hemisphere, showing surface and subsurface contours of constant pressure and density, and how they slope with respect to each other. Southern Hemisphere eddies have the signs of current and vorticity reversed. [Adapted from von Arx, W. S., *An Introduction to Physical Oceanography* (1962).]

or less; they have isopycnal, isothermal, and isohaline surfaces that dip deeply down in the middle; and they have anticyclonic, negative vorticity, as is shown on the left of Fig. 6.44. Cold core eddies have surface *depressions* of the same order; they have constant-property surfaces that rise upward; and they are cyclonic, low pressure systems. The balance of forces within them is described approximately by the gradient flow relationship (Eq. 6.93).

Such eddies are due to the unstable growth of Rossby waves to an extreme nonlinear limit. The Rossby (or planetary) wave is a propagating wave of potential vorticity, Π, and moves in the fluid as a circulation system that attempts to conserve that quantity. In the atmosphere, Rossby waves are best known via the alternating sequence of high and low pressure cells that make up the midlatitude weather systems, about and through which the jet stream flows (see Figs. 2.14 and 2.15b). In the ocean, they are most apparent through the sequence of meanders in the western boundary currents, which also have the alternating high and low pressure character of the atmospheric jets, but with much longer time scales (approximately 6 weeks versus 1 week) and smaller length scales (about 200 km versus 1000 km) than their atmospheric counterparts. However, Rossby waves in general have a wide range of scales in the ocean, as the dispersion equation below will reveal.

As these highs and lows oscillate about an equilibrium latitude, they take on or shed relative vorticity in an effort to balance their total angular momentum, as Fig. 6.45 illustrates. Figure 6.46 shows how the pair of vortices that constitute one wavelength of the Rossby wave add their circulations in the region between them and thus advance their phase to the west; this is a characteristic of all Rossby waves in a fluid not having a mean velocity. However, a moving fluid may advect the waves so that their phase speeds in an earth-fixed system can be in either direction, since the speeds are very low for the

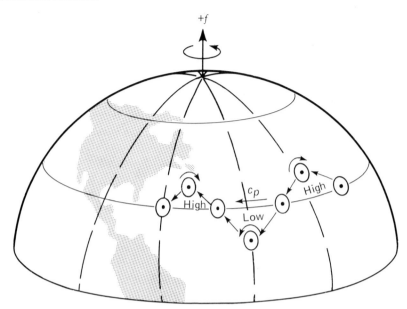

Fig. 6.45 Propagating Rossby wave of potential vorticity moving to the west at mid-latitudes, giving rise to a sequence of high and low pressure systems. In a current or wind moving to the east at $U > c_p$, the waves may appear to propagate to the east. [Adapted from von Arx, W. S., *An Introduction to Physical Oceanography* (1962).]

baroclinic modes. In the Gulf Stream, the eastward-flowing current apparently carries them in that direction at speeds of perhaps 0.05 m s^{-1}, while their intrinsic phase speeds might be of order 0.1 to 0.2 m s^{-1} to the west. Their group velocities in a stationary fluid are to the west for long

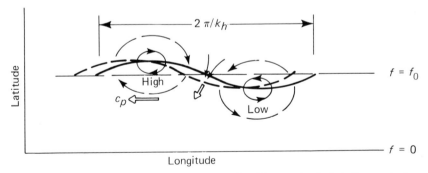

Fig. 6.46 Mid-latitude Rossby wave propagation. Phase velocity is to the west, since the flow in the region between high and low pressure cells has a westward component. Conservation of potential vorticity yields the restoring force required for an oscillation. (Adapted from Gill, A. E., *Atmosphere–Ocean Dynamics* (1982).]

wavelengths, and to the east for short wavelengths; since the group velocity is the speed of energy propagation, it is of somewhat greater interest than the phase speed. These dispersive characteristics will be discussed more fully ahead.

Figure 6.47 illustrates streamlines from a numerical model of the Gulf Stream that includes the nonlinear advective terms, and which shows unstable Rossby waves in the eastern extension of the current that bear much similarity to the observations. The calculation incorporates an instability mechanism, baroclinic instability, that causes amplification of the waves through the release of stored potential energy of the baroclinic density field.

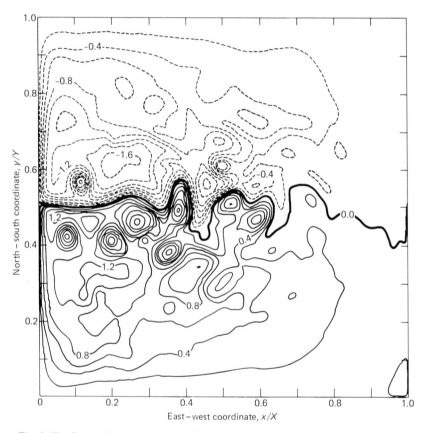

Fig. 6.47   Streamlines of unstable flow for a subtropical gyre of the same general configuration as for Figs. 6.30 to 6.32, from a nonlinear numerical model. Rossby waves and detached rings are visible above and below the current axis. [From Marshall, J., "A North Atlantic Ocean Circulation Model for WOCE Observing System Simulation Studies" (1986).]

The oscillations are propagating nonlinear Rossby waves having lengths of about 300 to 400 km, of the same magnitude as the ones observed.

We will now develop the equations for potential vorticity waves using the *quasi-geostrophic approximation*, which recognizes that the geostrophic balance is not perfect, and that the system tends to "hunt" about its equilibrium position. This hunting is the source of the Rossby waves in the western boundary currents (although not all Rossby waves are generated by instability—the wind contributes to their existence in considerable degree).

6.10   Quasi-Geostrophic Motion and Rossby Waves

It is apparent from the observed periods and wavelengths of Gulf Stream meanders that if they are indeed propagating waves, their dispersion characteristics put them into a quite distinct class from internal gravity waves or any other class that we have encountered to date. Recall that the lowest frequency of internal gravity waves is the local value of the Coriolis frequency, f. Because of the large spatial scale of Rossby waves, they are also called *planetary waves,* although in the case of the ocean, the boundedness of oceanic basins prohibits this appellation from being strictly correct, except perhaps in the Antarctic Circumpolar Current.

Rossby waves rely in part on the curvature of the earth (i.e., the variation of Coriolis force with north–south distance) for their existence. Therefore, we expect to find that the beta-plane approximation to that curvature will introduce into the relevant equations the parameter β as a leading quantity. However, the β effect is not the only source of variability in potential vorticity; the total fluid depth, $H + \xi$, also modifies the angular momentum of a fluid column, and thus variations in depth are quite equivalent to variations in Coriolis frequency in supporting Rossby waves. We therefore expect to find either the water depth or the equivalent layer depth, H_{eq}, appearing in the equations, which it does via the specification of the shallow water phase speed.

The nonintuitive character of planetary waves requires a careful mathematical treatment for the correct derivation of their characteristics, and we give here only an abbreviated version of the full development. The approach is to use the idea of quasi-geostrophic motion, in which the balance of forces is geostrophic to first order, but with small variations occurring about that balance. The equations of motion are the linear, inviscid, shallow water equations for a fluid steadily streaming toward the east on a beta plane. Thus we write, from the horizontal momentum equations:

$$\frac{\partial u}{\partial t} + U \frac{\partial u}{\partial x} - (f_0 + \beta y)v = -\frac{1}{\rho} \frac{\partial p'}{\partial x} = -f_0 v_g \,, \qquad (6.127)$$

and

$$\frac{\partial v}{\partial t} + U \frac{\partial v}{\partial x} + (f_0 + \beta y)u = -\frac{1}{\rho} \frac{\partial p'}{\partial y} = f_0 u_g \,. \qquad (6.128)$$

On the extreme right-hand side we have written the components of the horizontal pressure gradient in terms of the geostrophic speeds, with f_0 used instead of f. Also, we have included the effects of a steady horizontal eastward velocity, U, in the linearized convective derivative (cf. Eq. 5.160, for example). This allows for Rossby waves to appear as oscillations of an otherwise steady current system.

The next step is to decompose the total velocity, $\mathbf{u}$, into the first-order geostrophic velocity, $\mathbf{u}_g$, and a second-order *quasi-geostrophic part, $\mathbf{u}_a$,* according to:

$$u = u_a + u_g \qquad (6.129)$$

and

$$v = v_a + v_g \,. \qquad (6.130)$$

This is necessary because the time-dependence of the flow cannot be obtained from u_g alone because of the degeneracy of the geostrophic equations, even when they are time-dependent. In addition to the much-used hydrostatic assumption,

$$\nabla p' = \rho g \nabla \xi \,, \qquad (6.131)$$

we also invoke the integrated continuity equation in the form

$$\frac{D\xi}{Dt} + H_{eq} \left(\frac{\partial u}{\partial x} + \frac{\partial v}{\partial y} \right) = 0 \,. \qquad (6.132)$$

The interpretation of this relationship depends upon the vertical configuration that is being used; here we assume that $H_{eq} = \Delta\rho H_1/\rho$ is the upper layer *equivalent depth,* and are thus implicitly dealing with a reduced gravity model with a deep lower layer.

By taking the convective time derivatives of Eqs. 6.127 and 6.128 and eliminating u from the v equation and conversely, the resultant equations can be written as

$$\frac{D^2 u}{Dt^2} + f^2 u = ff_0 u_g - f_0 \frac{Dv_g}{Dt} \tag{6.133}$$

and

$$\frac{D^2 v}{Dt^2} + f^2 v = ff_0 v_g + f_0 \frac{Du_g}{Dt} . \tag{6.134}$$

We can neglect the rates of change of acceleration in these equations for slow motions; we next substitute Eqs. 6.129 and 6.130 for u_g and v_g into them. By keeping careful track of the order of the products of $(f_0 + \beta y)$ and $(u_a + u_g)$ (as may be done in a formal perturbation expansion) and retaining first-order terms, we may reduce these equations to the desired result for the quasi-geostrophic velocity:

$$f_0 u_a = -\beta y u_g - \frac{Dv_g}{Dt} = \frac{1}{\rho f_0} \left[\beta y \frac{\partial p'}{\partial y} - \frac{D}{Dt} \frac{\partial p'}{\partial x} \right] \tag{6.135}$$

and

$$f_0 v_a = -\beta y v_g + \frac{Du_g}{Dt} = \frac{1}{\rho f_0} \left[- \beta y \frac{\partial p'}{\partial x} - \frac{D}{Dt} \frac{\partial p'}{\partial y} \right] . \tag{6.136}$$

To this same order of approximation, the velocity components in the continuity equation, Eq. 6.132, should be u_a and v_a. By differentiating Eqs. 6.135 and 6.136 with respect to x and y, in that order, and substituting the derivatives in Eq. 6.132, along with the pressure-elevation relationships, an equation in ξ alone is obtained:

$$\frac{D}{Dt} \left(\frac{\partial^2 \xi}{\partial x^2} + \frac{\partial^2 \xi}{\partial y^2} - \frac{f_0^2}{c_h^2} \xi \right) + \beta \frac{\partial \xi}{\partial x} = 0 . \tag{6.137}$$

Here $c_h^2 = gH_{eq}$ is the square of the baroclinic speed for the equivalent depth.

Now an equation in one dependent variable (such as Eq. 6.137) is just what is needed to derive the dispersion relation, which, for harmonic space and time variations, can essentially be written by inspection of that equation. The result is

$$-i(\omega - Uk)(-k^2 - l^2 - f_0^2/c_h^2) + i\beta k = 0 , \tag{6.138}$$

or, solving for ω (where $\omega_D = \omega(\mathbf{k})$), we obtain

$$\omega = kU - \frac{\beta k}{k^2 + l^2 + f_0^2/c_h^2} . \tag{6.139}$$

This is the dispersion relationship for Rossby waves in a streaming fluid, and describes a number of interesting characteristics of those waves.

Neglecting for the moment the advective term kU, it is seen that for all positive wave numbers, the zonal phase speed, ω/k, is negative, indicating that Rossby waves propagate with advancing phase to the *west*. A plot of the dispersion relation is shown in Figs. 6.48a and 6.48b, using positive ω and negative k; the opposite choice is equally acceptable. Because of the variable Coriolis force, wave propagation is no longer isotropic in the horizontal plane, but differs in the meridional and the zonal directions. Hence the dispersion surface, $\omega(k,l)$, requires a three-dimensional representation, as indicated in Fig. 6.48. Figure 6.48a is a plot of $\omega(k,l = 0)$, and is a section or profile through the dispersion surface along $l = 0$. The scaling for frequency is in units of $\omega/(\beta c_h/f_0)$, while for wave number, it is in units of kc_h/f_0. There is a maximum frequency for Rossby waves of 0.5 (in these units) at a wave number of -1.0. The maximum phase velocity occurs for long wavelengths and has the value

$$(\omega/k)_{max} = -\beta c_h^2/f_0^2 = -\beta g H_{eq}/f_0^2 . \tag{6.140}$$

Thus near the equator, where f_0 is small, the waves have a larger phase velocity than closer to the poles, behaving as $1/f_0^2$. A typical mid-latitude phase speed for $H_{eq} = 0.4$ m is 0.6×10^{-2} m s^{-1}. Rossby waves in the vicinity of the Gulf Stream are thus very slow, indeed. The group velocity, $\mathbf{c}_g$, is

$$\mathbf{c}_g = \nabla_k \omega(\mathbf{k})$$

$$= \beta \left[k^2 + l^2 + \frac{f_0^2}{c_h^2} \right]^{-2} \left[\left(k^2 - l^2 - \frac{f_0^2}{c_h^2} \right) \hat{i} + 2kl\hat{j} \right] . \tag{6.141}$$

It is readily seen that for $l = 0$, $\mathbf{c}_g$ is zero at $kc_h/f_0 = -1$, while for $l \neq 0$, the zonal component vanishes along a hyperbola in the (k,l) plane given by

$$k^2 - l^2 = f_0^2/c_h^2 . \tag{6.142}$$

This is shown in the plan view of the dispersion surface, as illustrated by Fig. 6.48b. Here the contours of constant frequency are represented by cir-

(a)

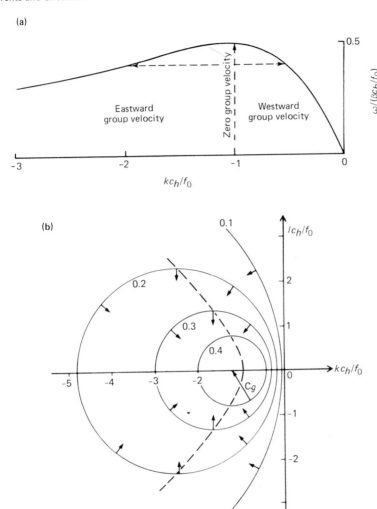

(b)

Fig. 6.48 Views of Rossby wave dispersion surface for midlatitude waves. (a) Profile view through $l = 0$. The maximum frequency also has zero group velocity. (b) Plan view. Circles are contours of constant frequency, $\omega/(\beta c_h/f_0)$. The group velocity has the direction given by small arrows. The dashed hyperbola separates eastward from westward group velocity. [From Gill, A. E., *Atmosphere–Ocean Dynamics* (1982).]

cles, on which are situated small arrows indicating the direction of the group velocity. Both phase and group velocities have components that are directed toward the *west* for wave vectors to the right of the dashed line representing Eq. 6.142, but the group velocity has components to the *east* for higher wave vectors, while c_p remains westward. Thus the direction of energy propagation, as given by Eq. 6.141, reverses sign for short enough wavelengths. In that region of wave number space, the Rossby wave is an example of a *backward wave*, i.e., one that has oppositely directed phase and group velocities.

If the advective speed, U, is now allowed to be nonzero, the Doppler shift that it induces can easily exceed the intrinsic westward phase speed with quite reasonable mean currents, and in an earth-fixed coordinate system, such waves appear to move in the direction of U. This seems to be the case in both the Gulf Stream and the jet stream.

Westward-propagating Rossby waves that encounter a reflecting barrier, say the edge of a continent, must undergo a transition to an eastward propagating wave after reflection. Since wave frequency is conserved in such a process, this means that the east–west wave number must change to one that allows eastward group propagation to occur, as indicated by the horizontal line in Fig. 6.48a. Thus Rossby waves significantly shorten their wavelengths after reflection from the western edge of an ocean basin, and lengthen them after reflection from the eastern edge. This is indicated in Fig. 6.49, which illustrates the wave vector transition process on the dispersion surface, at a sloping continental boundary.

The numerical differences in group speeds for eastward and westward propagating Rossby waves give rise to the alternative view of western boundary current intensification mentioned earlier. Figure 6.50 illustrates a vertical east–west section through a model oceanic basin with a thermocline that evolves under the influence of a zonal wind stress. At time $t = 0$, the wind is switched on, and the baroclinic response is shown at different subsequent times marked 1 to 15, which indicate thermocline depths. Under the influence of Ekman transport and downwelling, the thermocline deepens more or less uniformly until boundary effects are felt. Longer, higher speed Rossby waves move out from the eastern boundary toward the west, establishing a Sverdrup balance and a gently sloping thermocline in the process, as seen on the right. However, only slower, shorter waves may radiate from the western boundary, with the result that energy tends to accumulate there, increasing the slope of the thermocline and showing oscillations (cf. Fig. 6.47). The diagram may also be interpreted as a plot of the magnitude of the total eastward velocity, $u = u_a + u_g$ (see Eq. 6.129), as a function of x. Thus the accumulation of slow Rossby waves on the western boundary of an ocean also appears to explain the intensification of current and deepening of the thermocline there. Since they are waves of potential vorticity, this is equivalent to our earlier arguments on overall vorticity balance.

Rossby waves therefore appear to play a fundamental role in redistributing and dispersing large-scale time-varying energy in the ocean, with their effects not being confined to the subtropical gyres alone. However, in the

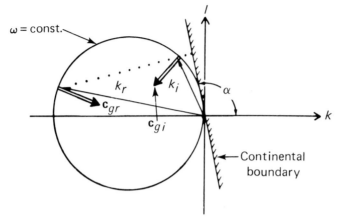

Fig. 6.49 Behavior of a Rossby wave vector after reflection at a continental boundary, as demonstrated using the dispersion surface (see Fig. 6.48b). Subscripts *i* and *r* indicate incident and reflected quantities. The difference wave vector is normal to the continental boundary. The frequency is conserved, but the nature of the Rossby dispersion relation requires a lengthening or shortening of the wave vector after reflections. [Adapted from Longuet–Higgins, M. S., *Proc. Royal Soc. London A* (1964).]

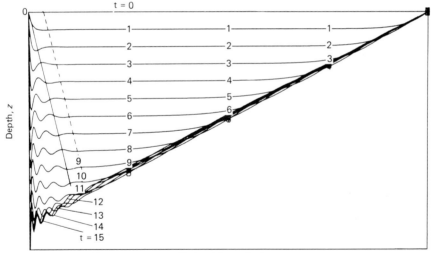

Fig. 6.50 Numerical evolution of the oceanic thermocline under the influence of wind stress. At $t = 0$, the wind is switched on. The thermocline deepens steadily under Ekman pumping, with a gentle east–west slope, until the boundary is felt. Faster, westward-traveling Rossby waves accumulate in the west, while eastward-traveling ones remain near the western boundary. [Adapted from Gill, A. E., *Atmosphere–Ocean Dynamics* (1982).]

equatorial zone, where their speeds are the highest, they find themselves in the company of other planetary waves peculiar to that region, toward which we now turn our attention.

6.11 Equatorial Waves and El Niño

The equatorial zone has an additional complement of waves beyond those that exist at higher latitudes, caused in the main by the vanishing of f_0 there. In principle, the condition of geostrophic balance, as specified by the right-hand side of Eq. 6.127, for example, fails at the equator. In practice, however, any wave motion having a modestly finite expanse across the equator instead senses a Coriolis force both above and below that line that turns it back toward $\Lambda = 0$, with the result that the equatorial zone acts as a waveguide for zonally propagating oscillations. The best known wave so trapped (and the only one that we will treat explicitly) is termed the *equatorial Kelvin wave,* which is an eastward-propagating excitation that is also dispersionless. The vorticity equations developed in the previous sections can be applied here, but with the replacement $f = \beta y$, since $f_0 = 0$ at the equator. Without repeating the calculus of that section, the relationships for the vertical displacement and horizontal velocity components can easily be verified by substitution to be:

$$\xi = \xi_0 \, \exp(-\tfrac{1}{2} \, \beta y^2/c_h) \, G(x - c_h t) \,, \tag{6.143}$$

$$u = \xi_0 \, (g/c_h) \, \exp(-\tfrac{1}{2} \, \beta y^2/c_h) \, G(x - c_h t) \,, \tag{6.144}$$

$$v = 0 \,, \tag{6.145}$$

and

$$\omega = kc_h \,. \tag{6.146}$$

These equations describe either a barotropic or a baroclinic wave of elevation that propagates in the x direction at a speed c_h and with an arbitrary shape factor, $G(x - c_h t)$, and with only a zonal (u) velocity. Its north-south profile is Gaussian, and it extends a characteristic distance, δ_e, on either side of the equator, with δ_e called the *equatorial radius of deformation:*

$$\delta_e \equiv (c_h/2\beta)^{1/2} = [(gH_{eq})^{1/2}/2\beta]^{1/2} \,. \tag{6.147}$$

A comparison of this relationship with Eq. 6.125 shows the difference in the parametric dependence of the equatorial and the Rossby radii of deformation, as well as the algebraic forms of the wave functions, i.e., Gaussian versus exponential.

Now the equatorial radius for *barotropic* motions, for which $H \approx 5000$ m (the actual oceanic depth), is of order 2000 km, which distance renders somewhat untenable the concept of an equatorially trapped wave. However, for baroclinic motions, $(gH_{eq})^{1/2} \approx 2$ m s^{-1}, for which the equivalent depth, H_{eq}, is of order 0.4 m. Then the characteristic distance, δ_e, is about 100 to 200 km, which implies that the equatorial waveguide for baroclinic Kelvin waves is narrow, being only a few degrees wide. The solutions (Eqs. 6.143 to 6.145) describe a zonally polarized wave of arbitrary east–west shape, $G(x - c_h t)$, propagating in the positive x direction at the velocity c_h without dispersion. The Kelvin wave is a unidirectional guided wave of considerable importance in equatorial dynamics.

Beyond the Kelvin wave, several other types of excitations occur in the tropics. Figure 6.51 shows a dispersion diagram for four types of equatorial waves, with frequency, ω, measured in units of $(2\beta c_h)^{1/2}$, and zonal wave number, k, in units of $(c_h/2\beta)^{-1/2}$. The Kelvin wave is represented by the straight line labeled as such. Trapped equatorial Rossby waves, which we

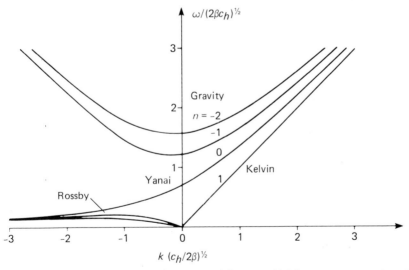

Fig. 6.51 Dispersion diagram for equatorial waves. Kelvin waves are eastward propagating dispersionless waves. Equatorial Rossby waves have an infinite number of north–south modes labled by $n = 1, 2, \ldots$. A mixed gravity–Rossby wave (or Yanai wave) has an index of $n = 0$. Higher frequency trapped internal gravity modes have negative indices. [From *Numerical Models of Ocean Circulation* (1975).]

have not treated here but which have properties analogous to those discussed in Section 6.10, appear as low-frequency, negative-k curves having modal numbers $n = 1, 2, \ldots$. The quantization occurs in the meridional direction, so that the wave number, $l_n^2 = (2n + 1)\beta/c_h$ is modally established. The equatorial Rossby wave dispersion relation is then (see Eq. 6.139 with $f_0 = 0$)

$$\omega_n = \frac{-\beta k}{k^2 + (2n + 1)\beta/c_h} . \qquad (6.148)$$

Internal gravity waves in the equatorial region occupy a higher range of frequencies than the others, but rather than being restricted to a low-frequency bound of $\omega = f_0$ for *all* modes, $n = -1, -2, \ldots$ (cf. Fig. 5.26), instead have that degeneracy removed and have a variety of frequencies at zero wave number. Another mode termed a *mixed gravity–Rossby wave* or a *Yanai wave* has only an eastward group speed, even for negative wave numbers.

On Fig. 6.52 are shown horizontal streamlines and current vectors for the Kelvin and Yanai waves for one zonal cycle, and extend over a distance of $4\delta_e$ north and south of the equator. Particles in the Kelvin wave move par-

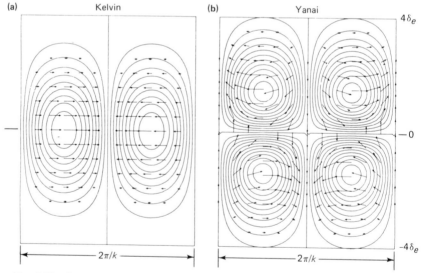

Fig. 6.52 Streamlines and current directions for (a) an equatorial Kelvin wave and (b) an equatorial mixed gravity–Rossby (Yanai) wave. One cycle of each is shown. Both waves have eastward phase and group velocities, but the Kelvin wave is symmetrical while the Yanai wave is antisymmetrical about the equator. [From Gill, A. E., *Atmosphere–Ocean Dynamics* (1982).]

allel to the x axis, while those in the Yanai wave alternately move anticyclonically and cyclonically in oval-shaped orbits.

These waves appear to play important roles in the progression of the phenomenon known as *El Niño,* which in Spanish means "The Child." The classic El Niño refers to a relatively sudden warming of near-surface waters off Peru and Chile that occurs quasi-periodically every several years approximately at Christmastime. The warming carries with it torrential rainfall and the near-disappearence of the anchovy fishery off the coast of equatorial South America at that time. Recent research suggests that the phenomenon is due to a chain of events that takes place over essentially the entire temperate and tropical Pacific and Indian Oceans. For reasons that are not altogether clear, the trade wind system over the tropical Pacific (shown in Fig. 2.15) weakens and even partially reverses during a few months in the autumn of the year preceeding an El Niño. This reversal of trade winds is associated with the *Southern Oscillation,* which is most apparent as a change in surface atmospheric pressure distribution between Australia and the southeastern Pacific. The steady surface stress of the trade winds, which had previously established a hydraulic pressure head by setting up water in the western Pacific by perhaps 1 m above equilibrium, then weakens quadratically with wind speed during the Southern Oscillation. When the trade winds rapidly decay under the influence of the changing pressure distribution, the potential energy of elevated water in the west is released, and a kind of pulse of water travels eastward across the Pacific, in part as a Kelvin wave whose speed of advance is approximately 2.5 m s^{-1}. The wave takes three to four months to cross the basin and carries with it a surge of warm, western Pacific surface water (cf. Fig. 2.18). Its arrival off the coast of South America results in a deepening of the thermocline from an initial depth of perhaps 40 m to 200 m or more, and a warming of the surface waters by some 2 to 4°C. Sea level in the western tropical Pacific drops for a period of months by 0.3 to 0.5 m, and rises in the east by roughly similar amounts. This is the main phase of the El Niño, which then slowly reverts back to more normal conditions over the next 12 to 18 months.

Figure 6.53 illustrates numerical calculations of the propagation of the trapped waves that in principle result from the release of the wind stress in the western Pacific. By some 35 days after the change in wind stress, the off-equator thermocline depth near $x = 9000$ km has risen by some 10 m, while on the equator near 6000 km, it has deepened by perhaps 20 m. By 69 days, the deepening has reached over 30 m off the coast of South America, where the incident Kelvin wave has now split into two poleward-propagating components along the edge of the simulated continent. In addition, a reflected westward-propagating equatorial Rossby wave develops that moves away with one-third the phase speed of the Kelvin wave. By 278 days,

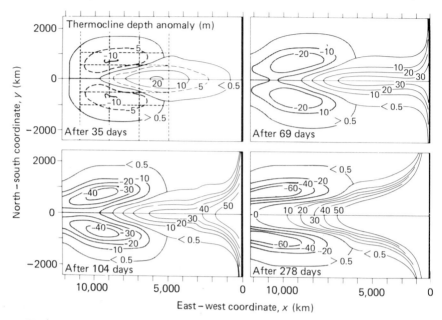

Fig. 6.53 Equatorial Kelvin and Rossby wave response, as exhibited by changes in the thermocline depth resulting from a wind stress anomaly in the Western Pacific. The Kelvin wave impacts on the coast by 69 days, and is reflected as a long, westward-traveling Rossby wave whose form is visible by 278 days. Positive anomalies indicate a deepening of the thermocline. [From McCreary, J. P., *FINE Workshop Proc.* (1978).]

the form of the planetary wave is fully visible, and a thermocline deepening in excess of 50 m has spread throughout much of the eastern Pacific. This simulates the fully developed phase of the El Niño.

Associated with the El Niño–Southern Oscillation cycle are global changes in short-term climate that manifest themselves in some locales as drought, in others as excessive rainfall, and in many locations around the world as seasonal-scale weather anomalies. While the ocean is in part responsible for these changes, the atmosphere is undoubtedly the agent that transmits the relatively rapid rainfall and temperature anomalies via teleconnections taking place through the general atmospheric circulation. The atmosphere–ocean system is strongly coupled on these time and space scales, and cause and effect are difficult to distinguish. In fact, both their dynamics and thermodynamics must be collectively studied in order to arrive at any reasonable understanding of the effects of this essentially single geophysical fluid system.

6.12   A Summary Dispersion Diagram

The myriad of waves discussed to this point are probably too varied to be summarized on one single dispersion diagram, but some vantage point for mid-latitude motions is provided by Fig. 6.54. This indicates, in a somewhat schematic form, the dispersion characteristics of planetary waves, tides, inertial oscillations, internal waves, surface gravity and capillary waves, and acoustic waves (these last to be studied in Chapter 7). Clearly the operative span of the wavelengths and frequencies is enormous, ranging from 10,000 km to millimeters, and from 1000 years to microseconds. However, there are significant spectral gaps between Rossby waves and gravity waves, and between gravity and acoustic waves as well. These gaps provide the justification for the separation of our study of geophysical fluid dynamics into its component parts, as we have done in Chapters 5, 6, and 7. There are interactions possible between these several types of excitations, of course, but they can be handled by a variety of perturbation techniques, for example by assuming that a higher-frequency wave moves *adiabatically* (with slow variation) in the current system of lower-frequency motions.

It should be noted that the wave numbers for the Rossby wave modes of Fig. 6.54 must be considered as negative (see Fig. 6.48); the properties of logarithmic plots prohibit access to the negative k axis in direct fashion.

Figure 6.55 is another summary diagram illustrating a "parameter space" that shows allowed and forbidden regions of wave propagation as a function of buoyancy frequency and Coriolis frequency. A wave *index of refraction, n_r,* is defined via

$$n_r \equiv kc_r/\omega , \qquad (6.149)$$

where c_r is a characteristic reference velocity for the type of wave under consideration, and $\omega/k = c_p$ is its phase speed. The small lemniscatelike figures in Fig. 6.55 are polar plots of $1/n_r$, which are thus proportional to the phase speed shown as a function of propagation angle off the vertical, θ (see Fig. 6.56). In Fig. 6.55, for example, internal wave frequencies are bounded above by N on the vertical axis and below by f_0 on the horizontal axis, and thus their propagation is confined to Region III in the upper left-hand quadrant of the parameter space. Within that space, their phase speeds vary as shown by the small dumbell figures, which describe speed variations as derivable from Eq. 5.146.

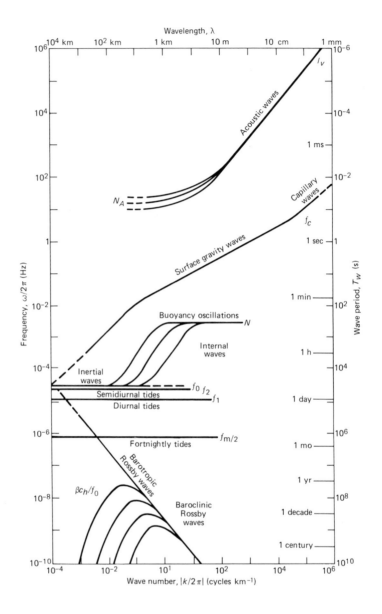

Fig. 6.54 Summary dispersion diagram for midlatitude oceanic waves, including Rossby, tidal, gravity, capillary, and acoustic waves in schematic form. The range of space and time shown includes most but not all of the relevant scales. Rossby wave numbers should be considered as negative. [Partly based on Hasselmann, K., *Prog. Oceanogr.* (1982).]

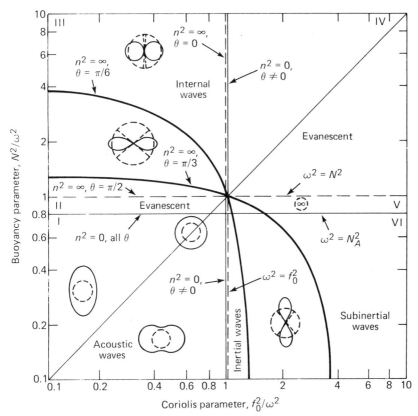

Fig. 6.55 Two-dimensional parameter "fluid" for a stratified, rotating, compressible fluid. The diagram is divided into six regions denoted by roman numerals, with the boundaries set (a) by resonances, where the index of refraction is infinite ($n^2 = \infty$), and (b) by cutoffs, where $n^2 = 0$. The locations of resonances depend on θ and are shown for $\theta_R = 0$, $\pi/6$, $\pi/3$, and $\pi/2$, while the locations of cutoffs are independent of angle. No wave propagation occurs in evanescent regions. In the propagating regions, the phase velocity varies with angle as shown by the small inset figures, which are phase velocity surfaces. The isotropic velocity of sound is indicated by the dotted circles and the resonance cones by the dotted lines passing through their centers. Various types of fluid waves exist in the regions as noted. [From Apel, J. R., *Johns Hopkins APL Technical Digest* (1986).]

6.13   Large-Scale Circulation

The relatively simple theoretical models discussed here have guided the development of more sophisticated ones, both analytical and numerical, to a considerable degree. With large computers, it has been possible to con-

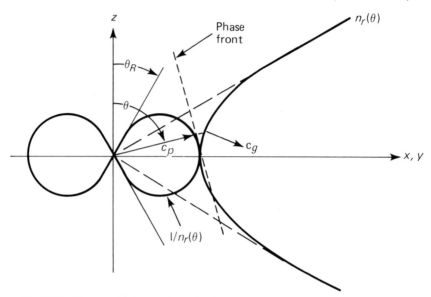

Fig. 6.56 Phase velocity surface, $1/n_r(\theta)$, and wave number surface, $n_r(\theta)$. The former gives the relative variation in phase velocity, c_p, as the angle from the vertical, θ, changes. The direction of the group velocity, $\mathbf{c}_g$, is specified by the normal to the wave number surface. [From Apel, J. R., *Johns Hopkins APL Technical Digest* (1986).]

struct so-called *general circulation models* of the world's oceans, starting with the primitive equations of Eqs. 4.51 to 4.60 (or their equivalent), and integrating them in space and time to obtain both the steady and time-varying flow conditions. One example is shown in Fig. 6.57, which gives the global pattern of streamlines (which are proportional to the mass transport) for the flow from a numerical model. Such models are at present capable of reproducing the general circulation reasonably well but, being limited in spatial and temporal resolution, are generally unable to define the mesoscale eddy field on a global scale; this is because of the limitations of computer speed and memory that exist even for the very largest machines available to date. However, eddy-resolving models for segments of a single ocean basin may be constructed, so that much can be learned thereby about the physics; an example is the one shown in Fig. 6.47.

Additionally, a central problem of large-scale oceanography is obtaining data on fine enough space and time scales to aid in model verification, or even in establishing the processes that govern certain situations. Global views of circulation, as represented by the surface setup shown in Fig. 6.58, for example, are almost always long-time composites of data taken over many years, and can only be regarded as relatively coarse approximations to the

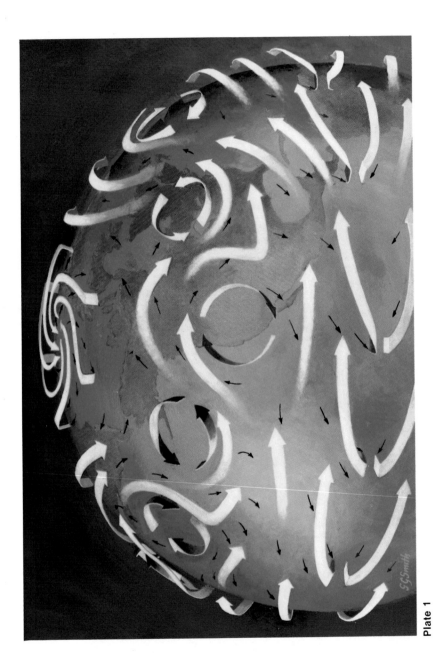

Plate 1

Fig. 2.14 Three-dimensional schematic view of major circulation features of the atmosphere. Combined Walker–Hadley cells are dominant features in the tropics, while the jet stream and high/low pressure systems characterize the temperate zones.

Plate 2

Fig. 5.32 Color photograph of the Strait of Gibraltar taken from the space shuttle *Challenger*, showing large internal solitary waves propagating eastward into the Mediterranean Sea. [Photo courtesy of the National Aeronautics and Space Administration and the Office of Naval Research.]

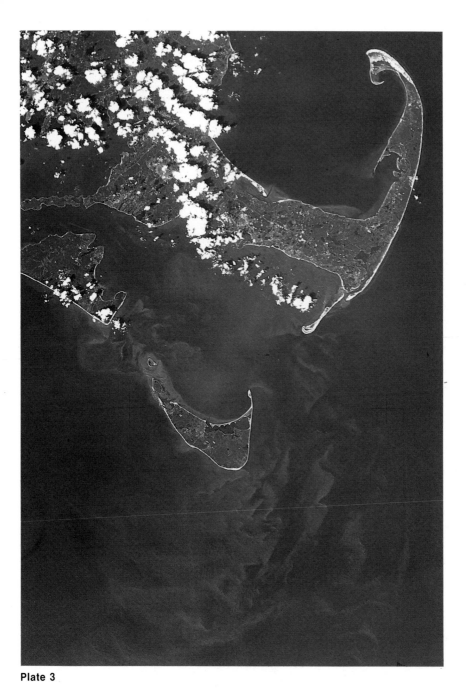

Plate 3

Fig. 8.28b Color photograph of the ocean south of Nantucket Island, made from Skylab. Shoals and internal waves are visible. [Figure courtesy of the National Aeronautics and Space Administration.]

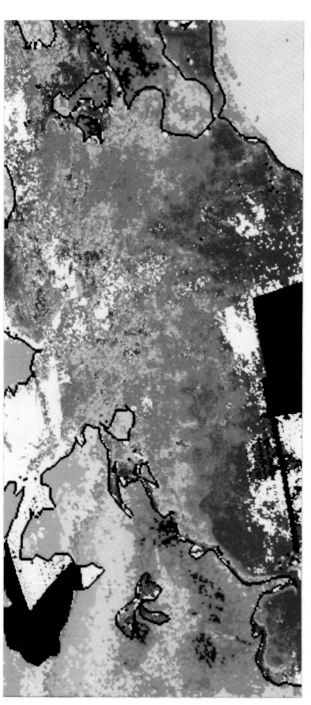
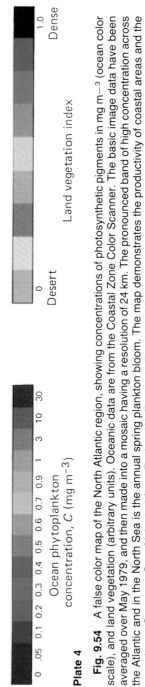

| 0 | .05 | 0.1 | 0.2 | 0.3 | 0.4 | 0.5 | 0.6 | 0.7 | 0.9 | 1 | 3 | 10 | 30 |

Plate 4

Ocean phytoplankton
concentration, C (mg m^{-3})

| 0 | | | | | | | | 1.0 |

Desert   Land vegetation index   Dense

Fig. 9.54 A false color map of the North Atlantic region, showing concentrations of photosynthetic pigments in mg m^{-3} (ocean color scale), and land vegetation (arbitrary units). Oceanic data are from the Coastal Zone Color Scanner. The basic image data have been averaged over May 1979, and then made into a mosaic having a resolution of 24 km. The pronounced band of high concentration across the Atlantic and in the North Sea is the annual spring plankton bloom. The map demonstrates the productivity of coastal areas and the near-desert Sargasso Sea. [Adapted from Esaias, W. E., et al., *Eos* (1986).]

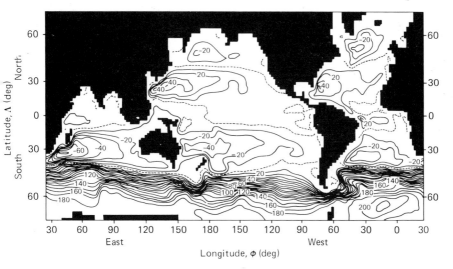

Fig. 6.57 Streamlines of the mass transport function from a numerical model of global ocean circulation. Such models of global ocean circulation do not generally resolve mesoscale features but are useful in simulating the large-scale circulation. [From Cox, M. D., *Proc. Symp. on Numerical Models of Ocean Circulation* (1975).]

actual state of affairs. While the general features of the oceanic circulation are visible in such composites, the study of dynamics and the resolution of mesoscale features is quite difficult over broad expanses of the sea.

Advanced instrumentation such as acoustic tomography (see Chapter 7), electromagnetic and optical sensors on satellites and aircraft, moored and drifting buoys, and expendable probes offer the possibility of global-scale observations of the sea. Many of these techniques are in the embryonic state and have not yet made significant contributions to our understanding of large-scale dynamics. Their promise is clear, nevertheless, and it behooves us to learn something of the physics that underlies such measurements. Beyond the instrumentation aspects, however, the study of acoustics, electromagnetics and optics of the sea is a desirable exercise in natural philosophy in its own right, and one that we now undertake.

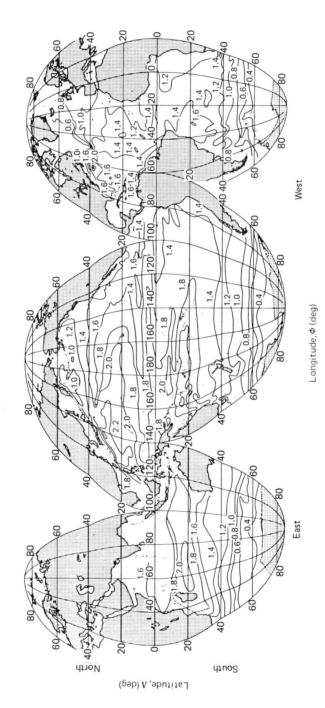

Fig. 6.58 Mean global sea surface elevations above an arbitrary equipotential surface close to the geoid, as calculated from a large number of density profiles. Major features of the ocean circulation are visible. [From Stommel, H., in *Studies on Oceanography* (1964).]

Bibliography

Books

Gill, A. E., *Atmosphere-Ocean Dynamics,* Academic Press, Inc., New York, N.Y. (1982).

Goldberg, E. D., I. N. McCave, J. J. O'Brien, and J. H. Steele, Eds., *The Sea,* Vol. 6, *Marine Modeling,* John Wiley and Sons, New York, N.Y. (1977).

LeBlond, P. H., and L. A. Mysak, *Waves in the Ocean,* Elsevier Scientific Publishing Co., Amsterdam, Netherlands (1978).

McCreary, J. P., "Eastern Tropical Ocean Response to Changing Wind Systems," in *Review Papers of Equatorial Oceanography — FINE Workshop Proceedings,* Nova N.Y.I.T. University Press (1978).

Neumann, G., and W. J. Pierson, Jr., *Principles of Physical Oceanography,* Prentice Hall, Englewood Cliffs, N.J. (1966).

Pedlosky, J., *Geophysical Fluid Dynamics,* Springer-Verlag, Berlin (1979).

Rhines, P. B., "Vorticity Dynamics of the Ocean General Circulation," *Ann. Rev. of Fluid Mech.,* Vol. 18, p. 433, Annual Reviews, Inc., Palo Alto, Calif. (1986).

Stommel, H., *The Gulf Stream,* 2nd ed., University of California Press, Berkeley, Calif. (1966).

Stern, M. E., *Ocean Circulation Physics,* Academic Press, Inc., New York, N.Y. (1975).

von Arx, W. S., *An Introduction to Physical Oceanography,* Addison-Wesley Publishing Co., Reading, Mass. (1962).

Warren, B. A., and C. Wunsch, *Evolution of Physical Oceanography,* The MIT Press, Cambridge, Mass. (1981).

Worthington, L. V., *On the North Atlantic Circulation,* The Johns Hopkins Oceanographic Studies, Vol. 6, The Johns Hopkins University Press, Baltimore, Md. (1976).

Yoshida, K. J., Ed., *Studies on Oceanography,* University of Washington Press, Seattle, Wash. (1964).

Journal Articles and Reports

Anderson, D. L. T., and A. E. Gill, "Beta Dispersion of Inertial Waves," *J. Geophys. Res.,* Vol. 84, p. 1836 (1979).

Apel, J. R., "A Linear Response Theory for Waves in a Geophysical Fluid," *Johns Hopkins APL Technical Digest,* Vol. 7, p. 42 (1986).

Bryan, K., and L. J. Lewis, "A Water Mass Model of the World Ocean," *J. Geophys. Res.* Vol. 84, p. 2503 (1979).

Cox, M. D., "A Baroclinic Numerical Model of the World Ocean: Preliminary Results," in *Numerical Models of Ocean Circulation,* National Academy Press, Washington, D.C. (1975).

Gustafson, T., and B. Kullenberg, "Untersuchungen von Traegheitssttromungen in der Ostsee," *Sven. Hydrogr.-Biol. Komm. Skr., Hydrogr.,* No. 13 (1936).

Hansen, D. V., "Gulf Stream Meanders Between Cape Hatteras and the Grand Banks," *Deep-Sea Res.,* Vol. 17, p. 495 (1970).

Hasselmann, K., "Ocean Model for Climate Variability Studies," *Prog. Oceanog.,* Vol. 11, p. 69 (1982).

Hawkins, J., and M. Lybanon, "Remote Sensing at NORDA," EOS *Trans. Amer. Geophys. Union,* Vol. 66, p. 482 (1985).

Leetmaa, A., and A. F. Bunker, "Updated Charts of the Mean Annual Wind Stress, Convergences in the Ekman Layers, and Sverdrup Transports in the North Atlantic," *J. Mar. Res.,* Vol. 36, p. 311 (1978).

Longuet-Higgins, M. S., "Planetary Waves on a Rotating Sphere," *Proc. Royal Soc. London A,* Vol. 279, p. 446 (1964).

Marshall, J., "A North Atlantic Ocean Circulation Model for WOCE Observing System Simulation Studies," Imperial College of Science and Technology, London, England (1986).

Maul, G., P. W. DeWitt, A. Yanaway, and S. R. Baig, "Geostationary Satellite Observations of Gulf Stream Meanders: Infrared Measurements and Time Series Analysis," *J. Geophys. Res.,* Vol. 83, p. 6123 (1978).

Munk, W. H., "On the Wind-Driven Ocean Circulation," *J. Meteorolo.,* Vol. 7, p. 79 (1950).

National Academy of Science, *Numerical Models of Ocean Circulation,* National Academy Press, Washington, D.C. (1975).

National Academy of Sciences, *Global Observations and Understanding of the Circulation of the Oceans,* National Academy Press, Washington, D.C. (1984).

Parker, C. E., "Gulf Stream Rings in the Sargasso Sea," *Deep-Sea Res.,* Vol. 18, p. 981 (1971).

Richardson, P. L., R. E. Cheney, and L. V. Worthington, "A Census of Gulf Stream Rings, Spring 1975," *J. Geophys. Res.,* Vol. 83, p. 6136 (1978).

Robinson, A. R., J. A. Carton, C. N. K. Mooers, L. J. Walstad, E. F. Carter, M. M. Rienecker, J. A. Smith, and W. G. Leslie, "A Real-Time Dynamical Forecast of Ocean Synoptic/Mesoscale Eddies," *Nature,* Vol. 309, p. 781 (1984).

Chapter Seven

Acoustical Oceanography

I must down to the seas again,
For the call of the running tide
Is a wild call and a clear call
That may not be denied...

Masefield, *Sea-Fever*

7.1 Introduction

Acoustical energy, yet another form of wave motion in the ocean, is of sufficient importance and varied enough characteristics to constitute a separate subject in our study of the physics of the sea. There are many applications of acoustical wave theory, with submarine signaling and detection, torpedo control, precision bathymetric measurements, sub-bottom profiling, plankton detection, and fish assay being the most prominent. More recent developments in acoustical remote sensing hold the promise of determining the subsurface temperature structure of large volumes of the sea. Acoustics enjoys another distinction from the remainder of the wave motions discussed to this point, in that signals may be generated, shaped, and detected by man in ways that are next to impossible with the other forms. Yet it should be recognized that essentially all of the physics of the subject is already contained in the mathematical statements made to date (with the exception of ionic contributions to the dissipative forces in the acoustic momentum equations).

7.2 Characteristics of Sound in the Sea

Sound waves propagate through the ocean extraordinarily well, especially at lower frequencies. Acoustical signals are attenuated to a much smaller degree than are optical and electromagnetic waves. For example, blue light might penetrate a hundred meters into the upper ocean, while radar wavelengths are effectively confined to a few centimeters of the surface (although at extremely low radio frequencies near a few kilohertz, the attenuated radio waves may reach depths of several tens of meters). At frequencies of a few hundred hertz, on the other hand, acoustical waves can propagate across an entire ocean basin before they are attenuated to levels that are masked by background noise. At the other end of the usable spectrum (at frequencies of the order of 1 MHz and above) the attenuation is large and acoustical signals have attenuation lengths of tens of meters or less. Within this range of more than five decades of frequency, a wide variety of effects comes into play in the propagation of sound waves.

In contrast to the other geophysical waves discussed to date, acoustical waves are *compressional* waves and are longitudinally polarized, i.e., their displacements, ξ, are parallel to their wave vectors, $\mathbf{k}$. To a first approximation, they are also *dispersionless*, in that their phase speed, c, is essentially independent of frequency at small amplitudes. This is not true of large amplitude waves, however, and acoustical radiation from sources such as underwater explosions and powerfully driven transducers will show dispersion in frequency and amplitude due to nonlinear or parametric effects. Nor are they dispersionless when propagation takes place in the various *waveguide modes* to be discussed ahead. There the geometry causes different frequencies that are close to the waveguide cutoff to propagate at different speeds, a phenomenon known appropriately enough as *geometric dispersion*. The speed of sound in the upper ocean is near 1500 m s^{-1}, and the frequency range of interest extends from a few hertz to beyond 10 MHz—wavelengths of roughly 1500 m to 0.15 mm. At the low frequency end, the spatial scales are the order of the wavelengths of internal waves and we can reasonably expect acoustical wave–internal wave interactions to occur. In the intermediate region, acoustical waves interact with surface waves after reflection or scattering from the ocean surface or with the bottom structure. At the higher end of the frequency range, one expects interactions with the so-called fine-structure and microstructure components of the oceanic density and temperature fields, said structure being brought about by such processes as salt-fingering, small-scale mixing, and other density- and instability-driven phenomena. Throughout the intermediate decades of frequencies, interactions with natural and man-made objects in the ocean, ranging from submarines and whales at one end to small fish and plankton at the other, strongly affect sound propagation. This entire class of interaction can be grouped under the subjects of *reflection, scattering,* and *absorption* of sound.

A less direct but nevertheless very important interaction of acoustic waves with the ocean comes from the gradual *refraction* of these waves by variations in the local speed of sound, the latter being due to temperature, pressure, salinity, and current variations that occur on spatial scales much longer than the acoustic wavelengths. Additionally, sound propagation across submarine ridges, sea mounts, and other bathymetric features can be altered by the processes of *diffraction* and *scattering* into the geometric "shadow zone" beyond the feature. Diffraction and refraction can be treated theoretically by a different set of analytical techniques from the like-scale interactions of the preceding paragraph; the techniques include WKB wave solutions, eikonal equations, ray-tracing methods, and the wave action equation (Section 8.9).

7.3 The Acoustic Wave Equation

We will now derive the acoustic wave equation and associated quantities from the hydrodynamic and thermodynamic equations developed earlier. The starting point for this is the set of relations for momentum, continuity, specific volume–pressure, and state (Eqs. 4.51–4.53, 4.54, 4.56, and 4.60, respectively). These are to be particularized for the situation at hand.

We will neglect the effects of background currents in what follows, although it is important to note that large-scale current systems can refract acoustic waves that traverse the current field over long distances. We also neglect the nonlinear term in the convective derivative, $\mathbf{u} \cdot \nabla \mathbf{u}$, thereby limiting ourselves to linear acoustic waves. We do not include the Coriolis acceleration, which is of miniscule direct consequence on these waves. Additionally, we take the viscous term to contain not only molecular damping, but add to it the previously neglected and important ionic contributions to acoustic attenuation. (A derivation of Debye attenuation is outlined in Chapter 8.) Finally, we make the small perturbation approximation for the pressure and density, *viz*:

$$p = p_0(z) + p'(\mathbf{x}, t) \tag{7.1}$$

and

$$\rho = \rho_0 + \rho'(\mathbf{x}, t). \tag{7.2}$$

Then the momentum equations can be rewritten as

$$\rho \frac{\partial \mathbf{u}}{\partial t} = -\nabla p' + [\mu \nabla^2 + B(\omega)]\mathbf{u}, \tag{7.3}$$

where the terms in the square brackets describe attenuation; the first of these is clearly due to viscosity, μ, while the right-hand-most term, $B(\omega)\mathbf{u}$, describes ionic absorption in a way that is discussed below. The continuity equation becomes

$$\frac{\partial \rho'}{\partial t} + \rho \nabla \cdot \mathbf{u} + \mathbf{u} \cdot \nabla \rho = 0 . \tag{7.4}$$

In these equations, the velocity, $\mathbf{u}$, is considered to be due to the acoustic pressure only; the influences of larger-scale velocities are introduced via their refractive effects, although it is possible to model acoustic–internal wave interactions, for example, by incorporating internal velocities and the buoyancy frequency into the basic equations at this point. We will not follow that route here.

The thermodynamic specific volume–pressure relation (Eq. 4.56) is simplified by the smallness of entropy and salinity changes, which, for acoustic waves, are essentially negligible; the entropy term describes heat flow and associated disorder. This is consistent with the experimental observation that the velocity of sound is given by the *isentropic* ($d\eta = 0$) change of pressure with respect to density, viz:

$$c^2 \equiv \left(\frac{\partial p}{\partial \rho} \right)_\eta ; \tag{7.5}$$

this is because the frequency of acoustic oscillations is so high that there is not time during one cycle for heat to flow and temperature to equilibrate, such as occurs during isothermal flows. Then the specific volume relationship (Eq. 4.56) simplifies to

$$\frac{\partial p'}{\partial t} = c^2 \frac{\partial \rho'}{\partial t} . \tag{7.6}$$

Proceeding further, the usual complications with the perturbation density and pressure terms arise. Expanding the continuity equation as well in terms of $\mathbf{u}_0 + \mathbf{u}'$ and $\rho_0 + \rho'$ and neglecting zero-order variations and products of first-order variables, the following relationship results among the first-order quantities:

$$\frac{\partial \rho'}{\partial t} + \rho \nabla \cdot \mathbf{u}' = 0 , \tag{7.7}$$

where we have neglected both the variation of density with depth as well as the products of the first-order velocity, $\mathbf{u}$, with other similar-order quantities. We will omit the primes on first-order quantities, except for p', for the remainder of this chapter.

Next we take the divergence of Eq. 7.3 (and again neglect second-order terms) to obtain

$$\rho \frac{\partial}{\partial t} \nabla \cdot \mathbf{u} = - \nabla^2 p' + [\mu \nabla^2 + B(\omega)] \nabla \cdot \mathbf{u} , \qquad (7.8)$$

where we have made use of the easily proven identity,

$$\nabla \cdot (\nabla^2 \mathbf{u}) = \nabla^2 (\nabla \cdot \mathbf{u}) . \qquad (7.9)$$

Next we substitute the velocity divergence from Eq. 7.4 and the relationship of Eq. 7.6 into Eq. 7.8 (and again neglect products of first-order terms) and rearrange the result to obtain the damped wave equation for the perturbation pressure:

$$0 = \nabla^2 p' - \frac{1}{c^2} \frac{\partial^2 p'}{\partial t^2} + \frac{1}{\rho c^2} [\mu \nabla^2 + B(\omega)] \frac{\partial p'}{\partial t} . \qquad (7.10)$$

The first two terms involve the pressure itself, while the other, describing attenuation due to viscosity and ion concentration, involves the time derivative of p'. When we write the time-dependence of p' as

$$p'(\mathbf{x},t) = p(\mathbf{x}) e^{-i\omega t} , \qquad (7.11)$$

the resultant space-dependent equation becomes

$$\left\{ \nabla^2 + \frac{\omega^2}{c^2} - \frac{i\omega}{\rho c^2} [\mu \nabla^2 + B(\omega)] \right\} p(\mathbf{x}) = 0 . \qquad (7.12)$$

Recall that the assumption of harmonic time-dependence for linear waves is no restriction, since an arbitrary temporal signal can be composed of harmonic waves via Fourier synthesis:

$$p(\mathbf{x},t) = \frac{1}{2\pi} \int_{-\infty}^{\infty} \tilde{p}(\mathbf{x},\omega) e^{-i\omega t} \, d\omega . \qquad (7.13)$$

If we further assume a plane wave solution to Eq. 7.12 behaving as $\exp(i\,\mathbf{k}\cdot\mathbf{x})$, the spatial attenuation may be derived by treating $|\mathbf{k}|$ as a complex quantity, $k = k_r + ik_i$, and solving for it after substitution of the plane wave into the differential equation. The result is, for small damping,

$$|\mathbf{k}| = k_r + ik_i \simeq \frac{\omega}{c} + i\left[\frac{\omega^2\mu}{2\rho c^3} - \frac{B(\omega)}{2\rho c}\right] = \frac{\omega}{c} + i\,\frac{\alpha(\omega)}{2}, \quad (7.14)$$

where $\alpha(\omega)/2$ is an abbreviation for the quantity in the square brackets, and is the reciprocal length for damping of the perturbation pressure. (There is no danger here of confusing the spatial damping factor, $\alpha(\omega)$, with the specific volume α, since we will not use the latter in this chapter.)

Equation 7.12 is to be solved subject to boundary conditions that will depend upon the geometry and the material from which the boundary is made. For acoustic waves in the ocean, the upper boundary surface, while nominally planar, is actually rough over a wide range of space and time scales, although it is also a very good reflector/scatterer of sound energy. The bottom, on the other hand, is stationary, but its composition ranges from hard rock to soft muds, with attendant variability in reflection, scattering, and absorption. Man-made objects generally present rigid boundaries, while fish are more pliable; indeed, the air bladders of certain fish are acoustically resonant over a range of frequencies, and their acoustical cross sections may be much larger than their geometrical ones (see Section 7.11).

From the general boundary conditions derived in Chapter 3, we observe that both the pressure and the normal components of velocity must be continuous across the interface. At the sea surface, the perturbation pressure must vanish (the total pressure equals the atmospheric pressure there), and the normal component of velocity must match that of the boundary, e.g., the normal velocity of the surface waves. This latter requirement has the effect of superimposing the Doppler frequencies of the surface wave spectrum on the surface-reflected signal and, indeed, observing that signal is one way of measuring the wave spectrum. A similar phenomenon results when the signal is reflected by a moving hard target; the Doppler shift due to the target's line-of-sight velocity is imposed on the frequency of the reflected signal.

The free surface conditions are therefore:

$$p'_b = 0 \qquad (7.15)$$

and

$$\mathbf{u}\cdot\hat{n} = \mathbf{u}_b\cdot\hat{n}. \qquad (7.16)$$

At the bottom or other similar surface, the normal component of **u** as well as $\partial \mathbf{u}/\partial t$ must vanish, which (by Eq. 7.3) implies that the normal pressure gradient must vanish:

$$\partial p'/\partial n = 0 \quad \text{at} \quad z = -H . \tag{7.17}$$

In this specification, we ignore the development of any type of viscous acoustic boundary layer, so that no conditions are imposed on the tangential velocity.

7.4 The Speed of Sound

Sound waves are intrinsically rapid oscillations in the density of an elastic medium brought about by pressure changes. Their speed can be calculated from the equation of state for seawater. From Eqs. 4.47 and 4.60 one readily obtains

$$\frac{1}{c^2} = \left(\frac{\partial \rho}{\partial p}\right)_\eta = -a_p \rho(s,T,p) \left[\frac{1 + \dfrac{p}{a_p} \dfrac{\partial a_p}{\partial p}}{1 - a_p p} \right], \tag{7.18}$$

because all of the pressure dependence is contained in the denominator of those relationships. Thus the speed of sound can be obtained from the accurate equations of state that have been developed. (Note that it is the ambient density and pressure that go into Eq. 7.18.) A somewhat different approach (direct measurement of c in the environment) has yielded equations for the speed of sound given in terms of salinity, temperature, and depth, z. This method is preferable and, indeed, is essential for the calculation of potential temperature, density, and buoyancy frequency, for example. A relationship thought accurate to some ± 0.2 m s^{-1} is given by Eq. 7.19, written in terms of departures of temperature (in degrees Celsius) from nominal values of $10°$ and $18°$C, and salinity from a reference value of 35 psu (formerly parts per thousand):

$$c = c(s,T,z) = c_0 + \alpha_0 (T - 10) + \beta_0 (T - 10)^2$$

$$+ \gamma_0 (T - 18)^2 + \delta_0 (s - 35)$$

$$+ \epsilon_0 (T - 18)(s - 35) + \zeta_0 |z| . \tag{7.19}$$

For pressures and temperatures close to the points of expansion, the numerical values of the coefficients are:

$$c_0 = 1493.0, \quad \alpha_0 = 3.0, \quad \beta_0 = -0.006, \quad \gamma_0 = -0.04,$$

$$\delta_0 = 1.2, \quad \epsilon_0 = -0.01, \quad \zeta_0 = 0.0164. \tag{7.20}$$

From these relationships, it is clear that the speed of sound depends essentially quadratically on temperature and linearly on salinity, with a single bivariate T–s term. The dependence on depth is important and dominates in deeper water, and represents an *increase* in sound speed at a rate of 1.64 m s^{-1} per 100 m depth. In shallow water, however, temperature dominates, and the character of the near-surface mixed layer and thermocline is very important in establishing the propagation of acoustical energy in the upper levels of the sea. On larger horizontal (x,y) scales, the horizontal variability in temperature and salinity will also cause the refraction of sound paths and will alter intensity levels significantly; the reasons for this are often found in the large lateral extent of such oceanic phenomena as internal, Rossby, and Kelvin waves, and boundary currents, rings, and eddies, which are important in conditioning acoustic propagation and signal variability via their effects on the velocity of sound. In the Arctic and Antarctic, salinity plays a more important role in establishing $c(\mathbf{x})$ than elsewhere, and near-surface propagation conditions in polar regions are quite different from those in more temperate zones. Additionally, in coastal areas, especially near regions of riverine input, salinity again influences the speed of sound.

Figure 7.1 shows a typical subtropical sound speed profile obtained from soundings in the Sargasso Sea as averaged over nine years. The solid line represents the mean, and the dashed lines 1 standard deviation about that mean, much of which is likely due to internal waves, especially in the main thermocline. The near-surface seasonal variability in the upper 300 m is also shown. An interesting and ubiquitous feature of deep-water sound speed profiles in temperate and tropical zones is the existence of a *velocity minimum* near 1200 to 1300 m depth. As will be shown ahead, this minimum leads to the development of a *deep sound channel* that traps acoustical energy and allows it to propagate over great distances. In subpolar waters, this channel shoals and on occasion disappears altogether, as illustrated in the profiles in Fig. 7.2. The surface mixed layer also establishes a shallow sound channel, as is discussed in Section 7.7.

The use of an adiabatically[1] varying sound speed to describe slow changes

---

[1] In this instance, "adiabatic" refers not to a thermodynamic variation but rather to a slow change of the motion that takes place as a result of gradual variations in the causative forces. See Section 8.9 for a discussion in terms of wave action.

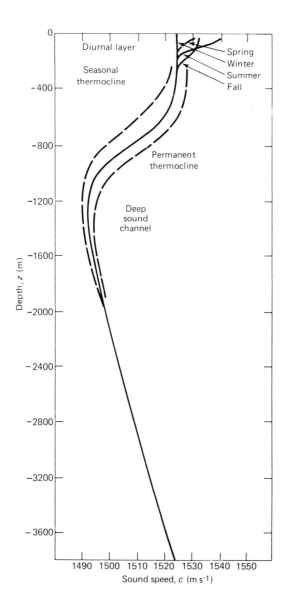

Fig. 7.1 Sound speed profile from a station near Bermuda as averaged over nine years. [Adapted from Jones, L. M., and W. A. Von Winkle, USN Underwater Sound Laboratory Report 632 (1965).]

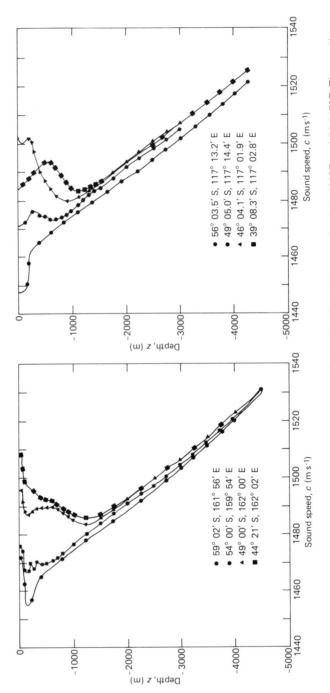

Fig. 7.2 Sound speed profiles along two meridians in the South Pacific at (a) approximately 160° E and (b) 117° E. The southern-most profiles are in sub-polar waters and show minimum speeds at very shallow depths. [From Denham, R. N., and A. C. Kibblewhite, *Antarctic Research Series 19* (1972).]

in acoustic propagation will be developed below in the context of the eikonal approximation, and will justify *a posteriori* the neglect of background vertical density variations in deriving the pressure wave equation. The slow variations in $\rho(z)$ are considered to be incorporated into the value of $c(z)$ instead. Such a treatment is acceptable as long as the fractional curvature of a wave front over a wavelength is very small. For low frequency acoustic waves propagating through an internal wave field, such may not be the case, and a somewhat different treatment may be required in which the density stratification and its variability are treated explicitly.

7.5 Mechanical Properties of Acoustic Waves

We earlier mentioned that acoustic waves are polarized longitudinally, or parallel to their direction of propagation, which is easily seen from Eq. 7.3 for the case of plane waves. In this case that relation becomes

$$-i\omega\rho\mathbf{u} = -ik p' - [\mu k^2 - B(\omega)]\mathbf{u} , \qquad (7.21)$$

which shows that $\mathbf{u}$ is parallel to $\mathbf{k}$. By forming the dot product of Eq. 7.21 with $\mathbf{k}$, and effecting a simple rearrangement, an important relationship between the perturbation pressure and the velocity is obtained:

$$p' = \{\rho c + i[\mu|\mathbf{k}| - B(\omega)/|\mathbf{k}|]\} |\mathbf{u}| , \qquad (7.22)$$

where the quantity in the curved braces is the *complex acoustic impedance, Z*:

$$Z \equiv \rho c + i[\mu|\mathbf{k}| - B(\omega)/|\mathbf{k}|] . \qquad (7.23)$$

$$= X + iY$$

Here we have defined the real part of the impedance as

$$X = \rho c . \qquad (7.24)$$

The similarity to electrical quantities is the source of the name: $V = IZ$, with pressure playing the role of electrical potential or voltage, and with electrical and fluid currents being analogous. The real part of the acoustic impedance, ρc, gives the proportionality factor between pressure and velocity for an acoustic wave; its value for seawater is approximately 1.56×10^6 kg $(m^2 s)^{-1}$.

The particle displacement in a sound wave, ξ, is related to the velocity by its time derivative:

$$\mathbf{u} = \frac{\partial \xi}{\partial t} . \tag{7.25}$$

A quantity called the *dilatation* or *condensation*, or fractional change in volume, is defined as

$$\delta = \frac{\rho'}{\rho} = -\nabla \cdot \xi . \tag{7.26}$$

From the continuity and pressure equations, one then obtains a relationship between pressure and particle displacement:

$$p' = -\rho c^2 \nabla \cdot \xi . \tag{7.27}$$

All of the quantities—$\mathbf{u}$, p', ξ, and δ—obey wave equations of the same form as Eq. 7.10.

Acoustic intensity, I, is defined as the rate at which energy is transmitted in the direction of propagation per unit area of wave front, i.e., the energy flux carried by the wave. This flux is the product of the force per unit area, or pressure, and the velocity (see Eq. 4.71):

$$\mathbf{I} = p \, \mathbf{u} \qquad W \, m^{-2} . \tag{7.28}$$

The average value of intensity for an undamped plane wave behaving as

$$\mathbf{u} = \mathbf{u}_0 \, \exp[i(\mathbf{k} \cdot \mathbf{x} - \omega t)] \tag{7.29}$$

and

$$p = p_0 \, \exp[i(\mathbf{k} \cdot \mathbf{x} - \omega t)] \tag{7.30}$$

can be written as

$$\langle I \rangle = \mathrm{Re} \left[\frac{1}{\lambda} \int_0^\lambda p u^* \, dx \right]$$

$$= \tfrac{1}{2} \, \mathrm{Re} \, (p u^*)$$

$$= \tfrac{1}{2} \, p_0 u_0 , \tag{7.31}$$

or

$$\langle I \rangle = \text{½} \, \rho c u_0^2 = \frac{p_0^2}{2\rho c} \, . \qquad (7.32)$$

Acoustic power fluxes are very small, with normal speech in air radiating at levels of order 10^{-5} W. A standard reference level for acoustic pressure in the sea is 1 μPa, or 10^{-6} N m^{-2}; levels are often quoted in decibels relative to this reference.

For wave forms other than plane waves, the expressions given for intensity in terms of the square of pressure are only approximate, and the actual waveform and its particle velocities must be determined in order for accurate values of intensity to be found.

7.6 A Simple Solution to the Wave Equation

As an example of a simple application of the wave equation to acoustic propagation in the sea, consider an undamped plane wave traveling in an ocean having both constant sound velocity and depth, and bounded above and below by smooth surfaces. While any function of the form $p = p(t + x/c)$ is a solution to Eq. 7.12 when damping is neglected, we shall confine our attention to plane waves for simplicity, and neglect refraction. Such a situation might arise on the continental shelf during the winter, when the wind mixing has rendered the water isothermal and isohaline, and where the depth dependence of speed plays a negligible role. Our assumed solution has the form

$$p(\mathbf{x},t) = P(z) \exp[i(\mathbf{k}\cdot\mathbf{x} - \omega t)] \, , \qquad (7.33)$$

which, after substitution into Eq. 7.12, becomes an ordinary differential equation for the amplitude function, $P(z)$:

$$\frac{d^2 P}{dz^2} + \left(\frac{\omega^2}{c^2} - k^2 - l^2 \right) P = 0 \, . \qquad (7.34)$$

Here the vertical wave number, m, is defined via the coefficient of P:

$$m = \pm \left(\frac{\omega^2}{c^2} - k^2 - l^2 \right)^{\text{½}} . \qquad (7.35)$$

A solution to this equation that satisfies the upper boundary condition, $p(0) = 0$, is

$$P(z) = p_0 \sin mz .\qquad(7.36)$$

Then the imposition of the lower boundary condition (Eq. 7.17) at the bottom, $z = -H$, results in a quantization of the vertical wave number into half-integer values according to

$$m_n = \pm (n - \tfrac{1}{2})\pi/H = \pm \left(\frac{\omega_n^2}{c^2} - k^2 - l^2\right)^{1/2}, \quad n = 1, 2, 3, \ldots .$$
$$(7.37)$$

where the infinite set of eigenfrequencies, ω_n, is given by

$$\omega_n^2 = c^2 \{k^2 + l^2 + [(n - \tfrac{1}{2})\pi/H]^2\} .\qquad(7.38)$$

Thus there exists a lowest radian frequency, ω_1, at which a plane wave may propagate in this shallow ocean. For very long horizontal wavelengths, this value is $c\pi/2H$, corresponding to one-quarter of a vertical wavelength, m_1, of acoustic energy fitting into the oceanic waveguide. It will be recognized that this condition is the one for an "open organ pipe" resonance, where reflection from dense to rare at the open end occurs at an antinode of the vibrating column; in this case, it occurs at the bottom. For a continental shelf depth of 100 m, say, the lowest permissible wave frequency, or cutoff frequency, $\omega_1/2\pi$, is approximately 3.75 Hz. All higher frequencies may propagate in this case, so that the depth limitation is not severe, even in a shallow sea. Figure 7.3 shows a schematic dispersion diagram for this case, with the abscissa being the horizontal wave number, $k_h = (k^2 + l^2)^{1/2}$, measured in units of $1/H$, and the ordinate is radian frequency. The first few quantized modes are shown, all of which asymptotically approach $\omega_n = k_h c$ as k_h becomes large; this relationship is the deep-water, constant-density dispersion relation for large horizontal wave numbers. These modes can be considered to be waveguide modes, with discrete values of the vertical wave number, m_n, being imposed by the horizontal boundaries, but with the horizontal wave vector, k_h, taking on those values required for propagation by the combination of the exciting frequency and the depth of the oceanic wave guide. Waves near the cutoff frequencies exhibit this property of *geometric dispersion*, with their phase speeds depending upon frequency, even though deep-ocean acoustic waves are themselves dispersionless.

We have included these upper frequency acoustical modes in the overall

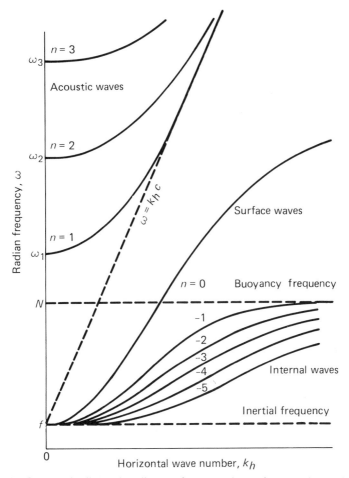

Fig. 7.3 Schematic dispersion diagram for acoustic, surface gravity, and internal gravity waves. The low-frequency cutoffs for acoustic modes are greatly reduced from their actual values for purposes of illustration. [Adapted from Eckart, C., *Hydrodynamics of Oceans and Atmospheres* (1960.)]

dispersion diagram for ocean waves shown on a different scale on Fig. 6.54. These diagrams imply that the lower frequency cutoffs are due to the shallow water waveguide effect, but in fact they may more nearly be due to the failure of the compressibility to provide a restoring force at frequencies near 1 to 10 Hz. This is the region of *infrasonics*, and is not a well-understood regime of sound waves in the sea.

In the presence of stratification (which has been neglected in the mathematics to this point), the propagation characteristics are more complicated. It can be shown, by taking into account the neglected vertical derivatives of density, that the resultant dispersion relation for combined internal and acoustic waves may be obtained from the simultaneous solution of the two equations that control the propagation characteristics. The first of these is a generalization of Eq. 7.37, written for the case of constant buoyancy frequency, N:

$$m = \pm \left[\left(\frac{\omega^2 - N^2}{\omega^2 - f^2} \right) \left(\frac{\omega^2 - f^2}{c^2} - k^2 - l^2 \right) \right]^{1/2} . \qquad (7.39)$$

This relationship shows the roles played by the Brunt-Väisälä and Coriolis frequencies as eigenfrequencies of two different types; i.e., one is a *resonance* at $\omega^2 = f^2$, and the other a *cutoff* at $\omega^2 = N^2$ (see Fig. 6.55). The second controlling equation (Eq. 7.40 below) arises from the imposition of boundary conditions as before, but now carried out with stratification present. It is an analogous relationship to the equation for surface gravity waves (Eq. 5.55) written without inclusion of capillary waves, and with the vertical wave number, m, replacing the horizontal one, k_h:

$$\omega^2 = N^2 - mg \tan mH . \qquad (7.40)$$

The simultaneous solution of these two equations establishes the eigenfrequencies as a function of wave number and depth. It can be seen from Eq. 7.39 that when $c \rightarrow \infty$, $\omega^2 \gg N^2$, and $\omega^2 \gg f^2$, m becomes imaginary. In that case, the horizontal boundary statement (Eq. 7.40) becomes the surface wave dispersion relationship given by Eq. 5.70. Figure 7.3 illustrates a schematic dispersion diagram for propagation of acoustic, surface, and internal waves in a stratified, compressible, bounded ocean.

The general solution of the guided wave problem posed above is the superposition of the allowed normal modes, weighted according to pressure amplitudes, $P_{0n}(\mathbf{k},\omega)$, which in turn depend upon the level of excitation by the source of the waves. Thus we may write the formal solution as the summation

$$p'(\mathbf{x},t) = \sum_{n=1}^{\infty} P_{0n}(\mathbf{k},\omega) \sin \left[(n - \tfrac{1}{2}) \pi z/H \right] \cos (\mathbf{k}_h \cdot \mathbf{x} - \omega_n t) . \quad (7.41)$$

An acoustic wave of arbitrary time dependence may be composed out of such normal modes by Fourier synthesis.

However, while this model is instructive, it is not a very realistic description of the actual state of affairs, and we must use other methods to more accurately describe sound propagation in the ocean.

7.7 Ray Tracing and the Eikonal Equation

The technique of *ray tracing* affords a very useful approach to the determination of propagation paths taken by acoustic signals, and can even give reasonable estimates of the sound intensity in regions free of caustics and other singularities. In order to understand the technique, however, we must first develop the *eikonal equation* of geometrical optics (*eikon*: Greek for *image*).

In a region where the speed of sound is constant, $c = c_0$, plane waves of the form of Eq. 7.30 are exact solutions to the dissipationless wave equation. In a region where the sound velocity varies slowly in all three dimensions of space as $c = c(\mathbf{x})$, it is reasonable to look for solutions that closely approximate plane waves, but which are slightly distorted or bent by the variation of the index of refraction, n. That quantity is defined as

$$n(\mathbf{x}) \equiv \frac{c_0}{c(\mathbf{x})} = \frac{k(\mathbf{x})c_0}{\omega} . \tag{7.42}$$

A more general definition of a *vector index of refraction* that accommodates the refraction due to a background current, $\mathbf{U}_0$, is given in terms of the wave vector $\mathbf{k}$ (see the discussion on the Doppler frequency following Eq. 5.161):

$$\mathbf{n}(\mathbf{x}) = \frac{\mathbf{k}(\mathbf{x})c_0}{\omega - \mathbf{U}_0 \cdot \mathbf{k}(\mathbf{x})} . \tag{7.43}$$

This form of the index of refraction is related to the slowness, s, of Eq. 5.62.

The eikonal equation may be obtained by first assuming that the phase and amplitude of a quasi-planar wave are slowly varying functions of space, and then deriving the differential equations for the phase function, $W(\mathbf{x})$ (which is the *eikonal*), and for the amplitude, $A(\mathbf{x})$. Thus we write the expression for our assumed quasi-plane wave solution:

$$p'(\mathbf{x},t) = A(\mathbf{x}) \exp\{ik_0[W(\mathbf{x}) - c_0 t]\} , \tag{7.44}$$

which we then substitute into an acoustic equation having an adiabatically varying speed of sound:

$$\nabla^2 p' - \frac{1}{c^2(\mathbf{x})} \frac{\partial^2 p'}{\partial t^2} = 0 \,. \tag{7.45}$$

The result of the differentiations of Eq. 7.45 in Eq. 7.44 is a complex expression,

$$0 = \left\{ \nabla^2 A - A \left[(\nabla W)^2 - \frac{c_0^2}{c^2} \right] \right\} + i \left\{ A \nabla^2 W + 2 \nabla A \cdot \nabla W \right\}, \tag{7.46}$$

whose real and imaginary parts must each be equal to zero. The assumption of slow variation is then seen to be equivalent to the assumption that the fractional curvature, $(1/A) \nabla^2 A$, is small over a wavelength, which means that the phase function, $W(\mathbf{x})$, must satisfy the *eikonal equation*:

$$(\nabla W)^2 \equiv \left(\frac{\partial W}{\partial x} \right)^2 + \left(\frac{\partial W}{\partial y} \right)^2 + \left(\frac{\partial W}{\partial z} \right)^2 = n^2 = \frac{c_0^2}{c^2} \,. \tag{7.47}$$

Constant values of $W(\mathbf{x})$ represent surfaces of constant phase, or *wave fronts*; the normals to the phase fronts, termed *rays*, are the paths of energy flux away from the source (Fig. 7.4). A bundle of adjacent rays is termed a *pencil of rays*, and within it, the acoustic energy is constant along the propagation path (neglecting dissipation; ray theory does not accommodate the processes of absorption, diffraction, or scattering).

The eikonal equation is time- and frequency-independent, and it leads to a set of first-order ordinary differential equations that describes the paths, $\mathbf{s}(x,y,z)$, of individual rays through the water, given an initial value of sound speed, c_0, or index of refraction, n_0. These equations can be simply integrated to obtain estimates of the progress of an acoustic signal through space. Let the ray $\mathbf{s}$ have direction cosines $\cos \alpha$, $\cos \beta$, and $\cos \gamma$, which are also the slopes of the ray (Fig. 7.5). These quantities satisfy the relationship

$$\cos^2 \alpha + \cos^2 \beta + \cos^2 \gamma = 1 \,. \tag{7.48}$$

It is a small exercise in analytic geometry to then show that the ordinary differential equations for the ray are

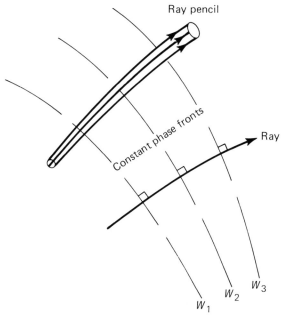

Fig. 7.4 Phase fronts and rays in a medium with a slowly varying speed of sound.

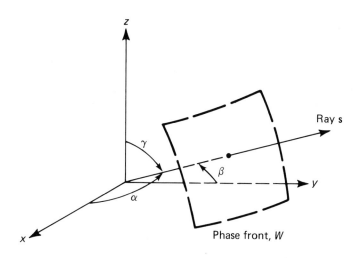

Fig. 7.5 Direction cosines giving the direction to a ray, **s**, and a phase front, W = constant.

$$\frac{d}{ds} (n \cos \alpha) = \frac{\partial n}{\partial x} ,$$

$$\frac{d}{ds} (n \cos \beta) = \frac{\partial n}{\partial y} ,$$

and

$$\frac{d}{ds} (n \cos \gamma) = \frac{\partial n}{\partial z} . \tag{7.49}$$

As an illustration of the application of the ray equations, let us assume that the index of refraction varies only in the vertical (which is the most important but not the sole variation), so that propagation is confined to the (x,z) plane. Then $\gamma = \pi/2 - \alpha$ and $\partial n/\partial x = 0$, so that

$$\frac{d}{ds} (n \cos \gamma) = \frac{\partial n}{\partial z} \tag{7.50}$$

and

$$\frac{d}{ds} (n \sin \gamma) = 0 , \tag{7.51}$$

and thus

$$n(z) \sin \gamma = \frac{c_0}{c(z)} \sin \gamma = \text{const} = \sin \gamma_0 . \tag{7.52}$$

By rearranging Eq. 7.52 slightly, we obtain the important equation:

$$\frac{c_0}{\sin \gamma_0} = \frac{c(z)}{\sin \gamma} . \tag{7.53}$$

This relationship is *Snell's Law*, and it can be used to construct the path of a ray starting at an angle γ_0 where the speed is c_0, and progressing through the medium where the speed is $c(z)$. From the definition of an element of arc, ds, it can be shown that the radius of curvature of a ray, R, is derivable from its curvature, $1/R$, via

$$\frac{1}{R} \equiv \frac{d\gamma}{ds} = - \frac{\sin \gamma_0}{c_0} \frac{dc}{dz} . \tag{7.54}$$

If it is assumed that a region of ocean has a linear velocity variation of the form $c(z) = c_0 + mz$ (where m is the slope, not the vertical wave number), the radius of curvature is then

$$R = -\frac{c_0}{\sin \gamma_0} \frac{1}{m} = \text{const.} ,\qquad(7.55)$$

which is a constant for a given ray. Hence R becomes the radius of a circle, and the path of a sound ray in a linearly varying sound profile is an arc of that circle. Equation 7.55 can be used to construct ray paths through an extended region of ocean by approximating the acoustic speed with a sequence of linear segments of varying slopes, m, and drawing short arcs of circles to define the ray, with $R < 0$ denoting a center *below* the initial angle. A simple example is shown in Fig. 7.6, and a more complicated one made up of segments of arcs is illustrated in Fig. 7.7. It can be seen in Fig. 7.7 that an acoustic speed profile that *decreases* with depth has the effect of refracting waves *downward*, while one that *increases* with depth bends rays *upward*. Such refraction is the origin of many of the vagaries of sound propagation in the sea.

It should be noted that ocean acoustics generally use the convention of *grazing angles*, $\gamma - \pi/2$ or $\theta - \pi/2$, rather than incidence angles, since most propagation occurs along nearly horizontal paths. However, to remain con-

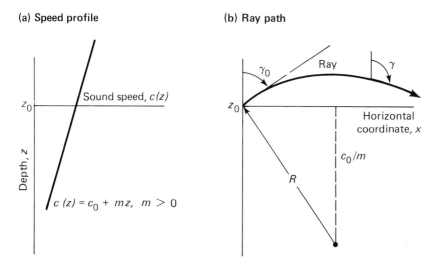

(a) Speed profile **(b) Ray path**

Fig. 7.6 (a) Sound speed profile varying linearly with depth, with speed decreasing in deeper water. (b) Resultant ray bending in circular arcs describes downward refraction. Opposite sign of speed gradient results in upward refraction.

(a) Speed profile   (b) Ray path

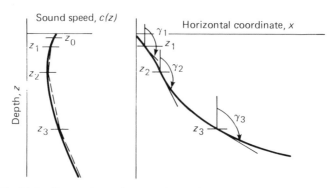

Fig. 7.7 (a) Continuously varying sound speed, $c(z)$, as approximated by a sequence of linear segments. (b) Ray paths composed of circular arcs starting at z_1 at an initial angle, γ_1.

sistent with the practice in the remainder of the book, we will continue with the incidence angle convention, unless otherwise noted.

The effects of the type of vertical variation in acoustic speed illustrated in Fig. 7.1 are shown in Fig. 7.8, where a source is assumed to be positioned in the sound channel at 1200 m depth, and to be radiating energy horizontally within angles of $\pm 15.2°$ from the horizontal, or over a beam width of approximately 30.4°. The rays oscillate about the axis of the deep sound channel, with upward-tending rays first bent toward the horizontal and then downward due to the temperature-dependent increase in sound speed toward the surface. Conversely, downward-tending rays subsequently bend upward, primarily because of the pressure-dependent increase in speed with depth (the term $\zeta_0 |z|$ in Eq. 7.19). Thus the sound channel serves as a waveguide for acoustic energy originating near its axis. Clearly, with this profile the water depth must exceed some 5 km if the $-15°$ ray is to avoid intersecting the bottom.

A sound source closer to the surface than the one in Fig. 7.8 will result in the kind of propagation shown in Fig. 7.9, under the usual conditions of mixed-layer/thermocline density structure; Fig. 7.9 shows how rays from a source at a depth of 15.8 m will behave in the vertical plane. Those rays leaving the source at angles greater than approximately $-1.76°$ will be refracted upward due to the pressure term in Eq. 7.19 and will be confined to the *near-surface sound channel* by a combination of upward refraction and downward surface reflection. These rays are often termed RSR, for refracted–surface reflected. Those leaving at steeper downward angles will be refracted toward greater depths. The net effect of these two factors is thereby to cre-

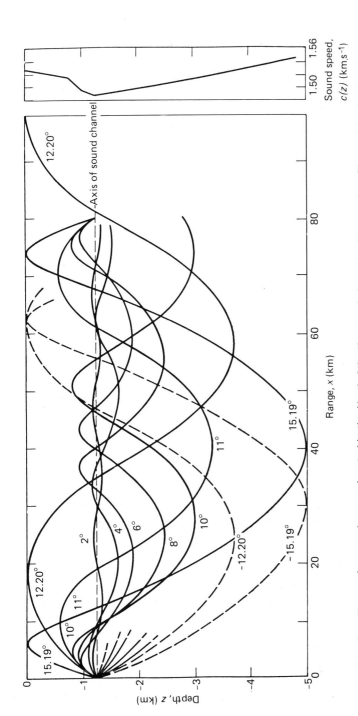

Fig. 7.8 Ray diagram for sound source located in the North Atlantic sound channel, with speed profile on right. Angles shown are measured with respect to horizontal, i.e., they are $\gamma - \pi/2$. [Adapted from Ewing, M., and J. L. Worzel, *Geo. Soc. Am. Memoir 27* (1948).]

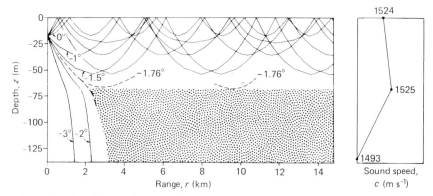

Fig. 7.9   Ray diagram for sound source located near $z = -16$ m, showing propagation and trapping of energy in the near-surface sound channel, and a shadow zone beneath 60 m. [Adapted from Urick, R. J., *Principles of Underwater Sound* (1982).]

ate an *acoustic shadow zone* below some 60 m depth and beyond a range of about 3000 m in this example, within which there are greatly reduced levels of insonification.

Another important effect of the increase of speed with depth is the occurrence of *surface convergence zones* that exist around a near-surface source at distances of approximately every 60 to 65 km. Near the surface, such zones should be considered as somewhat narrow, annular, concentric rings of increased intensity centered on the sound source, with amplification factors of 10 to 30 times normal propagation levels being common. Figure 7.10a illustrates rays emanating from an acoustic source at 91 m depth and shows three convergence zones having gradually increasing annular widths and therefore decreasing intensities. This phenomenon results in the propagation of detectable sound energy over very great distances (occasionally in excess of several hundred kilometers), provided the water is deep enough. For typical temperate-zone sound speed profiles, the water depth must exceed some 4800 to 5000 m and the near-surface temperature must be below roughly 23°C for convergence zone propagation to occur. However, the exact conditions are somewhat complicated and depend on several parameters. Shallow features such as the mid-ocean ridges interrupt this propagation to a considerable degree. Sources at greater depths result in less sharp focusing and more variable behavior, as Figs. 7.10b to 7.10d show.

7.8   Propagation Loss

The intensity of an acoustic wave varies inversely with distance for several reasons, the major ones being (1) spreading loss, (2) absorption by sea wa-

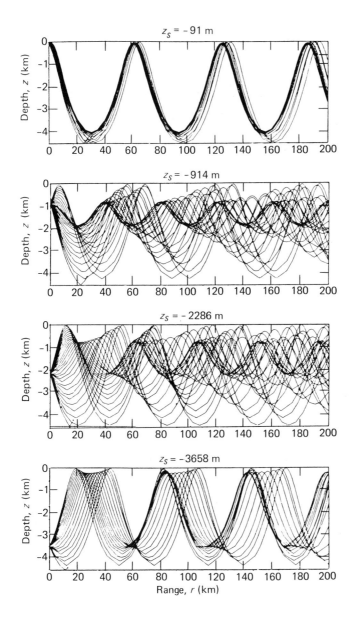

Fig. 7.10 Ray diagrams showing convergence zone phenomenon for sources at varying depths. [Adapted from Urick, R. J., *Principles of Underwater Sound* (1982).]

ter, (3) absorption by bottom materials, and (4) scattering by surface, volume, and bottom scatterers. A fifth factor, convergence zone focusing, can represent a local propagation *gain* (see Fig. 7.14 ahead).

The solution to the free-space wave equation in spherical coordinates shows that the pressure falls off inversely with the radial coordinate, r, with the intensity therefore varying as $1/r^2$. For a source, Q_0, in a spherically symmetric system, Eq. 7.45 may be written as

$$\frac{1}{r^2} \frac{\partial}{\partial r} \left(r^2 \frac{\partial p'}{\partial r} \right) - \frac{1}{c^2} \frac{\partial^2 p'}{\partial t^2} = 0 , \tag{7.56}$$

whose asymptotic solution is

$$p'(r) \sim \frac{-i\omega\rho}{4\pi r} Q_0 \exp[i(kr - \omega t)] , \tag{7.57}$$

with the associated average intensity

$$\langle I(r) \rangle = \frac{1}{2} \operatorname{Re} (p'u^*) = \frac{\rho\omega^2 Q_0^2}{32\pi^2 cr^2} , \tag{7.58}$$

where the quantity Q_0 is the volume flow of the medium at the surface of the source, with units of cubic meters per second, and is known as the *source strength*. This shows that in the absence of boundaries, inverse-square spreading loss obtains. However, this must be modified by the absorption losses as discussed in the context of Eq. 7.14, so that a frequency-dependent attenuation term, $\alpha(\omega)$, must be appended to the solution. Thus the form of the range-dependent intensity becomes

$$\langle I(r) \rangle = I_0 \left(\frac{r_0}{r} \right)^2 \exp[-\alpha(r - r_0)] , \tag{7.59}$$

where I_0 is the acoustic intensity at a reference distance of r_0, usually taken to be 1 m. In decibels, the propagation loss is given by $10 \log_{10} (I/I_0)$, or $20 \log (p/p_0)$, the latter because of the quadratic dependence of intensity on pressure (Eq. 7.32).

The form of the intensity attenuation coefficient, $\alpha(\omega) = 2k_i(\omega)$, as a function of frequency is complicated because of the varied contributions to it arising from several constituents of seawater. Attenuation is the loss of acoustic energy per unit path length due to both absorption and scattering. In addition to frequency, $\alpha(\omega)$ depends on temperature, pressure, salinity, and pH. In order of increasing frequency, the major contributions to $\alpha(\omega)$ are due to (1) volume scattering by inhomogeneities, and absorption by (2) boric acid, $B(OH)_3$, (3) magnesium sulfate, $MgSO_4$, and (4) molecular viscosity. Figure 7.11 summarizes its functional dependence on frequency for $T = 4°C$ and $p = 10^5$ Pa; in Fig. 7.11, the absorption and attenuation coefficients are given in decibels per kilometer, where the conversion factor from reciprocal meters to decibels per kilometer is $10^3 \times 10 \log e = 4.343 \times 10^3$. We will discuss the origins of each contribution briefly.

At low frequencies, say below 20 Hz, a frequency-independent attenuation coefficient, α_1, occurs and is approximated by

$$\alpha_1 \simeq 6.9 \times 10^{-7} \text{ m}^{-1}$$

$$\simeq 0.003 \text{ dB km}^{-1} . \tag{7.60}$$

Its origin is assumed to be scattering by inhomogeneities in the water column (discussed later) and leakage out of the deep sound channel. Next, at frequencies above roughly 100 Hz, molecular absorption from $B(OH)_3$ becomes dominant. The functional form of this absorption is somewhat like *Debye absorption* of electromagnetic radiation by a water molecule (see Chapter 8), and is of the form

$$\alpha_2(\omega) = \frac{A_2}{t_2} \frac{(\omega t_2)^2}{1 + (\omega t_2)^2} \quad \text{m}^{-1} . \tag{7.61}$$

This dependence on ω is quadratic for frequencies well below the *relaxation* frequency, $\omega_2 = 1/t_2$, but becomes constant for $\omega t_2 \gg 1$. The *relaxation time*, t_2, for boric acid is approximately 1.6×10^{-4} s at 4°C, but depends on absolute temperature, T, and salinity, s, approximately as

$$1/t_2 = 3.8 \, (s/35)^{1/2} \times 10^{(7 - 1051/T)} \quad \text{s}^{-1} . \tag{7.62}$$

At $s = 35$ psu and $T = 273 + 4$ K, this gives a value of $1/t_2 = 6.13 \times 10^3 \text{ s}^{-1}$. The coefficient A_2 is also slightly temperature- and pH-dependent.

An analogous behavior obtains for $MgSO_4$, with an expression of the form

$$\alpha_3(\omega) = \frac{P(p)}{t_3} \frac{A_3 S(s) \, (\omega t_3)^2}{1 + (\omega t_3)^2} \quad \text{m}^{-1} , \tag{7.63}$$

where the pressure-dependent factor is represented by $P(p) = 1 - 6.46 \times 10^{-9} p_0$, and where p_0 is the hydrostatic pressure in pascals. The relaxation time for this absorber is considerably shorter and its reciprocal is approximated by

$$1/t_3 = 1.38 \times 10^{(11 - 1520/T)} \quad s^{-1}, \tag{7.64}$$

which is equal to $4.5 \times 10^5 \ s^{-1}$ at 4°C. The salinity dependence is given by $A_3 S(s) = 3.73 \times 10^{-7} \ s$.

The final contribution to acoustic absorption, $\alpha_4(\omega)$, is due to viscosity (see Eq. 7.14). When due regard is made for the previously neglected contributions to dissipation arising from longitudinal strains of the form $\partial u_i / \partial x_i$ (see the discussion following Eq. 3.52), the net result is to multiply the shear coefficient of viscosity, μ, by a factor of 4/3. The form for viscous absorption then becomes $\alpha_4(\omega) = 4\mu\omega^2/3\rho c^3$. However, a more detailed study of molecular absorption by water reveals the need for three corrections to this expression. The first has to do with the existence of the two microstates of water cited in Section 4.2; the mechanical compression of an acoustic wave causes a small increase in the population of the more closely packed microstate and a concomitant apparent doubling of the effective coefficient of viscosity, so that the value of μ to be used in Eq. 7.65 is approximately 2×10^{-3} kg $(m \ s)^{-1}$. The second is a dependence on hydrostatic pressure, as given by $P(p)$ above. The third correction models the temperature-dependence of the observed viscous absorption in seawater by introducing the same relaxation time associated with $MgSO_4$, that is, t_3, as a multiplier to the theoretical formula. It is possible that the process at work here is hydration by the ionic components. The form for viscous absorption thus becomes

$$\alpha_4(\omega) = \frac{8}{3} \frac{\mu\omega^2}{\rho c^3} (\omega_3 t_3) P(p) \quad m^{-1}, \tag{7.65}$$

where $\omega_3 = 9 \times 10^5 \ s^{-1}$. Beyond these, there appears to exist another relaxation rate for viscosity, $1/t_4$, of about $3.9 \times 10^{12} \ s^{-1}$, which is the far-infrared region. Since this lies well above any acoustic frequencies of interest here, the low-frequency functional approximation of Eq. 7.65 is sufficient, with the increased numerical value of μ being used. However, for electromagnetic waves, this molecular relaxation is extremely important (see Chapter 8). Figure 7.11 illustrates the four contributions to acoustic attenuation separately, along with their sum. The general quadratic dependence of the sum $\alpha(\omega) = \Sigma \ \alpha_j(\omega)$ is clear.

Examples of propagation loss are shown in Figs. 7.12 and 7.13 for frequencies of 2 and 8 kHz, respectively, with various combinations of source,

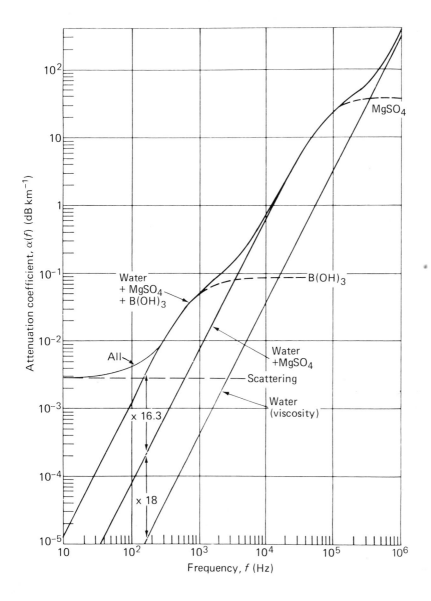

Fig. 7.11 Absorption or attenuation coefficients for seawater at $T = 4°C$ and at zero depth. The major contributions are due to boric acid, magnesium sulfate, and molecular viscosity. Scattering contributes an attenuation factor that is approximately constant. [Adapted from Fisher, F. H., and V. P. Simmons, *J. Acoust. Soc. Am.* (1977).]

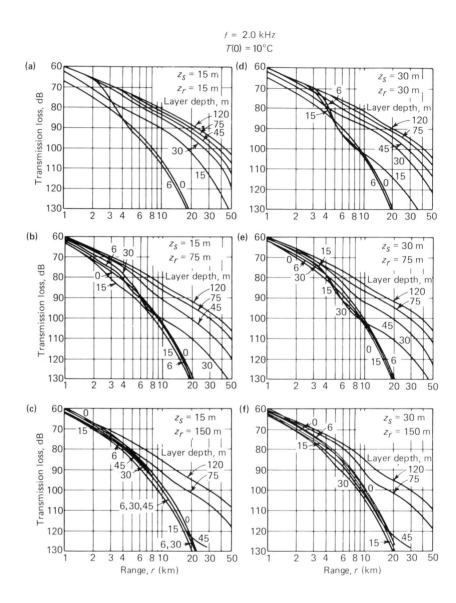

Fig. 7.12 Transmission loss for an acoustic signal at 2 kHz. The left-hand side, (a) to (c), is for the source at a depth of $z = -15$ m and the receiver at 15, 75, and 150 m; the parameter is mixed-layer depth. The right-hand side, (d) to (f), is for the source at 30 m and the receiver at 30, 75, and 150 m. Surface temperature $T_s = 10°C$. [Adapted from Condron, T. P., et al., *USN Underwater Sound Laboratory, TM-1110-14-55* (1955).]

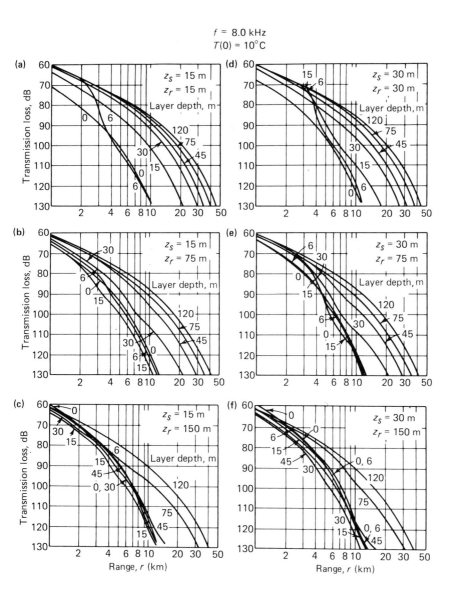

Fig. 7.13 Same as for Fig. 7.12, except at a frequency of 8 kHz. [Adapted from Condron, T. P., et al., *USN Underwater Sound Laboratory, TM-1110-14-55* (1955).]

receiver, and mixed layer depths. The losses are due to spreading, attenuation, and refraction. Zero dB is the loss at 1 m, which is the reference distance, r_0, of Eq. 7.59. Since this is a log-log plot, a straight line with a slope of -2 (20 dB per decade of range) represents inverse-square-law spreading. For deep mixed layers, most of the loss above that is due to absorption; however, for shallow layer depths, refraction and the resultant shadow zone play a much more important part in establishing propagation loss. No effect of convergence zones is visible in these plots because of the limited range on the graphs.

A diagram schematically summarizing the various factors contributing to overall transmission loss is given in Fig. 7.14, which is a log-log plot of intensity versus range. Inverse-square spreading sets the minimum loss, but is modulated somewhat by the process of interference between the directly transmitted and surface reflected waves at close ranges, a phenomenon termed the *Lloyd Mirror effect*. There are also increases at the convergence zones (CZ), two of which are shown. Absorption takes its toll with range at varying rates, depending on frequency, with the diagram shown implicitly being for low frequencies. Various kinds of ducting and bottom losses (to be discussed below) introduce further significant variations in transmission.

The determination of actual propagation loss in the ocean is a difficult and complicated undertaking that clearly requires detailed subsurface thermal information, bottom properties, and bathymetry for its correct derivation.

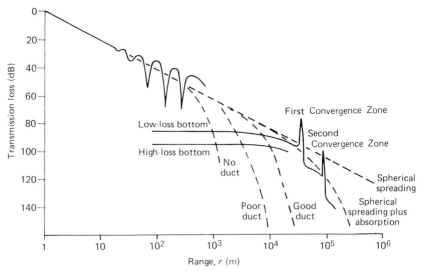

Fig. 7.14 Schematic transmission loss, illustrating contributions from various processes and conditions. [Adapted from Urick, R. J., *Sound Propagation in the Sea* (1982).]

7.9 Reflection and Scattering from Ocean Surfaces

The guided wave character of sound propagation in the sea is often estab-lished by reflections from the surface and the bottom. Each of these sur-faces has a somewhat different effect on the acoustic field and must be treated separately. We begin by describing generically the reflection process from a surface that is locally planar and time-invariant, and then generalize the basic reflection coefficients to other more complicated situations. When the surface becomes even slightly random, a statistical theory is essential to de-scribe the propagation; this topic falls under the rubric of *scattering in a ran-dom medium*, a rich and complicated subject that is touched on in Chapter 8 in the context of electromagnetic scattering. We will limit the discussion of scattering here to a few basic formulas and graphical results.

In order to derive the needed relationships for the *pressure reflection* and *transmission coefficients*, R_c and T_c, respectively, consider a plane wave in-cident on a flat surface dividing two regions of the oceanic environment. Each region has it own values of density, sound speed, and absorption, with the upper and lower regions being labled by subscripts 1 and 2, respectively. The incident and reflected rays are at angles with respect to the vertical of $\theta_i = \theta_r$: the refracted/transmitted ray is at an angle of θ_t (Snell's law). Fig-ure 7.15 shows two cases: (1) Figure 7.15a illustrates reflection in going from dense to rare, corresponding to the ocean–air interface; (2) Figure 7.15b shows the condition in going from rare to dense, corresponding to the water–bottom interface. In both cases the density and sound speed in the lower medium

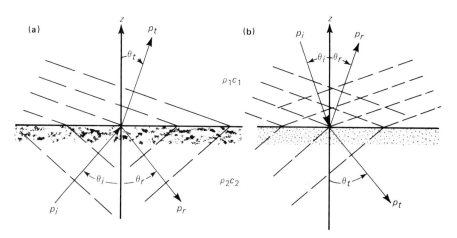

Fig. 7.15 Reflection and refraction at a plane interface: (a) ocean–air, and (b) bottom–ocean.

are greater than those in the upper (with the possible exception of propagation into soft muds; see Section 7.10).

In order to satisfy the boundary conditions, there must be both a transmitted and a reflected wave, with the horizontal components of their wave vectors being equal to that of the incident wave. However, the vertical components, m_1 and m_2, will be different. Additionally, the boundary condition requiring continuity of pressure across the interface (Eq. 3.122) demands that

$$p_i + p_r = p_t . \tag{7.66}$$

Also, continuity of the normal component of particle velocity further requires that

$$u_i \cos \theta_i + u_r \cos \theta_r = u_t \cos \theta_t . \tag{7.67}$$

Additional relationships between the p_j and the u_j come from Eq. 7.22 (with no loss term), and take the forms of:

$$p_i = \rho_1 c_1 u_i ,$$

$$p_r = \rho_1 c_1 u_r , .$$

and

$$p_t = \rho_2 c_2 u_t . \tag{7.68}$$

We now assume that the incident, reflected, and transmitted waves are all planar and have the form (supressing the time-dependence)

$$p_i = p_0 \exp[i(kx - m_1 z)] ,$$

$$p_r = R_c p_0 \exp[i(kx + m_1 z)] ,$$

and

$$p_t = T_c p_0 \exp[i(kx - m_2 z)] , \tag{7.69}$$

where R_c and T_c are the Rayleigh pressure reflection and transmission coefficients, respectively:

$$R_c = p_r / p_i$$

and

$$T_c = p_t/p_i \, . \tag{7.70}$$

The solutions obtained by eliminating the p_j and u_j give for these coefficients:

$$R_c = \frac{\rho_2 c_2 \cos \theta_i - \rho_1 c_1 \cos \theta_t}{\rho_2 c_2 \cos \theta_i + \rho_1 c_1 \cos \theta_t} \tag{7.71}$$

and

$$T_c = \frac{2\rho_2 c_2 \cos \theta_i}{\rho_2 c_2 \cos \theta_i + \rho_1 c_1 \cos \theta_t} \, , \tag{7.72}$$

as well as Snell's Law and the reflection angle conditions:

$$\frac{\sin \theta_i}{c_1} = \frac{\sin \theta_t}{c_2} \tag{7.73}$$

and

$$\sin \theta_i = \sin \theta_r \, . \tag{7.74}$$

The reflected and transmitted intensities depend on the absolute squares of R_c and T_c. Reflection from the sea surface results in a very small transmitted signal into the atmosphere (of the order of 4×10^{-6} or less) and the concomitant near-total reflection of acoustic energy back into the sea. This result can be obtained from Eq. 7.71 by substituting values for the acoustic impedance, ρc. At the bottom, on the other hand, there can be an appreciable transmitted/refracted intensity, up to angles of refraction approaching 90°; this is equivalent to total internal refraction in optics. The critical incidence angle, θ_{ci}, for a 90° transmitted angle is given by

$$\sin \theta_{ci} = \frac{c_1}{c_2} \, . \tag{7.75}$$

The totally refracted wave travels along the bottom as an evanescent edge wave whose amplitude falls off exponentially in both media.

Real surfaces are much more complicated than our planar geometry, of course, and many additional effects come into play. The first is the interference between the directly transmitted and the reflected waves mentioned above, a process termed the Lloyd Mirror phenomenon; this is indicated schematically in Fig. 7.16. The source and receiver are located as shown in Fig. 7.16a, with the reflected rays behaving as if they had originated at a virtual image source located above the surface. Since in the region close to the source, both the transmitted and reflected waves are reasonably self-coherent, at least at frequencies below a very few kilohertz, they set up an interference pattern whose behavior is illustrated in Fig. 7.16b, with deep nulls where destructive interference occurs. However, any appreciable random character to the surface will begin to fill in the valleys in the signal level, until the mirror effect disappears completely at higher sea states; it similarly disappears at frequencies above several kilohertz.

Another factor that significantly modifies the surface reflection coefficient is surface waves. As was discussed in Chapter 5, the amplitude distribution of surface waves is approximately Gaussian with a standard deviation equal to $\frac{1}{4} \xi_{1/3}$ (cf. Eq. 5.108). To a reasonable approximation, the slope distri-

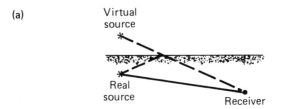

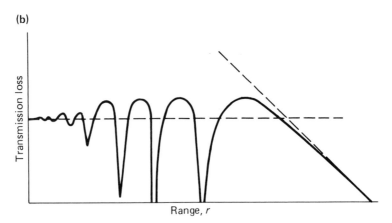

Fig. 7.16 Schematic of received signal intensity as a function of slant-range separating source and receiver. [Adapted from Urick, R. J., *Sound Propagation in the Sea* (1982).]

bution is similarly normal with a slope variance that increases with wind speed. Such a surface randomizes the local angles of reflection, spreading them out into a broadened pattern. The resultant scattering distribution in angle is described by the *scattering coefficient*, $S_c(\theta_i,\theta_r,\beta)$, which is a function of both incidence and scattering angles. It also depends on the rms slope of the surface wave slope distribution, σ_ζ; the latter is, of course, a function of wind speed. The scattering coefficient can also be considered as the scattering cross section per unit surface area and per unit solid angle, $d^2\sigma/dA\,d\Omega$.

From Eq. 5.95b, the surface wave slope variance is

$$\sigma_\zeta^2 = \langle \zeta^2 \rangle \equiv \langle (\nabla\xi)^2 \rangle \, , \qquad (7.76)$$

where the ensemble average is now taken over the entire Gaussian distribution of both gravity and capillary wave slopes. The *Cox and Munk slope distribution* (see Chapter 9) relates this quantity to surface wind speed, u_w, via

$$\sigma_\zeta^2 \equiv \tfrac{1}{2}\tan^2\beta = a + bu_w \, , \qquad (7.77)$$

where β is an rms slope angle, and the coefficients a and b depend on wind azimuth angle. Nominal values averaged over wind direction are $a \simeq 2.5 \times 10^{-3}$ and $b \simeq 1.97 \times 10^{-3}$, for u_w in m s^{-1}. By properly forming stochastic averages of the scattered intensity over the slope distribution function, the *bistatic scattering coefficient* may be obtained. The result is (cf. Eq. 8.118):

$$S_c(\theta_i,\theta_r,\beta) = F_1(\theta_i,\theta_r)\cot^2\beta\,\exp[-F_2(\theta_i,\theta_r)\cot^2\beta] \, . \quad (7.78)$$

Here the functions F_1 and F_2 are

$$F_1(\theta_i,\theta_r) = \left[\frac{1 + C_iC_r - S_iS_r}{C_i(C_i + C_r)}\right]^2 \left[\frac{C_i^2}{\pi(C_i + C_r)^2}\right] \, . \qquad (7.79)$$

and

$$F_2(\theta_i,\theta_r) = \frac{(S_i - S_r)^2}{(C_i + C_r)^2} \, , \qquad (7.80)$$

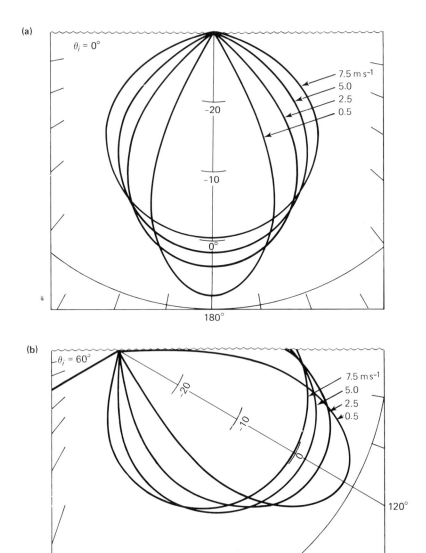

Fig. 7.17 Polar diagrams of the sea surface scattering coefficient in decibels for incidence angles of 0° and 60°, with wind speed as a parameter. [Adapted from Urick, R. J., *Sound Propagation in the Sea* (1982).]

where S_i, S_r, C_i, and C_r are abbreviations for $\sin \theta_i$, $\sin \theta_r$, $\cos \theta_i$, and $\cos \theta_r$. The term "bistatic" arises because the source and receiver are located at the two angles, θ_i and θ_r, with respect to the scattering area element. Figure 7.17 illustrates the calculated relative distribution of scattered energy with the angle θ_r, for normal incidence ($\theta_i = 0°$) and for $\theta_i = 60°$. Wind speed is the parameter, with speeds up to 7.5 m s^{-1} being shown. At higher wind speeds, the reflected energy is fairly widely dispersed in angle; hence the name "scattered" as a synonym for random, incoherent reflection.

At incidence angles beyond some 20 to 30°, there are very few specularly reflecting facets to contribute to the scatter, and another mechanism becomes dominant: *Bragg scatter*. In this case, the surface waves act as a kind of random diffraction grating with a wide range of "grating spacings" and "ruling directions." An incident acoustic wave selects out of the spectral distribution of spatial wave numbers and azimuthal directions those Fourier components that lead to constructive interference of the wave with itself. From the grating equation of optics, the relationship for constructive interference in the backscattered direction is

$$n\lambda = 2d \sin \theta_i \ , \qquad (7.81)$$

where $\lambda = 2\pi/k$ is the acoustic wavelength, $d = 2\pi/k_h$ is the surface wavelength, and n is the grating order. For first-order ($n = 1$) Bragg scatter from ocean waves, this relationship becomes

$$k_h = k_B = 2k \sin \theta_i \ , \qquad (7.82)$$

where k_B is the *Bragg surface wave number*. In addition, the surface wave crests must be nearly parallel to the acoustic wave crests. Figure 7.18 illustrates the Bragg condition for backscatter. As an example, a 10 kHz acous-

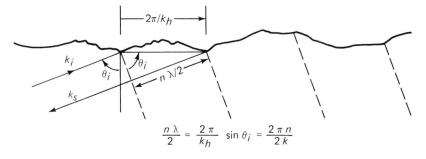

Fig. 7.18 Bragg scattering condition from a random sea surface for the case of $\theta_r = -\theta_i$, or backscatter. Constructive interference occurs when the distance between acoustic crests is an integral number of half wavelengths.

tic wave incident at $\theta_i = 60°$ scatters from ocean surface waves having approximately 0.09 m wavelengths, i.e., those in the ultragravity range.

Another physical process contributing to acoustic surface interactions is scattering from near-surface air bubbles in the upper 5 to 10 m of the mixed layer, especially at large incidence angles (low grazing angles). This source becomes increasingly important as the wind speed extends above white-capping speed—some 7 to 8 m s^{-1}.

Figure 7.19 shows the behavior of the *backscattering strength*, defined as

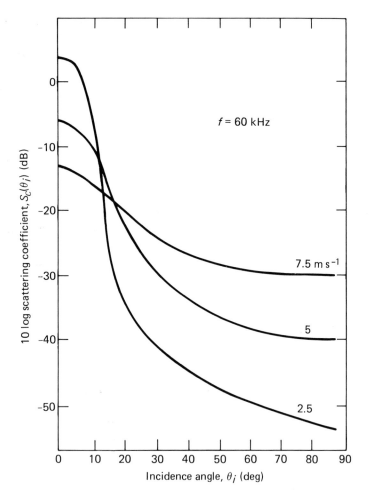

Fig. 7.19 Behavior of backscatter coefficient as a function of incidence angle and wind speed. Different scattering mechanisms dominate over different angular regions. [Adapted from Urick, R. J., *Sound Propagation in the Sea* (1982).]

$10 \log S_c(\theta_i, -\theta_i, \beta)$, over the entire range of incidence angles. In the region near vertical incidence, say out to $\theta_i \approx 25°$, backscatter from specular facets appears to dominate; this is the range wherein the Gaussian distribution of surface wave slopes has a significant population of facets oriented perpendicularly to the incident wave vector, which then act as small acoustic mirrors. In the range of angles from roughly 25 to 60°, Bragg scatter occurs from wave Fourier components meeting the grating condition. Finally, in the region from some 60 to 90°, the backscatter appears to be due to the near-surface bubble layer, which is relatively sensitive to wind speed but insensitive to angle. A certain amount of shadowing of incident acoustic waves by surface waves undoubtedly occurs here as well. Equation 7.78 is mainly applicable to the region of near-vertical incidence, the so-called *specular point* or *geometric optics* domain. (We will develop equations for specular point scattering and for Bragg scatter in the "slightly rough" region in the context of electromagnetic propagation in Chapter 8.)

Still another modification of scattering by real sea surfaces is the *Doppler effect*. If the acoustic source is of reasonably pure frequency, the surface wave velocity induces a range of Doppler shifts on the frequency of the scattered energy that is proportional to the surface velocity components aligned with the incident wave vector. Both the surface wave velocity and the mean current velocity (if one exists) contribute to the frequency displacement; the former introduces a *Doppler spreading*, $\langle \delta f \rangle$, about the source frequency, f_t, as indicated in Fig. 7.20. If the component of surface velocity along the line connecting a scattering element and the colocated source and receiver is $|\mathbf{u}| \cos \gamma$, then the Doppler shift induced by the scatterer due to its velocity, $\mathbf{u}$, is

$$\delta f = f_t \, \frac{2|\mathbf{u}|}{c} \cos \gamma . \tag{7.83}$$

A summation of all contributions from a random distribution of moving scatterers will yield the backscattered intensity at various Doppler-shifted frequencies, which are displaced from the central, transmitted frequency, f_t. This distribution will be a mapping of the wave height spectrum, $S(\omega/2\pi)$, into a *Doppler spectrum, $I_s(f_t + \langle \delta f \rangle)$*, with advancing waves contributing positive shifts and receding waves, negative shifts. The scattered intensity spectrum will appear as sidebands distributed about f_t, as suggested by Fig. 7.20. In addition, a *mean current* component along the line of sight will induce a Doppler frequency *shift* of the spectral components, Δf. Therefore the Doppler spectrum is a functional of the surface wave spectrum, transforming the latter in a convoluted fashion. It is possible to deconvolve the

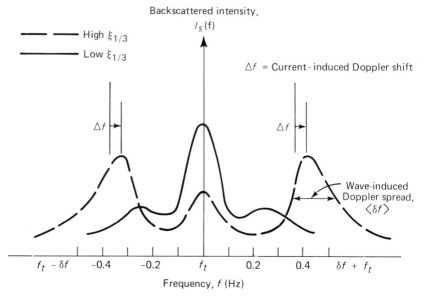

384

Fig. 7.20 Doppler spectra due to scattering from surface waves. Broadening is due to wave frequency distribution, while Doppler shift rises from a component of mean current along the line of sight.

acoustic signals to obtain reasonable estimates of both surface spectrum and current, however. Thus both wave spectra and mean current may be determined acoustically if the velocity of the source is known accurately.

7.10   Reflection, Refraction, and Scattering at the Sea Floor

From the brief description of the nature of the sea floor given in Chapter 2, it is clear that a much more diverse set of physical properties exists at the bottom than at the surface, and that a more comprehensive description is required to account for acoustic propagation in this regime. A reasonable approach must take into account the behavior of the reflection coefficient and scattering on a wide range of spatial scales; in addition, transmission and refraction occur in the bottom materials, with both longitudinally polarized acoustic waves and transverse *shear waves* being excited. These are attenuated by frequency-dependent absorption in the rocks, sands, and sediments.

Figure 7.21 is a simplified diagram showing rays for the various wave types just cited, with the phase front spacings proportional to their phase speeds; C and S denote compressional and shear waves, respectively. CR denotes

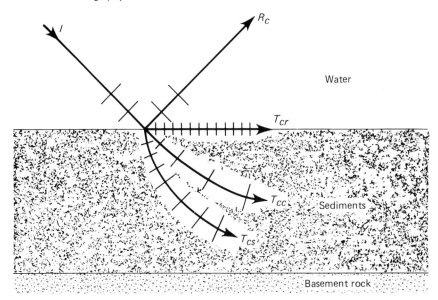

Fig. 7.21 Rays and wavefronts for water/sediment interface. *I*: incident ray. R_c: reflected ray. T_{cr}: refracted edge ray. T_{cc}: transmitted compressional (sound) wave. T_{cs}: transmitted shear wave. [Adapted from Urick, J. R., *Sound Propagation in the Sea* (1982).]

an edge wave known variously as the *Rayleigh, Stoneley,* or *Scholte* wave.

Reflection from the bottom exhibits many of the same features as from the surface, including interference between directly transmitted and reflected waves, scattering, and near-surface absorption. However, the penetration of the C and S waves into the bottom, which is accompanied by upward refraction and reflection from layers, brings an added dimension to the propagation. It also offers the possibility of probing the sub-bottom with low-frequency, high-power acoustic signals. This latter is illustrated in Fig. 7.22, which is a bottom profile made with a vertically propagating compressed pulse of acoustic energy swept between approximately 60 and 150 Hz, using a technique called *seismic reflection profiling*. The compressed pulse can in principle resolve sedimentary layers as thin as $c/2\Delta f \sim 8.3$ m in the vertical. On the left-hand side of Fig. 7.22 may be seen basement rock as deep as 1500 m below the overlying sedimentary layers. The variations in sediment layer thickness from left to right are thought to be due to variations in the speed of the bottom water and hence in the efficiency of scouring the sediments off the bedrock. The basement rock is quite rough and gives evidence of plate tectonic forces at work.

The densities of bottom materials and the speed of acoustic waves in them

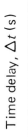

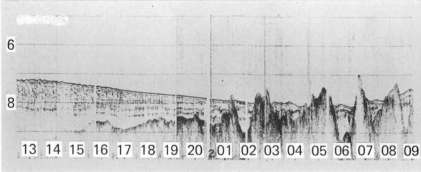

Time, t (h)

Fig. 7.22 Sub-bottom sedimentary layers and underlying bedrock mapped using seismic reflection profilometry. Sector shown is in the South Argentine Basin near the Falkland Escarpment, and was made with a low-frequency acoustic profiler. [From Ewing, J., and M. Ewing, *The Sea*, Vol. 4 (1970).]

are quite variable, with the *porosity* of the sediments, or the fraction of their volume occupied by water and gas bubbles, ranging from essentially zero to as great as 75%. Muds and silts have high porosity and low density and their compressional wave speeds can actually be a few percent less than the overlying water. Hard sands, which have low porosity and higher bulk density, may have speeds of 10 to 20% greater than seawater. Figure 7.23 shows the variation of c with porosity, Q. Table 7.1 gives measurements of acoustic constants for various types of near-surface sediments. As the depth of the sediments increases below the bottom, the included water is squeezed out by the overlying pressure and the speed increases rapidly, at least for the first kilometer or so. The associated refractive effects are large, so that sound energy entering sediments may emerge from the bottom after several kilometers of travel, as suggested by the ray diagrams on the right side of Fig. 7.24; the left side illustrates $c(z)$. During the sub-bottom propagation phase, sound is subject to additional frequency-dependent attenuation that is generally linearly proportional to frequency, as Fig. 7.25 shows. Clearly low-frequency acoustics is preferred for sub-bottom profiling of deep features; however, the vertical resolution is not high. Somewhat higher frequencies are preferred for probing shallow sediments. For a given frequency, attenuation in sediments is two to three orders of magnitude higher than in seawater.

The loss in transmitted energy after reflection from the sea floor is quite variable and depends on the geological province as well as incidence angles, frequencies, and roughness. Figure 7.26 suggests how the loss varies with incidence angle for two different regions of the North Atlantic—one a flat,

TABLE 7.1
Average Measured Constants, North Pacific Sediments*

Sediment type	Porosity, Q (%)	Density, ρ (10^3 kg m^{-3})	Sound speed, c (m s^{-1})
Continental terrace (shelf and slope) Sand			
Coarse	38.6	2.03	1836
Fine	43.9	1.98	1742
Very fine	47.4	1.91	1711
Silty sand	52.8	1.83	1677
Sandy silt	68.3	1.56	1552
Sand–silt–clay	67.5	1.58	1578
Clayey silt	75.0	1.43	1535
Silty clay	76.0	1.42	1519
Abyssal plain (turbidite)			
Clayey silt	78.6	1.38	1535
Silty clay	85.8	1.24	1521
Clay	85.8	1.26	1505
Abyssal hill (pelagic)			
Clayey silt	76.4	1.41	1531
Silty clay	79.4	1.37	1507
Clay	77.5	1.42	1491

*[From Hamilton, E. L., *Physics of Sound in Marine Sediments* (1974).]

sediment-filled plain and the other a sloping island rise. The sediment loss dominates when it is present in sufficient quantity.

Scattering from the bottom is similarly subject to wide variability; as for surface roughness, a statistical description is in order. The probability distribution function for the amplitude of sea floor undulations, $h(x,y)$, is assumed to be Gaussian (as was the equivalent surface wave distribution), and may be written

$$p(h)\ dh = \frac{1}{(2\pi\sigma_h^2)^{1/2}} \exp(-h^2/2\sigma_h^2)\ dh\ . \qquad (7.84)$$

Here σ_h is the rms value of the undulations, which are assumed to be statistically *homogeneous* on a local scale. Statistical homogeneity is the spatial equivalent to temporal *stationarity*, meaning that the probability distribution function is essentially the same throughout the local region of interest,

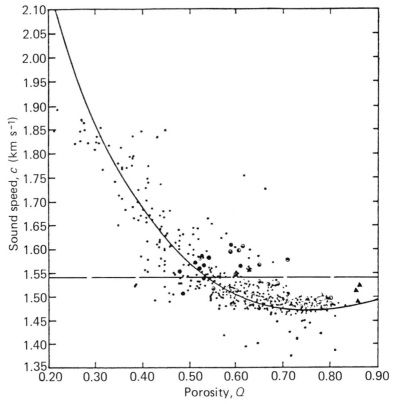

Fig. 7.23 Sound speed in sediments having included water, versus porosity. Measurements made on core samples. The speed of sound in water at $T = 4°C$ and $z = -4000$ m is 1547 m s^{-1}. [From Schreiber, B. C., *J. Geophys. Res.* (1968).]

no matter where its character is determined; such homogeneous character does not preclude the mean, standard deviation, or even higher-order moments from varying slowly in space, but only requires that they be fixed locally. The bottom roughness acts as a spatial Fourier analyzer, just as the surface roughness does for temporal fluctuations, and the bistatic scattering coefficient, $S_c(\theta)$ (which gives the scattered level of acoustic energy in terms of the Rayleigh reflection coefficient, R_c) is also Gaussian with the form

$$S_c(k, \theta - \theta_r) = |R_c|^2 \exp\{-[2k\sigma_h \tan(\theta - \theta_r)]^2\}, \qquad (7.85)$$

where R_c is given by Eq. 7.71. Thus the angular variation of the bottom-scattered signal is exponentially small in the quantity $\tan^2(\theta - \theta_r)$, with

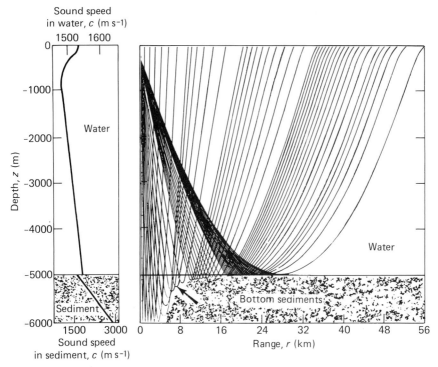

Fig. 7.24 Ray paths calculated from speed profile on left, for a source at 38 m depth. Steeper rays enter bottom, undergo total refraction, and emerge back into water column. Arrow shows first observable ray in practice. [Adapted from Christensen, R. E., et al., *J. Acoust. Soc. Am.* (1975).]

departures of the angle θ from the angle of specular reflection, θ_r. For slightly rough bottoms, σ_h is small and the reflection distribution is sharply peaked about the geometrical angle of reflection; for larger degrees of roughness, the distribution is broadened.

7.11 Reflection and Scattering by Bodies in the Water Column

Reflection and scattering from geometrically delimited objects in the water differ from the cases of reflection and scattering from the *distributed* surface and bottom, in that the dimensions of such objects are often on the order of the wavelength of the probing sound; large targets such as submarines and whales are exceptions, of course. For finite-sized targets, the concepts of the *scattering function* and *target cross section* are appropriate descriptors of the reflected energy. If I_i is the incident acoustic intensity,

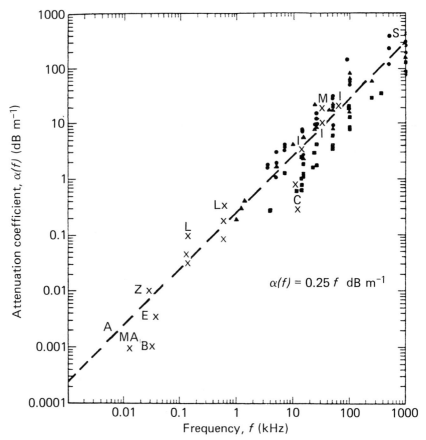

Fig. 7.25 Attenuation in natural saturated sediments and sedimentary strata. Data indicate a linear dependence on frequency. [Adapted from Hamilton, E. L., *Physics of Sound in Marine Sediments* (1974).]

each element of area dA on the target can be considered as a source emitting secondary Huygens wavelets. At a distance r from the target, the reflected intensity, dI_r, radiating from the element can be written as

$$dI_r = \frac{I_i}{r^2} \, S \, dA \, , \qquad (7.86)$$

where the scattering function, $S = S(k, \theta_r, \theta_i, \phi_i)$, depends on the polar angles of the incident and scattered sound and on frequency. (This function is essentially the same as the quantity S_c introduced in Eq. 7.78 for a *dis-*

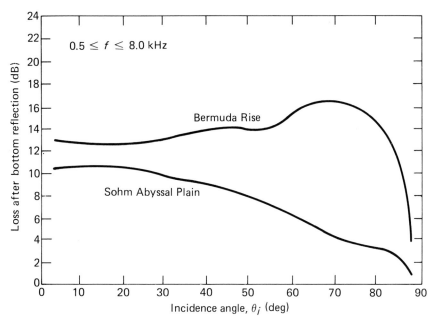

Fig. 7.26 Loss of acoustic intensity after scattering from the bottom in two physiographic provinces in the North Atlantic. Abyssal Plain is flat and sediment filled; Bermuda rise is steeper and rougher. [Adapted from Urick, R. J., *Sound Propagation in the Sea* (1982).]

tributed target.) The element dA also depends on the incidence angle geometry. The relationship of S to the differential cross section, $d^2\sigma$, is given by its product with dA and $d\Omega$, the element of solid angle centered on dA, thus:

$$d^2\sigma = S \, dA \, d\Omega . \tag{7.87}$$

Since the dimensions of the differential cross section are square meters, S is dimensionless; it may be thought of as the cross section per unit area per unit solid angle. Figure 7.27 shows the geometry of the scattering interaction, and indicates how the integrations to obtain the total cross section are to be carried out:

$$\sigma = \int_0^{4\pi} \int_A S \, dA \, d\Omega \tag{7.88}$$

$$= \frac{1}{I_i} \int_0^{4\pi} \int_A r^2 \, dI_r \, d\Omega . \tag{7.89}$$

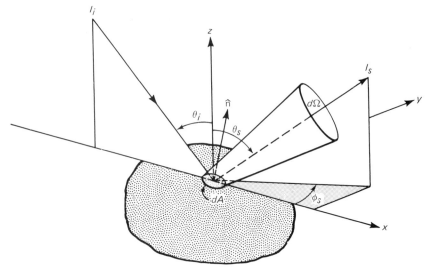

Fig. 7.27   Geometry of scattering from an elemental area dA into a solid angle, $d\Omega$.

Actual scattering calculations are difficult to carry out because of the complicated geometries and the multiple dependencies inherent in Eq. 7.88. For spheres (and a few other simple shapes), it is possible to evaluate the scattering function explicitly. The behavior of scattering from spheres is a prototype for the behavior of scattering from more complicated shapes, which often scatter much as do spheres when their dimensions are small compared with a wavelength. Rayleigh derived the scattering function for a sphere of radius R, reciprocal bulk modulus a_s, and density ρ_s; the sphere is assumed to be immersed in water whose analogous properties have subscripts w. In the limit of small kR, the Rayleigh formulation becomes

$$S(k,\theta) = \frac{(kR)^4}{\pi} \left[\frac{a_w - a_s}{3a_w} - \left(\frac{\rho_s - \rho_w}{2\rho_s + \rho_w} \right) \cos \theta \right]^2 . \quad (7.90)$$

For a rigid sphere, one has $a_w \gg a_s$; under this condition, the behavior of S with kR is shown in Fig. 7.28 for the condition of *backscatter*, i.e., $\theta = 180°$. Three regimes are apparent: (1) for wavelengths small compared with the radius, the scattering varies as k^4, or with wavelength as λ^{-4}; this is termed the *Rayleigh scattering region*, and in optics, such scattering is the explanation for the blueness of the sky; (2) for kR near unity, the *resonance regime* is reached $\vee$ 'ıerein the cross section oscillates above and below its asymptotic value due to constructive and destructive interference of waves

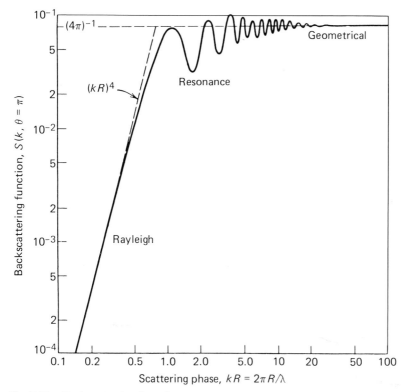

Fig. 7.28 Backscattering function, or normalized cross section per unit solid angle, for a rigid sphere of radius R as a function of scattering phase, kR. [Adapted from Stenzel, H., *Elektr. Nachr. Tech.*, Vol. 15 (1938).]

traversing the surface of the sphere; (3) for $kR \gg 1$, the *geometrical regime* occurs, wherein the wavelengths are small compared to the target radius and the cross section has the geometrical value of πR^2; here the scattering function has the value $1/4\pi$.

To complete the picture, Fig. 7.29 shows scattering at angles other than 180°. Here we have plotted the variation of S with angle as measured from the forward scattering direction, with the parameter kR ranging from much less than unity to 10. In the Rayleigh regime, $kR \ll 1$, and forward scatter is small, while scattering at angles greater than some 90° is relatively large. There is a null in the scattered radiation pattern at the angle determined by the vanishing of the quantity within the square brackets of Eq. 7.90. (For electromagnetic scattering, this angle is 90°.) For $kR \approx 1$, the scattering becomes more symmetrical in angle. As kR increases, the forward scatter becomes more peaked and may actually exceed the incident intensity; in addi-

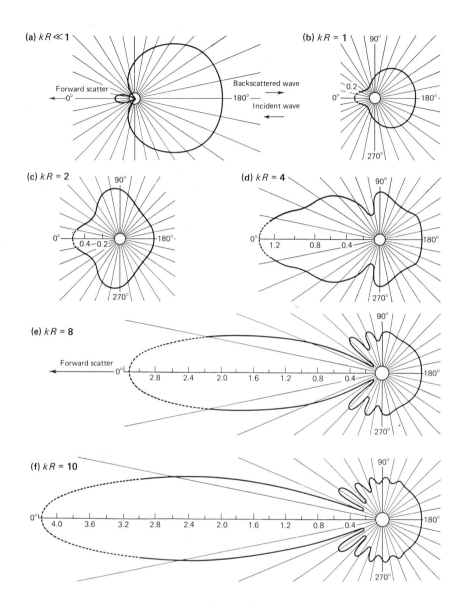

Fig. 7.29 Polar plots of pressure distribution of radiation scattered from a rigid sphere of radius R, as a function of scattering angle, θ, measured from direction of incident wave vector. Parameter is scattering phase, kR. Scattering amplitude in (a) for $kR \ll 1$ has been greatly enlarged over other cases. [Adapted from Stenzel, H., *Elektr. Nachr. Tech.*, Vol. 15 (1938).]

tion, for an appreciable range of angles around 180°, the backscatter appears to be uniform. This is the regime of *geometrical acoustics*, a name derived from the optical analogue and one in which the wavelength is small compared with the dimensions of the illuminated object.

With this description as a basis for understanding scattering from more complicated targets, we will briefly discuss three other examples: fish, plankton, and submarines.

Fish have air-filled sacks called *swim bladders* that are roughly prolate-ellipsoidal in shape and which can present a *resonant* target to sonar; under these conditions (all else being equal), their actual cross section can be two or more orders of magnitude greater than that of an equivalent rigid body, as shown in Fig. 7.30. This figure shows the normalized cross section of an

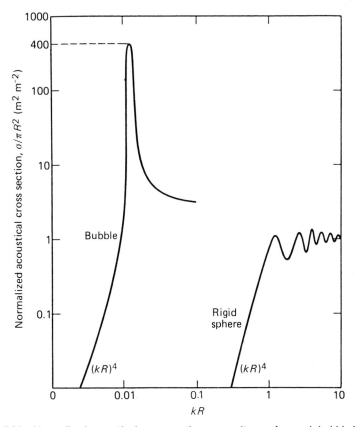

Fig. 7.30 Normalized acoustical cross sections per unit area, for an air bubble in water near the sea surface, and for a rigid sphere, both of radius R. Value of cross section for resonance peak corresponds to a natural resonance frequency of 52 kHz. [From Clay, C. S., and H. Medwin, *Acoustical Oceanography* (1977).]

air bubble in water having a natural resonance at 52 kHz at a pressure of 10^5 Pa, and compares it to that of a rigid sphere under the same conditions. This resonance makes the detection of fish with moderate- to high-frequency acoustic echo-sounding an altogether feasible undertaking. At high frequencies, the reflection may actually be due to body or skeletal formations.

Smaller planktonic forms scatter sound more nearly as Rayleigh scatterers, with their internal structures contributing increased cross sections over what would be expected from their geometrical size alone. The *deep scattering layer*, a stratum of plankton that rises at nightfall in response to reduced light levels and then sinks again at daybreak, is commonly observed with vertically viewing acoustic echo sounders. Another example is shown in Fig. 7.31, which illustrates the vertical motion of highly nonlinear internal waves as made visible by backscatter of 200 kHz acoustic signals from zooplankton that are neutrally buoyant and hence may undergo vertical excursions with the internal waves. Suspended sediments in the water also scatter acoustic energy at readily detectable levels, even though the intrinsic sizes of the sediment grains are extremely small compared to a wavelength. Apparently the sediments coagulate, or *flocculate* in seawater because of weak electrostatic molecular forces, thereby presenting acoustic targets of greatly increased effective size and scattering cross section.

Scattering from small discontinuities in the acoustic impedance of the water column is parameterized with a *volume scattering strength*, which is the normalized scattering intensity per unit volume, expressed in dB. Such scattering may be due to bubbles, density fluctuations, spatially distributed plank-

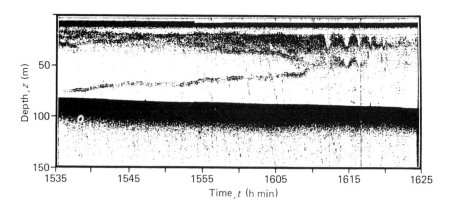

Fig. 7.31 Acoustic backscatter from plankton in continental shelf waters, as observed with a 200 kHz echo sounder. Nonlinear internal waves carry scattering layers through large vertical oscillations. The dark region near 80 to 90 m is due to reflection from the bottom.

ton concentrations, or unknown sources, and leads to *reverberation*, or drawn-out echos.

Submarines are large compared with all but the longest acoustic wavelengths and are therefore essentially geometric targets. Their *target strength* is an expression of their backscattering cross section per unit area and unit solid angle, and is expressed in dB. It is a strong function of azimuth angle, ϕ, with the broadside view presenting target strengths of some 25 dB due to specular reflection from the pressure hull, and with the bow or stern view being near 10 dB, due to the smaller geometrical size and wake absorption. Figure 7.32 illustrates this variation for frequencies in the region of 25 kHz.

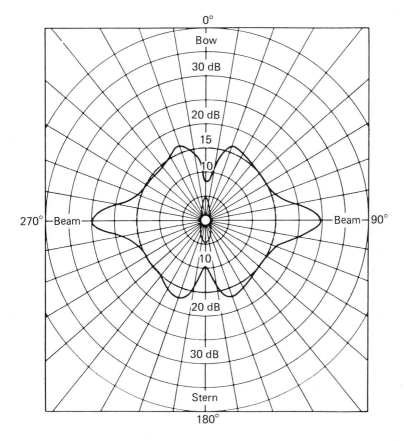

Fig. 7.32 Normalized backscatter cross section, or target strength of a submarine as a function of azimuth angle. Values are in dB relative to 1 m^2. [From Urick, R. J., *Principles of Underwater Sound* (1982).]

7.12   Acoustics and Geophysical Fluid Dynamics

The long distance transmission of acoustic energy through the ocean implies that an acoustic wave may sample a wide range of oceanic depths and horizontal paths during its transit. The signal received at the end of this transit has contained in its amplitude, frequency, and phase the integrated effects of sound velocity variations along the entire path. If the radiated signal characteristics are known sufficiently well, and if a wide enough variety of measurements is made, it is in principle possible to interpret the signal measurements to obtain information on the properties of the medium. This is the basis of acoustic *remote sensing*, and the interpretation of the observations is a problem in *inversion theory*.

From the discussions above, it is clear that a wide range of geophysical fluid flows can affect acoustic waves. Fine-structure, surface waves, internal waves, tides, mesoscale motions, and large-scale current systems all cause variations in transmission characteristics. In principle, their effects can be modeled by introducing their associated temperature, salinity, depth, and velocity dependencies into the vector index of refraction, **n**, *viz*:

$$\mathbf{n}(\mathbf{x},t) = \frac{\mathbf{k}c(\mathbf{x},t)}{\omega - \mathbf{U}_0(\mathbf{x},t)\cdot\mathbf{k}} , \qquad (7.91)$$

and then calculating the ray paths through the medium by a variational principle analogous to Fermat's principle in optics:

$$\delta \int_{\mathbf{x}_1}^{\mathbf{x}_2} \mathbf{n}\cdot d\mathbf{s} = 0 . \qquad (7.92)$$

Here the operation denoted by the variational operator, δ, is one of finding the path that makes the integral an extremum. In simplified form, this principle underlies the calculation of ray paths discussed in Section 7.7. An alternative but allied approach is to use the *wave action equation*, to determine intensity levels and their variations; this method is discussed in Section 8.9.

The concept of using acoustics to determine larger-scale dynamics is not entirely new, but its application to large volumes of the sea is relatively recent. A technique called *acoustic tomography* uses steady, low-frequency signals exchanged between several pairs of transmitter/receiver systems moored in the ocean. It derives its name from the tomographic technique used with x rays to probe human tissue via a 360° azimuthal scan of the body, followed by a Fourier inversion of the scattered x rays to obtain the distribution of scattering material. Figure 7.33 illustrates the site of an acoustic

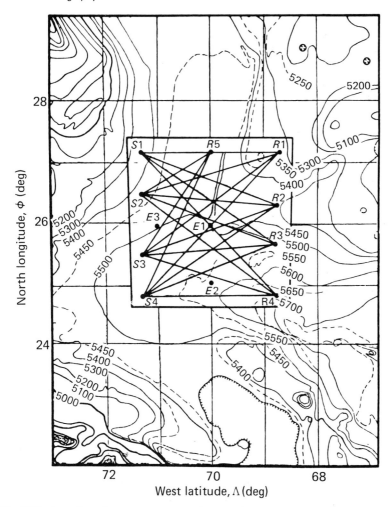

Fig. 7.33 Location of acoustic tomography experiment southwest of Bermuda. Four sources (*S*) transmit to each of five receivers (*R*) at hourly intervals. Environmental moorings at *E* are supplemented by in-water measurements. [From Munk, W. H., in *Oceanography: The Present and Future* (1983).]

tomography experiment southwest of Bermuda in which four sources, denoted by *S*, transmit to each of five receivers, *R*, for a total of 20 different *S–R* combinations. The time taken for an acoustic signal to propagate over a distance of the size of the 300 km square is approximately 200 s. However, for each horizontal path there are several multiple paths in the vertical that arrive at different angles off the vertical (see the right-handmost side of Fig.

7.34 near the R designator; this shows only downward-propagating rays). These multiple arrival angles imply that the propagation paths and hence the arrival times will differ over the same horizontal separation. The multiple paths, which may number from 10 to 20, allow discreet probing of the temperature profile of the ocean, since each traverses a different volume of the sea. The systematic variations in arrival times due to the various multiple paths through the mean vertical temperature profile may be as much as 3 to 4 s in the region shown. In the steady state, fluctuations in the signal travel times of order 100 ms due to tidal currents may easily be seen. Other geophysical fluid variations are also observed.

Now if a mesoscale dynamical system such as a cold core eddy (see Fig. 6.42, for example) moves through the array shown in Fig. 7.33, its temperature structure will sequentially affect all of the paths, both horizontal and vertical, past which it moves. Sophisticated inversion techniques may then be used to derive the thermal distribution from the multitude of measurements.

Although still in the experimental stage, acoustic tomography appears to offer a means for integral measurement of the temperature distribution in an oceanic volume, at least over linear dimensions of order 1000 km. Its extension to basin-wide scales (10^4 km) would require a large array of equipment and concomitant computer resources and analysis. While detectable acoustic signals from depth charges are known to have traveled 10,000 km, the extraction of intelligence from them is problematical.

7.13 Underwater Noise

No discourse on acoustical oceanography is complete without the mention of underwater ambient noise, i.e., undesired signals that emanate from a wide variety of sources at all ranges. Many noise sources are environmental, with wind, waves, rainfall, earthquakes, and bottom currents all contributing to the cacaphony of sound. Whales, porpoises, and certain kinds of pelagic species such as croakers and toadfish give off surprisingly large amounts of noise, much of which travels for appreciable distances. Anthropogenic noise originates from sources such as ship and submarine machinery, propeller motions, echo sounders, and underwater explosions, and also contributes to the ambient noise background. Lower-frequency sound may propagate for thousands of kilometers to arrive at a receiver at readily detectable levels; in fact, the presence of mid-ocean ridges located between source and receiver may effect low-frequency sound propagation more than any other factor entering into propagation loss. The presence of a multitude of sources implies that underwater noise is highly variable. At the highest frequencies,

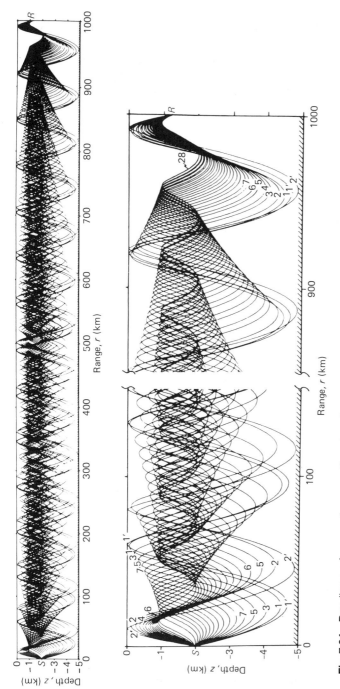

Fig. 7.34 Ray diagram for acoustic source, S, and receiver, R, separated by 1000 km. Lower panels are enlarged and show the first and last 100 km. Only rays that arrive at receiver from downward directions are plotted. [From Munk, W. H., in *Oceanography: The Present and Future* (1983).]

thermal noise sets an irreducible background. Figure 7.35 presents a schematic frequency spectrum for noise which indicates that below perhaps a few hundred hertz, man-made noise generally dominates, while in the kilohertz region, environmental noise is the major contributor. All of this assumes that the platform from which the noise observations are made is essentially si-

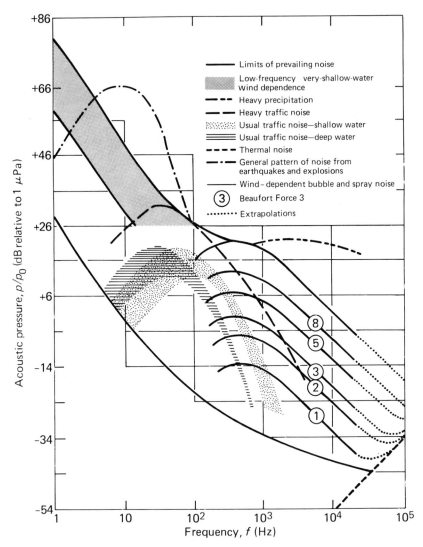

Fig. 7.35 Noise pressure levels in a 1 Hz band in dB (relative to 1 μPa) versus frequency. Minimum and maximum noise levels vary widely, with both natural and man-made sources contributing. [Adapted from Wenz, G. M., *J. Acoust. Soc. Am.* (1962).]

lent; however, very often *self-noise* from equipment or the flow of water about the platform is the limiting noise source against which detection of acoustic signals in the sea must be made.

Bibliography

Books

Beckmann, P., and A. Spizzichino, *The Scattering of Electromagnetic Waves from Rough Surfaces*, The Macmillan Co., New York, N.Y. (1963).

Brekhovskikh, L. M., *Waves in Layered Media*, Academic Press, Inc., New York, N.Y. (1960).

Clay, C. S., and H. Medwin, *Acoustical Oceanography*, John Wiley & Sons, New York, N.Y. (1977).

Denham, R. N., and A. C. Kibblewhite, in *Antarctic Oceanology II: The Australian-New Zealand Sector*, D. E. Hayes, Ed., *Antarctic Research Series*, Vol. 19, Am. Geophys. Union, Washington, D.C. (1972).

Eckart, C., *Hydrodynamics of Oceans and Atmospheres*, Pergamon Press, Inc., New York, N.Y. (1960).

Ewing, J., and M. Ewing, in *The Sea*, Vol. 4, Part 1, *New Concepts of Sea Floor Evolution*, A. E. Maxwell, Ed., John Wiley & Sons, New York, N.Y. (1970).

Flatté, S. M., Ed., *Sound Transmission Through A Fluctuating Ocean*, Cambridge University Press, Cambridge, England (1979).

Hamilton, E. L., "Geoacoustic Models of the Sea Floor," in *Physics of Sound in Marine Sediments*, L. Hampton, Ed., Plenum Press, New York, N.Y. (1974).

Morse, P. M., *Vibration and Sound*, McGraw-Hill Book Co., New York, N.Y. (1948).

Morse, P. M., and K. U. Ingard, *Theoretical Acoustics*, McGraw-Hill Book Co., New York, N.Y. (1968).

Munk, W. H., "Acoustics and Ocean Dynamics," in *Oceanography: The Present and Future*, P. G. Brewer, Ed., Springer-Verlag, New York, N.Y. (1983).

Officer, C. B., *Introduction to the Theory of Sound Transmission*, McGraw-Hill Book Co., New York, N.Y. (1958).

Tolstoy, I., and C. S. Clay, *Ocean Acoustics*, McGraw-Hill Book Co., New York, N.Y. (1966).

U.S. Navy, *Physics of Sound in the Sea*, U.S. Government Printing Office, Washington, D.C. (1969).

Urick, R. J., *Principles of Underwater Sound*, 3rd ed., McGraw-Hill Book Co., New York, N.Y. (1982).

Urick, R. J., *Sound Propagation in the Sea*, Peninsula Publishing, Los Altos, Calif. (1982).

Journal Articles and Reports

Christensen, R. E., J. A. Frank, and W. H. Geddes, "Low-Frequency Propagation via Shallow Refracted Paths through Deep Ocean Unconsolidated Sediments," *J. Acoust. Soc. Am.*, Vol. 57, p. 1421 (1975).

Cox, C. S., and W. H. Munk, "Measurements of the Roughness of the Sea Surface from Photographs of the Sun's Glitter," *J. Opt. Soc. Am.* Vol. 44, p. 838 (1954).

Condron, T. P., P. M. Onyx, and K. R. Dickson, *U.S. Navy Underwater Sound Laboratory Tech. Memo 1110-14-55* (1955).

Ewing, M., and J. L. Worzel, "Long-Range Sound Transmission," in *Geological Society of America Memoir 27* (1948).

Fisher, F. H., and V. P. Simmons, "Sound Absorption in Sea Water," *J. Acoust. Soc. Am.*, Vol. 62, p. 558 (1977).

Jones, L. M., and W. A. Von Winkle, U.S. Navy Underwater Sound Laboratory Report 632 (1965).

Schreiber, B. C., "Sound Velocity in Deep Sea Sediments," *J. Geophys. Res.*, Vol. 73, p. 1259 (1968).

Schulkin, M., and H. W. Marsh, "Absorption of Sound in Sea-Water," *J. Brit. IRE*, Vol. 25, p. 493 (1963).

Stenzel, H., *Elektr. Nachr. Tech.*, Vol. 15, p. 71 (1938). English translation in DRL Technical Report 159, Defense Research Laboratory, Univ. of Texas, Austin, Texas (1959).

Wenz, G. M., "Acoustic Ambient Noise in the Ocean: Spectra and Sources," *J. Acoust. Soc. Am.*, Vol. 34, p. 1936 (1962).

Chapter Eight

Electromagnetics and the Sea

The toughest job in Washington
is being able to tell the difference between
the tides, the waves, and the ripples.

Califano

8.1 Introduction

In this chapter we will examine electromagnetics as it applies to the sea, the interest in which arises from a number of problems occupying those who study the ocean. The areas to be examined include:

1. Interaction of radar and radio waves with the sea surface and near surface,
2. Emission of radio and microwave energy from the sea surface, and
3. Induced voltages caused by seawater moving in a magnetic field.

Interest in such problems stems in part from the capability of remotely measuring a number of physical properties of the ocean using electromagnetic radiation, thereby learning about certain limited but nevertheless important oceanic processes; the operational needs for communication and surveillance in the marine environment; and the propagation of infrared and optical wavelength energy in the sea (although the study of optics will, in the main, be deferred until Chapter 9).

The physical properties or processes that can be observed via electromagnetic radiation (albeit with varying accuracies and coverages) include the following: ocean surface wind stress (i.e., magnitude and direction), surface wave spectra, internal wave surface signatures, surface geostrophic current values, sea surface topography, sea surface temperature, and sea ice cover. Many of these have been alluded to in the discussions of the previous chapters. In addition, hydrodynamic transport rates in limited passages can be inferred from the electromagnetic potentials induced in electrodes and cables mounted on the sea floor.

8.2 Maxwell's Equations and Constitutive Relations

The starting point for our discussion must perforce be Maxwell's equations and their associated *constitutive relationships* giving the electromagnetic properties of seawater, e.g., dielectric constant and conductivity. We write those equations (using MKSA units) in terms of the electric field, **E**; the polarization field, **P**; the electric displacement, **D**; and the magnetic flux density, **B**; and use the fields **H** and **M** only as necessary in order to account for polarization influences and boundary conditions. The governing equations are then

$$\nabla \cdot \mathbf{E} = \frac{1}{\epsilon_0} \rho_{total} = \frac{1}{\epsilon_0} (\rho_{true} - \nabla \cdot \mathbf{P}) , \tag{8.1}$$

$$\nabla \cdot \mathbf{B} = 0 , \tag{8.2}$$

$$\nabla \times \mathbf{E} = -\frac{\partial \mathbf{B}}{\partial t} , \tag{8.3}$$

and

$$\nabla \times \mathbf{B} = \mu_0 \left[\mathbf{j}_{true} + \rho_{true} \mathbf{u} + \nabla \times (\mathbf{P} \times \mathbf{u}) + \frac{\partial}{\partial t} (\epsilon_0 \mathbf{E} + \mathbf{P}) \right] , \tag{8.4}$$

where ϵ_0 and μ_0 are the electric permittivity and magnetic susceptibility, respectively, of free space (see Appendix Four for numerical values). When applied to the ocean, the interpretation of these equations is, approximately:

1. The source of **E** field divergence is the *total* electric charge density, ρ_{total}, which is composed of *true* charges such as ions in water, plus both permanent and induced *polarization charges* described by **P**, the electric dipole moment per unit volume;
2. There is no divergence of magnetic flux, **B**, in the absence of magnetic charges, and lines of magnetic force are therefore closed;
3. Rotation or curl of **E** is induced by time changes in **B**;
4. Rotation or curl of **B** has sources in the *true* current density, $\mathbf{j}_{true}$ + $\rho_{true}\mathbf{u}$ (the second term due to ionic charge motions advected by the moving fluid at velocity **u**); in the curl of the advected electric dipole

moment per unit volume, $\mathbf{P} \times \mathbf{u}$; and in time changes of the electric and the polarization fields, whose sum is the electric displacement. No term appears in the magnetization field, $\mathbf{M}$, since seawater is essentially nonmagnetic.

The solutions to Maxwell's equations are enormously varied and the equations cover all aspects of electromagnetism, with no known exception.

Seawater is a moderately good conductor of electricity, with a conductivity, σ, and a dielectric "constant," ϵ, that are frequency-, temperature-, and salinity-dependent. It is also electrically strongly polarizable, as will be discussed below. At optical frequencies, the ocean is much more transparent than at radio or radar frequencies. This makes the *optics* of the sea take on a quite different flavor than for lower frequency electromagnetics. Its study is therefore taken up separately in Chapter 9.

The influences of the polarization charge density, $\nabla \cdot \mathbf{P}$, can be swept into the dielectric function, $\epsilon(\omega)$, by rewriting Eq. 8.1 in a form involving the electric displacement vector, $\mathbf{D}$, which is solely due to the true (ionic) charge density:

$$\nabla \cdot \mathbf{D} = \rho_{true} \; . \tag{8.5}$$

Now the relationship between $\mathbf{E}$ and $\mathbf{P}$ for a linearly polarized medium is

$$\mathbf{P} = \epsilon_0 \chi \mathbf{E} \; , \tag{8.6}$$

where χ is the *electric susceptibility*. Then $\mathbf{D}$ is related to $\mathbf{E}$ and $\mathbf{P}$ via

$$\mathbf{D} = \epsilon_0 \mathbf{E} + \mathbf{P} = \epsilon_0 (1 + \chi)\mathbf{E} = \epsilon \mathbf{E} = \epsilon_0 \epsilon_r \mathbf{E} \; . \tag{8.7}$$

Thus $\mathbf{D}$, $\mathbf{E}$, and $\mathbf{P}$ are all proportional, with the constants of proportionality describing the electrical response of the medium to the vacuum field, $\mathbf{E}$. Equation 8.7 defines the *relative dielectric function*, $\epsilon_r(\omega)$ (sometimes called the specific inductive capacity, κ); this function is complex for a lossy dielectric such as seawater. Its multiplication by the permittivity of free space, ϵ_0, converts ϵ_r to the *absolute dielectric function*, $\epsilon(\omega)$. In terms of the displacement vector, Eq. 8.4 for a *stationary* medium becomes

$$\nabla \times \mathbf{B} = \mu_0 \left(\mathbf{j}_{true} + \frac{\partial \mathbf{D}}{\partial t} \right) \; . \tag{8.8}$$

Now the true current density in a stationary medium is related to $\mathbf{E}$ by Ohm's Law, with the conductivity, σ, being the constant of proportionality:

$$\mathbf{j}_{true} = \sigma \mathbf{E} \ . \tag{8.9}$$

(See Eq. 8.200 for a generalization of Ohm's Law when the effect of a magnetic field becomes appreciable.) As with the dielectric constant, the conductivity is a complex function of frequency, temperature, and salinity. However, its intrinsic frequency dependence is small for wavelengths longer than submillimeter waves, and the numerical values of σ given below are considered adequate for most electrolytes in the range of oceanic salinities (30 to 38 psu), and for frequencies extending to well above 100 GHz.

If we now assume harmonic time-dependence for all quantities (as is appropriate for radiation fields), we may write Maxwell's equations for a nonmoving ocean as

$$\epsilon \nabla \cdot \mathbf{E} = \rho_{true} \ , \tag{8.10}$$

$$\nabla \times \mathbf{B} = \mu_0 (\sigma - i\omega\epsilon) \ \mathbf{E} \ , \tag{8.11}$$

$$\nabla \cdot \mathbf{B} = 0 \ , \tag{8.12}$$

and

$$\nabla \times \mathbf{E} = i\omega \mathbf{B} \ , \tag{8.13}$$

with the induced fields, $\mathbf{D}$ and $\mathbf{P}$, no longer appearing explicitly, their effects having been incorporated into σ and ϵ, and with the vacuum fields, $\mathbf{B}$ and $\mathbf{E}$, being the ones of interest. We must know how σ and ϵ depend on the material properties in order to do this. For such an analysis, plane wave solutions are most convenient.

By further assuming plane, polarized wave solutions to Maxwell's equations of the forms

$$\mathbf{E} = \mathbf{E}_0 \ \exp[i(\mathbf{k} \cdot \mathbf{x} - \omega t)] \tag{8.14}$$

and

$$\mathbf{B} = \mathbf{B}_0 \ \exp[i(\mathbf{k} \cdot \mathbf{x} - \omega t)] \ , \tag{8.15}$$

and substituting these into Eqs. 8.1 to 8.4, one obtains a relationship between $\mathbf{k}$ and ω that involves ϵ and σ:

$$\mathbf{k}^2 = \omega^2 \mu_0 \left[\epsilon + i \ \frac{\sigma}{\omega} \right] = \omega^2 \mu_0 \epsilon_0 \left[\epsilon_r + i \ \frac{\sigma}{\epsilon_0 \omega} \right] \ . \tag{8.16}$$

This relationship is clearly complex and therefore describes both wave propagation and dissipation.

Recalling that the free-space velocity of light, c_0, is

$$c_0^2 = \frac{1}{\mu_0 \epsilon_0} , \qquad (8.17)$$

then the ω–$\mathbf{k}$ relationship of Eq. 8.16 becomes

$$\frac{\mathbf{k}^2}{\omega^2} = \frac{1}{c^2} = \frac{1}{c_0^2} [\epsilon_r + i\sigma/\epsilon_0\omega] , \qquad (8.18)$$

which implies that the velocity of light in seawater, c, is less than in free space, since $\epsilon_r \geq 1$. Then in seawater, the complex velocity is

$$c = \frac{\omega}{k} = \frac{c_0}{\sqrt{\epsilon_r + i\sigma/\epsilon_0\omega}} , \qquad (8.19)$$

and the index of refraction is

$$n = \frac{c_0}{c} = \sqrt{\epsilon_r + i\sigma/\epsilon_0\omega} . \qquad (8.20)$$

Thus ϵ_r and σ establish both the propagation characteristics and the absorption of waves in the sea. We require equations describing their values as a function of the material parameters.

8.3 Dielectric Function and Electrical Conductivity

The Debye formulation of the dielectric function of a polar molecule is a reasonable one for *pure* water, with the polar contribution to the relative dielectric constant, $\epsilon_r = \epsilon/\epsilon_0$, being derived from a fairly simple relaxation theory. The functions found are then to be supplemented with a conductivity due to the ion concentrations in the oceanic electrolyte. The Debye formulation treats only the contribution to the total polarization arising from the *permanent* dipole moment (see Section 4.1) and does not attempt to describe the origins of the far-infrared value of $\epsilon_{ir} = 4.9$, or of the optical value, $\epsilon_{opt} = n_{opt}^2 = (1.34)^2 = 1.80$. The infrared and optical values are established by *induced* electric dipole moments and depend on details of both electronic and molecular rotational/vibrational transitions. Often ϵ_{ir} is denoted by ϵ_{∞}, but this notation can lead to a misinterpretation as being the same as ϵ_{opt}. Figure 8.1 illustrates the variation of the real and imaginary parts of ϵ_r over a range of microwave frequencies, and demonstrates its reduction from the low-frequency static values to intermediate levels as ω increases, as well as showing the effects of salinity at $s = 35$ psu. Figure 8.2 illustrates the real and imaginary parts of the index of refraction for pure water (where $n = \sqrt{\epsilon_r}$), ranging from the near-infrared to wavelengths of 100 m.

The physics of the polarization process is found in the tendency of *unbonded* polar molecules to line up with the incident electric field under its influence, which ordering is partially resisted by hydrogen bonding and by random thermal motions. Because of its large permanent dipole moment of 1.84 debye, water responds to an applied static field, $\mathbf{E}_s$, with a very large polarization field, $\mathbf{P}_s$; hence the high value for ϵ_s of 81. In addition, induced atomic and electronic polarizabilities result in a field denoted by $\mathbf{P}_{ind}$ and contribute the values cited for ϵ_{ir} and ϵ_{opt} above. The total polarization, $\mathbf{P}$, is the sum, $\mathbf{P}_s + \mathbf{P}_{ind}$.

If the applied electric field varies in time, the permanent dipoles attempt to follow it, but are resisted by molecular forces. The major resisting force is thought to reside in the partial ordering of the H_2O molecules according to the hydrogen-bonded structure of Fig. 4.1b. The applied field leads to the breakage of a certain fraction of these bonds and to their transition from the ice-like to the gas-like microstate. These unbonded polar molecules are then free to contribute to the electric polarization, with the energy that goes into breaking of bonds appearing as heat. A high frequency acoustical field appears to cause the same effect (see Section 7.8). The removal of the field results in a *relaxation* of the polarization as the hydrogen bonding recovers to its equilibrium level, with the relaxation time termed t_r. For pure water at $T = 20°C$, $t_r \approx 6.0 \times 10^{-11}$ s.

To derive the Debye expression for ϵ_r, it is assumed that the polar contribution, $\mathbf{P}_{pol}$, gives rise to ϵ_r according to Eq. 8.7:

$$\mathbf{D}_{pol} = \epsilon_0 \epsilon_r \mathbf{E} = \epsilon_0 \mathbf{E} + \mathbf{P}_{pol} , \qquad (8.21)$$

while that from the induced component contributes ϵ_{ir} (neglecting for now the optical contribution):

$$\mathbf{D}_{ind} = \epsilon_0 \epsilon_{ir} \mathbf{E} = \epsilon_0 \mathbf{E} + \mathbf{P}_{ind} . \qquad (8.22)$$

The total polarization gives ϵ_s:

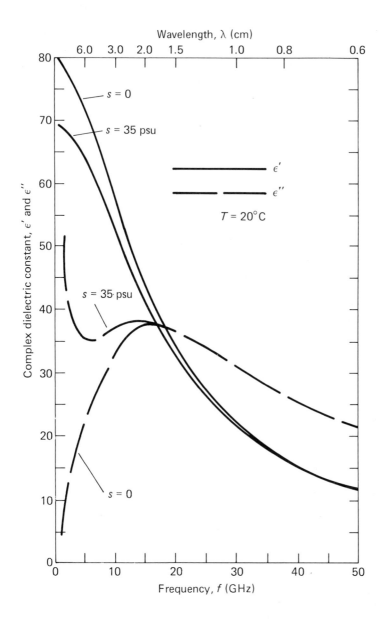

Fig. 8.1 Real and imaginary parts of the relative dielectric function at microwave frequencies, for two salinities. This basic function in part sets reflection, absorption, transmission, and emission coefficients. [Adapted from Wilheit, T. T., *The Nimbus-5 User's Guide* (1972).]

$$\mathbf{D} = \epsilon_0 \epsilon_s \mathbf{E} = \epsilon_0 \mathbf{E} + \mathbf{P} . \tag{8.23}$$

On the application of a time-varying field, $\mathbf{E} = \mathbf{E}_0 \exp(-i\omega t)$ (where the frequency ω is significantly below infrared frequencies), the orientation of the polar moments and production of induced moments approach their equilibrium values, accompanied by a certain amount of bond breakage. It is assumed that $\mathbf{P}_{pol}$ changes at a rate set by the departure of the instantaneous polarization from the equilibrium value, divided by the relaxation time:

$$
\begin{aligned}
\frac{d\mathbf{P}_{pol}}{dt} &= \frac{\mathbf{P} - (\mathbf{P}_{ind} + \mathbf{P}_{pol})}{t_r} \\
&= \frac{\epsilon_0 (\epsilon_s - \epsilon_{ir})}{t_r} \mathbf{E}_0 \, e^{-i\omega t} - \frac{\mathbf{P}_{pol}}{t_r} .
\end{aligned}
\tag{8.24}
$$

The solution to Eq. 8.24, assuming that all polarization fields follow the driving field, is

$$\mathbf{P}_{pol} = \frac{\epsilon_0 (\epsilon_s - \epsilon_{ir})}{1 - i\omega t_r} \mathbf{E}_0 \, e^{-i\omega t} , \tag{8.25}$$

where we have used Eqs. 8.21 to 8.23. From Eq. 8.21, the relative dielectric function due to the polar molecules is then

$$\epsilon_r(\omega) = \epsilon_{ir} + \frac{\epsilon_s - \epsilon_{ir}}{1 - i\omega t_r} . \tag{8.26}$$

After rationalization and separation into real and imaginary parts, we obtain for the real part

$$\mathrm{Re}\, \epsilon_r(\omega) = \epsilon_{ir} + \frac{\epsilon_s - \epsilon_{ir}}{1 + (\omega t_r)^2} , \tag{8.27a}$$

and for the imaginary part

$$\mathrm{Im}\, \epsilon_r(\omega) = \frac{\epsilon_s - \epsilon_{ir}}{1 + (\omega t_r)^2} \, \omega t_r . \tag{8.27b}$$

These are the forms of the Debye equations, and describe the propagation of electromagnetic fields below infrared frequencies in fresh water quite adequately. Figure 8.2 shows the form of the actual complex index of refraction, $n = n_r + in_i$, as a function of wavelength, with the Debye contribution shown dashed. The infrared molecular bands between wavelengths of

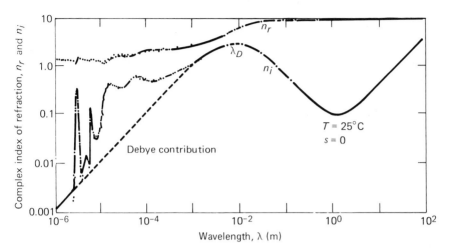

Fig. 8.2 Real and imaginary parts of the index of refraction for pure water from near-infrared to microwave lengths. Their character at mid- to very long IR wavelengths is set by Debye relaxation effects; at near-IR, the molecular polarizability establishes *n* as well. [From Ray, P. S., *Appl. Optics* (1972).]

10^{-3} and 10^{-6} m lead to significant departures from this simple theory, and the ionic conductivity makes another contribution.

The addition of salts to pure water complicates the behavior of ϵ_r significantly. The reduction in ϵ_r from the fresh water values, as shown in Fig. 8.1, is due to the process of *hydration* of the ionic components. Each ion orients approximately five to six water molecules around itself in a kind of sheath that partially immobilizes the molecules and prevents them from fully contributing to the polarization. The hydration also reduces the relaxation time by factors of order 10% for ordinary salinities.

The salt ions constitute the vast majority of the charge carriers contributing to the conductivity and they add additional dielectric and absorption terms to the complex index of refraction, according to Eq. 8.20. The DC conductivity, σ_s, is 4.29140 S m^{-1} (siemens per meter, or mhos per meter) at $T = 15.0°C$ and $s = 35.0$ psu. As the frequency varies at constant T and s, this value is assumed to hold approximately for frequencies up to those in the vicinity of the ion plasma frequency, ω_p, as we will now discuss.

An analogous simple theory (developed by Drude) of ionic conduction in a time-varying field yields a complex, frequency-dependent conductivity, $\sigma(\omega)$, given by

$$\sigma(\omega) = \frac{Ne^2 t_c}{\epsilon_r m (1 - i\omega t_c)} , \qquad (8.28)$$

where t_c is the *ion collison time*, N is the ion number density, e is its charge, and m is its mass. This form arises from a consideration of the equation of motion of an electric charge in a sinusoidally varying electric field. The electrical force on the charge is $e\mathbf{E}/\epsilon_r(\omega)$, but it is damped due to collisions with other entities in the fluid at a rate proportional to its velocity and inversely proportional to the collision time, t_c. Newton's equation is then

$$m\frac{d^2\mathbf{x}}{dt^2} = (e\mathbf{E}_0/\epsilon_r) \, e^{-i\omega t} - \frac{m}{t_c}\frac{d\mathbf{x}}{dt} , \qquad (8.29)$$

whose particular solution is

$$\mathbf{x} = \frac{-e\mathbf{E}/\epsilon_r}{m\omega(\omega + i/t_c)} . \qquad (8.30)$$

The current density, $\mathbf{j}$ (with units of amperes per square meter) is proportional to the number of charges per unit volume, N, their charge state, e, and their velocity, $d\mathbf{x}/dt$:

$$\mathbf{j} = Ne\, d\mathbf{x}/dt = \sigma\mathbf{E} \qquad \text{A m}^{-2} . \qquad (8.31)$$

Substitution of the time derivative of Eq. 8.30 into Eq. 8.31 yields the expression for conductivity above.

The complex form of σ can be separated into its real and imaginary parts, with the static conductivity for seawater being written as

$$\sigma_s = \sum_j \frac{N_j e_j^2 t_{cj}}{\epsilon_r m_j} , \qquad (8.32)$$

where the index j ranges over all ionic species present, both positive and negative. From the value of $\sigma_s = 4.29$ S m^{-1}, Eq. 8.32 yields a species-averaged collision time, t_c, of about 5×10^{-12} s. It should be noted that t_c depends on both T and s, while N depends only on s.

Useful regression equations for the static conductivity as a function of temperature and salinity are given by Eqs. 8.33 to 8.36:

$$\sigma_s(T,s) = \sigma(25,s) \, e^{-\beta\Delta} , \qquad (8.33)$$

$$\Delta = 25 - T , \qquad (8.34)$$

$$\beta = \beta(\Delta, s)$$

$$= 2.033 \times 10^{-2} + 1.266 \times 10^{-4} \Delta + 2.464 \times 10^{-6} \Delta^2$$

$$- s(1.849 \times 10^{-5} - 2.551 \times 10^{-7} \Delta$$

$$+ 2.551 \times 10^{-8} \Delta^2) , \tag{8.35}$$

and

$$\sigma(25, s) = s[0.182521 - 1.46192 \times 10^{-3} s$$

$$+ 2.09324 \times 10^{-5} s^2 - 1.28205 \times 10^{-7} s^3]. \tag{8.36}$$

(The inverse relationship giving salinity as a function of conductivity and temperature is available as an adjunct to the UNESCO equation of state; see Chapter 4.)

For time-varying fields, we may write the complex conductivity as

$$\sigma(\omega) = \sum_j \text{Re } \sigma_j + i \text{ Im } \sigma_j$$

$$= \sum_j \frac{\sigma_{sj}}{1 + (\omega t_{cj})^2} + i\omega\epsilon_0 \frac{\omega_{pj}^2 \, t_{cj}^2}{1 + (\omega t_{cj})^2} , \tag{8.37}$$

where ω_{pj} is the radian *ion plasma frequency*:

$$\omega_{pj} = \left(\frac{N_j e_j^2}{\epsilon_0 \, \epsilon_r \, m_j} \right)^{1/2} . \tag{8.38}$$

The plasma frequency due to the lightest ion is a *cutoff* frequency for ion oscillations in a charged medium, much as the buoyancy frequency of Eq. 5.30 is a cutoff frequency for internal waves (in fact, both depend on the square root of a density), except that ω_p is the *lowest* frequency that can propagate in an electrical conductor. Thus for a salinity of $s = 35$ psu, the concentration of Na^+ ions is $N \approx 2 \times 10^{26}$ ions m^{-3}, $\epsilon_r \approx 6.2$, and the Na^+ plasma frequency is $f_p \approx 250$ GHz. The sodium ion plasma frequency is presumably the frequency at which seawater ceases to be mainly a conductor and becomes primarily a dielectric. However, this is a region of wavelengths where not enough is known of the properties of electrolytes to say that the physics is completely understood.

Even very fresh water generally has enough dissolved salts to be a conductor at ordinary radar frequencies. For example, a lake with a salinity of 10^{-2} times oceanic (or $s = 0.36$ psu) has a plasma frequency of approximately 25 GHz, and thus is a conductor up to the low range of millimeter wavelengths. The surface of such a lake will reflect microwave radiation very nearly as well as the sea surface, and its conductivity will be close to σ_s.

Rewriting Eq. 8.18 to include the real and imaginary parts of both the dielectric function and the conductivity then gives an important relationship for the complex index of refraction:

$$\frac{\mathbf{k}^2 c_0^2}{\omega^2} \equiv n^2 = \epsilon_r + i\sigma/\epsilon_0\omega$$

$$= \left\{ \epsilon_\infty + \frac{\epsilon_s - \epsilon_{ir}}{1 + (\omega t_r)^2} - \frac{(\omega_p t_c)^2}{1 + (\omega t_c)^2} \right\}$$

$$+ i \left\{ \frac{(\epsilon_s - \epsilon_{ir})\omega t_r}{1 + (\omega t_r)^2} + \frac{\sigma_s}{\omega\epsilon_0[1 + (\omega t_c)^2]} \right\}$$

$$\equiv \epsilon' + i\epsilon'' . \tag{8.39}$$

This equation defines the total complex relative dielectric function, with the quantity ϵ_∞ being the short-wavelength dielectric constant. The real and imaginary parts of Eq. 8.39 are denoted by ϵ' and ϵ'', in conformity with the conventional usage; thus the complex ionic conductivity is subsumed into the complex dielectric function. It is worth noting that the general form of Eq. 8.39 is similar to that for acoustic absorption, Eqs. 7.61 to 7.64, except for the ω^2 in the numerator of the acoustic attenuation. Indeed, the same relaxation process described by t_r may be at work in both acoustics and electromagnetics.

Values of ϵ' and ϵ'' as a function of temperature at two different frequencies are presented in Tables 8.1 and 8.2. The variation with nominal temperature is significant.

TABLE 8.1

Values of ϵ' and ϵ'' for Distilled Water as a Function of Temperature ($f = 1.43$ GHz).

Temperature, T (°C)	ϵ'	ϵ''
5	84.54	10.742
10	82.96	8.888
20	79.56	6.285
30	76.14	4.783

TABLE 8.2
Values of ϵ' and ϵ'' for Distilled Water as a Function of Temperature
$(f = 2.653 \text{ GHz})$.

Temperature, T (°C)	ϵ'	ϵ''
5.5	80.52	20.03
15	79.57	13.75
24	77.44	10.18
30	75.88	8.61

[From Klein, L. A., and C. T. Swift, *IEEE Trans. Antennas and Propag.* (1977)].

The index of refraction for electromagnetic radiation, n, is related to the dielectric function and the conductivity via the first line of Eq. 8.39, *viz*:

$$n^2 = \epsilon' + i\epsilon'' . \tag{8.40}$$

Therefore, n is also complex:

$$n = n_r + in_i . \tag{8.41}$$

The wavelength dependencies of n_r and n_i in pure water illustrated in Fig. 8.2 show that n_r ranges from near 1.33 at visible wavelengths ($\lambda < 1$ μm) to close to 9 at radio wavelengths. The effects of molecular absorption are most apparent in the behavior of n_i at wavelengths shorter than 1 cm, beyond the Debye relaxation wavelength of some 0.02 m. Similar data for seawater do not appear to be available.

8.4 Reflection at a Plane Surface

The interaction of electromagnetic radiation with the ocean surface is most readily studied by first examining the behavior of plane wave solutions to Maxwell's equations at flat surfaces, and then building up rough surfaces out of small planar segments. We shall find that the reflection and refraction processes depend on ϵ and σ in very important ways.

Assume that a plane, polarized wave of the form $\exp[i(\mathbf{k} \cdot \mathbf{x} - \omega t)]$ is incident on a horizontal surface of the ocean at an angle of incidence, θ_i. Let $\hat{n}$ be the outward normal vector to the surface, and let the angles of reflection and refraction be θ_r and θ_t. Because of the transverse polarization of plane electromagnetic waves, two cases must be distinguished: (1)*Transverse* or *horizontal polarization* (with respect to the **E** vector), with the **E** field at right angles to the plane containing **k** and $\hat{n}$, the so-called *plane of incidence*, and with the **B** field in that plane; this is illustrated in Fig. 8.3a. (2) *Parallel* or *vertical polarization*, with **E** *in* the plane of incidence, in which event **B** is parallel to the surface; here "vertical" is something of a misnomer, since the **E** field has both horizontal and vertical components; this is illustrated in Fig. 8.3b. The terms "vertical" and "horizontal" are in common usage in the literature in spite of the seeming ambiguity.

If the ocean were in fact an excellent conductor, the penetration of the electromagnetic field beneath the surface would be very slight because the ionic and dipolar reactions to the electromagnetic force would always induce a surface charge density, Σ, and a surface current density, **K**, of just the correct amount required to completely shield out the fields. The surface boundary conditions on the normal and tangential fields for the perfect conductor case impose the following relationships:

1. Normal component of **E**:

$$\hat{n} \cdot \epsilon \, \mathbf{E}_a = \Sigma . \tag{8.42}$$

2. Tangential component of **B**:

$$\hat{n} \times \mathbf{B}_a = \mu_0 \mathbf{K} . \tag{8.43}$$

3. Normal component of **B**:

$$(\hat{n} \cdot \mathbf{B})_a = (\hat{n} \cdot \mathbf{B})_w . \tag{8.44}$$

4. Tangential component of **E**:

$$\hat{n} \times \mathbf{E}_a = \hat{n} \times \mathbf{E}_w . \tag{8.45}$$

Here the subscripts a and w refer to air and water, respectively. If the fields in water, $\mathbf{E}_w$ and $\mathbf{B}_w$, were truly exactly zero just beneath the surface, the boundary conditions of Eqs. 8.44 and 8.45 would require that they also be zero in air immediately above the surface. This situation is shown in Fig. 8.4a, which illustrates the behavior of the parallel ($\parallel$) and perpendicular ($\perp$) components of **E** and **B** near the ocean surface.

An instructive example is the one of finite conductivity, $\sigma = \sigma_s$, and zero dielectric loss. We first solve Maxwell's equations for the fields in air, assuming that the conductivity is infinite. Then we apply the boundary conditions to those equations in the conductor, using the actual value of σ and

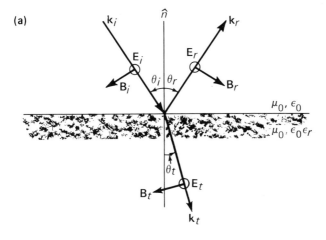

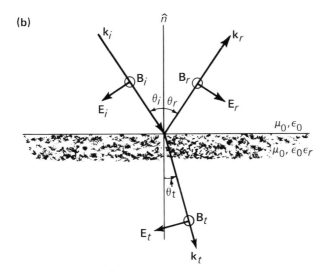

Fig. 8.3 Nomenclature for reflection at a smooth horizontal surface. (a) Transverse or horizontal polarization, with electric field at right angles to plane of incidence; (b) parallel or vertical polarization, with **E** field in plane of incidence.

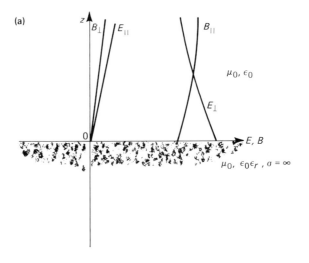

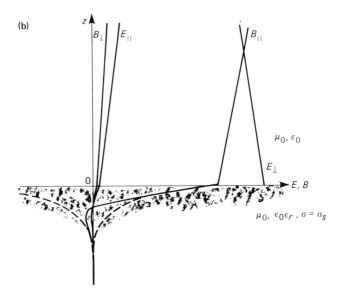

Fig. 8.4 (a) Electric and magnetic fields at the surface of a perfect conductor. Charge mobility completely shields out the interior fields, which requires that $E_\parallel$ and $B_\perp$ also be zero just outside the surface. (b) Fields near the surface of a good but not perfect conductor. Within the material, $B_\parallel$ is the dominant source of energy. Now small $E_\parallel$ and $B_\perp$ exist across the boundary. The dashed curves indicate the exponential envelope of oscillating field amplitudes. [Adapted from Jackson, J. D., *Classical Electrodynamics* (1975).]

taking account of the fact that the spatial variations in the z direction are much more rapid than the variations along the surface. Thus in Eqs. 8.10 to 8.13, we can then replace the del operator with

$$\nabla \simeq \hat{n} \frac{\partial}{\partial z} , \qquad (8.46)$$

where z is upward, as previously. We may also neglect the displacement current term in Eq. 8.11, $\partial(\epsilon \mathbf{E})/\partial t$, because of the assumption of high conductivity. Under these conditions, Maxwell's equations in seawater reduce to

$$\mathbf{E}_w \simeq \frac{1}{\mu_0 \sigma_s} \hat{n} \times \frac{\partial \mathbf{B}_w}{\partial z} , \qquad (8.47)$$

$$\mathbf{B}_w \simeq -\frac{i}{\omega} \hat{n} \times \frac{\partial \mathbf{E}_w}{\partial z} , \qquad (8.48)$$

and

$$\hat{n} \cdot \mathbf{B}_w \simeq 0 . \qquad (8.49)$$

Equations 8.47 and 8.48 can be combined to obtain a differential equation for the tangential magnetic flux density in water, $\hat{n} \times \mathbf{B}_w$:

$$\frac{\partial^2}{\partial z^2} (\hat{n} \times \mathbf{B}_w) + \frac{2i}{\delta^2} (\hat{n} \times \mathbf{B}_w) \simeq 0 . \qquad (8.50)$$

Here the quantity δ is the high-conductivity *skin depth* for electromagnetic radiation:

$$\delta \simeq (2/\mu_0 \sigma_s \omega)^{1/2} . \qquad (8.51)$$

The solutions for the real fields in the water are

$$\mathbf{B}_w = \mathbf{B}_{\parallel} e^{-|z|/\delta} \cos (z/\delta - \omega t) , \qquad (8.52)$$

and

$$\mathbf{E}_w \simeq \left(\frac{\mu_0 \omega}{2 \sigma_s} \right)^{1/2} (\hat{n} \times \mathbf{B}_{\parallel}) e^{-|z|/\delta}$$

$$\times [\cos (z/\delta - \omega t) + \sin (z/\delta - \omega t)] . \qquad (8.53)$$

Thus the fields beneath the surface exhibit the properties of (1) rapid exponential decay within the skin depth with oscillations superimposed; (2) a phase difference between $\mathbf{E}_w$ and $\mathbf{B}_w$; and (3) the magnetic field much larger than the electric field, with both $\mathbf{E}_w$ and $\mathbf{B}_w$ being almost exactly parallel to the surface and propagating normal to it. There is nevertheless a second-order tangential electric field, $\mathbf{E}_\parallel$, just *outside* the surface that, from the boundary condition of Eq. 8.45, must be the same as Eq. 8.53 evaluated at $z = 0$. To the same order, there must be a small normal component in air, $\mathbf{B}_\perp$, which may be obtained from Eq. 8.13; in water, $\mathbf{E}_\perp$ is zero. The fields are plotted schematically in Fig. 8.4b.

The order of magnitude of δ is $250/\sqrt{f}$, which for typical marine radar frequencies ($f \approx 10^{10}$ Hz), is 0.025 m, while for extremely low frequencies ($f \approx 10^3$ Hz), it is approximately 7.9 m. The dimension of the skin depth for ELF waves forms the basis for communications with shallowly submerged receivers. It is worth noting that at such low frequencies the speed of light in seawater (see Eq. 8.19) is very low; at $f = 10^3$ Hz, $c_r \approx c_0 (\epsilon_0 \omega / 2\sigma_s)^{1/2} \approx 2.5 \times 10^4$ m s^{-1}, or about 16 times the speed of sound.

In the case of a moderately good conductor such as the sea, our neglect of the displacement current is not completely justified, and the perfect-conductor skin depth is not accurate. Under these conditions, correct values for the wave number and absorption coefficient in water may be obtained from Eq. 8.39. These are:

$$k_r = \frac{\sqrt{\epsilon'}\,\omega}{c_0} \left[\frac{1}{2}\left(1 + \sqrt{1 + \tan^2\gamma} \right) \right]^{1/2} \tag{8.54}$$

and

$$k_i = \frac{\sqrt{\epsilon'}\,\omega}{c_0} \left[\frac{1}{2}\left(-1 + \sqrt{1 + \tan^2\gamma} \right) \right]^{1/2}, \tag{8.55}$$

where $\tan\gamma$ is the *loss tangent*:

$$\tan\gamma \equiv \frac{\epsilon''}{\epsilon'}. \tag{8.56}$$

Here ϵ' and ϵ'' are given by Eq. 8.39, or are obtained from experimental data.

We now return to the question of reflection from the sea surface. The intensity of radiation, or energy flux per unit area, is given by the Poynting vector, $\mathbf{E} \times \mathbf{H}$, whose time-averaged value is $\langle \mathbf{S} \rangle$:

$$\langle \mathbf{S} \rangle = \frac{1}{2}\,\mathrm{Re}\,(\mathbf{E} \times \mathbf{B}^*)/\mu_0$$

$$= \frac{1}{2}\,\mathrm{Re}\left[\sqrt{\epsilon/\mu_0}\,\mathbf{E}_0^2\,(\mathbf{k}/k) \right]$$

$$= \frac{1}{2} \text{ Re } \left[\epsilon \ \mathbf{E}_0^2 \ (c_0 / \sqrt{\epsilon_r}) \right] \quad \text{W m}^{-2} , \qquad (8.57)$$

where, for a plane wave, $\mathbf{B} = \mathbf{k} \times \mathbf{E}/\omega$. Thus the energy propagates in the same direction and at the same velocity as the fields, with the speed of energy transport being the real part of $c_0 / \sqrt{\epsilon_r}$.

By solving the electromagnetic boundary problem much as was done for the acoustic problem (see Section 7.9), the *Fresnel reflection coefficients* are obtained, which are analogous to the acoustic Rayleigh pressure coefficients; however, the added complication of a polarization dependence exists for electromagnetic waves. The reflection and transmission coefficients for the fields are defined as the ratio of the reflected or transmitted field to the incident field. Thus for horizontal, h, or vertical, v, polarizations, the field reflection coefficients are:

$$R_h = \frac{E_{rh}}{E_{ih}} \qquad (8.58)$$

and

$$R_v = \frac{E_{rv}}{E_{iv}} . \qquad (8.59)$$

Analogous formulas may be derived for the transmission coefficients, $T_h(\theta)$ and $T_v(\theta)$. However, it is easy to show that

$$T_h = 1 + R_h . \qquad (8.60)$$

The reflected *energy flux* in an electromagnetic wave is given by the reflected Poynting vector, so that the *power reflection coefficients*, r_h and r_v, are

$$r_h = |R_h|^2 \qquad (8.61a)$$

and

$$r_v = |R_v|^2 . \qquad (8.61b)$$

After some algebra, these are found to be

$$r_h = |R_h|^2 = \frac{(p - \cos \theta)^2 + q^2}{(p + \cos \theta)^2 + q^2} \qquad (8.62)$$

and

$$r_v = |R_v|^2 = \frac{(\epsilon' \cos \theta - p)^2 + (\epsilon'' \cos \theta - q)^2}{(\epsilon' \cos \theta + p)^2 + (\epsilon'' \cos \theta + q)^2} , \qquad (8.63)$$

where the auxiliary functions p, q, g, and h are

$$p = \frac{1}{\sqrt{2}} (g + h)^{1/2} , \qquad (8.64)$$

$$q = \frac{1}{\sqrt{2}} (g - h)^{1/2} , \qquad (8.65)$$

$$g = [(\epsilon' - \sin^2 \theta)^2 + \epsilon''^2]^{1/2} , \qquad (8.66)$$

and

$$h = \epsilon' - \sin^2 \theta . \qquad (8.67)$$

In these equations, both the real and imaginary parts of the dielectric function, ϵ' and ϵ'', must include the conductivity contributions given in Eq. 8.39.

The *power transmission coefficients,* t_h and t_v, are more complicated than the absolute squares of T_h and T_v; however, conservation of energy requires that

$$r_{v,h} + t_{v,h} = 1 , \qquad (8.68)$$

so that knowledge of r suffices to find t.

Fresnel power coefficients are plotted versus incidence angle in Fig. 8.5 for a radar wavelength, λ, of 0.03 m, using the complex dielectric function discussed above. At normal incidence (nadir), the two coefficients become identical, as they must; the polarization distinction disappears when $\theta_i = 0$. Off nadir, however, their behavior is quite different, with r_v dipping to very small values near 83°, which is the *Brewster angle*, θ_B, for seawater at this frequency, before rapidly rising to 100% at grazing angles. However, r_h climbs monotonically toward unity reflection, the so-called "dielectric reflection" condition. Recall that the Brewster angle is that angle for which the vertically polarized reflectance vanishes (for nonabsorbing materials), or is a minimum for absorbing ones. Only vertically polarized radiation has this minimum, and for a dissipationless dielectric ($\epsilon'' = 0$), $r_v = 0$ when

$$\tan^2 \theta_B = \epsilon' \approx 81 , \qquad (8.69)$$

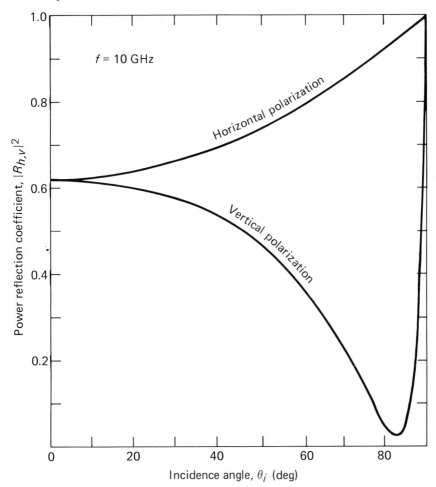

Fig. 8.5 Squares of Fresnel coefficients versus incidence angle, for a wavelength of 0.03 m. Reflectance is the same for both polarizations at normal incidence; vertically polarized component has a minimum at Brewster angle of 83.7° for seawater. [From Stewart, R. H., *Methods of Satellite Oceanography* (1985).]

or at an angle

$$\theta_B = \tan^{-1}(9) = 83.7°, \tag{8.70}$$

when $T = 5.5°C$ and $f = 2.7$ GHz (see Tables 8.1 and 8.2). Thus randomly polarized radiation incident on the sea at angles near the Brewster angle will be reflected largely as horizontally polarized energy.

8.5 The Radar Equation for a Distributed Surface

The *radar equation* is a relationship among (1) the focused, transmitted power radiated by a source of electromagnetic energy located at a distance r_1 from a distributed surface; (2) the reflective/scattering properties of that surface; and (3) the scattered power incident on a receiver at a distance r from the surface. If transmitter and receiver are colocated, the geometry is termed *monostatic*; if separated, it is *bistatic*. We shall treat the latter instance, since it includes the monostatic backscattering geometry as a special case.

Figures 8.6a and 8.6b illustrate the bistatic geometry. The transmitter radiates at an instantaneous power level P_t, which is focused into a beam confined approximately to a solid angle, Ω_t, by the antenna, whose gain is G_t. The transmitter antenna gain varies significantly with the antenna's polar angles, θ_t and ϕ_t, as described by its *beam diffraction pattern,* but in the forward direction, is very approximately the reciprocal ratio of the solid angle of the beam to the total angle subtended by a sphere, 4π. Thus

$$G_t \approx \frac{4\pi}{\Omega_t} \ . \tag{8.71}$$

The transmitted power radiates into the complete sphere, so that at a distance r_1 from the source, the Poynting vector is (cf. Eq. 8.57)

$$\mathbf{S}_t = \frac{P_t G_t (\theta_t, \phi_t)}{4\pi r_1{}^2} \hat{r}_1 \ , \tag{8.72}$$

where $\hat{r}_1$ is a unit vector in the direction of propagation. At the surface of the scatterer, the beam illuminates an area with a varying illumination pattern described by $L^2(x', y')$, that depends upon the projection of the antenna diffraction pattern on the surface.

Each scattering element on the surface, s_i, scatters energy as described by its differential cross section per unit area of its surface per unit solid angle, $d^2\sigma/d\Omega \, dA$; this quantity depends on the incidence angle, θ_i, and the scattering angles, θ_s and ϕ_s. Thus we write

$$\frac{d^2\sigma}{d\Omega \, dA} \equiv \frac{\sigma^0 (\theta_i, \theta_s, \phi_s, x', y')}{4\pi} \ , \tag{8.73}$$

where σ^0 is the *normalized scattering cross section,* i.e., the cross section per unit area of surface. Because of variations in surface scattering properties, it will in general be a function of position (x', y') on the surface; in fact, it is this local variation of scattering cross section that allows a radar image

(a)

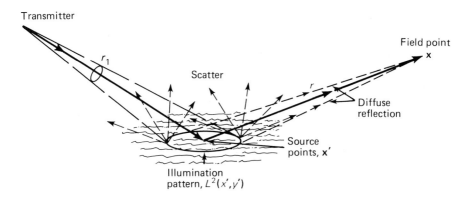

(b)

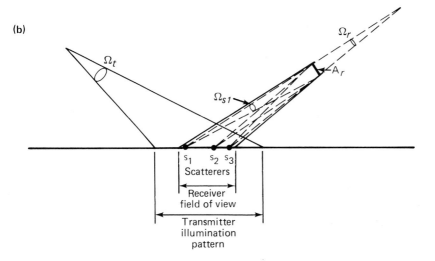

Fig. 8.6 (a) Schematic of bistatic reflection and scatter from sea surface. Illuminated region, *L*, scatters energy at all angles. A receiver at **x** senses energy from a distributed region; for a constant beam width, the area viewed increases as r^2, while scatter from each point decreases at the same rate, so that the signal is independent of distance as long as the illumination is constant. (b) Bistatic geometry. Scatterers at s_i re-radiate Huygen's wavelets into the receiver aperture, A_r.

such as that shown in Fig. 5.28 to be constructed. As will be demonstrated in the next section, each scatterer acts as a point source of secondary Huygens wavelets whose re-radiated power spreads geometrically as $1/4\pi r^2$; thus a receiver viewing the illuminated area with a receiving aperture beam width of solid angle, Ω_r, will intercept energy from all scatterers within its field of view. Each scatterer in the field will contribute that amount of energy which it re-radiates into the solid angle, Ω_s, subtended by the antenna when viewed from the scatterer's position. The total received power is the summation from the individual scattering elements.

The power contributed by all scatterers lying in the ith element of area on the surface, $\Delta A_i'$, is then the product of the incident flux and the differential cross section:

$$\Delta P_i = \frac{P_t G_{ti}}{4\pi r_{1i}^2} \sigma_i^0 \, \Delta A_i' \, . \tag{8.74}$$

This is radiated into the sphere surrounding the element, and at a distance r at angles θ_{si} and ϕ_{si}, the power per unit area per unit solid angle, or *radiance*, is

$$\Delta S_i = \frac{\Delta P_i}{4\pi r_i^2} = \frac{P_t G_{ti} \sigma_i^0}{(4\pi)^2 \, r_{1i}^2 \, r_i^2} \, \Delta A_i' \, . \tag{8.75}$$

The total scattered power per unit area per unit solid angle at the distance r is the sum over all scatterers in the illuminated area:

$$S(r) = \sum_i \Delta S_i = \sum_i \frac{P_t G_{ti} \sigma_i^0}{(4\pi)^2 \, r_{1i}^2 \, r_i^2} \, \Delta A_i' \, , \tag{8.76}$$

or in the limit of a continuous distribution, the sum goes to

$$S(r) = \frac{P_t}{(4\pi)^2} \iint\limits_{A'} \frac{G_t \sigma^0}{r_1^2 \, r^2} \, dA' \, . \tag{8.77}$$

A receiver antenna having an effective capture area of A_r at this distance will receive the power, P_r, given by

$$P_r(r) = \iint\limits_{A_r} S \, dA_r \simeq \frac{P_t A_r}{(4\pi)^2} \iint\limits_{A'} \frac{G_t \sigma^0}{r_1^2 \, r^2} \, dA' \, , \tag{8.78}$$

where the right-hand-most expression assumes that the power flux is essentially uniform over the antenna area. Equation 8.78 is the bistatic radar equation for a distributed target. From it, we see that the properties of the distributed surface are all incorporated into the cross section, with the remaining factors being instrumental or geometric in nature. Understanding electromagnetic measurements of the ocean in considerable degree then devolves to understanding the factors that influence σ^0. The next section will concentrate on developing that comprehension.

One item is worth discussing in Eq. 8.78: With a distributed target having statistically homogeneous scattering properties, increasing the distance of observation, r, does not generally decrease the received power significantly. This is because as the receiver draws away from the surface, even though the power from an individual scatterer may fall off as $1/r^2$, the area viewed with an antenna of constant beam width will increase quadratically, as will the number of scatterers within the field of view, so that, to a first approximation, the signal remains constant. This means that, if care is taken, measurements made at close distances may be applied to observations at other distances.

If the transmitter is at sufficiently large distances so that r_1 is essentially constant over the area of illumination, the incident wave may be considered as locally planar, and the quantity $(P_t/4\pi r_1^2)$ as the incident power flux, S_i. The gain G_t may be described by the normalized illumination function $L^2(x',y')$, so that the radiance at a distant field point, $\mathbf{x}$, may be written as

$$ S(\mathbf{x}) = S_i \int\int_{-\infty}^{\infty} L^2(x',y') \ \frac{\sigma^0}{4\pi r^2} \ dA' \qquad \text{W m}^{-2}\ \text{sr}^{-1}\ . \qquad (8.79) $$

This form has the advantage of avoiding reference to transmitter and receiver properties, and will be used next in a more rigorous development of scattering from the sea.

8.6 Scattering of Electromagnetic Radiation

When an electromagnetic wave is incident on a material surface such as the ocean, it is in general partly reflected and partly transmitted. The reflection and transmission coefficients for plane waves on planar surfaces represent the simplest mathematical description of these processes. Real surfaces, however, are almost always much more complicated than that, and the formulation of the problem of reflected/scattered fields mirrors those difficulties.

The scattered field observed at some distant point from the scattering surface may be considered as the coherent sum of fields radiated from all elemental areas distributed over the illuminated surface. Each small surface element, dA', may be characterized by a differential cross section per unit area per unit solid angle subtended by the element. This quantity, $d^2\sigma/dA\ d\Omega$, was also encountered in Chapter 7 in the context of acoustic scattering. The problem of calculating the scattered energy then divides into the two tasks of calculating the scattered electromagnetic fields from each element of area, and then adding up the re-radiation from every scatterer for every point in the surrounding space. At the basis of this approach is the concept of Huygen's wavelets, which are assumed to re-radiate as spherical waves from every illuminated point, s_i, on the surface (see Fig. 8.6). Their coherent sum (as weighted by the distribution of scattering sources, each with its own constituent parameters and geometrical orientation) gives the desired field strength. The mathematical formulation of the source of such wavelets is found in the Green function for the radiation field, $G(\mathbf{x} - \mathbf{x}', t - t')$, whose source terms are delta function charges and currents located at point $\mathbf{x}'$ at time t', and which contribute to the field at point $\mathbf{x}$ at time t.

The approach most often used in the scattering problem is due to Kirchhoff, and was originally applied to the diffraction of waves by various geometrical configurations; however, it is equally applicable to scattering. It has found its way into the theory for scattering of acoustic energy as well, and in principle it is also useful for calculations of scattering of other types of waves in fluids.

The route to be followed in arriving at the final formulas for the scattering cross section is long and arduous, and the results themselves are imperfect. Because this is an area of active research, the best course of action in an exposition at this level is to lay out the basics of the problem, next outline the approach, and then cite some useful but incomplete formulas. The reader is referred to the bibliography for more complete developments. (See especially Jackson (1975), Valenzuela (1978), and Holliday et al. (1986).)

Vector Wave Equations

In the derivations leading up to Eq. 8.16, it was assumed that the electromagnetic field could be described by plane waves, and that the result of the del operator working on a field was to generate $i\mathbf{k}$. However, the wavelike character of the fields is independent of the plane wave assumption, as may be shown in a brief derivation. First, assume $\mathbf{u} = \mathbf{0}$ in Eq. 8.4; then take the curl of that relationship and substitute Ohm's law and the definition of the displacement, $\mathbf{D} = \epsilon \mathbf{E}$, into the result. In terms of the absolute dielectric function, there results

$$\nabla \times \nabla \times \mathbf{B} = \mu_0 \sigma \nabla \times \mathbf{E} + \mu_0 \epsilon \frac{\partial}{\partial t} \nabla \times \mathbf{E} . \tag{8.80}$$

An expansion of the double curl shows that the space derivatives may be rewritten as

$$\nabla \times \nabla \times \mathbf{B} = \nabla(\nabla \cdot \mathbf{B}) - \nabla^2 \mathbf{B} . \tag{8.81}$$

Since, from Maxwell's equations, $\nabla \cdot \mathbf{B} = 0$, Eq. 8.80 becomes

$$\nabla^2 \mathbf{B} = \mu_0 \sigma \frac{\partial \mathbf{B}}{\partial t} + \mu_0 \epsilon \frac{\partial^2 \mathbf{B}}{\partial t^2} , \tag{8.82}$$

where we have used Faraday's Law (Eq. 8.3) to eliminate $\mathbf{E}$. This equation shows that the magnetic (as well as the electric) field obeys a wave equation having a damping term represented by the first-order time derivative. If we now confine our attention to the medium above the surface of the sea, for which $\sigma \simeq 0$, the free-space wave equation becomes

$$\nabla^2 \mathbf{B} - \frac{1}{c^2} \frac{\partial^2 \mathbf{B}}{\partial t^2} = 0 , \tag{8.83}$$

with the wave velocity given by $1/(\mu_0 \epsilon_0)^{1/2}$ in the limit of small conductivity. The additional assumption of harmonic time-dependence, when substituted in Eq. 8.83, yields the *Helmholtz equation:*

$$\nabla^2 \mathbf{B} + \mathbf{k}^2 \mathbf{B} = 0 . \tag{8.84}$$

The scalar components of both the magnetic and the electric fields obey similar relationships in the region above the sea surface, as long as the atmosphere is essentially nonconducting.

Green's Theorem and the Scalar Kirchhoff Integral Formula

The next step in arriving at the Kirchhoff scattering theory is contained in an offspring of the divergence theorem, which asserts that for any vector field $\mathbf{A}$ defined in a volume V bounded by a closed surface S, the result of integrating the divergence of $\mathbf{A}$ over the volume is given by the integral of the normal component of $\mathbf{A}$ evaluated over the bounding surface. Using the same notation as in previous chapters, the divergence theorem is:

$$\iiint_V \nabla \cdot \mathbf{A} \, d^3x = \oiint_S \mathbf{A} \cdot \hat{n}_0 \, d^2x , \tag{8.85}$$

where $\hat{n}_0$ is an *outward*-directed normal to the surface. For our problem, let $\mathbf{A} = B \nabla G$, where B is an arbitrary component of the magnetic flux density. (Here we elect to deal with the magnetic rather that the electric field since, for a perfect conductor, the more general result given by Eq. 8.92b below is simpler when phrased in terms of $\mathbf{B}$ than of $\mathbf{E}$.) In addition, we choose G to be a scalar function describing the potential of an outgoing wave radiating from an infinitesimal current source, i.e., a Huygen's wavelet. Thus

$$G(R) = \frac{e^{ikR}}{4\pi R} , \tag{8.86}$$

where

$$R = |\mathbf{x} - \mathbf{x}'| \tag{8.87}$$

is the distance between the *source point* located at $\mathbf{x}'$ and the *field point* located at $\mathbf{x}$. The potential, G, is the Green function for the problem of radiation in free space. Substituting these definitions into the divergence theorem, we obtain

$$\iiint\limits_V (\nabla B \cdot \nabla G - B\nabla^2 G)\, d^3\mathbf{x} = \oiint\limits_S B\, \hat{n}_0 \cdot \nabla G\, d^2\mathbf{x} , \tag{8.88}$$

where the normal derivative is $\partial G/\partial n_0 = \hat{n}_0 \cdot \nabla G$. Next we interchange B and G in the definition of $\mathbf{A}$ and subject the new product to the same procedure; then subtract the equation so obtained from Eq. 8.88. The result is termed *Green's second theorem* for scalar fields:

$$\iiint\limits_V (B\nabla^2 G - G\nabla^2 B)\, d^3\mathbf{x} = -\oiint\limits_S \left(B\, \frac{\partial G}{\partial n} - G\, \frac{\partial B}{\partial n} \right) d^2\mathbf{x} . \tag{8.89}$$

Here the sign of the right-hand side is now negative because the direction of the unit normal has been reversed to point *into* the interior of the bounding surface, i.e., $\hat{n} = -\hat{n}_0$ (see Fig. 8.7).

The wave equation given by Eq. 8.84 is correct if there are no magnetic currents in the interior of the integration volume. However, if there are sources of the field within the volume, the magnetic field obeys an inhomogeneous wave equation, with the right-hand side describing the source function. For the Green function, the source term is a delta function describing infinitesimal charges or currents located at position $\mathbf{x}'$; for the problem at hand, these are either the sources of the incident field, B_i, or are those induced by it on the bounding surface:

$$\nabla^2 G + \mathbf{k}^2 G = -\delta(\mathbf{x} - \mathbf{x}') . \tag{8.90}$$

To convince one's self of this property of the function, differentiate Eq. 8.86 twice and add $\mathbf{k}^2$ times G to it; the result is zero, unless $\mathbf{x} = \mathbf{x}'$, in which case the function itself is singular. This, of course, is a reflection of the character of the delta function.

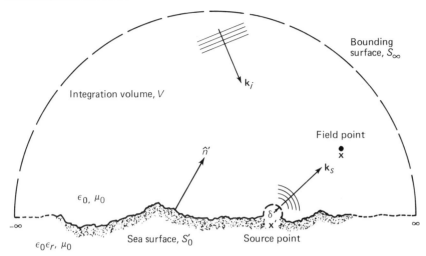

Fig. 8.7 Integration volume for evaluating scattering integrals, with the rough sea surface containing scattering sources, and the hemisphere at infinity originating the incoming radiation. Sources are excluded by integrating each around a small hemisphere, δ, and letting it shrink to zero. [Adapted from Valenzuela, G. R., *Boundary Layer Meteorology* (1978).]

Next we substitute Eqs. 8.84 and 8.90 into the left-hand side of Green's theorem to obtain

$$\iiint\limits_{V} B\delta(\mathbf{x} - \mathbf{x}')\, d^3\mathbf{x} = -\oiint\limits_{S} \left(B\,\frac{\partial G}{\partial n} - G\,\frac{\partial B}{\partial n} \right) d^2\mathbf{x} \ . \tag{8.91}$$

The volume integration is zero unless the observation point coincides with the source point, in which case the delta function picks out the field value at that point. This integration must be done carefully, but it yields the value of the total magnetic field at point $\mathbf{x}'$ in terms of the integral of the surface current distribution over other points, $\mathbf{x}''$:

$$B(\mathbf{x}') = \oiint\limits_{S''} \left(B\,\frac{\partial G}{\partial n''} - G\,\frac{\partial B}{\partial n''} \right) d^2\mathbf{x}'' \ . \tag{8.92a}$$

It is worth emphasizing that the values of B on both sides of Eq. 8.92a represent the *total* field, which is considered to be the sum of the incident field, B_i, and the scattered field, B_s. The dummy integration variable is $\mathbf{x}''$, with inward normal $\hat{n}\,''$, while the field point is $\mathbf{x}\,'$. These last three quantities will be relabeled below with one less prime for simplicity.

The vector equivalent of Eq. 8.92a can be derived from the vector form of the divergence theorem and in the limit of perfect conductivity can be shown to be:

$$\mathbf{B(x)} = \oiint_{S'} [(\hat{n}\,' \times \mathbf{B}) \times \nabla\,'G(\mathbf{x} - \mathbf{x}\,')]\, d^2\mathbf{x}\,' .\tag{8.92b}$$

The relationships given by Eqs. 8.92a and 8.92b are exact integral equations for the magnetic field in the interior of the volume, given in terms of the values on the closed bounding surface.

The closed surface invoked in Eqs. 8.92a and 8.92b is next divided into two parts, with one being a large, concave surface capping the integration domain at very great distances and denoted by S_∞, and the other being an approximately level surface denoted by S_0 (see Fig. 8.7). The sources of the incident magnetic field are assumed to lie on S_∞, so that the integration over the large cap gives just B_i. On that same surface, the scattered field is vanishingly small, since it satisfies the *radiation condition* and therefore recedes at least as fast as $1/R$ as the surface becomes very large. However, on the level surface, S_0, the currents induced at point $\mathbf{x}\,'$ are due to both the imposed incident field and the scattered field from other induced sources located elsewhere on S_0, say at $\mathbf{x}\,''$. Thus the quantity that appears in the surface integrals over S_0 is the *total field*, B_t. For the scalar case, we may therefore write Eq. 8.92a (with one prime dropped) as

$$B_t(\mathbf{x}) = B_i(\mathbf{x}) + \iint_{S_0'} \left[B_t(\mathbf{x}\,')\frac{\partial G}{\partial n'} - G\frac{\partial B_t(\mathbf{x}\,')}{\partial n'} \right] d^2\mathbf{x}\,'$$

$$= B_i(\mathbf{x}) + B_s(\mathbf{x}) ,\tag{8.93}$$

where B_i results from the integral over S_∞ and where the integral for the *scattered field*, B_s, is over the *open* surface, S_0', only. This equation is the scalar Kirchhoff integral formula, with the division into incident and scattered fields made explicit.

There are mathematical inconsistencies in the scalar Kirchhoff formula related to our inability to specify values for both B_t and $\partial B_t/\partial n'$ on the surface (i.e., mixed Dirichlet and Neumann boundary conditions). However, these troubles are circumvented to a reasonable degree by the use of the vector formulation (especially Eq. 8.92b) evaluated over similar bounding surfaces, and with the same division of the fields into incident and scattered quantities. In the vector relationship of Eq. 8.92b, derivatives of $\mathbf{B}_t$ do not appear, so in some sense it appears equivalent to the scalar relationship of Eq. 8.92a with $G(\mathbf{x} - \mathbf{x}') = 0$ on the surface.

A more complete discussion of the inconsistencies in the scalar Kirchhoff theory (see Jackson (1975), p. 429, for example) leads to the conclusion that, within the domain wherein the theory is valid at all, essentially equivalent results are obtained for the scattered field whether or not one approximates the integrand of Eq. 8.92a only by the first term, or only by the second term, or by one-half their sum. If the source and observation points are far from the scattering surface, the relative amplitudes of the fields resulting from all three approximations are very nearly the same. Because the more exact vector theory of Eq. 8.92b contains only the term involving the normal derivative of the magnetic flux density, we shall choose the scalar equivalent to it such that the Green function vanishes on the surface (the Dirichlet boundary condition).

With this choice, the values of the total magnetic field in the integrand of Eq. 8.92a describe equivalent induced surface currents on the horizontal bounding surface of the type given by Eq. 8.43. If now the observation point, $\mathbf{x}$, is moved down to a source point, $\mathbf{x}'$ located *on* the horizontal surface, an integral equation is obtained for the equivalent surface current density:

$$B_t(\mathbf{x}') = B_i(\mathbf{x}') + \iint\limits_{S_0'} B_t(\mathbf{x}'') \frac{\partial}{\partial n''} G(\mathbf{x}' - \mathbf{x}'') \, d^2\mathbf{x}'' \,. \qquad (8.94)$$

Evaluating the integral around the singularity presented by $\mathbf{x}' = \mathbf{x}''$ yields two contributions, one equal to $\frac{1}{2} B_t(\mathbf{x}')$, which arises from the singularity, and the other a principal value integral that avoids the point $\mathbf{x}' = \mathbf{x}''$ by redefining the surface S_0 to contain an infinitesimal hemisphere that excludes the singularity (see Fig. 8.7). Clearly the surface current contribution from the singularity is just one-half the left-hand side, so that solving for $B_t(\mathbf{x}')$ and multiplying by 2 gives an integral equation for the surface current density involving (1) the incident field on the surface and (2) the integral of the total field, also over that surface. The first approximation to its solution is to take $B_t(\mathbf{x}') = 2B_i(\mathbf{x}')$, which neglects the contribution from the distant induced currents on the local current. We then return to the relationship for the field *off* the surface (Eq. 8.93) and, using this approximation to the integrand, obtain

$$B_t(\mathbf{x}) = B_i(\mathbf{x}) + 2 R_{v,h} \iint\limits_{S_0'} B_i(\mathbf{x}') \frac{\partial}{\partial n'} G(\mathbf{x} - \mathbf{x}') \, d^2\mathbf{x}'$$

$$= B_i(\mathbf{x}) + B_s(\mathbf{x}) \,, \qquad (8.95)$$

where $R_{v,h}$ are given by Eqs. 8.58 and 8.59. The reflection coefficients take into account the properties of the medium. The effect of the first approximation has been to set the total field under the integral equal to twice the incident field. Thus this admittedly sketchy derivation of the *Kirchhoff approximation* yields the result that the value of the scattered field at the surface is equal to that of the incident wave, so that in the integrand, the total field is approximated by twice the incident field.

The vector theory improves on the scattering calculations considerably and yields more rigorous results, but at the price of significantly greater complication. Nevertheless, we shall continue to use the scalar formulation ahead because the results are somewhat simpler and easier to understand. In the past, the vast majority of scattering calculations carried out for environmental problems have made use of the scalar theory, which appears to have given reasonable results. However, its deficiencies make the results somewhat suspect, so the vector theory is probably required in any realistic treatment of electromagnetic scattering in the natural environment, especially if polarization effects or higher-order approximations are important.

Scattering from the Sea Surface in the Kirchhoff Approximation

We now apply the scalar Kirchhoff formula to the ocean, which may be assumed to be instantaneously frozen on the time scales of the short pulses used in radar. We furthermore assume that the ocean is a good conductor (which it is not, but the inclusion of finite values of σ adds an additional complication that further obscures an already difficult calculation without greatly improving the physics). We bound the volume of integration by the actual, roughened sea surface along the lower portion (over which we will perform the integrations), and by the large hemisphere capping the ocean, as illustrated schematically in Fig. 8.7. (The curvature of the earth must be taken into account if the segment of the ocean surface of interest is large enough.)

To proceed further, additional simplifications are required. First, it is generally assumed that the observation point is located in the *far radiation field*, so that near-field effects are negligible. The far-field approximation allows the Green function to be rewritten using the geometry illustrated in Fig. 8.8. This figure defines the coordinate system to be used in the evaluation of the surface integral, Eq. 8.95. The incidence and scattering angles, θ_i and θ_s, respectively, are defined with respect to the vertical, and $\mathbf{k}_i$ and $\mathbf{k}_s$ are the associated wave vectors. The coordinates of the sea surface are $[x',y',\xi(x',y')]$; those of the field point are $\mathbf{x}$, with $r = |\mathbf{x}|$, and with the distance between source and field being $R = |\mathbf{x} - \mathbf{x}'|$. From Fig. 8.8, clearly

$$R = r - \mathbf{k}_s \cdot \mathbf{x}'/k . \tag{8.96}$$

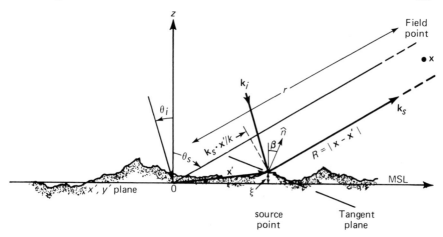

Fig. 8.8 Geometry for evaluating Eq. 8.94 in the far-field approximation. Incident energy is scattered from source point, x', where sea elevation is ξ, toward field point at x. [Adapted from Beckmann, P., and A. Spizzichino, *The Scattering of Electromagnetic Waves from Rough Surfaces* (1963).]

The far-field approximation consists of neglecting the difference between R and r in the denominator of Eq. 8.86, and then writing the phase of the exponential using Eq. 8.96:

$$G(R) = \frac{e^{ikR}}{4\pi R} \simeq \frac{e^{ikr}\, e^{-i\mathbf{k}_s \cdot \mathbf{x}'}}{4\pi r} \tag{8.97}$$

and

$$\hat{n}' \cdot \nabla' G(R) \simeq -i\mathbf{k}_s \cdot \hat{n}'\, G(R) \,, \tag{8.98}$$

which neglects a factor in $i/k_s R$ in comparison with unity. We may then write Eq. 8.94 with the far-field approximation taken outside the integral, and with the surface integral evaluated on the open surface defined by the instantaneous sea elevations, $\xi(x',y')$:

$$B_s(\mathbf{x}) \simeq \frac{e^{ikr}}{2\pi r}\, R_{v,h} \iint\limits_{S_0'} B_i\, \hat{n}' \cdot \nabla' G\, d^2\mathbf{x}' \,. \tag{8.99}$$

It is usually assumed that the incident field is a plane wave, which neglects the curvature of the wavefront illustrated in Fig. 8.6, not a serious issue. Supressing the time-dependence of the plane wave, we write for the incident field

$$B_i(\mathbf{x}') = B_0\, e^{i\mathbf{k}_i \cdot \mathbf{x}'} , \tag{8.100}$$

where the spatial coordinate $\mathbf{x}'$ is an integration variable.

Now the integral of Eq. 8.99 is taken over the illuminated ocean area only, and must be weighted by the illumination pattern of the antenna, as shown in Fig. 8.6, if the limits of integration are to extend to very great distances. We model this with the illumination distribution, $L(x',y')$. After the substitution of Eq. 8.100 and the normal gradient, Eq. 8.98, into Eq. 8.99, we obtain for the scattered field at an observation point:

$$B_s(\mathbf{x}) \simeq \frac{-iB_0 e^{ikr}}{2\pi r} R_{v,h} \int\!\!\!\int_{-\infty}^{\infty} L(x',y')\, \exp[i(\mathbf{k}_i - \mathbf{k}_s)\cdot\mathbf{x}']\, \mathbf{k}_s\cdot\hat{n}'d^2\mathbf{x}' . \tag{8.101}$$

The geometry of the problem must now be made explicit. Figure 8.9 illus-

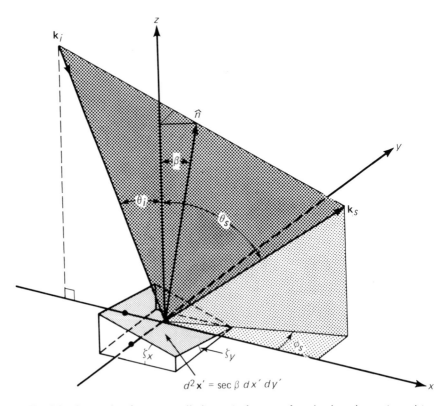

Fig. 8.9 Scattering from a small element of sea surface having slopes ζ_x and ζ_y, and a total tilt angle of β.

trates a small element of sea surface of area $d^2\mathbf{x}'$ tilted at an angle β, and its unit normal, $\hat{n}'$. The slopes of the element are

$$\zeta_x' = \partial\xi/\partial x' \qquad (8.102a)$$

and

$$\zeta_y' = \partial\xi/\partial y' \, , \qquad (8.102b)$$

where $\xi(x',y')$ is the surface elevation with respect to mean sea level, which is taken to be the (x',y') plane. The tangent of the angle β is the overall slope of the surface element. Both the incident and scattered wave vectors, $\mathbf{k}_i$ and $\mathbf{k}_s$, respectively, as well as $\hat{n}'$, lie in the plane of reflection, but the scattering process has rotated the vector $\mathbf{k}_s$ out of the (x',z') plane by an azimuth angle, ϕ_s.

Figure 8.10 shows the differential geometry of the scattering area. From it and Fig. 8.9, the following relationships may be deduced:

$$\mathbf{k}_i = k(\hat{i}\sin\theta_i - \hat{k}\cos\theta_i)$$

$$= \mathbf{k}_{ih} + \mathbf{k}_{iz} \, , \qquad (8.103)$$

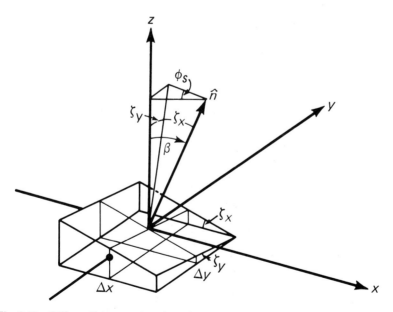

Fig. 8.10 Differential geometry of an element of scattering area. Incident and scattered waves obey a local law of reflection.

$$\mathbf{k}_s = k[\sin \theta_s (\hat{i} \cos \phi_s + \hat{j} \sin \phi_s) + \hat{k} \cos \theta_s] , \qquad (8.104)$$

$$\hat{n}' = \sin \beta (\hat{i} \cos \phi_s + \hat{j} \sin \phi_s) + \hat{k} \cos \beta \qquad (8.105)$$

$$= \frac{-\nabla_h' \xi + \hat{k}}{[1 + |\nabla_h' \xi|^2]^{1/2}} ,$$

$$\nabla_h' \xi = \tan \beta (\hat{i} \cos \phi_s + \hat{j} \sin \phi_s) \qquad (8.106)$$

$$= \hat{i} \frac{\partial \xi}{\partial x'} + \hat{j} \frac{\partial \xi}{\partial y'} ,$$

and

$$d^2\mathbf{x}' = \sec \beta \, dx' \, dy' . \qquad (8.107)$$

The second expression for the unit normal in Eq. 8.105 comes from consideration of the components of slope in terms of tan β.

These transformations reduce the S_0' integration from one over the sea surface to one over the level (x',y') plane. Substituting them into Eq. 8.101 yields

$$B_s(r) = \frac{ik_z B_0 e^{ikr}}{2\pi r} R_{v,h} \int\limits_{-\infty}^{\infty}\!\!\int L(x',y')$$

$$\times \left[1 - a \frac{\partial \xi}{\partial x'} - b \frac{\partial \xi}{\partial y'} \right] e^{-i\kappa \cdot \mathbf{x}'} \, dx' \, dy' , \qquad (8.108)$$

where the coordinate of the scattering point on the free surface is

$$\mathbf{x}' = x'\hat{i} + y'\hat{j} + \xi(x',y')\hat{k} . \qquad (8.109)$$

The *difference wave vector, κ,* is

$$\kappa \equiv \mathbf{k}_s - \mathbf{k}_i \equiv \kappa_h + \kappa_z . \qquad (8.110)$$

The difference wave vector may be decomposed into horizontal (h) and vertical (z) components, a move that will be useful in the next step in developing an algorithm for the scattered field.

The remaining coefficients incorporate the scattering angles:

$$a = \tan \theta_s \cos \phi_s \qquad (8.111)$$

and

$$b = \tan \theta_s \sin \phi_s \ . \tag{8.112}$$

Additional, more dominant angular dependencies are carried by the exponential term in Eq. 8.108, as we shall see below.

The Random Sea Surface

We now take account of the randomness of the sea surface, using the concepts of ensemble averaging and Gaussian distributions of both height and slope (see Eqs. 5.97 and 7.76). One may elect to treat either the field quantities or the Poynting vector (see Eq. 8.57) as a random variable to be averaged over the distribution function; we will take the ensemble average of the magnetic field as an example, because the results are somewhat simpler. However, it is the absolute square of the field that is actually required, as discussed below. In Eq. 8.108, the independent random variables imposing fluctuations on B_s are the height, ξ, and the slope, $\nabla_h \xi$. The ensemble average or expectation value, denoted by $\langle B_s \rangle$, is formed by multiplying it by the probability distribution function, $p(\xi) \ d\xi$ or $p(\zeta_x, \zeta_y) \ d\zeta_x \ d\zeta_y$, and integrating over all possible values of the random variables. We will indicate this expectation operation by

$$EX(F) \equiv \langle F \rangle_\xi = \int_{-\infty}^{\infty} F(\xi) \ p(\xi) \ d\xi \ , \tag{8.113}$$

for an arbitrary function, F. Then Eq. 8.108 may be written as

$$\langle B_s(r) \rangle = \frac{ik_z B_0 e^{ikr}}{2\pi r} R_{v,h} \int\int_{-\infty}^{\infty} L(x', y') \exp[-i\kappa_h \cdot (x'\hat{i} + y'\hat{j})]$$

$$\times \left\langle \left[1 - a\frac{\partial \xi}{\partial x'} - b\frac{\partial \xi}{\partial y'} \right] \exp\left[-i\kappa_z \xi(x',y') \right] \right\rangle dx' \ dy' \ , \tag{8.114}$$

where the order of integration over the horizontal plane and the random variable has been interchanged. Also, we have indicated that the ensemble averaging of both height and slope may be carried out collectively; however, they are related as given by Eq. 8.102. Integration by parts converts them into an expression containing heights only, as Eq. 8.116 below will show.

While Eq. 8.114 gives an indication of how the stochastic averaging proceeds for the field strengths, it is the ensemble average of the absolute square (i.e., the power) that is generally measured by a receiver that makes observations over several seconds. We will therefore outline the calculation of σ^0 next.

Cross Section of the Sea Surface

The general bistatic scattering cross section per unit area, $\sigma^0(\theta_i,\theta_s,\phi_s)$, is defined as the normalized energy flux scattered by a unit area of the sea from incidence angle θ_i into angles (θ_s,ϕ_s), and observed at large distances; the normalization is done by dividing by the incident power. From the earlier definitions of power density, the cross section may be written as

$$\frac{d^2\sigma}{d\Omega \, dA} \equiv \frac{1}{4\pi} \, \sigma^0(\theta_i,\theta_s,\phi_s) \equiv \frac{r^2}{A_{eff}} \, \langle |B_s|^2 \rangle \, \bigg/ \, |B_0|^2 \, . \qquad (8.115)$$

Here A_{eff} is the effective area illuminated and is related to $L^2(x',y')$. As defined, the quantity σ^0 is dimensionless, and is often quoted numerically in dB relative to 1 m^2 of ocean surface.

Using this definition of cross section and then performing the ensemble averaging indicated by Eq. 8.113 on the absolute square of Eq. 8.108, one obtains an integral approximation to the *backscatter* cross section for a Gaussian sea as given by (see Holliday et al. (1986)):

$$\sigma^0(\theta_i, -\theta_i, 0) \equiv \sigma^0(\theta_i) = \frac{k^4}{\pi k_z^2} |R_{v,h}|^2 \int\!\!\!\int_{-\infty}^{\infty} L(x',y') \, \exp(-i \, 2\mathbf{k}_h \cdot \mathbf{x}')$$

$$\times \exp\{-4k_z^2\sigma_\xi^2[1 - \rho(\mathbf{x}')]\} \, dx' \, dy' \, . \qquad (8.116)$$

The angular dependences here are carried by the wave vector, which has been resolved into horizontal and vertical components (Eqs. 8.103 and 8.104, respectively). For backscatter, $\mathbf{k}_{sh} = -\mathbf{k}_{ih}$ and $k_{sz} = -k_{iz}$. The new function that appears here, $\rho(\mathbf{x}')$, is the normalized *spatial autocorrelation function* for the sea surface, and is analogous to the temporal autocorrelation function of Eq. 5.104, with the normalization factor being the variance of the sea-height distribution, σ_ξ^2 (see Eq. 8.161 ahead). This function describes how ocean waves are correlated with themselves at increasing distances of separation, $\mathbf{x}'$, and is the *inverse* Fourier transform of the two-dimensional surface wave spectrum, $S(\mathbf{k})$; it arises when the square of the field is taken and then ensemble averaging is carried out. These two-dimensional spectral quantities are simple generalizations of the temporal functions discussed in Section 5.12. The dominant behavior of the integral comes from the exponential terms; the Gaussian-like factor is due to the random distribution of sea heights and describes *specular-point scatter* at near-vertical incidence, while the factor in $i2\mathbf{k}_h \cdot \mathbf{x}'$, together with the correlation function, describes *Bragg scatter*, which is more important at larger incidence angles.

Equation 8.116 represents a formal but useful solution for the random scattered electromagnetic power in the Kirchhoff approximation. To obtain more explicit formulas requires additional approximations and calculations, and we will content ourselves with descriptions of the three most often used results: (1) the *physical optics* or *specular-point* approximation, which is valid for small incidence angles, θ_i; (2) the *small-perturbation* or *Bragg scatter* approximation, which is applicable when the surface roughness is small compared with a wavelength of electromagnetic radiation and which holds at large incidence angles; and (3) the *two-scale* or *tilted-plane* approximation, in which the large waves are assumed to tilt the small-scale Bragg scatterers, with each wave scale having its own probability distribution function. A fourth approximation (which we will not discuss) is the so-called *wedge scattering* model, which attempts to account for the enhanced scatter that is sometimes observed on the forward face of waves by modeling the steep crests with a wedgelike shape (cf. Fig. 5.4). In addition, at very large incidence angles, wave shadowing and diffraction effects occur that are not at all treated via the Kirchhoff approximation.

Specular Point Scattering

If the ocean waves have radii of curvature, R_x and R_y, that are large compared with the wavelength of the electromagnetic waves (cf. Eq. 3.122), they act as small flat tangent planes located at each scattering point on the surface. The condition that this approximation should hold is

$$(R_x^2 + R_y^2)^{1/2} \, |\mathbf{k}| \cos^3 \theta_i \gg 1. \tag{8.117}$$

Clearly short electromagnetic wavelengths, large radii, and small incidence angles act to make this condition valid. The backscatter cross section for this case ($\theta_s = -\theta_i$) is found to be, for a Gaussian sea,

$$\sigma_{v,h}^0 (\theta_i) = \frac{|R_{v,h}(0)|^2}{2\sigma_u \sigma_c} \sec^4 \theta_i \, \exp(-\tan^2 \theta_i / 2\sigma_u^2) . \tag{8.118}$$

Here the Fresnel coefficients (Eqs. 8.62 and 8.63) are evaluated at zero angle due to the specular point assumption. The quantities σ_u and σ_c are the standard deviations of the anisotropic slope distribution in the upwind (u) and crosswind (c) directions; these are dependent on wave height and length in addition to direction (see Eqs. 9.29 and 9.30 and Fig. 9.15). The slopes in question are those of the waves contributing to the scatter. Figure 8.11

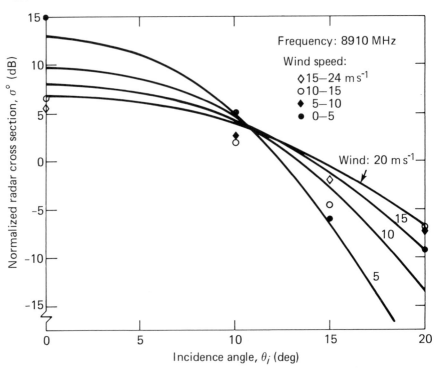

Fig. 8.11 Behavior of normalized backscatter cross section for near-nadir angles, with the wind speed as the parameter. [From Barrick, D. E., *IEEE Trans. Antennas and Propag.* (1968).]

shows the dependence of this formulation of $\sigma^0(\theta_i)$ on incidence angle and wind speed, for a radar frequency of 8910 MHz ($\lambda = 0.0337$ m), compared with experimental data. The agreement is reasonable when corrections are made to the numerical values of the Fresnel coefficients for surface organic films, or surfactants. The wind speed dependence of σ^0 is the basis for measuring oceanic surface winds, an example of which is shown in Fig. 2.15b.

Small Perturbation (Bragg) Theory

When the electromagnetic wavelength is long and the surface heights small, i.e., $|k\xi|^2 \ll 1$, the surface is considered to be only slightly rough, and perturbation methods can be used to evaluate the scattering integrals. In the backscatter geometry, the appearence in the integrals of delta functions in **k** space has the effect of picking out of the two-dimensional surface wave spectrum those waves that meet the condition of Bragg scatter, wherein only water waves of wave vector, $\mathbf{k}_w$, having components satisfying

$$(\mathbf{k}_w)_x = k_B = 2k \sin \theta_i \qquad (8.119)$$

and

$$(\mathbf{k}_w)_y \equiv 0 \qquad (8.120)$$

contribute appreciably to the returned power. Thus backscattering occurs as with a diffraction grating, but only from waves traveling parallel or antiparallel to the line of sight. This is the same condition that was cited for acoustic wave scattering from the underside of the sea surface (see Eq. 7.82). For grazing incidence angles ($\theta_i \approx 90°$), the water waves causing electromagnetic Bragg scattering are those waves whose lengths are one-half the radar wavelength (cf. Fig. 7.18) and propagating along the line of sight, either approaching or receding. In this case, the theory yields for the backscatter cross section

$$\sigma_{ij}^0(\theta_i) = 8\pi k^4 \cos^4 \theta_i |g_{ij}(\theta_i)|^2$$

$$\times [S(2k \sin \theta_i, 0) + S(-2k \sin \theta_i, 0)]. \qquad (8.121)$$

The quantities $S(\pm 2k \sin \theta_i, 0)$ are values of the two-dimensional wave number spectrum evaluated at the positive and negative Bragg wave vectors. The coefficients $g_{ij}(\theta_i)$ are modifications of the Fresnel coefficients and depend upon polarization, with the indices (i, j) denoting (h, v). For horizontal transmission and horizontal reception, they are

$$g_{hh}(\theta_i) = \frac{\epsilon_r - 1}{[\cos \theta_i + (\epsilon_r - \sin^2 \theta_i)^{1/2}]^2}, \qquad (8.122)$$

while for vertical transmission and reception,

$$g_{vv}(\theta_i) = \frac{(\epsilon_r - 1)[\epsilon_r(1 + \sin^2 \theta_i) - \sin^2 \theta_i]}{[\epsilon_r \cos \theta_i + (\epsilon_r - \sin^2 \theta_i)^{1/2}]^2}. \qquad (8.123)$$

Figure 8.12 gives the angular, polarization, and conductivity dependencies of these coefficients.

The small perturbation results are probably most applicable to longer radio wavelengths propagating at large incidence angles, in excess of perhaps 25 to 30°. However, somewhat surprisingly, they seem to hold for radar wavelengths as short as approximately 0.010 m (or even shorter) in moderate sea states, for reasons that are not altogether clear.

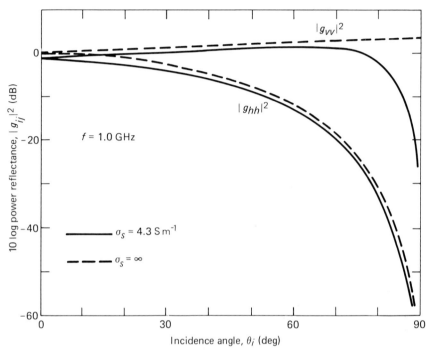

Fig. 8.12 Dependence of Bragg scattering coefficients (Eqs. 8.122 and 8.123) as a function of incidence angle, for a slightly rough sea, with polarization and conductivity as parameters. [From Wright, J. W., *IEEE Trans. Antennas and Propag.* (1968).]

Composite Surface Scattering

An attempt to include both the specular point and Bragg contributions has been incorporated into the two-scale or composite surface model. In this case, the ocean is assumed to scatter electromagnetic waves from a large number of slightly rough patches riding on longer gravity waves; the tilting of the gravity waves adds an additional incidence angle contribution, θ', and azimuth angle deviation, ϕ', to the overall slope distribution. The cross sections averaged over the slightly rough patches are similar to those given by Eqs. 8.121 to 8.123, but with the angles supplemented by the gravity wave tilts. Averaging over another probability distribution for long wave slopes then gives the complete ensemble-averaged cross section. The results are complicated; however, they predict reasonable values of the cross sections, a certain amount of depolarization, and scattering from waves traveling at angles other than just the angle of illumination.

Figure 8.13 illustrates data and calculations of backscatter cross sections

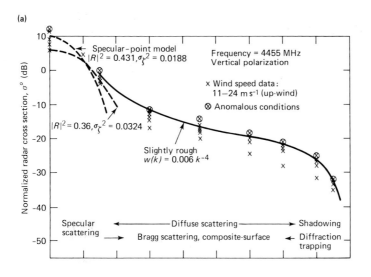

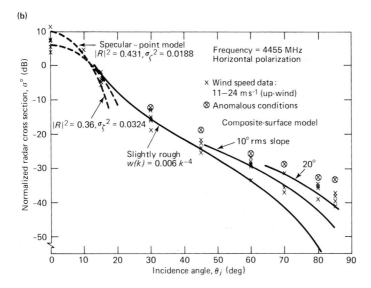

Fig. 8.13 Observed and theoretical values of backscatter coefficients versus incidence angle, for a frequency of 4455 MHz: (a) Vertical polarization; (b) horizontal polarization. Regions of applicability of various approximations are shown. [Adapted from Valenzuela, G. R., *Boundary Layer Meteorology* (1978).]

at both polarizations, for a frequency of 4455 MHz. It is seen that the horizontally polarized scattered power falls off more steeply with increasing angles than does vertically polarized power. The different regimes of approximation are shown along the bottom. All wind speeds for which measurements were made (11 to 24 m s^{-1}) have been lumped together. If the data had been stratified according to speed, the cross section would have shown the same sort of behavior at larger angles as that seen in Fig. 8.11; the general dependency is much as shown on Fig. 7.19 for the acoustic case.

Bistatic Cross Sections

The graphical examples given above represent monostatic (backscatter) cross sections. However, the general case of bistatic cross sections of the ocean is more complicated, and relatively little work has been done on the subject. The formulation for the magnetic field given by Eq. 8.114 is valid for this case, as is the definition of cross section in Eq. 8.115. Figure 8.14 shows the behavior of a calculated bistatic behavior for σ^0 as a function of azimuth angle, for the case of equal incidence and scattered angles of 49.5°, and a rough ocean surface for which $\sigma_\zeta = 0.4$; the illumination direction is $\phi_s = 0°$. For identical polarizations, the calculation predicts appreciable forward scatter over a sector of order $\phi_s \approx \pm 25°$, and a null in the scattered energy near $\pm 60°$. There is rather large cross-polarized forward scattering at angles where the like-polarized scatter is a minimum. The backscatter case corresponds to $\phi_s = 180°$, where the cross sections are identical.

8.7 Radar Scatter and Geophysical Fluid Dynamics

While the theory of electromagnetic scatter from the ocean is of scientific interest in its own right, it is the application of these measurements to other problems that has stimulated much of the interest in the subject. Buried in the functional dependencies of dielectric constant, angular variations, polarization sensitivities, and rms roughness parameters is significant information on some of the basic geophysical fluid processes discussed in earlier portions of this book. We shall now give several applications of the topic to wind, wave, and surface signature observations.

Wind Stress Measurements

Figure 8.11 gives a simplified view of the variation of backscatter coefficient with wind speed over a limited range of parameters. Figure 8.15, a cross-plot of the speed dependence, shows a sequence of measurements made with

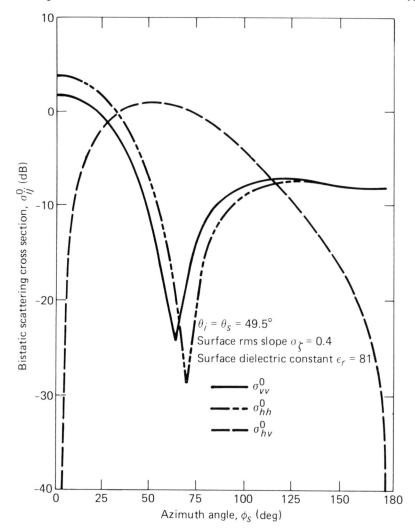

Fig. 8.14 Bistatic scattering coefficient for a very rough dielectric ($\epsilon' = 81$) versus azimuth angle, for incidence and reflection angles of 49.5°. Roughness induces considerable cross-polarized component. [From Ulaby, F. T. et al., *Microwave Remote Sensing,* Vol. II (1982).]

vertically polarized radiation at 13.96 GHz from a radar scatterometer on an aircraft, along with best-fit linear dependencies on u_w at each incidence angle. Since this is a log-log plot, the straight lines indicate that at any fixed incidence angle, the cross section depends on speed raised to some power, at least over a limited range of speeds. Thus σ^0 can be modeled as

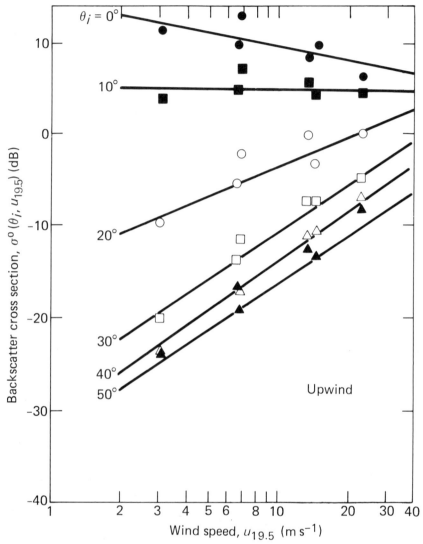

Fig. 8.15 Experimental variation of backscatter coefficient at 13.96 GHz versus wind speed, with incidence angle as a parameter, for upwind azimuth. Straight-line fits indicate a power-law variation over limited range shown (see Fig. 8.16). [From Jones, W. L. et al., *IEEE J. Oceanic Engr.* (1977).]

$$\sigma^0(\theta_i, u_w) = a u_w^{\eta(\theta_i)} \ . \tag{8.124}$$

Here the wind speed is that measured at 19.5 m height and corrected to neutral stability conditions, and $\eta(\theta_i)$ is the power to which u_w is raised. Fig-

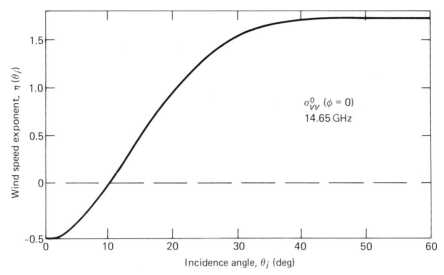

Fig. 8.16 Exponent of power law for wind speed-backscatter relationship versus incidence angle. This is essentially a crossplot of slopes of the straight lines of Fig. 8.15 versus θ_i. [Adapted from Ulaby, F. T. et al., *Microwave Remote Sensing*, Vol. II (1982).]

ure 8.16 shows the variation in this exponent, which depends mainly on incidence angle and which ranges between approximately -0.5 at vertical incidence to $+1.7$ at $50°$, over a range of wind speeds from 3 to 25 m s^{-1}. Although some differential sensitivity to wind speed appears to exist above 25 m s^{-1}, the saturation of the surface wave spectrum and the increases in spray, foam, and white water at higher winds mitigate against such simple relationships continuing to hold. The angular variability of σ^0 with incidence angle is incorporated into $\eta(\theta_i)$, and is a different way of modeling the basic cross section change shown in Fig. 8.13, for example. There is also an azimuthal dependence in backscatter due to the differences in upwind/ crosswind/downwind roughness, which Eq. 8.118 includes via the rms values of slope. This is observed in empirical data, as is illustrated in Fig. 8.17; the azimuth angle, ϕ, in Fig. 8.17 is now the radar illumination direction measured from the upwind direction. Algebraic models of the backscatter cross section generally use a three-term cosine distribution to describe this, so that the total functional form of σ^0 is

$$\sigma_{jj}^0 (\theta_i, \phi, u_w) = a_{jj} u_w^{\eta(\theta_i)} (1 + b \cos \phi + c \cos 2\phi) . \quad (8.125)$$

Here the indices $(jj) = (hh)$ or (vv) indicate a dependence on polarization as well.

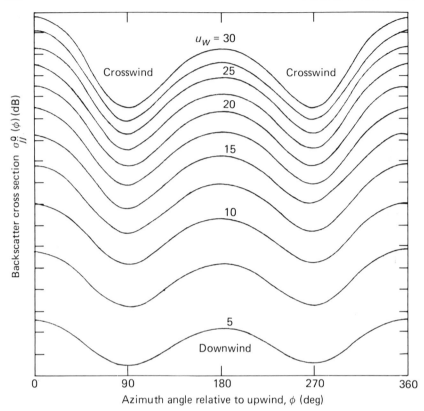

Fig. 8.17   Dependence of C-band backscatter versus azimuth angle, with wind speed as a parameter. Units are 1 dB, with only relative changes shown; absolute values of σ^0 depend on incidence angle much as shown on Fig. 8.13. [From *ERS-1 Technical Annex* (1986).]

The data shown have all been for frequencies near 15 GHz, but there is frequency variability in σ^0 as well. For near-vertical incidence, it may be found in the reflection coefficients via their dependence on ϵ_r, and perhaps implicitly in the rms values for the slope distribution (cf. Eq. 8.118). For the Bragg regime, Eq. 8.121 indicates that in addition to the dielectric constant, both the projection of the wave vector on the surface, and the surface wave energy at the Bragg wave numbers contribute in perhaps more important ways than ϵ_r.

There is also an implicit temperature dependence of σ^0 in both the dielectric constant and in the surface tension; the latter will be most important for short radar wavelengths, where capillary waves serve as both Bragg and specular point scatterers. The presence of surfactants (which are ubiquitous in the marine environment) changes the surface tension and short-wave damping constants greatly. These unmodeled physical effects introduce much variability in the scattering process and lead to a degree of uncertainty in their interpretation.

The speed and angle dependencies of σ^0 are the basis for the measurement of wind velocity over the ocean with a radar scatterometer on spacecraft or aircraft. With this device, one narrow fan beam is used to illuminate a very elongated segment of ocean at an azimuth angle of about 45° to the side of the flight path; over this segment, a range of incidence angles from about 25° to 60° is covered. The absolute value of cross section is measured as a funtion of θ_i, using Doppler filtering to differentiate between ranges, and hence angles. As the platform moves, a swath of surface is covered, as shown in Fig. 8.18. A second fan beam viewing at 135° from the flight path (or 90° from the first beam) covers the same swath a short time later as the flight

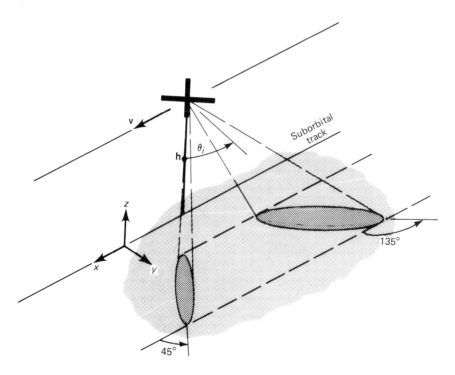

Fig. 8.18 Geometry of swath for two-fan-beam scatterometer, with beams at 45° and 135° off flight direction. Absolute backscatter measurements at angles from wind of ϕ and $\phi + 90°$ give wind solutions.

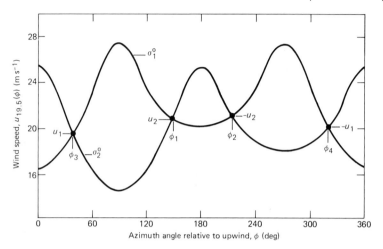

Fig. 8.19 Graphical solutions for wind velocity from a two-fan-beam scatterometer, showing two possible speeds and four possible directions at intersections. [From Stewart, R. H., *Methods of Satellite Oceanography* (1985).]

moves along a constant direction, and repeats the same kind of measurements. On the other side of the platform, this process is duplicated. The net effect of this is that for each resolution element illuminated on the sea surface, the backscatter is measured at two azimuth angles separated by approximately 90°, so that the curves of Fig. 8.17 are sampled at two unknown angles with respect to the wind direction. As shown in Fig. 8.19, for each resolution cell the two observed values of $\sigma°$ may belong to a considerable range of wind speeds and directions, but only two speeds and four angles represent simultaneous solutions for the wind velocity. Thus a pair of scattering measurements has a fourfold ambiguity in the determination of velocity, although only two values of speed exist. If wind direction can be even coarsely inferred from other sources such as large-scale meteorological analyses, the ambiguities can be resolved. The scatterometer wind field of Fig. 2.15b was determined with a two-antenna system on the Seasat spacecraft. However, the addition of a third antenna and the accompanying paired measurements greatly ameliorates the problem, so that it is estimated that less than 10% of the total number of observations would be ambiguous with such an arrangement.

It is likely that the backscattering actually depends on the wind stress rather than the velocity, since the small-scale roughness is a function not only of the wind speed but also the conditions of air–sea temperature difference (cf. Section 2.6). For purposes of describing the wind-driven circulation, it is the stress that is of importance, of course, and its direct measurement on a global scale is a promising development. More research is needed before the issue may be decided.

Radar Altimeter Measurements

The radar altimeter is a vertically viewing, narrow pulse, pencil beam device that determines three parameters that can be used to make a variety of geophysical measurements. The parameters are (1) time delay between pulse transmission and return of the backscattered energy, to an accuracy of approximately 0.1 ns; (2) the shape of the returned pulse, especially its leading edge; and (3) the absolute value of the backscatter coefficient. These measurements are shown schematically in Figs. 8.20a and 8.20b. The time delay, when coupled with a knowledge of the velocity of propagation through the ionosphere and the wet troposphere, can be converted to altitude measurements having a precision of order of a few centimeters. The shape of the pulse can be used to determine the significant wave height, $\xi_{1/3}$, over a range of that quantity from 0.5 up to perhaps 20 m. The peak value of σ^0 can be used to measure wind speed through a knowledge of its dependence on surface roughness.

Assume that the electromagnetic index of refraction can be determined for the ionosphere (e.g., through ionosondes) and the moist atmosphere (via surface pressure and integrated water vapor measurements). If $c(z)$ is the resultant profile for the speed of light, then the altitude to the surface, $\mathbf{h}(x,y)$, can be determined from the round-trip time delay, T_d, via

$$T_d = 2 \int_0^{\mathbf{h}} \frac{dz}{c(z)} . \tag{8.126}$$

The reflecting ocean surface is not smooth, and the mean level at which electromagnetic reflection occurs lies slightly below mean sea level (msl) because of the asymmetry of wave profiles, when measured over the first Fresnel zone, as shown in Fig. 8.20. (The first Fresnel zone is the region of reflecting surface illuminated by the pulse of spatial thickness, $\delta z = ct_p/2$, when the combined curvatures of the wave front and earth's surface result in the spherical wavefront departing from the surface by the amount δz (see Fig. 8.20b).

If the altimeter is on a spacecraft whose orbit may be determined to submetric accuracies, these radar measurements can be put to use in the determination of several oceanic and geodetic quantities. Figure 8.21 illustrates the geometry, with the radius vector, $\mathbf{R}(x,y)$, from the satellite to the earth's center of mass indicated, as well as the altitude to the surface, $\mathbf{h}(x,y)$. The difference between these two is the elevation of the sea surface topography,

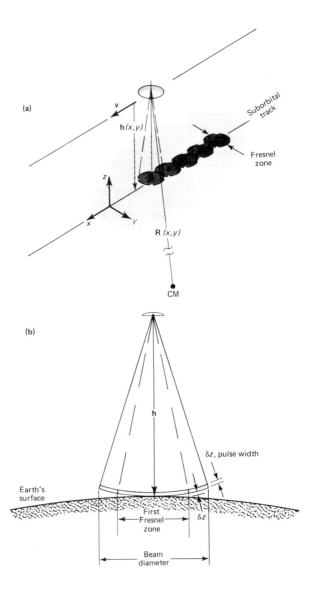

Fig. 8.20 Geometry of altimetric measurements of sea surface. (a) Fresnel zone pattern along subsatellite path; (b) Illumination of curved surface of the earth by a thin shell of radiation of thickness δz, which defines first Fresnel zone. Beam diameter is usually greater than first zone.

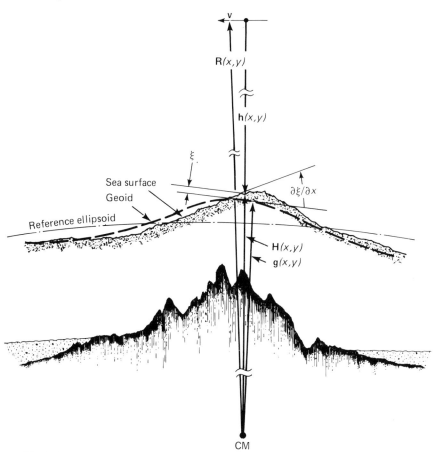

Fig. 8.21 Measurement of sea surface setup, ξ, relative to geoidal heights, $\mathbf{g}(x,y)$, from satellite altimeter. Satellite position with respect to center of mass, $\mathbf{R}(x,y)$, comes from orbit determination; height above sea level, $\mathbf{h}(x,y)$ derives from altimeter. Differences give surface elevations, or potential energy, and surface slopes give kinetic energy of geostrophic flow. Local geoidal heights are correlated with elevations of dense bottom materials (see Figs. 3.8 and 3.9).

$\mathbf{H}(x,y)$, as averaged over the Fresnel zone. Further averaging takes place along the suborbital track if time integration is used. The departure of the actual sea surface topography from the equipotential surface (i.e., the geoid) can be determined if the geoidal heights, $\mathbf{g}(x,y)$, are known independently (see Fig. 3.7, for example); this departure is termed the sea surface elevation, or setup, $\xi(x,y)$.

If additional corrections for tides, the inverse barometer effect, wind-stress setup, and other nongeostrophic distortions can be made, the residual signal can be attributed to geostrophic flow, both baroclinic and barotropic (see Fig. 6.23 and Eqs. 6.84 and 6.85). The slope (with respect to the geoid) of the surface along the suborbital path is a measure of the component of surface current at right angles to the path, and the elevations are measures of the gravitational potential energy stored in surface setup and setdown. Such observations can be assembled over several days on a near-global basis with measurements from a satellite in a high-inclination orbit, during which the surface projections of the orbit are painted out, with the oblique orbit crossings giving two components of the surface current vector. From these, the surface geostrophic current field can be determined on a spatial grid with dimensions of the order of a hundred kilometers and an averaging time of several days. Longer-term records, when subjected to narrowband filtering, can determine those tidal components having amplitudes in excess of several centimeters.

Even if the geoid is not known with sufficient accuracy to measure the departures of the surface from it, a satellite placed in an exactly repeating orbit (to within the diameter of the Fresnel zone) can nevertheless determine the *time variability* of setup along the suborbital tracks. Furthermore, even if the orbit does not repeat exactly, the temporal changes of elevation at the points where the ascending and descending orbits cross over each other may be determined. Figure 8.22 illustrates the observation of the Gulf Stream boundary and the migration of a cold-core eddy out from under the spacecraft's path over a period of 17 days, as determined during a repeat-orbit phase of Seasat.

The accuracy with which these quantities must be known is very high. The vertical component of the orbit, the geoid height, and the distance between satellite and sea level must be determined in a way that yields an overall rms height error of order 0.01 m over the dimensions of a baroclinic Rossby radius, if mid-latitude geostrophic currents are to be found to within about 0.20 m s^{-1}. In spite of these seemingly impossible requirements, present indications are that they may be met with analysis of large amounts of data, mainly through the averaging that is possible with literally millions of satellite observations. The determination of the geoid itself is assisted by the altimeter measurements, but such a geoid is contaminated by steady state current setup. Therefore, independent gravity observations on a global scale are necessary to arrive at the steady state current field. The geoid measurements can be met with a high precision gravity-sensing satellite system via the perturbations introduced into the satellite's orbit by small surface gravity anomalies, of order a few milligals (see Section 3.6).

Beyond currents, tides, and and other surface distortions, the measurement of significant wave height and wind speed are readily accomplished with an altimeter; Figs. 8.23 and 8.24 illustrate the principles behind this. In Fig.

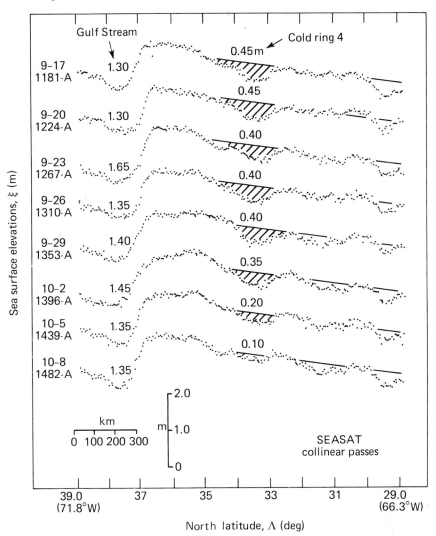

Fig. 8.22 Seasat height measurements for eight repeat passes over a three week period. Cold ring near 33.5°N (see Figs. 6.42 and 6.43) slowly moved away from subsatellite path. Gulf Stream boundary is near 37 to 38°N. [From Cheney, R. E., and J. G. Marsh, *J. Geophys. Res.* (1981).]

8.23, the spherical shell of the short radar pulse is seen incident on a sea surface having significant wave height $\xi_{1/3}$. The backscattered pulse will be broadened out by the distribution of reflecting heights presented by the surface, and when the backscatter is measured over several hundred pulses, a

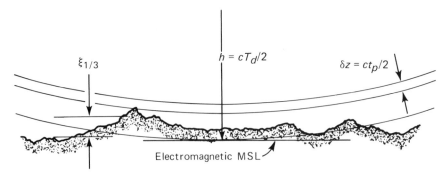

Fig. 8.23 Schematic of altimeter pulse incident on rough sea surface. The reflected pulse is broadened by the distribution of waves and can be used to measure the significant wave height if the spatial pulse width, $ct_p/2$, is smaller than $\xi_{1/3}$. Electromagnetic sea level usually lies slightly below mean sea level because of the asymmetry in scattering from wave crests versus troughs.

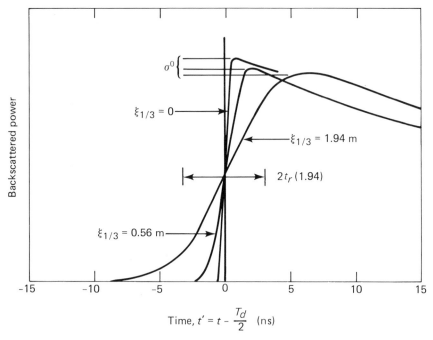

Fig. 8.24 Broadening of the leading edge of an altimeter pulse due to the distribution of reflecting wave heights. Rise time measures significant wave height, while σ° measures wind speed at nadir. Three rise times are shown, corresponding to significant wave heights of 0, 0.56, and 1.94 m. [Adapted from Walsh, E. J. et al., *Boundary Layer Met.* (1978).]

smooth envelope to the return is seen that slowly changes with geographical location as the sea state evolves. The returned pulse shape is a convolution of the outgoing pulse, the delta-function response of the sea surface, and the distribution of surface elevations. If the sea surface is Gaussian, the leading edge of the average return pulse involves an error function, erf(x):

$$\langle P_r(t) \rangle = P_0 \left[\tfrac{1}{2} + \text{erf}\left(\frac{t - T_d/2}{t_r} \right) \right] . \qquad (8.127)$$

This is displayed in Fig. 8.24, which shows three pulse shapes corresponding to significant wave heights of 0, 0.56, and 1.94 m. The apparent rise time of the leading edge of the pulse, t_r, is determined from

$$t_r^2 = t_p^2 + \frac{\xi_{1/3}^2}{c^2} \ln 2 , \qquad (8.128)$$

where t_p is the radar pulse width. Both P_0 and $\langle P_r(t) \rangle$ can be measured and the backscatter cross section can be determined (Eq. 8.118); also $\xi_{1/3}$ may be measured from the rise time.

The value of $\langle P_r \rangle / P_0$ at the maximum of the pulse may be related to the wind speed along the subsatellite track by using the functional behavior shown in Fig. 8.11 at $\theta_i = 0°$. The wind speed determinations of Fig. 5.16a were made simultaneously with the significant wave height measurements of Fig. 5.16b using the radar altimeter data from Seasat in conjunction with such methods. A useful relationship between backscatter cross section at or near nadir and the wind speed at 19.5 m has been derived from a comparison of satellite altimeter and scatterometer observations, and surface buoy measurements, and is given by

$$\sigma^0(\theta_i = 0, u_w) = a u_w^\eta , \qquad (8.129)$$

where $a = 31.77$ and $\eta = -0.468$. This is applicable to radars having frequencies near 15 GHz for all areas of the ocean, over a range of wind speeds from approximately 3 to 15 m s^{-1}; it is altogether possible that the upper range is considerably higher, but the functional form has not been verified at higher speeds.

Radio Frequency Scattering

The basic mechanisms for electromagnetic scattering discussed above are also at work in the case of high frequency radio scatter, where by high frequencies we mean those in the general range of 3 to 30 MHz. Usually the radio wavelengths are of the same order as the dominant oceanic wavelengths, and the incidence angles are near grazing, i.e., 80 to 90°. There the operative scattering mechanism is almost exclusively Bragg scatter, with strong return coming from those ocean waves lying within the illuminated area and having wavelengths meeting the condition: $\lambda_w = \lambda_r/2$. Thus a 3 MHz radio wave would be scattered from ocean waves of 50 m length propagating either parallel or antiparallel to the radio wave vector.

Figure 8.25 is a schematic illustration of radio wave probing of the ocean surface via ionospheric propagation of a pulsed Doppler radio transmission, which may reach out to distances in excess of 4000 km. The illumination may also be by direct transmission, which is useful for local measurements out to perhaps 50 km. The areas covered from a transmitter located in central California are shown in Fig. 8.26. Figure 8.27 shows a Doppler spectrum obtained at 13 MHz, with the sidebands at $\pm \delta f_B$ due to advancing and receding waves, respectively, as well as a small mean Doppler shift induced by ocean currents. The measurement technique is analogous to the acoustic scatter case of Fig. 7.20.

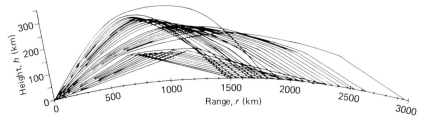

Fig. 8.25 Ray path diagram of forward-scattered high frequency radar propagation using ionospheric propagation to reach out to 1500 to 3000 km. Sea state measurements are possible with such an arrangement. [From Barrick, D. E., *Marine Technical Society J.* (1973).]

8.8 Synthetic and Real Aperture Imaging Radars

High resolution radars are capable of providing *images* of the sea surface that contain interesting but incompletely understood data on fluid processes in and on the sea. Images from aircraft and spacecraft have shown signatures of surface and internal waves, shallow (and not so shallow) bathymetric features, wind speed patterns, and distributions of oil and other surfactants. The frequencies of such radars range from approximately 1.4 to 35 GHz, and the attenuation lengths are therefore of order centimeters or less (see Eqs. 8.51 and 8.55). It is clearly the case that the mechanisms leading to surface signatures of subsurface features are not ones that involve penetration

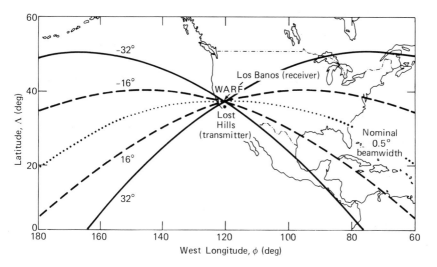

Fig. 8.26 Ocean areas covered by wide aperture Doppler radar located in central California. Both eastern Pacific and Gulf of Mexico coverage is obtained. [From Barnum, J.R. et al., *IEEE Trans. Antennas and Propag.* (1977).]

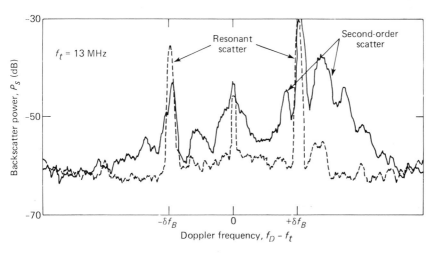

Fig. 8.27 HF Doppler spectrum at 13 MHz showing two sidebands at the displaced Bragg frequency, $\pm \delta f_B$, which is 0.37 Hz for 11.5 m ocean waves. First-order peaks are displaced slightly toward higher frequencies due to mean currents. [From Stewart, R. H., *Methods of Satellite Oceanography* (1985).]

of electromagnetic radiation into the sea. Instead, the features being imaged must be caused by subsurface dynamics that lead to changes in radar cross section, or to other modulations of image intensity. There exists a reasonable theory that accounts for the variations of σ^0 as a function of spatial position on scales that are as small as the resolution element of the radar—typically tens of meters. The mechanisms at work are both hydrodynamic and electromagnetic.

We will first present samples of imagery of the sea surface showing hydrodynamic signatures, and next briefly describe the functioning of the two major types of imaging radars used in obtaining the images, the synthetic aperture radar—SAR—and the real aperture radar—RAR, or, as it is more commonly known, the side-looking airborne radar, SLAR. Finally, in Section 8.9, we will outline a first-order theory that appears to account for the observations in terms of (1) current-induced modulations of the surface wave spectrum, and (2) in the case of SAR, additional modulations of the image intensity due to velocity-induced Doppler shifts.

Figure 8.28a is a SAR image of the area south of Nantucket Island, Massachusetts, taken from the Seasat satellite in August 1978. The resolution of the image is approximately 25 m and the radar wavelength is 0.214 m. The features in the image are essentially maps of shallow (≈ 20 m depth) bathymetry, internal waves, and regions of smooth, low-backscatter ocean due to stable, cooler unwelled water east of Nantucket. The image was made near the maximum in tidal flow, a fact that is important to the mechanisms responsible for the imaging of the bottom; another image of the same region made near slack tide does not show bathymetric features. Figure 8.28b (Plate 3) is a color photograph of the same region taken from the Skylab manned spacecraft several years earlier, and clearly shows white sand in the shallows via upwelling light reflected from the bottom. The correlation between the patterns of backscatter variations and depth is obvious. Figure 8.29 is an image from Seasat showing what appears to be a packet of moderately coherent internal waves associated with a subsurface bathymetric feature at a depth of approximately 500 m. Figure 8.30 shows long wavelength storm waves from hurricane Josephine taken from an imaging radar on the space shuttle *Challenger* in October 1984; the lower part of the figure is the Fourier transform of the image, with certain corrections applied to render it an approximation to the wave height spectrum.

From these (and other) images, it is indubitably the case that imaging radar may mirror subsurface features under certain limited circumstances, e.g., low wind speed, strong current shear, and shallow water. Surface waves are imaged if large and long enough, although the high-fidelity determination of their spectral content is not always possible, as we will touch on below.

It is first necessary to understand how these instruments reach the fine

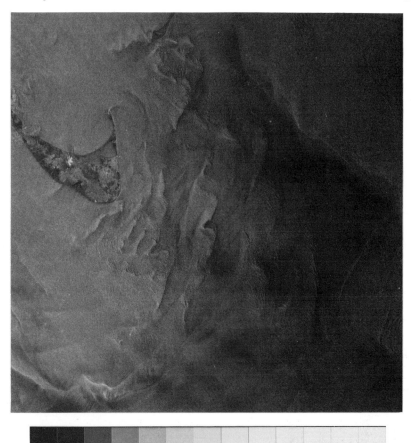

Fig. 8.28a Seasat synthetic aperture radar image of the ocean south of Nantucket Island made during flood tide. Tidal flow over bathymetric features makes shoals visible; internal waves and upwelling are also imaged. Compare with the Skylab color photograph of the same region [Fig. 8.28b; Plate 3]. [Digital image courtesy of The Johns Hopkins University Applied Physics Laboratory; data courtesy of Jet Propulsion Laboratory, California Institute of Technology (1978).]

spatial resolution that enables them to discern small-scale oceanic features. Both the SAR and the RAR achieve resolution in cross-track or *slant range*, r, with a short pulse length, as shown in Fig. 8.31; the range resolution, δr, is $ct_p/2$, where t_p is the pulse length. This resolution projects into a *ground range* resolution, $\delta y = (ct_p/2) \csc \theta_i$. Thus ground range resolution is poor

Fig. 8.29 Seasat SAR image of internal waves over the Wyville Thompson Ridge southeast of Faeroe Islands; ridge depth is 500 m or more. [Figure courtesy of Jet Propulsion Laboratory, California Institute of Technology (1978).]

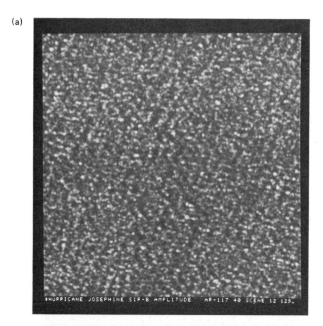

(a)

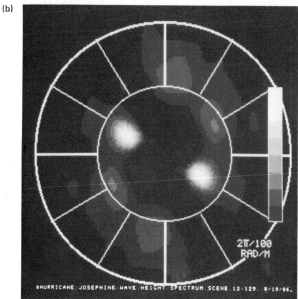

(b)

Fig. 8.30 (a) Shuttle SIR-B image of 300 m long storm waves in hurricane Josephine during October 1984; the area shown is 6.4 km on a side. (b) Digital Fourier transform of the image, corrected to yield the height variance spectrum. Wave vector origin is at the center, and the outer ring corresponds to $k = 2\pi/100$ rad m^{-1}. [Figure courtesy of The Johns Hopkins University Applied Physics Laboratory (1986).]

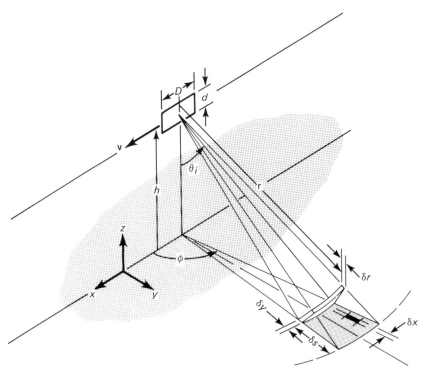

Fig. 8.31 Viewing geometry of a synthetic aperture radar. Range direction is *y*; azimuth direction is *x*. [Adapted from Stewart, R. H., *Methods of Satellite Oceanography* (1985).]

at small incidence angles. In the azimuth coordinate, ϕ, the two types of radars differ significantly in the method used to achieve resolution. The RAR uses a combination of large antenna dimension, D, high frequencies, and low altitudes, h, in arriving at the *azimuth* (along-track) *resolution, δx,* and hence is not useful from spacecraft if high resolution is required. The SAR makes use of the principle of *aperture synthesis* to construct an effective antenna dimension in the direction of motion that is many times its physical dimension, hence its name. It accomplishes this by transmitting a sequence of highly repeatable, stable signals and coherently adding up the scattered returns over an integration time t_i (typically a very few seconds in the case of the ocean). During this *coherent integration* process, the platform has moved a distance $D' = vt_i$, which is the synthetic aperture length. The effective radian beam width is approximately the reciprocal of the number of electromagnetic wavelengths, D'/λ, contained in the synthetic aperture and hence is very narrow, with the result that the resolution in the along-track direction is very high. If the real beam width is chosen so that a point target is illuminated for just the duration of the integration time, it can be shown that the azimuth resolution, δx, is just one-half the real aperture dimension, D. Thus the SAR has a surface resolution of

$$\delta x = D/2 \qquad (8.130a)$$

and

$$\delta y = (ct_p/2) \csc \theta_i \, . \qquad (8.130b)$$

A typical aircraft SAR may have a resolution of 3×3 m^2, while spacecraft radars are closer to 8×25 m^2. The image shown in Fig. 5.28 is an example of the former; those in Figs. 8.28 to 8.30 lie between 25 and 40 m resolution.

The processes of constructing an image from the two types of imaging radars are quite different, as might be expected. Images are made up of discrete picture elements (pixels). Both SAR and RAR assign pixel positions in the range (r) dimension proportional to the time delay to the ground point in question. The RAR assigns the x location in proportion to the platform velocity. The SAR assigns the other coordinate according to the Doppler shift of the received signal, which is due to a combination of platform velocity and, in the case of an orbiting satellite, the appropriate component of the earth's surface rotational velocity. Thus the (x,y) coordinates in a SAR image are established by loci of constant time delay and constant Doppler shift, a nonorthogonal coordinate system. It is easy to show that the latter are hyperbolas (on a planar surface) given by

$$\left(\frac{x}{h}\right)^2 \left[\left(\frac{\delta f_{D0}}{\delta f_D}\right)^2 - 1\right] - \left(\frac{y}{h}\right)^2 = 1 \, , \qquad (8.131)$$

where δf_{D0} is the Doppler-shift in frequency in the direction of platform motion, and δf_D is the (lesser) Doppler shift elsewhere in the plane. This is shown in Fig. 8.32, which demonstrates the nonorthogonal coordinate system in terms of lines of constant range delay (circles) and constant Doppler frequency (hyperbolas), and illustrates a rectangular pixel at one location. The transmitted frequency, f_t, is Doppler-shifted by the platform's motion at velocity v by the value

$$\delta f_D = \frac{2f_t v}{c} \frac{x}{(x^2 + y^2 + h^2)^{1/2}} \, . \qquad (8.132)$$

The maximum Doppler frequency clearly comes from the direction of the positive x axis; the frequency shifts are positive in the approaching direction,

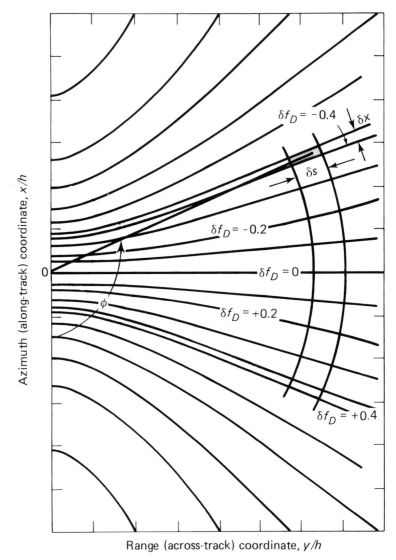

Range (across-track) coordinate, y/h

Fig. 8.32 SAR coordinate system, defined by hyperbolas of constant Doppler shift, and circles of constant range delay. Platform motion is toward the bottom of figure.

negative in the receding, and zero along the direction orthogonal to the flight path.

It is also clear that *any* relative motion between transmitter and target will induce a Doppler shift, and that because of the kind of mapping illustrated in Fig. 8.32, the positional assignment of a *moving* target will be incorrect in the along-track or azimuth direction. This leads to the "moving-train-off-the-track" phenomenon in SAR imagery, and is very important for understanding SAR images of the moving ocean, especially rapidly propagating surface waves. A point target moving with a *radial* velocity component, v_r, will be displaced from its true position by an amount Δx in the *azimuth* coordinate given by

$$\Delta x = \frac{r}{v}\, v_r\ , \tag{8.133}$$

where, as before, v is the platform velocity. Thus the larger the range or the slower the platform speed, the worse the error. This leads to distortions of surface waves in SAR images such as Fig. 8.30 and hence disfigures their apparent spectral content. In addition, the image of a moving target is defocused because of the processing method used to construct the picture from the time history of the received radar signal.

Surface Wave Spectra

In light of the mismappings of moving oceanic targets that occur with SAR, it is somewhat surprising that fast-traveling gravity waves are imaged at all. However, this is the case, down to a point—said point being dependent on the radar and platform as well as the wave velocity and amplitude. From Eq. 8.133, it may be seen that the azimuth shift is proportional to the range-to-platform-velocity ratio as well as the target's radial, line-of-sight velocity component. A platform such as the space shuttle has an r/v value of approximately 35 s, and a steep incidence angle of some 20 to 30°. This means that the radial velocity is very nearly the same as the vertical component of whatever wave velocity it is that is responsible for the Doppler shift. There are three candidate velocities; in increasing order of magnitude, they are the orbital (surface current) velocity, $\mathbf{u}$; the wave group velocity, $\mathbf{c}_g$; and the wave phase velocity, c_p. Opinion is divided, with the present experimental evidence supporting (but not unequivocally) the *orbital* velocity hypothesis, but this issue is still open. Let us assume that the orbital velocity is the operative one. A deep-water gravity wave has a vertical component of current given by Eq. 5.57; i.e., $w_0 = gk_h\xi_0/\omega$, where k_h is the horizontal wave number for the gravity wave. Now the oscillating character of the wave orbital velocity will lead to a bunching of the backscattered intensity due to the Doppler displacement and hence an enhancement of the wave image in some regions of the wave phase and a diminution in other regions. Using this velocity in Eq. 8.133 and assuming that the displacement, Δx, can be no more than one-fourth of the azimuth component of wavelength before significant distortion of the image has occurred, we obtain an estimate of the shortest azimuth-traveling wave that can be imaged with moderate fidelity in an SAR:

$$\lambda_{min} = (\sqrt{2\pi g}\ \ 4r\xi_0/v)^{2/3} \ . \tag{8.134}$$

Here we have used the deep water gravity wave dispersion relation for ω/k_h. Equation 8.134 suggests that a wave having an amplitude of 1 m and a wavelength of about 100 m is near the limit of along-track, azimuthal resolution, with longer or lower waves meeting that condition being visible. This estimate is borne out by experiment to a reasonable degree, with the implications being that long storm waves and swell may in fact be imaged by synthetic aperture radar.

In the range coordinate, little or no problem with velocity bunching exists. Here the wavelength resolution is set by the ground range resolution, δy, with the theoretical Nyquist requirement of two samples per wavelength needed to resolve a periodic signal actually being more nearly four to five samples in practice. A SAR with a 25 m resolution would then be able to discern 100 to 125 m waves; again, this agrees with experience.

Images of the type shown in Fig. 8.30 can be fast Fourier transformed digitally to obtain a two-dimensional *intensity* spectrum, $I(\mathbf{k})$, that shows periodicities clearly associated with the waves (see Fig. 8.30b). There is a question here as to the relationship between the intensity spectrum and the ocean wave height or slope spectrum. No satisfactory theoretical answer exists at present, but experimental observations suggest that with appropriate corrections, the SAR image spectrum is quite close to the ocean wave *slope* spectrum, $S_\zeta(\mathbf{k}_h)$. Since the slope, ζ, is the gradient of the surface height, the variance spectrum associated with the slope is readily shown to be $\mathbf{k}^2$ times the height spectrum. Denoting the wave number in the range (y) direction by k_r and that in the azimuth (x) direction by k_a, we may write

$$S_\zeta(\mathbf{k}_h) = S_\zeta(k_r,k_a) = (k_r^2 + k_a^2)\ S_\xi(\mathbf{k}_h) \ . \tag{8.135}$$

Thus transforming from height to slope in the Fourier domain is a simple algebraic operation. The slope spectrum emphasizes higher frequencies compared with the height spectrum.

Corrections must be applied to the raw intensity spectrum for instrument illumination, finite resolution, image speckle, and broadband noise, all of which yield a function, $\hat{I}(k_r,k_a)$, that can be compared with the actual two-dimensional wave slope spectrum of the same region of ocean obtained simultaneously by other means. Figure 8.33 illustrates a Fourier transform of an ocean wave image obtained from the space shuttle radar operating at a

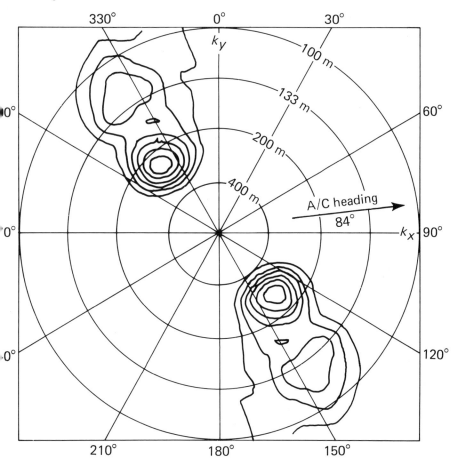

Fig. 8.33 Surface slope wave number spectrum, $S_\xi(k_x,k_y)$, from SIR-B during a storm off Chile, contoured in eight intensity intervals. [From Beal, R. C., et XI al., *Science* (1986).]

radar wavelength of 0.23 m. The image has been processed as discussed above, and shows what is believed to be a slope spectrum of 300-m-long storm waves off Chile having rms wave heights, σ_ξ, of approximately 1.1 m (or significant wave heights of $\xi_{1/3}$ = 4.5 m), and rms wave slopes, σ_ζ, of approximately 0.023. The spectrum indicates a direction of wave travel of about 135° true (the ambiguity in direction present in any Fourier transform was resolved with other data), and shows intensity, or slope spectrum, contoured in eight intervals. Division by $\mathbf{k}^2$ then gives the height spectrum, $S_\xi(\mathbf{k}_h)$. Figure 8.34 shows a slope spectrum obtained nearly simultaneously using an airborne

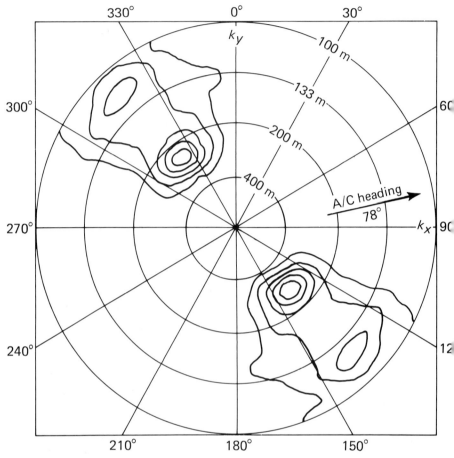

Fig. 8.34 Surface slope wave number spectrum from a radar ocean wave spectrometer taken by an aircraft 1 hr later than that in Fig. 8.33. [From Beal, R. C., et XI al., *Science* (1986).]

radar ocean wave spectrometer; the agreement is considered satisfactory, indicating that synthetic aperture radar most likely yields slope spectra for wavelengths of $2\pi/k_a \gtrsim 100$ m.

The system transfer function from wave height spectrum to image spectrum imposed by the nature of the SAR imaging process is assumed to be linear, algebraic, and to be described by a *modulation transfer function,* $R_{mtf}(k_r, k_a)$:

$$R^2_{mtf} = k_h k_a{}^2 g(r/v)^2 \ (\sin^2 \theta_i \ \sin^2 \phi_i + \cos^2 \theta_i)$$
$$+ k_r{}^2 (4 \cot \theta)^2 / |g_{ii}(\theta_i)|^2 . \tag{8.136}$$

As before, θ_i and ϕ_i are the incidence and azimuth angles of the radar relative to the nadir and flight direction, respectively; $g_{ii}(\theta)$ are the reflection coefficients given by Eqs. 8.122 and 8.123, and $\mathbf{k}_h = k_r \hat{r} + k_a \hat{a}$ is the wave vector of the surface wave. Then the assumed transformation is

$$\hat{I}(\mathbf{k}) = S_{\hat{\zeta}}(\mathbf{k}_h) = R_{mtf}^2 \, S_{\xi}(\mathbf{k}_h) \; . \tag{8.137}$$

The function R_{mtf}^2 varies in wave number with angle and the magnitude of the wave vector, and has the effect of amplifying high frequencies by the factor k_r^2 in the range direction and by k_a^3 in the azimuth direction. Dividing the image spectrum by the modulation transfer function thus approximately corrects for the influence of tilt modulation in the range direction and velocity bunching in the azimuth direction. Care must be exercised near the origin of the wave vector.

Backscatter Maps of the Ocean

From the above discussions, it is clear that changes in image or backscatter intensity distributions in general originate from at least three causes: (1) modulation by surface hydrodynamic processes, (2) changes in the mean tilt over long gravity waves (see Eq. 8.118), and (3) the velocity bunching effect arising from misplaced pixels in the SAR Doppler coordinate. Of the three, it is thought that hydrodynamic modulation is the most pervasive, except for large surface waves, where both tilting and bunching lead to additional modulations, as just described. Thus an ocean feature shows image contrast with respect to its surroundings if, over the pixels delineating it, there is (1) a differential in backscatter cross section from neighboring elements, or (2) there is a differential in Doppler shift (however, this latter is generally a factor that reduces image fidelity and contrast). Thus moving ocean features are imaged by SAR to the extent that a group of scatterers has a coherent, ensemble-averaged component of motion over the resolution cell of the radar that contributes to the mean Doppler shift.

Refer now to the general equation for backscatter (Eq. 8.116) and to its approximations in the limits of near-specular angles of incidence (Eq. 8.118), and in the Bragg region (Eq. 8.121). In each of these, properties of the wave height spectrum are involved in some important way or other – in Eq. 8.116 via the Fourier transform of the exponentiated autocorrelation function, in Eq. 8.118 by way of the rms values of the slope distribution, and in Eq. 8.121 through the spectrum itself evaluated at the Bragg wave vector. To a reasonable approximation, therefore, variations in the surface wave spectrum, $\delta S(\mathbf{k})$, are mirrored as variations in the radar cross section; radar images or maps of the latter will represent the distribution of modulations in the surface wave spectrum at selected frequencies, if properly interpreted. (In a synthetic aperture radar, the additional complications of velocity bunching or smearing will be imposed on the spatial variations as changes in image intensity and target geometry.) Thus for small changes in $S(\mathbf{k})$ due to surface geophysical processes, we expect the correspondence:

$$\frac{\delta \sigma^0(\mathbf{x},t)}{\sigma^0} = \frac{\delta S(\mathbf{k};\mathbf{x},t)}{S_{eq}(\mathbf{k})} = \frac{S(\mathbf{k};\mathbf{x},t) - S_{eq}(\mathbf{k})}{S_{eq}(\mathbf{k})} \ , \tag{8.138}$$

where $S_{eq}(\mathbf{k})$ is the equilibrium spectrum. We then need to know how hydrodynamic effects modulate the surface wave spectrum.

A theoretical method of treating such hydrodynamic interactions is contained in the equations for *conservation of wave action* (or, more properly, *non*conservation of action). Because of its specific applicability to the ocean imaging problem, we will outline the derivation of the equations here and refer the reader to the bibliography for more detailed expositions.

8.9 The Wave Action Equation

It should be emphasized that the material in this section is quite general, and could equally well have been incorporated after Section 5.12, for example. However, its development has been delayed until this point, first to provide additional motivation for its inclusion, and secondly to draw on other material presented subsequent to that section.

The conservation equations for mechanical systems—mass, linear and angular momentum, and energy—may be derived most generally from variational calculations involving Lagrangian or Hamiltonian mechanics. As a side benefit, these calculations make clear the times at which those conservation laws fail, or become in need of modification (e.g., when friction is present, or when system parameters are changing in space or time). For example, in the theory of small oscillations, it is found that the sum of the kinetic and potential energies of the oscillatory system is not conserved if the restoring force is made to slowly vary; such a situation might arise when a small mass is connected to springs that are slowly being stretched while the mass oscillates. As the spring tension increases, the oscillatory energy of the mass increases, as does its frequency of oscillation. Nevertheless, the Lagrangian formulation shows that in spite of this slow variation, the ratio of energy (as averaged over a period) to the frequency of that period remains approximately a constant, A. This quantity, $A = E/\omega$, is termed the *action* and is an *adiabatic invariant* of the system (see footnote 1 in Section 7.5).

Similarly, if the system is a continuous medium whose properties depend on time, the energy per unit volume is no longer a constant; if it depends on space, the momentum per unit volume is no longer conserved. (We will derive an equation below for the total rate of change of energy in such a circumstance.) However, the wave *action density* is conserved in both of these cases, and a conservation equation may be derived for this quantity that describes the interplay between the variables of interest in physical space, in wave number space, and in time. It will be seen that the action equation is a powerful analytical tool. However, in order to arrive at it, we first need a few preliminary kinematic statements about waves.

Wave Kinematics

Slow changes in the properties of the medium lead to concomitant slow changes in the propagation of waves, an example of which is the refraction of acoustic waves by temperature and depth variations. In Section 7.7, the eikonal equation was derived for rays and phase fronts in a slowly varying medium by assuming a quasi-plane wave as the solution to the wave equation, and then deriving the relationships obeyed by the amplitude and phase functions. In the present case, we will allow for slow changes in both $\mathbf{x}$ and t, and write for the instantaneous amplitude

$$\xi(\mathbf{x},t) = a(\mathbf{x},t) \, e^{iW(\mathbf{x},t)} \; . \tag{8.139}$$

The phase function, $W(\mathbf{x},t)$, is assumed to have the form of a local wave vector, $\mathbf{k}(\mathbf{x},t)$, and a local frequency, $\omega(\mathbf{x},t)$ (which must also satisfy the wave dispersion equation). Thus we take

$$W(\mathbf{x},t) = \mathbf{k}(\mathbf{x},t)\cdot\mathbf{x} - \omega(\mathbf{k};\mathbf{x},t)t \; . \tag{8.140}$$

The nonuniformity of the medium will be built into the problem by including in the frequency (and, via the dispersion equation, in the wave vector) a dependence on our generalized fluid parameter, $\gamma(\mathbf{x},t)$, which in turn might represent a variable depth or changing ambient density gradient, for example. We also explicitly allow for a variable background current, $\mathbf{U}(\mathbf{x},t)$. All of these quantities are assumed to change slowly over a wavelength or a period.

From the definition of the phase function, one may calculate the wave vector and radian frequency via

$$\mathbf{k} = \nabla W \tag{8.141}$$

and

$$\omega = -\frac{\partial W}{\partial t} .$$

(8.142)

By cross-differentiation, it is immediately obvious that

$$\frac{\partial \mathbf{k}}{\partial t} + \nabla \omega = 0 ,$$

(8.143)

which is a kinematic relationship that is a conservation equation for the *density of waves*. If no new waves are being created by a local disturbance, for example, Eq. 8.143 states that the local rate of change of wave vector is due to the spatial convergence of the frequency, i.e., the number of wave crests passing a point per unit of time.

If the medium is moving with a background velocity $\mathbf{U}(\mathbf{x},t)$, it is clear that waves at the intrinsic or Doppler frequency,

$$\omega_D = \omega - \mathbf{U} \cdot \mathbf{k} ,$$

(8.144)

are being advected past a fixed observer at the apparent frequency, ω. (We recall from Eqs. 5.160 and 5.161 that the Doppler frequency is the one that satisfies the dispersion relationship, and that ω is the apparent frequency entering into the oscillatory solutions.) Thus an alternative writing of the formula for conservation of wave density is

$$\frac{\partial \mathbf{k}}{\partial t} = - \nabla (\omega_D + \mathbf{U} \cdot \mathbf{k}) .$$

(8.145)

Now the apparent frequency is

$$\omega = \mathbf{U} \cdot \mathbf{k} + \omega_D [\mathbf{k}(\mathbf{x},t), \gamma(\mathbf{x},t)] ,$$

(8.146)

since the intrinsic frequency varies with the properties of the medium. With such implicit variablity, the Doppler frequency will change locally according to the calculus of implicit derivatives. From Eq. 8.141, $\nabla \times \mathbf{k} = \mathbf{0}$, since the curl of the gradient of a scalar vanishes identically, and we may write

$$\nabla \omega_D = (\nabla_{\mathbf{k}} \omega_D)^{\dagger} \cdot (\nabla \mathbf{k}) + \frac{\partial \omega_D}{\partial \gamma} \nabla \gamma ,$$

(8.147)

where the dagger (†) indicates a matrix transpose, and where, in various equivalent notations,

Electromagnetics and the Sea

$$(\nabla_\mathbf{k} \omega_D)^\dagger \cdot (\nabla \mathbf{k}) = \sum_{i,j=1}^{2} \frac{\partial \omega_D}{\partial k_j} \frac{\partial k_j}{\partial x_i} \hat{\imath}_i$$

$$= \left[\begin{array}{cc} \dfrac{\partial \omega_D}{\partial k} & \dfrac{\partial \omega_D}{\partial l} \end{array} \right] \left[\begin{array}{cc} \dfrac{\partial k}{\partial x} & \dfrac{\partial k}{\partial y} \\[2mm] \dfrac{\partial l}{\partial x} & \dfrac{\partial l}{\partial y} \end{array} \right] . \tag{8.148}$$

The $\mathbf{k}$ derivative of ω_D is just the group velocity:

$$\mathbf{c}_g = \nabla_\mathbf{k} \omega_D (\mathbf{k}) . \tag{8.149}$$

By substituting Eq. 8.147 into 8.145 and performing the differentiations on the convective term, we obtain for the total rate of change of the wave vector:

$$\frac{\partial k_i}{\partial t} + \sum_j (c_{gj} + U_j) \frac{\partial k_i}{\partial x_j} = - \frac{\partial \omega_D}{\partial \gamma} \frac{\partial \gamma}{\partial x_i} - \sum_j k_j \frac{\partial U_j}{\partial x_i} . \tag{8.150}$$

In a similar fashion, by differentiating Eq. 8.146 with respect to time, we obtain

$$\frac{\partial \omega}{\partial t} + \sum_j (c_{gj} + U_j) \frac{\partial \omega}{\partial x_j} = \frac{\partial \omega_D}{\partial \gamma} \frac{\partial \gamma}{\partial t} + \sum_j k_j \frac{\partial U_j}{\partial t} . \tag{8.151}$$

The left-hand sides of Eqs. 8.150 and 8.151 may be considered as total derivatives following the motion along spatial trajectories, or rays, defined by

$$\frac{d\mathbf{x}}{dt} = \mathbf{c}_g + \mathbf{U} , \tag{8.152}$$

and a trajectory in wave-number space given by

$$\frac{d\mathbf{k}}{dt} = - \mathbf{k}^\dagger \cdot \nabla \mathbf{U} - \frac{\partial \omega_D}{\partial \gamma} \nabla \gamma . \tag{8.153}$$

These curves are known as *characteristics* or *rays* and define points of *stationary phase* in four-dimensional phase space, (x,y,k,l), along which energy (and action) flow. In configuration space, the propagation velocity is $\mathbf{c}_g + \mathbf{U}$. All of this is highly reminiscent of the problem of acoustic propagation in a slowly varying medium, and is in fact a generalization of the simple ray concepts that includes time as well as space variations, and which takes into account both frequency and wave vector changes. In a homogeneous, time-independent medium for which γ and $\mathbf{U}$ are constant, the equations for ω and $\mathbf{U}$ reduce to

$$\frac{d\omega}{dt} = 0 \tag{8.154}$$

and

$$\frac{d\mathbf{k}}{dt} = \mathbf{0} . \tag{8.155}$$

In such a medium, the frequency and wave vector are constant in time along a characteristic.

An Energy Equation

With this amount of kinematics behind us, we may turn our attention to energy variation. Let $\mathcal{E} = dE/dA$ be the averaged perturbation wave energy density per unit area (both kinetic and potential), as in Eq. 5.90. If energy is conserved in the system, the equation for its conservation is of the same general type as the other conservation equations of Chapter 3 (e.g., Eq. 3.76). Because we know that the propagation speed is $\mathbf{c}_g + \mathbf{U}$, we may write

$$\frac{d\mathcal{E}}{dt} = \frac{\partial \mathcal{E}}{\partial t} + \nabla \cdot (\mathbf{c}_g + \mathbf{U})\mathcal{E} = 0 . \tag{8.156}$$

If instead it is the action density per unit area that is the conserved quantity, we will have

$$\frac{\partial}{\partial t}\left(\frac{\mathcal{E}}{\omega_D}\right) + \nabla \cdot \left[(\mathbf{c}_g + \mathbf{U})\frac{\mathcal{E}}{\omega_D}\right] = 0 . \tag{8.157}$$

When the medium is nonuniform, there will be an interchange between the waves and the background velocity, in much the same way that there are exchanges between turbulence and a mean flow. We may derive the energy rate equation for this from the action conservation relationship by expanding the derivatives in Eq. 8.157 and rewriting it so that the left-hand side deals with energy changes only:

$$\frac{\partial \mathcal{E}}{\partial t} + \nabla \cdot [(\mathbf{c}_g + \mathbf{U})\mathcal{E}] = \frac{\mathcal{E}}{\omega_D}\left[\frac{\partial \omega_D}{\partial t} + (\mathbf{c}_g + \mathbf{U}) \cdot \nabla \omega_D\right] . \tag{8.158}$$

With the substitution of Eqs. 8.147 and 8.151, Eq. 8.158 becomes

$$\frac{\partial \mathcal{E}}{\partial t} + \nabla \cdot [(\mathbf{c}_g + \mathbf{U})\mathcal{E}] =$$

$$- \frac{\mathcal{E}}{\omega_D} \left[\mathbf{k} \cdot (\mathbf{c}_g \cdot \nabla)\mathbf{U} - \frac{\partial \omega_D}{\partial \gamma} \left(\frac{\partial \gamma}{\partial t} + \mathbf{U} \cdot \nabla \gamma \right) \right] , \quad (8.159)$$

which is the desired rate equation. Clearly the energy interacts with *derivatives* of the mean flow and the generalized parameter describing the variable medium. This equation finds utility in describing propagation in variable flows or over changing bottom depths, for example.

The Action Spectrum

Let $S_{eq}(\mathbf{k}) = S_{eq}(k,l)$ be the two-dimensional equilibrium height spectrum defined via the height autocorrelation operation,

$$C_\xi(\mathbf{x}') = \frac{1}{4XY} \int_{-X}^{X} \int_{-Y}^{Y} \xi(\mathbf{x}) \, \xi^*(\mathbf{x} - \mathbf{x}') \, d^2\mathbf{x} , \quad (8.160a)$$

and its Fourier transform,

$$S_{eq}(\mathbf{k}) = \int\!\!\!\int_{-\infty}^{\infty} C_\xi(\mathbf{x}') \, e^{-i\mathbf{k} \cdot \mathbf{x}'} \, d^2\mathbf{x}' . \quad (8.160b)$$

The inverse Fourier transform of $S_{eq}(\mathbf{k})$ gives an alternative expression for the autocorrelation function. If we normalize C_ξ by dividing it by the height variance, σ_ξ^2, we obtain the *spatial correlation function*, $\rho(\mathbf{x})$, in terms of the spectral transform:

$$\rho(\mathbf{x}) \equiv \frac{C_\xi(\mathbf{x})}{\sigma_\xi^2} = \frac{1}{(2\pi\sigma_\xi)^2} \int\!\!\!\int_{-\infty}^{\infty} S_{eq}(\mathbf{k}) \, e^{i\mathbf{k} \cdot \mathbf{x}} \, d^2\mathbf{k} . \quad (8.161)$$

This form also arose in Eq. 8.116.

Similar operations may be used to obtain the equilibrium action spectrum, $N_{eq}(\mathbf{k})$. The relationship between the two is, from the definition of action,

$$N_{eq}(\mathbf{k}) = (\rho g + k^2 \tau_s) \, S_{eq}(\mathbf{k})/\omega_D(\mathbf{k}) , \quad (8.162)$$

where

$$\alpha = \int\int N_{eq}\,(\mathbf{k})\,d^2\mathbf{k} \tag{8.163}$$

is the total action per unit area. (The capillary term involving τ_s appears because $S(\mathbf{k})$ is defined in terms of a height variable only, via Eqs. 8.160 and 8.161; see Eq. 5.94, for example.)

If now the slow variations of the medium are introduced, both S and N become functions of $\mathbf{x}$ and t as well. The integrations of Eqs. 8.160 and 8.161 may no longer range over all space, but are restricted to the domain where the wave field is locally homogeneous. However, the wave height spatial correlation function almost always decays to very small values within lag lengths of a very few spatial wavelengths, so that the integrand of Eq. 8.160b is effectively negligible beyond those dimensions. We may then define local energy and action spectra that change on space and time scales that are long compared with their correlation scales; these are nonequilibrium quantities, by their very definitions. Thus for the action spectrum, we write

$$N(\mathbf{k};\mathbf{x},t) = (\rho g + k^2 \tau_s)\,S(\mathbf{k};\mathbf{x},t)/\omega_D\,(\mathbf{k};\mathbf{x},t)\ , \tag{8.164}$$

where $\omega_D(\mathbf{k};\mathbf{x},t)$ is the quantity appearing in Eq. 8.146.

Given the functional dependencies of $N(\mathbf{k};\mathbf{x},t)$, its total time derivative must be

$$
\begin{aligned}
\frac{d}{dt}N(\mathbf{k};\mathbf{x},t) &= \frac{\partial N}{\partial t} + \nabla_{\mathbf{k}}N\cdot\frac{d\mathbf{k}}{dt} + \nabla N\cdot\frac{d\mathbf{x}}{dt}\\[2mm]
&= \frac{\partial N}{\partial t} - \left(\mathbf{k}^{\dagger}\cdot\nabla\mathbf{U} + \frac{\partial\omega_D}{\partial\gamma}\nabla\gamma\right)\cdot\nabla_{\mathbf{k}}N\\[2mm]
&\quad + (\mathbf{c}_g + \mathbf{U})\cdot\nabla N\ .
\end{aligned}
\tag{8.165}
$$

Here we have used the equations for the total time derivatives of $\mathbf{k}$ and $\mathbf{x}$ (Eqs. 8.152 and 8.153), since the changes of $\mathbf{k}$ and ω must occur along the characteristics.

Equation 8.165 is a portion of our desired result. It states that the total change in the action spectrum is due to (1) intrinsic, local time changes of N, (2) changes due to variations in the properties of the medium interacting with the $\mathbf{k}$ space gradient of action, and (3) movements of $\mathbf{x}$ space gradients at the combined group and convective velocities. This is both a dynamic and a kinematic result. It asserts that while the total action in four-dimensional phase space and time is conserved (a dynamical statement), there are changes in the local spectral level due to interactions with currents (**U**), current gradients or shears (∇**U**), and other gradients in the medium ($\nabla\gamma$). These interactions modify the spectral distribution in **x** and t by the processes of convection and refraction in **x** space, and by the redistribution of energy among wave vectors in **k** space (all of which are essentially kinematical in nature). It is a conservation law of considerable importance whose underlying physical basis is a variant of Liouville's theorem on the constancy of the phase space distribution function, in the absence of collisions (or other interactions involving rearrangements). A calculation that forms the proper statistical moment of the Liouville equation should yield the conservation law for action above.

Wind-Wave Growth and Damping

The analogue to collisions in the Liouville equation are macroscopic processes contributing to the nonconservation of action density. In situations where there are sources or sinks of wave energy, Eq. 8.165 must be modified by including descriptors of those quantities on the right-hand side, which will be abbreviated as $(dN/dt)_{inter}$. The nature of the terms depends on the problem of interest, of course. Since we have been motivated in this development by the study of radar scatter from a wind-roughened surface, our right-hand side should clearly include a source/sink term for wind waves, as well as any other mechanisms leading to rearrangements of waves not already included on the left. Now there appears to exist a global equilibrium frequency spectrum for wind waves, $S_{eq}(\omega)$, of the general form of Eq. 5.109, and the source/sink term should model the tendency for the spectrum to grow toward $S_{eq}(\omega)$ if equilibrium has not been reached, or to decay toward it if there exists a higher level of wave energy than that maintained by the wind. A rigorous derivation of the form of the source/sink term involves complicated scattering integrals describing forcing by wind pressure, turbulent eddies and other stresses, interactions between wave vector components that scatter action into and out of wave number intervals, and ultimately molecular dissipation. As is often done in these cases, this highly complicated sequence of events is modeled to first order by a simple relaxation time approximation, much as was used in the Debye/Drude models of the dielectric and conductivity functions (Eqs. 8.24 and 8.29). The most rudimentary function consistent with our requirement for an approach to equilibrium from above or below is of the form

$$\left(\frac{dN}{dt}\right)_{inter} = -\frac{N(\mathbf{k};\mathbf{x},t)}{t_r(\mathbf{k})}\left[\frac{N(\mathbf{k};\mathbf{x},t)}{N_{eq}(\mathbf{k})} - 1\right]. \qquad (8.166)$$

This term is zero if $N = N_{eq}$; if $N > N_{eq}$, it represents a decay back toward equilibrium, while if $N < N_{eq}$, it acts as a source term. The relaxation time, $t_r(\mathbf{k})$, is a function of the wave vector and wind velocity, and describes the fact that longer waves grow, interact, and decay more slowly than shorter ones. It should be emphasized that $1/t_r(\mathbf{k})$ is the net relaxation rate due to *all* interactions, not just wind–wave growth and turbulent decay; a host of physics is implicitly encapsulated into this single term. Figure 8.35 shows the current best estimates for $1/t_r$ as a function of $\mathbf{k}$; wind speed, u_w; and wind direction, ϕ; relative to $\mathbf{k}$.

A Solution for the Action Spectrum

With this form of interaction term on its right-hand side, Eq. 8.165 becomes an inhomogeneous, first-order, time-dependent ordinary differential equation for N, to be integrated along the characteristics given by Eqs. 8.152 and 8.153. The transformation

$$Q(\mathbf{k};\mathbf{x},t) = 1/N(\mathbf{k};\mathbf{x},t) \tag{8.167}$$

reduces it to

$$\frac{dQ}{dt} + \frac{Q}{t_r} = \frac{Q_{eq}}{t_r} . \tag{8.168}$$

An integrating factor is readily found for this and results in the solution

$$\frac{Q(\mathbf{k};\mathbf{x},t)}{Q_{eq}(\mathbf{k})} = 1 + N_{eq}(\mathbf{k}) \int_{-\infty}^{t} \sum_{i,j=1}^{2} \left(k_j' \frac{\partial U_j}{\partial x_i'} + \frac{\partial \omega_D}{\partial \gamma} \frac{\partial \gamma}{\partial x_i'} \right) \frac{\partial Q_{eq}}{\partial k_i'}$$

$$\times \exp\left[- \int_{-t'}^{t} \frac{dt''}{t_r(\mathbf{k}'')} \right] dt' . \tag{8.169}$$

Having this solution in hand, we retreat backward through the transformations from Q to N to S, using Eq. 8.164 to recover the modified wave spectrum:

$$\frac{S(\mathbf{k};\mathbf{x},t)}{S_{eq}(\mathbf{k})} = \frac{Q_{eq}(\mathbf{k})}{Q(\mathbf{k};\mathbf{x},t)} . \tag{8.170}$$

The dynamics described by the integrand of Eq. 8.169 are approximately as follows: Waves starting with initial vector **k**′ at time *t*′ propagate from an ambient region (while decaying toward equilibrium) into a region of current shear, ∇**U**, or gradient in property, ∇γ; here they interact with the gradients, undergo refraction and changes in velocity, and by time *t* have their wave vector changed to **k** and their propagation path bent toward new directions.

Such a prescription is quite useful in the study of the interaction of surface waves with a bounded current system such as the Gulf Stream, where in principle it can describe the enhanced sea state occurring near the current boundary. Additionally it can account for long/short surface wave interactions in terms of the effects of long wave currents and shear on the shorter components. We will illustrate a third example: the interaction between internal wave currents and short surface waves, as shown by the oceanic inter-

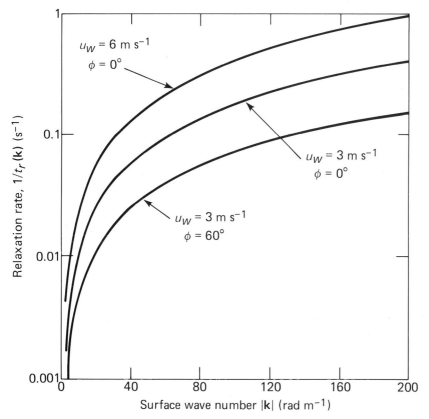

Fig. 8.35 Surface wave relaxation rate, $1/t_r(\mathbf{k})$, for two wind speeds and two angles. [Adapted from Hughes, B. A., *J. Geophys. Res.* (1978).]

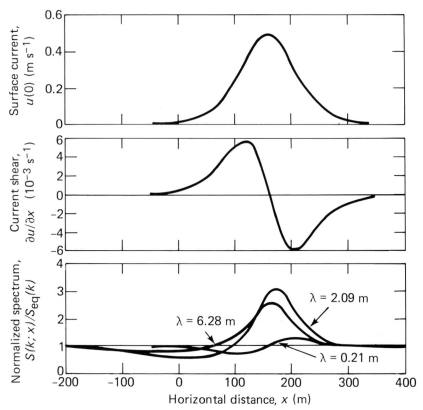

Fig. 8.36 Computed perturbation to surface wave spectrum due to internal wave currents and current shear. Maximum current derivative is 6×10^{-3} s^{-1}. [From Thompson, D. R., *Johns Hopkins APL Tech. Digest* (1985).]

nal wave image of Fig. 5.28, in order to make the abstract calculations somewhat more compelling.

Direct measurements of the surface currents of the internal waves, $\mathbf{U}(\mathbf{x},t)$, and the surface wave spectrum allow calculation of the quantities $\partial U_j/\partial x_i$ and $\partial Q_{eq}/\partial k_i$; these, plus numerical values for t_r obtained from wind observations and from the functional form shown on Fig. 8.35, complete the data required to evaluate Eq. 8.169 as a function of surface wave vector and wave position through the internal wave field. Figure 8.36 illustrates the calculated spectral perturbations over one internal wave cycle, for three values of surface wavelengths greater than or equal to the radar Bragg wavelength of 0.21 m; the largest perturbation occurs to surface waves near 2 m length. This perturbation then forms the basis for the subsequent calculation of radar backscatter modulation as given by Eq. 8.116. The results are shown

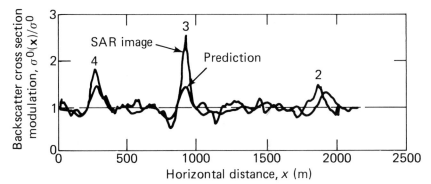

Fig. 8.37 Comparison of calculated and observed radar backscatter modulations from internal waves off New York. Numbers 2, 3, and 4 identify the three brightest waves visible in Fig. 5.28, starting in the center and proceeding downward. [From Gasparovic, R. F., et al., *Johns Hopkins APL Tech. Digest* (1985).]

in Fig. 8.37 as traces taken along a narrow sector from top to bottom of the image of Fig. 5.28. The agreement between calculated and observed backscatter is considered to be good. Similar computations have been made for backscatter modulations arising from narrow, shallow shoals in the North Sea, where the currents, $U(x,t)$, are tidal in nature, and comparable agreement has been obtained.

8.10 Emission from the Sea Surface

As with all material media, the ocean emits and absorbs radiation as a result of its thermodynamic temperature, T, due to the thermal agitation of both bound and free electrons in the medium. Measurements of the intensity, wavelength dependence, and polarization of this radiation can be related to a number of physical and chemical properties of the sea surface, although the problem of determining those characteristics from emission measurements is, in many cases, a difficult one to solve practically.

In this section we will discuss a few properties of thermal radiation and then apply our understanding to the determination of thermodynamic temperature, surface wind speed, salinity, and as a by-product of the methods used to measure the oceanic variables, to the inference of properties of the intervening atmosphere, such as water vapor and liquid water.

Electromagnetic Fields and Radiant Quantities

The relationships between the electromagnetic fields, **E** and **B** (or **D** and **H**) and the fundamental radiant variables such as *radiance, irradiance*, and their spectral decompositions is a worthwhile one to make clear, if only briefly. The fields considered in the earlier portions of this chapter were assumed to be coherent, monochromatic, and polarizable, whereas thermal radiation is incoherent, generally not polarized, and possessed of a continuous distribution of frequencies. However, they are linked by the field energy densities, the Poynting vector, and the electronic density of states for the distribution of oscillators. We will only cite a few formulas and direct the reader to texts such as Sommerfeld's excellent book on thermodynamics for a more extensive discussion.

The Poynting vector, $\mathbf{S}$, is in general defined as

$$\mathbf{S} = \mathbf{E} \times \mathbf{H} , \tag{8.171}$$

and has the dimensions of watts per square meter of wavefront. For a harmonic time dependence (but arbitrary space dependence), the time average of this quantity over a cycle is obtained by the prescription

$$\langle \mathbf{S} \rangle = \tfrac{1}{2} \ \mathrm{Re} \ (\mathbf{E} \times \mathbf{H}^*) . \tag{8.172}$$

In a lossless medium characterized by ϵ and μ, Eq. 8.172 becomes

$$\langle \mathbf{S} \rangle = \tfrac{1}{2} \ \sqrt{\epsilon/\mu} \ \ |\mathbf{E}|^2 \, \mathbf{n} , \tag{8.173}$$

where $\mathbf{E}$ is the amplitude of the electric field and $\mathbf{n}$ is a unit vector in the direction of the propagation vector.

The electric and magnetic fields possess energy densities, w_e and w_m, whose time-averaged sum is

$$
\begin{aligned}
\langle w \rangle &= \langle w_e \rangle + \langle w_m \rangle \\
&= \tfrac{1}{4} \ [\mathbf{E} \cdot \mathbf{D}^* + \mathbf{B} \cdot \mathbf{H}^*] \\
&= \tfrac{1}{4} \ [\epsilon |\mathbf{E}|^2 + \mu |\mathbf{H}|^2] \\
&= \tfrac{1}{2} \ \epsilon |\mathbf{E}|^2 \qquad \mathrm{J \ m}^{-3}
\end{aligned}
\tag{8.174}
$$

for harmonic waves. Thus the relationship between energy density and energy flux, or radiance, is

$$\langle \mathbf{S} \rangle = \langle w \rangle c = \tfrac{1}{2} \ \sqrt{\epsilon/\mu} \ \ |\mathbf{E}|^2 \, \mathbf{n} , \tag{8.175}$$

$$c = \frac{\mathbf{n}}{(\epsilon\mu)^{1/2}} . \tag{8.176}$$

A conservation equation for energy density may be written linking these quantities; for a conducting but nonmoving medium with conductivity σ, it may readily shown to be

$$\frac{\partial\langle w \rangle}{\partial t} + \nabla \cdot \langle \mathbf{S} \rangle = \frac{1}{c}\frac{\partial\langle S \rangle}{\partial t} + \nabla \cdot \langle \mathbf{S} \rangle$$

$$= -\langle \mathbf{j} \cdot \mathbf{E} \rangle = -\langle (\sigma - i\omega\epsilon)\mathbf{E}^2 \rangle . \tag{8.177}$$

The left-hand side is amenable to the same kind of interpretation as the fluid conservation equations in terms of the increase of energy density being due to the convergence of energy flux, but with a sink term on the right due to work done on charge carriers by the electric field. (The magnetic field dissipates no power since the magnetic force is at right angles to the velocity.) Equation 3.79 is an example of such a conservation equation.

The radiance associated with a small radiating surface is the energy it emits in a direction (θ,ϕ) per unit of surface area normal to the direction in question, per unit of time, per unit solid angle. For plane waves, radiance is undefined because of the solid angle singularity, but for fields behaving as $1/r$ (or more rapidly), the radiance, $L(\theta,\phi)$, emitted from an element of area dA may be written as

$$L \equiv d^2P/d\Omega \, dA \cos\theta = \langle \mathbf{S} \rangle \cdot \mathbf{n} = \tfrac{1}{2} \, \mathrm{Re} \, \langle \mathbf{E} \times \mathbf{H}^* \rangle \cdot \mathbf{n} , \tag{8.178}$$

where we have assumed that the element of area is a *Lambertian radiator* that emits equally well in all directions, so that the effective area of the radiator is its projection in the direction of interest, $dA \cos\theta$. Equation 8.178 is valid for the far radiation field and for sufficiently small elements of area such that each may be considered an infinitesimal source. Thus the Poynting vector for a spherical wave constitutes the radiance when viewed in the far zone. (The differential cross section of Eq. 8.115 is the equivalent to the normalized radiance for scattered radiation.)

One more elemental differential radiation quantity may be defined, the *spectral radiance*, which is the radiance per unit interval of radian frequency, $d\omega$. If the electromagnetic spectral flux density, $\mathbf{S}_\omega$, is defined such that

$$d\mathbf{S} = \mathbf{S}_\omega \, d\omega \tag{8.179}$$

is the radiance emitted in a bandwidth $d\omega$ centered at ω, then the spectral radiance may be written as

$$L_\omega = d^3 P/d\Omega \, dA \, \cos\theta \, d\omega$$

$$= \langle \mathbf{S}_\omega \rangle \cdot \mathbf{n}$$

$$= \tfrac{1}{2} \, \text{Re}(\mathbf{E} \times \mathbf{H}^*)_\omega \cdot \mathbf{n} \qquad \text{W m}^{-2} \, \text{sr}^{-1} \, \text{s} \, . \qquad (8.180)$$

The spectral radiance per unit wavelength interval, L_λ, may similarly be defined, with the relationship between it and L_ω being the requirement for equal power in a unit band:

$$L_\lambda \, d\lambda = L_\omega \left| \frac{d\omega}{d\lambda} \right| d\lambda \, . \qquad (8.181)$$

Here $|d\omega/d\lambda| = \omega^2/2\pi c$ is the Jacobian of the transformation.

The *irradiance, E,* is generally a *received* energy flux, with upward and downward fluxes considered separately. The *upward irradiance, E_u,* is the radiant flux incident on an infinitesimal element of horizontal surface from below, for example. If the radiator were an *extended source,* such as the sea surface, the irradiance could be integrated over the lower hemisphere to obtain the power incident on a unit horizontal area of the receiver located above the surface from all angles:

$$E_u = \int_0^{2\pi} \int_{\pi/2}^{\pi} L(\theta,\phi) \, \cos\theta \, \sin\theta \, d\theta \, d\phi \qquad \text{W m}^{-2} \, . \qquad (8.182)$$

There are numbers of other radiant functions of interest (several of which are discussed in Chapter 9), but the spectral radiance is the most fundamental.

Blackbody Radiation

In this section we denote blackbody radiant quantities with an asterisk. The spectral radiance of a blackbody, L_ω^*, is a universal function of the temperature and frequency, the distribution for which may be derived by considering the emission from a collection of quantum oscillators in thermal equilibrium, using the methods of statistical mechanics. From such a calculation, one obtains the radiance per unit radian frequency as given by the Planck function:

$$L_\omega^* = \frac{\hbar \omega^3}{4\pi^3 \, c_0^2} \, \frac{1}{\exp(\hbar\omega/k_B T) - 1} \qquad \text{W m}^{-2} \, \text{sr}^{-1} \, \text{s} \, . \qquad (8.183)$$

Here h is Planck's constant (the quantum of action; $h = 6.626 \times 10^{-34}$ J s); $\hbar = h/2\pi$; k_B is Boltzmann's constant ($k_B = 1.381 \times 10^{-23}$ J K^{-1}); where the velocity of propagation in the medium is and c_0 is the speed of light in vacuum ($c_0 = 2.998 \times 10^8$ m s^{-1}). In terms of wavelength, the spectral radiance is

$$L_\lambda^* = \frac{2hc_0^2}{\lambda^5} \frac{1}{\exp(hc_0/\lambda k_B T) - 1} \qquad \text{W m}^{-2}\text{ sr}^{-1}\text{ m}^{-1}, \quad (8.184)$$

where the final m^{-1} in the units for L_λ indicates a differential unit of wavelength. Given the size of optical wavelengths, a more practical unit for $d\lambda$ is 1 μm, in which event, L_λ^* is to be multiplied by 10^{-6}. An integration of Eq. 8.184 over all wavelengths gives the radiance, L^*:

$$L^* = \int_0^\infty \frac{2hc_0^2}{\lambda^5} \frac{d\lambda}{\exp(hc_0/\lambda k_B T) - 1}$$

$$= \frac{2\pi^4 k_B^4 T^4}{15 c_0^2 h^3}$$

$$\equiv bT^4, \qquad (8.185)$$

indicating that the total power emitted by a blackbody increases as the fourth power of temperature, a very sensitive function of T. A further integration of the power radiated into the outward hemisphere per unit projected area of a Lambertian surface yields the radiant exitance, M^*:

$$M^* = bT^4 \int_0^{2\pi} \int_0^{\pi/2} \cos\theta \sin\theta \, d\theta \, d\phi$$

$$= \pi b T^4$$

$$= \sigma_{SB} T^4 \qquad \text{W m}^{-2}, \qquad (8.186)$$

where σ_{SB} (yet another sigma!) is the Stefan–Boltzmann constant, whose theoretical value is 5.670×10^{-8} W m^{-2} K^{-4}. At a temperature of 300 K, each square meter of the surface of a blackbody therefore emits 459.3 W. It also absorbs exactly this same amount from its surroundings, if it is in radiative equilibrium with them. This is a form of Kirchhoff's law, which states that at every wavelength, the spectral emissivity, e_λ, is equal to the spectral absorptivity, a_λ. The emissivity is the ratio of the spectral radiance actually emitted by a body, $L_\lambda(\theta,\phi)$, to that emitted by a blackbody, and hence is a number less than or equal to unity because a blackbody radiates at the maximum rate thermodynamically possible:

$$L_\lambda(\theta,\phi) = e_\lambda(\theta,\phi) \, L_\lambda^* \; . \tag{8.187}$$

Thus the emission may be found from the product of the spectral emissivity and the Planck radiance function. (The emissivity of seawater ranges from 0.99 at far-infrared wavelengths to of order $\frac{1}{3}$ to $\frac{1}{2}$ at microwave frequencies; see below.) Kirchhoff's law states that if a material is a good absorber of radiation at any frequency, then it will also be a good emitter, provided that its temperature is high enough for its thermal excitation to provide it with energy to emit. For example, the ocean is a very good absorber at both ultraviolet and thermal infrared (10 μm) wavelengths (see Chapter 9), but it only emits sensible radiation at the infrared end because the Planck distribution for 300 K contains too little energy at ultraviolet wavelengths to detect readily.

We now relate the emissivity to the Fresnel power reflection coefficients of Eqs. 8.61 to 8.67. If the ocean is irradiated with energy at IR or longer wavelengths, a portion is reflected or scattered, and the remainder is transmitted below the surface. Since the spatial absorption coefficient (Eq. 8.55) is very large at these wavelengths, essentially all of the energy transmitted into the sea will be absorbed and almost none will be scattered back across the surface or be reflected from the bottom. Hence for this case, the sum of reflected and absorbed energy must be equal to the incident energy; in normalized units, the specific coefficients obey the relationship

$$r_\lambda + a_\lambda = 1 \; . \tag{8.188}$$

In more general cases when there is incomplete attenuation or subsurface scattering, this relationship must be replaced with one that takes account of transmissivity, t_λ:

$$r_\lambda + a_\lambda + t_\lambda = 1 \; . \tag{8.189}$$

In Eqs. 8.188 and 8.189, polarization indices (h,v) must be appended if that parameter is of importance.

Looking back on Eqs. 8.61, it may be seen that the Fresnel reflection coefficients are the quantities needed to calculate the emissivity; making its polarization indices explicit, we write

$$e_{\lambda,h,v} = a_{\lambda,h,v} = 1 - r_{\lambda,h,v} = 1 - |R_{h,v}|^2 \; . \tag{8.190}$$

The wavelength, polarization, and angular dependencies of the emissivity are just those derived from the complex relative dielectric function, $\epsilon' + i\epsilon''$, which is an additional justification for the time spent on understanding that

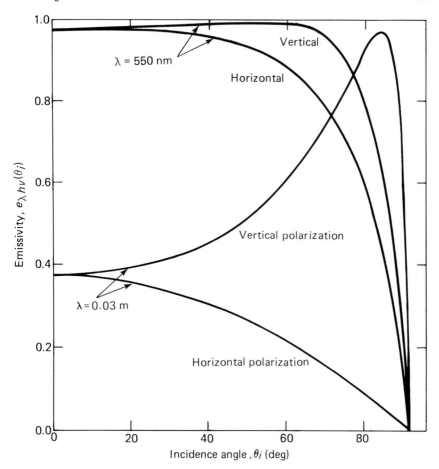

Fig. 8.38 Spectral emissivity of a flat sea at $\lambda = 0.03$ m and 550 nm, versus incidence angle, with polarization as a parameter. [Adapted from Stewart, R. H., *Methods of Satellite Oceanography* (1985).]

function earlier. Figure 8.38 shows the emissivity at a microwave and a visible wavelength. It is only near the Brewster angle that the emission at 10 GHz approaches that of a blackbody. Over a moderate range of incidence angles from zero to perhaps 30°, the emissivity for either polarization at this frequency does not depart appreciably from 0.38. Nevertheless, while the near-constancy of e_λ in this range of angles is a general property, the numerical value of that constant is dependent on frequency, as the illustration suggests.

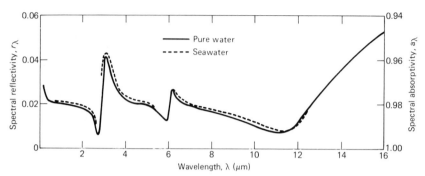

Fig. 8.39 Spectral reflectance for fresh and salt water over the range from visible to infrared wavelengths, for radiation incident on a plane water surface. [From Friedman, D., *Appl. Optics* (1969).]

Infrared Emission

Figure 8.39 shows the reflectivity, r_λ, for both fresh and seawater over the visible, near-IR, and thermal-IR regions of wavelengths; for the infrared portion of this regime, Eq. 8.190 holds well. It is clear that smooth water absorbs on the average some 98% of the solar infrared radiation falling on it at *normal* incidence and is therefore effectively black at these wavelengths; however, at other angles, the amount of absorbed radiation varies, as shown by Fig. 8.38. In the ocean, the situation is much more complicated, with roughness variations, the oblique incidence angle of sunlight at higher latitudes, and the existence of a broad source of diffuse skylight all modifying the absorption of solar radiation considerably. We will return to this question in Chapter 9.

At thermal infrared wavelengths, especially near 10 μm, the emissivity is about 99% and it depends very little on near-nadir viewing angles and surface roughness. This has made that wavelength regime a useful one for determining the surface or "skin" temperature of the sea, provided that accurate atmospheric corrections can be made. Because of the very small attenuation length at infrared frequencies (see Eq. 8.55), the emitted IR radiation comes from a layer just a few molecular diameters in thickness, and this layer may be cooler (in the case of evaporation) or warmer (in the case of intense heating) than the so-called bulk temperature immediately beneath the surface. Departures from the subsurface temperature of as much as $\pm 2°C$ are possible; nevertheless, under nominal oceanic conditions, the infrared temperatures are probably within $0.5°C$ of the actual temperatures. The required atmospheric corrections generally come from the same multispectral radiometer used to make the temperature measurements, with a channel operating at the secondary atmospheric window of 3 to 5 μm (see Fig. 2.2) being the source used for the correction. However, at this wavelength, the absorptivity of water is lower, and additionally the reflected solar radiance is not negligible, in contrast to the case at 10 μm, so that such corrections are not as accurate for daytime data as for nighttime observations. Figure 2.9 shows a medium resolution IR image of central North America made with a satellite multispectral imaging radiometer; Figure 6.41 illustrates a thermal IR image of the Gulf Stream that has been corrected for the atmospheric effects and has been computer-enhanced to bring out oceanic features.

Microwave Brightness Temperature

At the mean temperature of the earth, the maximum in the blackbody distribution curve occurs near $\lambda = 10$ μm. At longer wavelengths, i.e., microwave lengths greater than perhaps 3 mm, one may approximate the Planck function with the *Rayleigh–Jeans law*. In Eq. 8.183, if $\hbar\omega/k_B T \ll 1$, an expansion of the exponential yields that law in terms of the blackbody radiance:

$$
L_\omega^* \simeq \frac{\omega^2 k_B T}{4\pi^3 c_0^2} \ .
\tag{8.191}
$$

It is interesting that Planck's constant has disappeared from this relationship, indicating that quantum effects are not important in radiation at microwave frequencies at terrestrial temperatures. The equation demonstrates the useful result that the long-wavelength blackbody radiance is linearly proportional to thermodynamic temperature. This proportionality has led to the concept of *brightness temperature, T_b*, which is defined as the product of the thermodynamic temperature and the spectral emissivity:

$$
T_b(\lambda,\theta,\phi) \equiv e_\lambda(\theta,\phi) T \ .
\tag{8.192}
$$

If the emissivity is known and the radiance can be measured accurately, this relationship can be used to determine T. For example, over a narrow bandwidth, $\Delta\omega/2\pi$, centered at ω_0, the integrated microwave radiance may be approximated by

$$
\begin{aligned}
L(\theta,\phi) &\simeq L_{\omega_0}^* \, e_{\omega_0}(\theta,\phi) \, \Delta\omega/2\pi \\[2mm]
&= k_B T_b(\omega_0,\theta,\phi) \frac{\omega_0^2}{4\pi^3 c_0^2} \, \Delta\omega \\[2mm]
&= \frac{2k_B T_b(\lambda_0,\theta,\phi)}{\lambda_0^2} \, \Delta f \ ,
\end{aligned}
\tag{8.193}
$$

where we have used the electromagnetic dispersion relationship to write the power in terms of the wavelength of observation, λ_0. This equation is widely used in radiometry. When applied to a radiometer receiver viewing a distributed target through an intervening atmosphere, several other considerations must be taken into account; these are often lumped into terms described by an *apparent temperature* (total radiation incident on the antenna), and an *antenna temperature* (apparent temperature convolved with antenna response and efficiency); these three temperatures are not synonymous.

Recall that the dielectric function, and through it, the emissivity, are functions of the thermodynamic temperature, although not sensitive ones (see Tables 8.1 and 8.2). In addition to the temperature, $e_\lambda(\theta,\phi)$ depends on all of the other parameters appearing in the Fresnel coefficients, including angles,

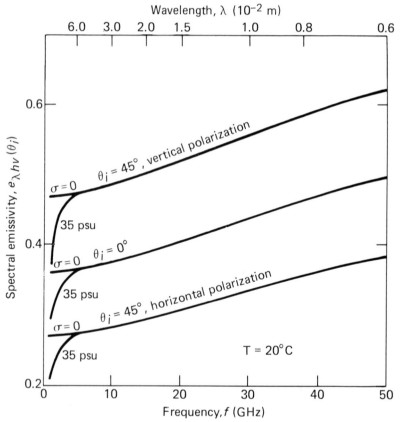

Fig. 8.40 Microwave spectral emissivity of fresh and seawater at 20°C, for two incidence angles and polarizations. Salinity has an effect only below $\lambda \approx 0.05$ m. [From Wilheit, T. T., *Boundary Layer Met.* (1978).]

frequency, polarization, and conductivity. Figure 8.40 illustrates the smooth-surface wavelength dependence of e_λ for incidence angles of $0°$ and $45°$, for both polarizations. It may be seen that at wavelengths longer than about 0.05 m, there is some salinity sensitivity, but the errors intrinsic to radio-metric measurements appear to limit this to an accuracy of about $\delta s \approx 1$ psu.

Beyond the flat-surface emission, there is also a dependence (whose origins are not completely understood) on surface roughness and on surfactants, foam, bubbles, and white water. Small-scale roughness leads to enhanced emissivity at microwave frequencies, first by virtue of a greater effective area of radiator, and secondly by the appearance of white water and foam, which occurs above some 6 to 8 m s^{-1} wind speed and which is thought to be due

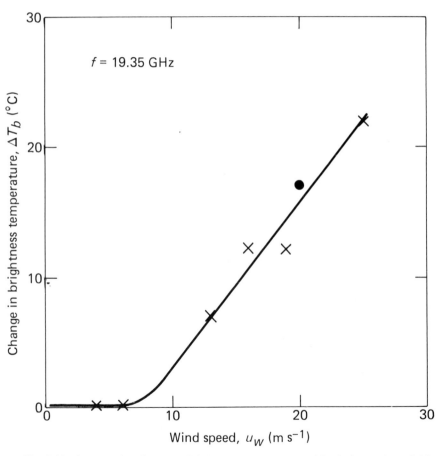

Fig. 8.41 Increase in microwave brightness temperature with wind speed, at 19.35 GHz. [From Wilheit, T. T., *Boundary Layer Met.* (1978).]

to improved radiative impedance matching. The resultant increase in microwave brightness temperature varies roughly linearly with surface wind speed from near 7–8 to above 20 m s^{-1}. Figure 8.41 shows this dependence as a function of wind speed, for vertical incidence. This is a microwave phenomenon; at infrared wavelengths, the emissivity is high enough so that roughness has little effect.

The atmosphere also emits and absorbs infrared and microwave radiation, with the associated radiant quantities showing a complicated dependence on concentrations of water vapor, liquid water, rainfall rate, and aerosols, as well as on frequency, polarization, and angle. While a discussion of such effects is beyond the scope of this book, they are nevertheless important in making radiometric measurements of the sea. The reader is referred to the texts by Stewart and by Ulaby et al. for a discussion of the atmospheric physics and engineering considerations.

Radiometry and Ocean Observations

An antenna viewing the sea through the atmosphere receives radiation from all objects within its field of view (including thermal radiation from land and spurious signals from transmitters), as weighted by the antenna beam pattern. The approach used to determine geophysical variables such as T, u_w, s, and liquid and vaporous water concentrations is to make a number of simultaneous, colocated measurements of the radiation from all of these quantities using a multispectral, dual-polarized scanning radiometer viewing at a constant incidence angle. This is indicated schematically in Fig. 8.42, which shows the radiometer beam pattern scanned about a constant angle, a method called a conical scan, and one often used on spacecraft. In each radiometer channel, the received power is attributed to the sum of contributions from the geophysical variables being viewed, with each contribution assumed to be linearly dependent on its associated variable, and with various weights that are set by a combination of theory and prior experimentation. Thus, if the power received in Channel i is denoted by P_i and is attributed to all geophysical variables of relevance, G_j, we may write

$$P_i = \sum_j b_{ij} G_j \ , \tag{8.194}$$

or in matrix notation,

$$\mathbf{P} = \mathbf{BG} \ . \tag{8.195}$$

The nonsquare matrix **B** contains the power weighting functions, which are usually specified as a function of geographical position and season. For the ocean, the four variables that may be determined readily are

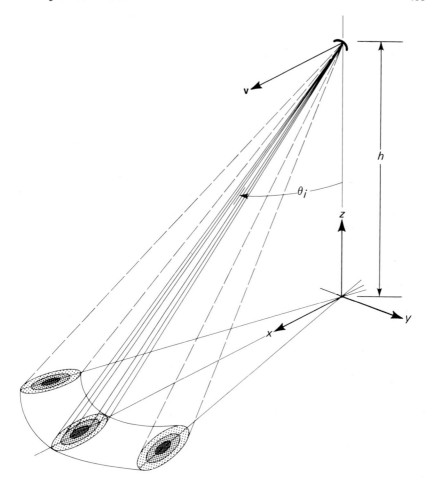

Fig. 8.42 Viewing geometry of three-frequency scanning microwave radiometer. Conical scan gives constant incidence angle as platform moves, and zig-zag pattern gives contiguous coverage.

$$\mathbf{G} = \begin{bmatrix} T \\ u_w \\ V \\ L \end{bmatrix}, \qquad (8.196)$$

where V is the vertically integrated (columnar) concentration of atmospheric water vapor in kilograms per square meter, and L is the equivalent in liquid water. (We are assuming here that the radiometer is located entirely above the sensible atmosphere.) To solve for the geophysical variables in terms of the observations, an inversion of Eq. 8.195 is effected:

$$\mathbf{G} = \mathbf{B}^{-1}\mathbf{P} \ . \tag{8.197}$$

Usually there are more radiometer channels used than there are independent

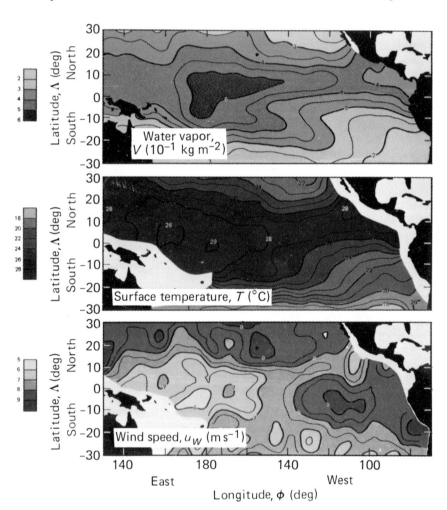

Fig. 8.43 Geophysical fields derived from scanning microwave radiometer on Nimbus-7 satellite. (a) Columnar water vapor in units of 10^{-1} kg m^{-2}; (b) surface temperature in °C; (c) Wind speed in m s^{-1}. [From Liu, W. T., *Air-Sea Interaction with SSM/I and Altimeter* (1985).]

variables, so that the solution is not a straightforward matrix inversion; however, this overdetermination of **G** leads to greater stability and accuracy to the solutions.

Figure 8.43 illustrates columnar water vapor, V, surface temperature, T, and wind speed, u_w, as obtained over the tropical Pacific from a 10-channel microwave radiometer on the satellite Nimbus-7. The maps clearly show the wet Intertropical Convergence Zone just north of the equator, as well as the cooler upwelling area off South America. The resolution of these maps is rather coarse, with 50 to 250 km footprints obtaining because of limitations on the size of scanning antennas on spacecraft. Nevertheless, for large-scale studies of the ocean surface, these data are quite useful.

8.11 Induced EMF's and Currents

The motion of conducting seawater through the earth's magnetic field induces electrical fields and currents in the water whose magnitudes are measures of the velocity or the volume transport, depending on the arrangement of the measurement system. This phenomenon is similar to the dynamo effect, wherein a current flowing through a material in a magnetic field that is at right angles to the flow will result in the accumulation of charges of opposite signs on the two sides of the conductor paralleling the flow. The associated electric field is directed across the material at right angles to both the flow and the magnetic field, and a transverse current may be drawn through an external circuit.

In the ocean, this effect has been exploited in a variety of instrumentation configurations. For example, a device called the electromagnetic velocity profiler is used to determine the relative variation of a horizontal current, $\mathbf{u}_h(z)$, with depth. The profiler is allowed to free-fall through the water column while spinning. Pairs of electrodes measure the induced motional electric currents, which are related to the baroclinic component of the horizontal water current; the vertically averaged barotropic component, $\langle \mathbf{u}_h \rangle$, must be determined by other means (e.g., a three-beam acoustic Doppler echo-sounder viewing the sea floor). A second arrangement uses a pair of electrodes towed behind a ship; the component of water current (at the depth of the towed system) orthogonal to the ship's velocity vector is measured by this means. A third application, which we will discuss as an example of the method, is found in the determination of the integrated volume transport across some horizontal distance, W, typically a region of current flow near a shore station. An undersea cable or wire is laid on the sea floor (communications cables serve well) and the voltage developed between the near and far ends of the conductors is measured as a function of time. As we will show later, this voltage is proportional to the volume flow of water passing over the cable, provided certain ancillary information may be obtained, and corrections for geomagnetic noise may be made.

The theory for the effect in the oceanic electrolyte parallels that for the dynamo. If a charge, q, moves at a macroscopic velocity, $\mathbf{u}$, in a magnetic field whose magnetic induction is $\mathbf{B}$, and an electric field, $\mathbf{E}$, the charge will be subject to the Lorentz force, $\mathbf{F}$, where

$$\mathbf{F} = q(\mathbf{E} + \mathbf{u} \times \mathbf{B}) \ . \tag{8.198}$$

In the absence of other influences such as collisions, this force will cause a spiraling of the charges about the magnetic field, with positively charged ions rotating to the right and negatively charged ones to the left; superimposed is an acceleration along the direction of the electric field. The net result of these two forces, if steady, is to cause a drifting of positive charges to the left and negative charges to the right of the velocity vector, $\mathbf{u}$, in the same kind of orbits shown in Fig. 6.7. (In that case of inertial oscillations, the Coriolis force replaces the magnetic force, and the horizontal pressure gradient replaces the electric potential gradient, or electric field, of Eq. 8.198. There are no oppositely signed inertial oscillations in the same hemisphere, thankfully.)

A generalized Ohm's Law replaces Eq. 8.9 under these circumstances, and this may be derived by a slight extension to the Drude theory leading to Eq. 8.28. Thus if $d\mathbf{x}/dt$ and $d^2\mathbf{x}/dt^2$ are the microscopic velocity and acceleration of an ion of mass m and charge e, and if the ion is subject to a collisional damping force at an average collision frequency, $1/t_c$, its equation of motion is

$$m\, d^2\mathbf{x}/dt^2 = e(\mathbf{E}'/\epsilon_r + \mathbf{u} \times \mathbf{B}) - (m/t_c)\, d\mathbf{x}/dt \ . \tag{8.199}$$

Here the dielectric effect on the field has been taken into account by dividing $\mathbf{E}'$ by ϵ_r, the relative dielectric function. For fields that vary slowly compared with the time needed to establish charge equilibrium (i.e., the relaxation time), a quasi-steady state results that represents a balance between the average forcing by the electromagnetic fields and the average effects of damping. Setting Eq. 8.199 equal to zero in the steady state and defining the macroscopic, averaged current density as in Eq. 8.31, we obtain

$$\mathbf{j} = \langle Ne\, d\mathbf{x}/dt \rangle = \langle Ne^2 t_c/m \rangle (\mathbf{E} + \mathbf{u} \times \mathbf{B})$$

$$= \sigma(\mathbf{E} + \mathbf{u} \times \mathbf{B}) \ . \tag{8.200}$$

The electric field $\mathbf{E} = \mathbf{E}'/\epsilon_r$ is that actually developed in seawater, and σ is the low-frequency conductivity. This formula replaces Eq. 8.9 when magneto–ionic effects are important, as they are here. In addition to this modification to the conductivity, the polarization field of seawater is altered from that given by Eq. 8.6 to

$$\mathbf{P} = \epsilon_0 \chi (\mathbf{E} + \mathbf{u} \times \mathbf{B}) \ . \tag{8.201}$$

Also, the advection of the polarization appears to a fixed observer to be a magnetization, or magnetic dipole moment per unit volume, $\mathbf{M}$, where

$$\mathbf{M} = \mathbf{P} \times \mathbf{u} \ . \tag{8.202}$$

The curl of the induced magnetization vector acts as an additional source of electrical current, $\mathbf{j}_m$, given by

$$\mathbf{j}_m = \nabla \times \mathbf{M} = \nabla \times (\mathbf{P} \times \mathbf{u}) \ . \tag{8.203}$$

It is thought that this effect is weak in the ocean.

The form of Eq. 8.200 indicates that the electric current density has a component at right angles to both the water current and the magnetic field. Our interpretation of this is that the ions, which migrate transverse to the flow direction under the driving influence of the magnetic force, compose a current that brings about a charge separation, and hence an electric field that opposes further migration. This field is always opposite to the direction of $\mathbf{u} \times \mathbf{B}$. However, if there are return paths for the electric charges (such as exist in the motionless water or the conducting earth), a continuous flow of current may exist. (This current would set up a secondary magnetic field, which, for nominal current densities, is rather small.) Thus Eq. 8.200 reflects a balance among advection, charge separation, electric field development, and current flow.

In order to apply these concepts to the measurement of oceanic current, we particularize them to the horizontal flow of seawater in the geomagnetic field. To a first approximation, the earth's magnetic field is modeled by a tilted geomagnetic dipole, as shown in Fig. 8.44, where the dipole axis is inclined at an angle of 11.5° from the spin axis; the south pole of the dipole intersects the earth's surface at a geographical latitude of approximately 78.5°N, 69°W, near Thule, Greenland. (The actual "north" Magnetic Pole, or point where the field lines are vertical, is close to 75°N, 100°W; the lines of magnetic force point into the ground there, toward the south pole of the imagined interior dipole.) An expression for the magnetic induction, written using spherical coordinates (r,θ), is

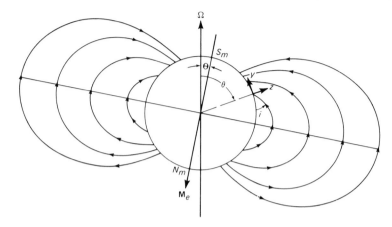

Fig. 8.44 Lines of force of the earth's dipole field approximation, which is tilted at an angle of about 11.5°. The dip angle, i, is the inclination that the local field makes at the earth's surface.

$$\mathbf{B}(r,\theta) = -\frac{\mu_0 M_e}{4\pi r^3} [2 \cos{(\theta - \Theta)}\hat{i}_r + \sin{(\theta - \Theta)}\hat{i}_\theta] . \qquad (8.204)$$

Here $\mu_0 = 4\pi \times 10^{-7}$ H m^{-1}, $\Theta = 11.5°$ is the tilt angle of the dipole axis with respect to the spin axis, and $M_e = 8.02 \times 10^{22}$ A m^2 is the earth's dipole moment. Using the same local tangent-plane approximation as previously used for fluid flow, the field components in magnetic north and magnetic vertical at the earth's surface are

$$B_y = B_0 \sin{(\theta - \Theta)} \qquad (8.205)$$

and

$$B_z = -2B_0 \cos{(\theta - \Theta)} , \qquad (8.206)$$

where $B_0 = \mu_0 M_e / 4\pi R_e^3 = 3.122 \times 10^{-5}$ T (or 0.3122 G) is the theoretical value for the surface field at the magnetic equator. (T = tesla = weber per square meter; G = gauss.) Here the negative sign indicates that in the Northern Hemisphere, the field lines are directed northward and downward. The angle i in Fig. 8.44 is the dip angle, i.e., the angle that the field line at the earth's surface makes with respect to the local horizontal. (In the literature on geomagnetism, the x, y, and z components of $\mathbf{B}$ are called X, Y, and Z, and are quoted with respect to a local right-handed geographical tan-

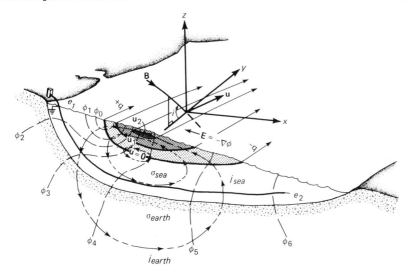

Fig. 8.45 Schematic of current flow in a confined channel, showing isotachs labeled by values of velocity, u; isopotentials labeled by ϕ; and flow lines of current density, j. An undersea cable senses the potential difference, $\phi_6 - \phi_1$, which is proportional to the volume transport of seawater flowing past the span of the cable. [Partially based on von Arx, W. S., *An Introduction to Physical Oceanography* (1961).]

gent plane wherein the positive direction for x is northward, y is eastward, and z is downward. We shall not make use of this convention here.)

With this description in hand, we can return to the oceanography. Figure 8.45 shows an ocean current system, $\mathbf{u}_h$, flowing slightly east of north in the Northern Hemisphere, with the local line of force, **B**, intersecting the current at the dip angle. An application of Eq. 8.200 to this geometry gives

$$-\partial\phi/\partial x = E_x = -vB_z + j_x/\sigma_s \qquad (8.207a)$$

and

$$-\partial\phi/\partial y = E_y = -uB_z + j_y/\sigma_s , \qquad (8.207b)$$

where the oceanic conductivity is taken as the static value, σ_s (see Eq. 8.33). The induced electric fields are assumed to be electrostatic, i.e., varying so slowly in time that no radiation results and no inductive effects exist; this means that they may be derived from the scalar electric potential, ϕ. Equations 8.207 indicate that it is only the vertical component of the earth's field that interacts with the horizontal flow to induce the electric field and current density.

Figure 8.45 also shows a schematic of the fluid and electromagnetic fields, and illustrates velocity isotachs, equipotentials, and induced and return current densities. The latter quantities penetrate into the conducting earth as well as the ambient, nonmoving seawater; the currents are carried by the salt ions in the water and free electrons in the solid earth. Figure 8.45 suggests that the charge separation across the current region resulting from the transverse drift is the source of the opposing electric field, and that this field sends a return electric current back through both the motionless portions of the water column and the solid material.

In the geographical configuration shown, the potential difference, $\phi_6 - \phi_1$, developed between two electrodes e_2 and e_1 (which might be the inner and grounded conductors of an open-ended submarine cable, for example) is the line integral of the electric field strength along the width, W, of the cable span. From Fig. 8.45 and Eq. 8.207a, for an east–west cable,

$$\phi_6 - \phi_1 = \int_0^W E_x(x)\, dx$$

$$= -\int_0^W B_z \langle v^*(x) \rangle\, dx - \int_0^W [j_x(x)/\sigma_s]\, dx . \quad (8.208)$$

Here the northward average velocity, $\langle v^*(x) \rangle$, can be shown to be a weighted average of the actual velocity component, $v(x,z)$, over the vertical coordinate, where the weighting function is the conductivity, $\sigma_s(x,z)$, of the entire region carrying the current. Since the vertical integral of the conductivity gives the total conductance, G, of the current paths, we have

$$G(x) = \int_{-H}^0 \sigma_s(x,z)\, dz + \int_{-R_e}^{-H} \sigma_{earth}(x,z)\, dz$$

$$= G_{sea} + G_{earth} \quad \text{S} . \quad (8.209)$$

With this notation, the conductivity-weighted velocity is

$$\langle v^*(x) \rangle = 1/(G_{sea} + G_{earth}) \int_{-H}^0 \sigma_s(x,z)\, v(x,z)\, dz . \quad (8.210)$$

Now the total volume transport, V, of the current system out to the distance W is the integral of the northward velocity over x and z:

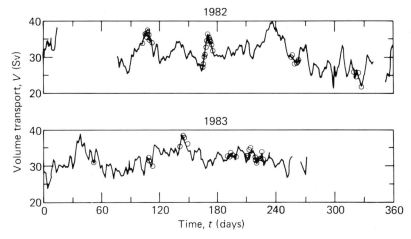

Fig. 8.46 Time variations of the volume transport of the Florida current near Miami as observed over two years, derived using motionally induced voltages on a submarine cable. Circles are transport measurements from vertically profiling moored current meters. [From Larsen, J. C., and T. B. Sanford, *Science* (1985).]

$$V(W) = \int_0^W \int_{-H}^0 v(x,z)\, dz\, dx \quad \text{Sv}\ . \qquad (8.211)$$

Thus the observed voltage difference, $\phi_6 - \phi_1$, is related to the transport and can be used to measure that quantity if the magnetic field, the earth's conductance, and the ocean conductivity are known, and if the effects of nonlocal currents are negligible. The solid earth conductance and magnetic field strength can be determined for the geographical region of interest; the oceanic conductivity can be calculated with sufficient accuracy from temperature profiles and point salinity observations, and thus the required factors derived to yield the transport from the voltage measurement. Corrections for ionospherically induced geomagnetic noise are also needed.

Figure 8.46 shows electromagnetic observations of the total transport of the Florida Current near Miami over two years, along with intermittent values inferred with vertically profiling current meters, which have been used to calibrate the earth conductance. The agreement between the two methods is excellent, with the data showing that the mean flow of the Gulf Stream past Florida is close to 30 Sv, but that fluctuations up to 15 Sv (50% of the mean) occur with periods of 30 to 60 days. These may be related to meteorological forcing upstream (say, between the Mid-Atlantic Ridge and the Gulf of Mexico) or to other, more distant factors responsible for short-term climate variations.

Bibliography

Books

Bass, F. G., and I. M. Fuks, *Wave Scattering from Statistically Rough Surfaces*, Pergamon Press, Oxford, England (1979).

Beckmann, P., and A. Spizzichino, *The Scattering of Electromagnetic Waves from Rough Surfaces*, Pergamon Press, Oxford, England (1963).

Born, M., and E. Wolf, *Principles of Optics*, 5th ed., Pergamon Press, Oxford, England (1975).

Derr, V. E., Ed., *Remote Sensing of the Troposphere*, U.S. Government Printing Office, Washington, D.C. (1972).

Elachi, C., *Spaceborne Radar Remote Sensing Applications and Techniques*, IEEE, New York, N.Y. (1988).

Goldstein, H., *Classical Mechanics*, Addison Wesley Publishing Co., Reading, Mass. (1950).

Gower, J. F. R., Ed., *Oceanography from Space*, Plenum Press, New York, N.Y. (1981).

Gower, J. F. R., Ed., *Passive Radiometry of the Ocean*, R. Reidel Publishing Co., Dordrecht, The Netherlands (1980).

Hasted, J. B., *Aqueous Dielectrics*, Chapman and Hall, London, England (1973).

Jackson, J. D., *Classical Electrodynamics*, 2nd ed., John Wiley and Sons, New York, N.Y. (1975).

Kinsman, B., *Wind Waves*, Dover Publications, New York, N.Y. (1965).

Lathi, B. P., *An Introduction to Random Signals and Communication Theory*, International Textbook Co., Scranton, Pa. (1968).

LeBlond, P. H., and L. A. Mysak, *Waves in the Ocean*, Elsevier Scientific Publishing Co., Amsterdam, The Netherlands (1978).

Long, M. W., *Radar Reflectivity of Land and Sea*, Lexington Books, Lexington, Mass. (1975).

Maul, G. A., *Introduction to Satellite Oceanography*, R. Reidel Publishing Co., Dordrecht, The Netherlands (1985).

Panofsky, W. K. H., and M. Phillips, *Classical Electricity and Magnetism*, Addison Wesley Publishing Co., Reading, Mass. (1955).

Phillips, O. M., *The Dynamics of the Upper Ocean*, 2nd ed., Cambridge University Press, Cambridge, England (1977).

Robinson, I. S., *Satellite Oceanography*, John Wiley and Sons, Ltd., West Sussex, England (1985).

Saltzmann, B., Ed., *Satellite Oceanic Remote Sensing; Advances in Geophysics*, Vol. 27, Academic Press, Inc., Orlando, Fla. (1985).

Sommerfeld, A., *Thermodynamics and Statistical Mechanics*, Academic Press, Inc., New York, N.Y. (1956).

Stewart, R. H., *Methods of Satellite Oceanography*, University of California Press, Berkeley, Calif. (1985).

Ulaby, F. T., R. K. Moore, and A. K. Fung, *Microwave Remote Sensing*, Vol. I (1981), Vol. II (1982), Addison Wesley Publishing Co., Reading, Mass. Vol. III (1986), Artech House, Inc., Dedham, Mass.

Whitham, G. B., *Linear and Nonlinear Waves*, John Wiley and Sons, New York, N.Y. (1979).

Journal Articles and Reports

Barnum, J. R., J. W. Maresca, and S. M. Serebreny, "High-Resolution Mapping of Oceanic Wave Fields with Skywave Radar," *IEEE Trans. Antennas and Propag.,* Vol. AP-25, p. 128 (1977).

Barrick, D. E., "Rough Surface Scattering Based on the Specular Point Theory," *IEEE Trans. Antennas and Propag.,* Vol. AP-16, p. 449 (1968).

Barrick, D. E., "The Use of Skywave Radar for Remote Sensing of Sea States," *Marine Tech. Soc. J.* Vol. 7, p. 29 (1973).

Beal, R. C., et XI al. "A Comparison of SIR-B Directional Ocean Wave Spectra with Aircraft Scanning Radar Spectra," *Science,* Vol. 232, p. 1531 (1986).

Cheney, R. E., and J. G. Marsh, "Seasat Altimeter Observations of Dynamic Topography of the Gulf Stream Region," *J. Geophys. Res.,* Vol. 86, p. 473 (1981).

European Remote Sensing Satellite-1 Announcement of Opportunity – Technical Annex, European Space Agency, Paris (1986).

Friedman, D., "Infrared Characteristics of Ocean Water (1.5–15 μm)," *Appl. Optics,* Vol. 8, p. 2073 (1969).

Gasparovic, R. F., J. R. Apel, A. Brandt, and E. S. Kasischke, "Synthetic Aperture Radar Imaging of Internal Waves," *Johns Hopkins APL Tech. Digest,* Vol. 6, p. 338 (1985).

Gower, J. F. R., and J. R. Apel, Eds., *Opportunities and Problems in Satellite Measurements of the Sea,* UNESCO Technical Papers in Marine Science 46, UNESCO, Paris (1986).

Holliday, D., G. St-Cyr, and N. E. Woods, "A Radar Ocean Imaging Model for Small to Moderate Incidence Angles," *Intl. J. Remote Sensing,* Vol. 7, p. 1809 (1986).

Hughes, B. A., "The Effect of Internal Waves on Surface Wind Waves, 2. Theoretical Analysis," *J. Geophys. Res.,* Vol. 83, p. 455 (1978).

Jones, W. L., L. C. Schroeder, and J. L. Mitchell, "Aircraft Measurements of the Microwave Scattering Signature of the Ocean," *IEEE J. Oceanic Engr.,* Vol. OE-2, p. 52 (1977).

Klein, L. A., and C. T. Swift, "An Improved Model for the Dielectric Constant of Sea Water at Microwave Frequencies," *IEEE Trans. Antennas and Prop.* Vol. AP-25, p. 104 (1977).

Larsen, J. C., and T. B. Sanford, "Florida Current Volume Transports from Voltage Measurements," *Science,* Vol. 227, p. 302 (1985).

Liu, W. T., cover illustration, *Air-Sea Interaction with SSM/I and Altimeter,* Jet Propulsion Laboratory (1985).

Ray, P. S., "Broadband Complex Refractive Indices of Ice and Water," *Appl. Optics,* Vol. 11, p. 1836 (1972).

Sanford, T. B., "Motionally Induced Electric and Magnetic Fields in the Sea," *J. Geophys. Res.* Vol. 76, p. 3476 (1971).

Thompson, D. R., "Intensity Modulations in Synthetic Aperture Radar Images of Ocean Surface Currents and the Wave/Current Interaction Process," *Johns Hopkins APL Tech. Digest,* Vol. 6, p. 346 (1985).

Valenzuela, G. R., "Theories for the Interaction of Electromagnetic and Ocean Waves – A Review," *Boundary Layer Met.,* Vol. 13, p. 61 (1978).

Walsh, E. J., E. A. Uliana, and B. S. Yaplee, "Ocean Wave Heights Measured by a High-Resolution Pulse-Limited Radar Altimeter," *Boundary-Layer Met.,* Vol. 13, p. 263 (1978).

Wilheit, T. T., "The Electrically Scanning Microwave Radiometer (ESMR) Experiment," in *The Nimbus-5 User's Guide*, R. R. Sabatini, Ed., NASA Goddard Space Flight Center, Greenbelt, Md. (1972).

Wilheit, T. T., "A Review of Applications of Microwave Radiometry to Oceanography," *Boundary-Layer Met.*, Vol. 13, p. 277 (1978).

Wright, J. W., "A New Model for Sea Clutter," *IEEE Trans. Antennas and Propag.*, Vol. AP-16, p. 217 (1968).

Chapter Nine

Optics of the Sea

J'ai vu le soleil bas, taché d'horreurs mystiques
Illuminant de longs figements violets,
Pareils à des acteurs de drames très antiques,
Les flots roulant au loin leurs frissons de volets.

Rimbaud, *Le Bateau Ivre*

9.1 Introduction

Marine optics is concerned with the physical behavior of light on the surface and within the volume of the sea; how light energy enters into other physical, chemical, and biological processes in the ocean; and the transmission of information via optical signals above and through seawater. This last includes studies in other disciplines of oceanography via optical methods. The sources of light include solar radiation, biological and chemical species, and anthropogenic sources such as coherent and incoherent light controlled by man.

Although optical radiation is electromagnetic and governed by the same kinds of equations discussed in the previous chapter, in practice it is convenient to treat it as a special case of electromagnetics. Because of the wavelength-dependent absorption by seawater and its constituents, the wavelengths of interest are essentially confined to the range from the near-ultraviolet (say $\lambda \approx 300$ nm) to the near-infrared (extending to perhaps 1 μm). This is the domain of classical optics.

As was discussed in Chapter 2, the diurnally and seasonally varying solar radiation in this wavelength regime is either reflected from the surface or transmitted beneath it; in the latter case, the radiation is scattered or absorbed by seawater and the suspended and dissolved materials in its upper layers. Here the energy undergoes transmutations of a variety of types, including absorption (and therefore thermalization), scattering by particulates in suspension, conversion to chemical energy by photosynthesis, and re-emission as fluorescent radiation by chlorophyll-a, phycoerythrin, and other optically active pigments in the ocean. These volume interactions take place in the *euphotic zone*, i.e., the upper sunlit reaches of the sea; the bottom of the euphotic zone is commonly defined as that depth where the energy has fallen to 1% of the value immediately below the surface. The lower limit of the euphotic zone varies greatly in time and space, but in the most transparent waters (such as the Sargasso Sea), it may lie near a depth of 150 m, although it is more typically near 100 m in the open sea. It is in this layer that the vast majority of primary production of biological material in the world's oceans takes place (the chief exception being isolated thermosynthetic organisms that grow near deep sea heat sources). Because of the value of the optical index of refraction ($n \simeq 1.34$) and the variation of solar incidence angles with time of day, latitude, and season, the transmission of light into the ocean also depends strongly on geography and time of year. As will be more fully explained ahead, it is mainly the biological populations and their waste products that introduce the large observed variability in the optical properties of the open sea; the intrinsic optical characteristics of pure seawater are very nearly constant, or depend only weakly on salinity, temperature, and depth. An exception to the biologically induced changes occurs near shore, where land sources such as suspended sediments and humic substances also modify the optics. The sediment effects are most pronounced near major rivers such as the Yangtze, Amazon, and Mississippi Rivers, for example. Another class of terrigeneous materials that varies in the sea and is optically active includes dissolved organic compounds, also known as *gelbstoffe* (yellow substance) or *gilvin*. The five known factors that contribute to optical properties are then (1) pure seawater, (2) chlorophyll and other similar pigments, (3) yellow substance, (4) detritus, and (5) suspended materials, both planktonic and sedimentary. It is fortunate for the study of the optics of the sea that the effects of these substances are largely separable (although they are closely linked in the marine ecosystem).

In the discussion to follow, we will first define the intrinsic physical quantities such as irradiance, absorption, etc., that characterize the light field, and next establish the optical properties of pure seawater. Then the major effects of actual ocean water with its wide range of constituents will be outlined, first concentrating on surface reflection, scattering, and transmission, and then examining the volumetric effects. For shallow water, bottom reflectivity must also be considered. We conclude with an application of the theory to the remote measurement of the concentrations of certain constituents of seawater via optical observations.

9.2 Optical Fields and Parameters

In this section we will define some of the more commonly used optical fields and parameters without immediately attempting to relate them to oceanographic problems of interest. Such an approach slightly disconnects the physics from the discussion, but has the advantage of grouping the quantities in one place and listing them side by side for the sake of comparison; it also allows subsequent discussions to proceed with less diversion.

Although optical radiation is electromagnetic in nature, its generally incoherent, wideband character (lasers excepted) has led to physical descriptions phrased not so much in terms of electric and magnetic fields as in terms of intensity, radiance, irradiance, and a wide variety of absorption, scattering, and attenuation coefficients. These quantities are usually further detailed via their spectral distributions, although the spectrally integrated energy is also of interest in studies of oceanic solar heating and photosynthesis, for example.

Radiant Fields

Spectral (or *monochromatic*) *radiance* is the most fundamental and detailed radiant quantity of interest in environmental optics. It may be defined either for emitted radiation (as was the case for blackbody radiance, as given by Eq. 8.184), or for radiation incident on a surface, termed *field radiance,* with the direction of propagation of the radiant flux distinguishing the transmitted and received cases. Figure 9.1a shows a portion of an extended source considered to be emitting radiant energy, dQ, per unit time, dt, in a unit spectral bandwidth, $d\lambda$, per unit area of its surface, dA, into a solid angle, $d\Omega$. This quantity is the *spectral radiant emittance*, and may be written as

$$L_\lambda(z,\theta,\phi) = d^4Q/dt\ dA\ \cos\theta\ d\Omega\ d\lambda \qquad \text{J s}^{-1}\text{ m}^{-2}\text{ sr}^{-1}\text{ nm}^{-1}, \qquad (9.1)$$

where the element of radiating area has a projection in direction θ of $dA\cos\theta$. Next consider a surface element (Fig. 9.1b) of area dA, located at depth z and illuminated by a flux of energy of bandwidth $d\lambda$, coming from the direction (θ,ϕ). During time dt, the amount of energy

$$d^4Q = L_\lambda(z,\theta,\phi)\ dt\ dA\ \cos\theta\ d\Omega\ d\lambda \qquad \text{J} \qquad (9.2)$$

falls on the element from the solid angle, $d\Omega$. Here $L_\lambda(z,\theta,\phi)$ is the *field spectral radiance* at depth z associated with an extended source located elsewhere, and has the units of watts per square meter per steradian per nanometer. When further distinguished as to polarization, it constitutes the most detailed, fundamental descriptor of the radiant field that is generally required. The pres-

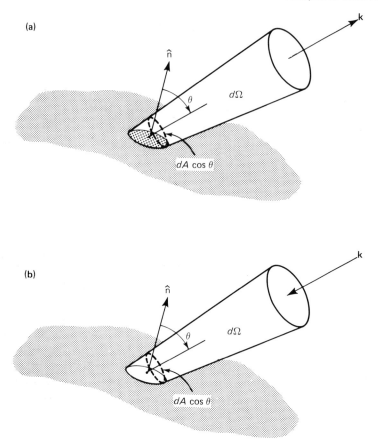

Fig. 9.1 Definitions of radiance as power per square meter per steradian, for a projected surface area varying as cos θ: (a) Radiance emitted from a surface. (b) Radiance incident on a surface.

ence of the cos θ factor in Eqs. 9.1 and 9.2 is important to observe. It arises because a fixed element of area has a cosine projection when viewed away from the normal, $\hat{n}$.

The most commonly measured quantities are the various *spectral irradiances*, which are differentiated from one another by (1) the range of solid angles over which they are defined, (2) the angular response of the receiver (e.g., cosine versus omnidirectional character), and (3) the spectral interval under observation. The simplest of these is the *spectral scalar irradiance, $E_{\lambda 0}(z)$*, which is the band-limited power intercepted by a unit area of a small object (such as a photosynthetic particle) that is capable of collecting energy from all directions. From the definition of field spectral radiance (Eq. 9.2) this quantity is

$$E_{\lambda 0}(z) = \int_{4\pi} L_\lambda(z,\theta,\phi)\, d\Omega \qquad \text{W m}^{-2}\,\text{nm}^{-1}. \qquad (9.3)$$

Scalar irradiance is illustrated in Fig. 9.2a, which indicates (via the lengths of arrows) a nonuniform angular distribution of radiance, $L_\lambda(\theta,\phi)$, incident on a small collector at the center. This kind of variable light distribution with direction is typical of the subsurface optical field, with the downwelling light more intense than the upwelling light.

When integrated over 4π steradians as well as over all wavelengths present, the scalar irradiance, $E_0(z)$, at depth z gives the total flux of light energy at depth z:

$$E_0(z) = \int_0^{2\pi}\int_0^{\pi}\int_0^{\infty} L_\lambda(z,\theta,\phi)\, d\lambda\, \sin\theta\, d\theta\, d\phi \qquad \text{W m}^{-2}. \qquad (9.4)$$

The absorption characteristics of seawater effectively limit the wavelengths of interest to the range between 300 and 1000 nm near the surface and to even more narrow intervals at depth. We shall discuss this aspect of the subsurface light field at greater length in Section 9.6.

The underwater light field due to natural illumination is generally homogeneous in the horizontal on a local scale, but varies in the vertical relatively rapidly. At any depth, then, there is a difference between the intensity of the *downwelling* light and the *upwelling* light because of (1) the surface distribution of the light source, and (2) the attenuation of that source due to oceanic absorbers and scatterers acting on the intensity distribution down to the depth in question. The net effect is measured by the difference between the down-

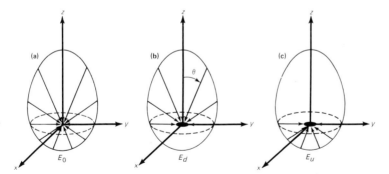

Fig. 9.2 Schematic of three types of nonisotropic irradiances: (a) Scalar irradiance, E_0, incident on a small omnidirectional detector such as a photosynthetic cell. (b) Downward irradiance from the upper hemisphere, E_d, through a small aperture. (c) Upward irradiance from the lower hemisphere, E_u, through a small aperture. E_u and E_d measure upwelling and downwelling light, respectively.

welling and the upwelling spectral irradiances. The *downward spectral irradiance*, $E_{\lambda d}(z)$, is the energy per unit time passing through a horizontal aperture of area dA (therefore a *cosine collector*) from above, integrated over the upper hemisphere (see Fig. 9.2b); it thus measures the *downwelling* light flux per unit wavelength. Since the projected area of the horizontal aperture is $dA \cos \theta$ in the direction (θ, ϕ), this quantity is calculated by integrating the radiance distribution in ϕ from 0 to 2π and in θ from 0 to $\pi/2$.

$$E_{\lambda d}(z) = \int_0^{2\pi} \int_0^{\pi/2} L_\lambda (z,\theta,\phi) \cos \theta \sin \theta \, d\theta \, d\phi . \tag{9.5}$$

Similarly, the *upward spectral irradiance*, $E_{\lambda u}(z)$, is defined as the cosine-weighted radiance incident on a horizontal aperture dA from *below*, integrated over the lower hemisphere (Fig. 9.2c):

$$E_{\lambda u}(z) = - \int_0^{2\pi} \int_{\pi/2}^{\pi} L_\lambda (z,\theta,\phi) \cos \theta \sin \theta \, d\theta \, d\phi . \tag{9.6}$$

(The negative sign is required because $\cos \theta$ is negative over the lower hemisphere.) This quantity is therefore a measure of the *upwelling* light in any wavelength interval, which is light that had initially propagated to deeper depths than the level of observation, but which has been scattered back into the upper hemisphere.

The *net downward irradiance* is an important quantity that is related to the rate at which energy is absorbed in seawater. Its spectral decomposition is denoted by $E_{\lambda_{net}}(z)$ and is defined as the difference between the upward and downward spectral irradiances:

$$E_{\lambda_{net}}(z) = E_{\lambda d}(z) - E_{\lambda u}(z) \qquad \text{W m}^{-2} \text{ nm}^{-1}$$

$$= \int_0^{2\pi} \int_0^{\pi} L_\lambda (z,\theta,\phi) \cos \theta \sin \theta \, d\theta \, d\phi . \tag{9.7}$$

As will be shown ahead during the discussion of radiative transfer, the vertical derivative of the net irradiance gives the divergence of the radiative flux, $\nabla \cdot \langle \mathbf{S} \rangle$, which was introduced in the internal energy rate equations (Eq. 4.40) and the relationships that follow it (see also Eq. 8.177). By integrating over all wavelengths, we obtain

$$\nabla \cdot \langle \mathbf{S} \rangle = \frac{\partial}{\partial z} E_{net} = \frac{\partial}{\partial z} (E_d - E_u) \qquad \text{W m}^{-3}$$

$$= \int_0^\infty \int_0^{2\pi} \int_0^\pi \frac{\partial}{\partial z} L_\lambda(z,\theta,\phi) \cos\theta \sin\theta \, d\theta \, d\phi \, d\lambda , \qquad (9.8)$$

which gives the rate at which radiant energy is deposited in a unit volume of ocean due, in the main, to absorption by water, dissolved salts, particulates, and photosynthetic molecules.

By integrating the radiance over the entire area of the source or receiver, one obtains the *spectral radiant intensity,* I_λ, whose units are watts per steradian per nanometer, and which gives the total power per unit bandwidth radiated into a solid angle:

$$I_\lambda(z,\theta,\phi) = \int_A L_\lambda(z,\theta,\phi) \, dA . \qquad (9.9)$$

This quantity makes no reference to source or receiver dimensions and is therefore useful for specifying point source strengths.

Attenuation Coefficients

In addition to the basic Fresnel coefficients for reflection and transmission, optics makes use of a variety of other absorption, scattering, and attenuation coefficients, where *attenuation* means the net reduction in radiance, irradiance, or intensity resulting from both absorption and scattering. In the absence of fluorescence and bioluminescence, essentially all photons penetrating beneath the surface of the sea are ultimately either absorbed or scattered back out across the air–sea interface; thus absorption and scattering are the only two quantities needed to express the total attenuation of light intensity beneath the sea.

Experimentally, it is found that the radiance and the various irradiances all decay approximately exponentially with increasing depth, once a region immediately beneath the surface has been passed; in the latter region, various effects such as spatially varying attenuation distort the local radiance distributions somewhat away from exponential. A radiant field that is missing the usual inverse-square geometrical spreading superimposed on absorption may be approximated by a plane wave, and this characterizes the illumination of the sun. A *diffuse attenuation coefficient* for the scalar, upward, downward, and net irradiances may then be defined via the local logarithmic derivative of the parent quantity. Thus for the scalar irradiance, for example, the diffuse attenuation coefficient is

$$K_{\lambda 0}(z) = -\frac{d}{dz} \ln E_{\lambda 0} = -\frac{1}{E_{\lambda 0}} \frac{dE_{\lambda 0}}{dz} . \qquad (9.10)$$

This relationship may be integrated to give the variation in scalar irradiance between any two depths, z and z_1:

$$E_{\lambda 0}(z) = E_{\lambda 0}(z_1) \exp\left[-\int_{z_1}^{z} K_{\lambda 0}(z') \, dz'\right]$$

$$\simeq E_{\lambda 0}(z_1) \exp[-\bar{K}_{\lambda 0}(z - z_1)] . \tag{9.11}$$

The second expression assumes that the diffuse attenuation coefficient is approximately a constant, $\bar{K}_{\lambda 0}$; while for some waters this may be approximately true, it is often found that layers of highly absorbing phytoplankton at various depths in the euphotic zone greatly modify the diffuse coefficient locally. Similar relationships hold for diffuse attenuation coefficients governing the decay of each of the quantities in Eqs. 9.3 to 9.7, which coefficients will be denoted by the same subscript used to distinguish between the radiant field quantities themselves.

Since the diffuse coefficients are defined in terms of the fields and their derivatives, to some degree they must therefore depend on the distribution of sources (e.g., whether the sky is clear or overcast), especially near the surface. However, they are set primarily by the intrinsic optical properties of the ocean; for this reason, they are often termed *apparent optical properties.* In order to characterize the medium independently of the source, it is preferable instead to use absorption and scattering functions that depend only on the medium and the substances in it, and not on the source geometry or the illumination characteristics of the radiant fields. To accomplish this, three new *intrinsic optical properties* of the medium are defined—beam absorption, beam scattering, and beam attenuation coefficients. These describe the loss of intensity from a parallel, monochromatic beam of light (such as might be provided by a laser, for example) as it traverses the medium.

The *monochromatic beam absorption coefficient* is the fractional loss in power, as given by the change in the monochromatic radiant *flux,* $d\Phi_\lambda$, of a small, parallel beam, due to *absorption* by the medium over an infinitesimal increment of path, ds; the flux is the rate of energy transport by the beam and is expressed in watts. (Here "absorption" refers to loss of energy due to dissipative mechanisms.) Thus

$$a_\lambda \equiv -\left(\frac{1}{\Phi_\lambda} \frac{d\Phi_\lambda}{ds}\right)_{abs} \qquad \text{m}^{-1} . \tag{9.12}$$

From its geometrical definition, it is clear that the beam absorption coefficient is essentially the absorption cross section per unit volume of water. The

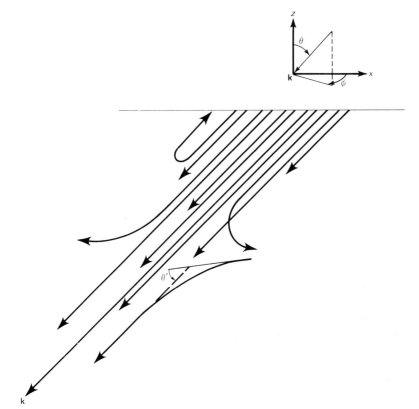

Fig. 9.3 Schematic of attenuation of a parallel photon flux in traversing an absorbing and scattering medium. The diagram shows both the absorption (via decrease in flux density) and scattering out of and into beam (curved lines). The irradiance varies approximately as an exponential function of depth.

absorption process is suggested in Fig. 9.3 via the decrease in the undiverted photon flux density as the light traverses the fluid.

The *monochromatic beam scattering coefficient* describes the loss of flux due to redirection of photons out of the beam by the process of scattering; it is indicated in Fig. 9.3 by way of the outwardly curved lines, and is defined as

$$b_\lambda \equiv -\left(\frac{1}{\Phi_\lambda} \frac{d\Phi_\lambda}{ds} \right)_{scat} \qquad \mathrm{m}^{-1} \, . \qquad (9.13)$$

Since more flux is generally scattered out of the beam than into it, the net coefficient is positive; numerically, it is typically a few percent of the absorption cross section, but can be considerably larger under some circumstances (see Table 9.4). Most scattering takes place at relatively small angles from the original direction.

Since absorption and scattering include all attenuation processes, by definition, the *total monochromatic beam attenuation coefficient* must then be

$$c_\lambda \equiv a_\lambda + b_\lambda \ . \tag{9.14}$$

This quantity is the analogue of the diffuse irradiance attenuation coefficient of Eq. 9.10, except it is defined for a beam of parallel radiation rather than diffuse irradiance.

Returning to the beam scattering coefficient: This quantity represents the net scattering of the photon beam taking place over all polar angles, (θ, ϕ). However, a more detailed descriptor is often needed to represent the distribution of intensity as a function of angle. The *monochromatic volume scattering function, $\beta_\lambda(\theta)$*, represents the intensity, $dI_\lambda(\theta, \phi)$, scattered out of a small volume, d^3x, of a monochromatic beam of irradiance E_λ and into the angular range between θ and $\theta + d\theta$, while traversing path length ds:

$$\beta_\lambda(\theta) \equiv \frac{1}{E_\lambda} \frac{dI_\lambda(\theta, \phi)}{d^3x} \qquad \mathrm{m}^{-1} \, \mathrm{sr}^{-1} \ . \tag{9.15}$$

Because it is defined in terms of the ratio of intensity to irradiance, the volume scattering function picks up the solid angle discriminant, $d\Omega$, of the intensity. This function is a very important one in optics since it gives the detailed geometric description of the scattering process as a function of θ. Thus $\beta_\lambda(\theta) \sin \theta \, d\theta \, d\phi$ is the loss of intensity by scattering into the solid angle $d\Omega$ centered on the volume element d^3x, and the integral over all angles must therefore be the beam scattering coefficient, b_λ:

$$b_\lambda = \int_0^{2\pi} \int_0^{\pi} \beta_\lambda(\theta) \sin \theta \, d\theta \, d\phi$$

$$= 2\pi \int_0^{\pi} \beta_\lambda(\theta) \sin \theta \, d\theta \ . \tag{9.16}$$

For a random distribution of isotropic scatterers, the distribution is clearly symmetrical about the direction of incidence, as suggested in Fig. 9.4.

A review of the discussion of electromagnetic scattering from a rough surface given in Section 8.5 leads to the conclusion that $\beta_\lambda(\theta)$ is equivalent to the differential cross section per unit volume per steradian per unit wavelength;

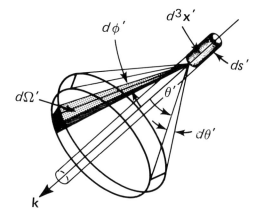

Fig. 9.4 Scattering out of a parallel beam by scatterers in a small volume element, d^3x, while the beam traverses a distance ds'. The volume scattering function, $\beta(\theta')$, measures the intensity in solid angle, $d\Omega' = \sin\theta'\, d\theta'\, d\phi'$, and is azimuthally symmetric about the incidence direction.

clearly the optical target being considered here is not a differential element of surface but of volume. Thus we have

$$\frac{d^2\sigma(\lambda,\theta)}{d\Omega\, d^3x} = \beta_\lambda(\theta) \; . \tag{9.17}$$

As with the irradiance, one may define beam scattering functions for forward or backward scattering, or for scattering into any finite interval of interest, by the appropriate angular ranges of integration.

In the discussions to follow, we will have occasion to deal with many of the quantities defined by Eqs. 9.2 to 9.16, which constitute the most important radiant functions in the extensive repertoire of optics. A few more specialized definitions may be required from time to time; these will be derived as needed.

9.3 Surface Illumination: Sun, Sky, and Artificial

As we did in Chapter 6 in the study of air–ocean interaction, we will attempt to describe the sequence of cause and effect taking place during the propagation of light energy from just above the sea surface, across the interface, and into the upper layers of the medium. The first entry in this odyssey is the description of the nature of surface illumination, of which there are at least three important sources: direct solar radiation, skylight, and anthropogenic.

Direct Solar Illumination

Figure 9.5 shows the surface spectral irradiance of the sun as observed through a standard air mass containing 0.01 m of precipitable water. This graph is a more detailed version of the data shown in Fig. 2.2 (on the right side). The modifications of the incident solar spectral irradiance from the near-blackbody illumination at the top of the atmosphere to that shown in Fig. 9.5 are due to (1) atmospheric scattering, both Rayleigh and aerosol, which removes an appreciable part of the short-wavelength, blue component from the direct solar beam, and (2) absorption, chiefly by ozone, water vapor, and

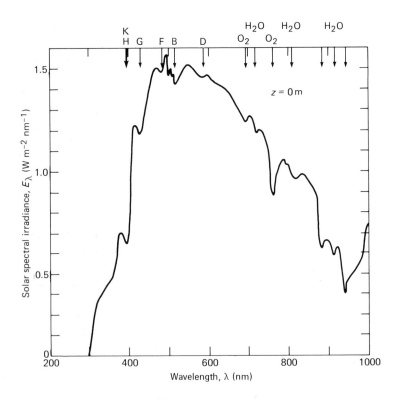

Fig. 9.5 Downwelling spectral irradiance from direct solar illumination, as viewed through a standard sea level air mass, 3.5 mm of ozone, and 10 mm of precipitable water. Solar Fraunhofer lines (K, ..., D) and absorption bands from O_2 and H_2O may be seen at positions of arrows. [From Jerlov, N. G., *Marine Optics* (1976).]

oxygen. Attenuation by these molecules imposes additional fine structure on the solar spectrum above and beyond the modulation from the solar Fraunhofer lines. The spectral integral of this quantity gives the clear-sky insolation at the surface of the earth with the sun in the zenith, a value of approximately 1000 W m^{-2}.

Typical diurnal variations in solar illumination at the ground near a latitude of 36°S are shown in Fig. 9.6, for dates close to the winter and summer solstices and at the equinoxes, under conditions of varying amounts of cloud cover occurring during the day. An overcast sky of 100% cloud cover may reduce the surface insolation to perhaps 10 to 20% of the clear-sky irradiance. Over 90% of the total energy is absorbed by the ocean at these latitudes, and the time integrals of the curves of Fig. 9.6 give the daily heating pulse deposited into the euphotic zone.

Beyond the daily variability, insolation is a function of latitude and season as well, due to the obliquity of the polar regions and the 23.5° tilt of the earth's axis with respect to the ecliptic. Calculated values of daily direct insolation in the Northern Hemisphere throughout the year (which values ignore the effects of the atmosphere) are shown in Fig. 9.7; latitude is the parameter on the graph. The ordinate is in units of megajoules per square meter per day, where 50 MJ day^{-1} is equal to 578.7 W. The irradiance data of Fig. 2.5 represent the annual average of these curves. It is seen that the daily heating rate is quite uniform near the equator; north of the Arctic Circle it may exceed the equatorial value in the summer, but becomes essentially zero in the winter. Clouds, atmospheric scattering, absorption, and skylight modify these values significantly.

Sky Illumination

An example of the spatial distribution of clear-sky radiance throughout the celestial hemisphere is given in Fig. 9.8; the values give the contributions from all wavelengths. This plot shows relative *sky radiance* or *skylight*, for the sun near 53° zenith angle (angle from the local vertical) and 0° azimuth angle; the radiance is normalized to unity at the zenith. In a clear sky, the direct solar illumination is confined to the sun's disk, which occupies about a 0.5° angle. However, molecular and aerosol scattering lead to diffuse light emanating from the entire sky in quantities that vary significantly with direction, but which in their spatially integrated intensity amount to perhaps 15 to 20% of the direct illumination when the sun is above some 30° of the horizon. At approximately 90° zenith angle and 180° azimuth angle from the sun, there is a region of skylight that is relatively highly polarized due to Rayleigh scattering. No simple analytical expression for this radiance distribution appears to be available.

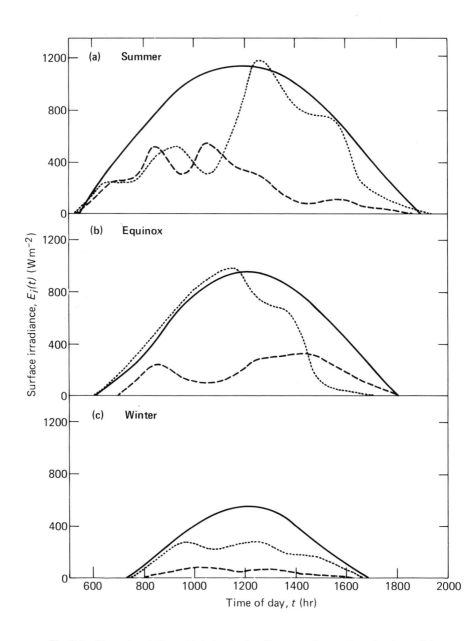

Fig. 9.6 Diurnal variation of total solar irradiance under varying cloud conditions, at three times of year. The dashed curves show the effects of cloud cover. Geographical coordinates: 149°27′E, 35°49′S, 770 m elevation. (a) Summer solstice, (b) Equinox, (c) Winter solstice. [Adapted from Kirk, J. T. O., *Light and Photosynthesis in Aquatic Ecosystems* (1983). Originally due to F. X. Dunin.]

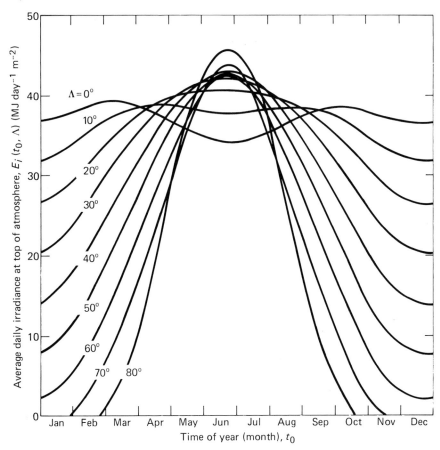

Fig. 9.7 Theoretical insolation at the top of the atmosphere as a function of the annual cycle and Northern Hemisphere latitude. Tropics are relatively uniformly illuminated, while 80°N is in darkness for 4 months. [From Kirk, J. T. O., *Light and Photosynthesis in Aquatic Ecosystems* (1983). Based on data from K. Ya. Kondratyev (1954).]

In the presence of clouds, sky radiances are much more complex. While dense clouds smooth out the illumination somewhat, even for a heavily overcast sky, it is found that the distribution of radiance in zenith angle is not at all uniform, but is given by a cardioid of the form

$$L(\theta_z) = 3E_d(0)(1 + 2\cos\theta_z)/7\pi, \qquad (9.18)$$

where $E_d(0)$ is the downwelling irradiance on a zenith-facing surface. Under broken clouds, the distribution varies discontinuously and little can be said

Zenith angle, θ_Z (deg)

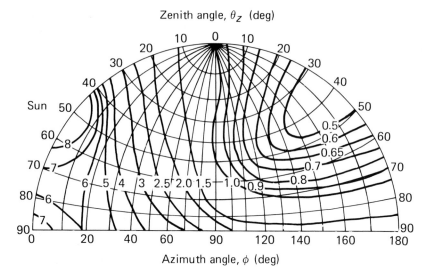

Azimuth angle, ϕ (deg)

Fig. 9.8 Radiance distribution for a clear sky with the sun at 53° zenith angle and 0° azimuth angle (normalized to unity at the zenith). [Adapted from Jerlov, N. G., *Marine Optics* (1976).]

about it in general. Mariners and aviation weather forecasters often report cloud cover in terms of tenths of the visible sky occupied by clouds, and such data form useful adjuncts to more comprehensive cloud cover distribution data available from satellites.

These spatial and temporal variations mean that the amount of solar energy incident on the sea has a wide range of values, with concomitant implications for the subsurface illumination field and heat input to the upper ocean. Such heating variations play an important role in large-scale ocean dynamics, an example of which is the El Niño phenomenon (Section 6.11).

Artificial Illumination

Measurement of light absorption and scattering using narrowband artificial light sources is the common approach to determining the inherent optical properties, a_λ and b_λ. Monochromatic, highly collimated laser sources are ideal for this purpose, and have been used to considerable advantage over incandescent light sources shaped by lenses and filters. When operated with sufficient power and with tuned, filtered receivers, lasers can be used in full daylight. A range of sources is available whose combined outputs span the full spectrum of wavelengths of oceanic interest. As will be seen ahead, blue light having a wavelength near 472 nm can penetrate most deeply in clear oceanic waters, while in coastal waters, green light near 550 nm is minimally attenuated.

However, values for a_λ and b_λ determined in this fashion do not yet seem to be available for the full range of oceanic water types. The classification of optical water types given in Section 9.8 is made on the basis of many years of measurements, using a variety of methods. Improved observations with modern instrumentation are required for more accurate understanding of the optical properties of seawater.

9.4 Surface Interactions: Reflection, Scattering, and Refraction

Our earlier discussions of the Fresnel coefficients illustrated the importance of those quantities in establishing the reflection, scattering, and transmission of light across the surface. The values of $|R|^2$, σ^0, and $|T|^2$ in turn are functionals of the complex dielectric function, or equivalently, the complex index of refraction. Table 9.1 gives the dependence of the real part of n at $\lambda = 589.3$ nm as a function of temperature and salinity. Over the ranges of nominal values of these quantities, the index of refraction does not depart from a value of 1.34 by more than 0.003. For many purposes, therefore, a constant value for the real part of n will suffice, particularly for estimating refractive effects. Nevertheless, even though the wavelength dependence of the index is not pronounced, its variation is important in calculations of scattering, for example. Table 9.2 lists values of n for seawater at $T = 4°$ and $20°$, and shows that

TABLE 9.1

Refraction Anomaly, $(n - 1.30000) \times 10^5$.
for $\lambda = 589.3$ nm, versus Salinity and Temperature

Salinity, s (psu)	Temperature, T (°C)			
	0	10	20	30
0	3400	3369	3298	3194
5	3498	3463	3390	3284
10	3597	3557	3482	3374
15	3695	3652	3573	3464
20	3793	3746	3665	3554
25	3892	3840	3757	3644
30	3990	3934	3849	3734
35	4088	4028	3940	3824
40	4186	4123	4032	3914

[Adapted from Sager, G., *Beit. Meeresk.* (1974).]

TABLE 9.2
Variation of Refractive Index of Seawater
$s = 35$ psu, $p = 1.013 \times 10^5$ Pa.

Wavelength, λ (nm)	Refractive Index, n $T = 4°C$	Refractive Index, n $T = 20°C$
400	1.35121	1.34994
450	1.34716	1.34586
500	1.34425	1.34295
550	1.34206	1.34077
600	1.34035	1.33904
650	1.33894	1.33765
700	1.33776	1.33644

[Adapted from Austin, R. W., and G. Halikas, "The Index of Refraction of Seawater" (1976).]

the dispersion of the index amounts to about 4% over the range of wavelengths listed.

In general, n is a function of salinity, temperature, and pressure, as well as of wavelength. The partial derivatives of n at near-surface pressures, at $T = 30°C$, and at $s = 35$ psu, as averaged over the wavelength interval between 400 and 700 nm, are approximately as follows: $\partial n/\partial s = 1.84 \times 10^{-4}$ (psu)$^{-1}$; $\partial n/\partial T = -1.27 \times 10^{-4}$ (°C)$^{-1}$; and $\partial n/\partial p = 1.41 \times 10^{-10}(Pa)^{-1}$.

It should also be noted that the wavelengths cited are those *in vacuo*. The speed of light in seawater is near 2.24×10^8 m s^{-1}, which means that wavelengths in water are approximately three-fourths of those in air ($n \simeq 4/3$). The color of light, on the other hand, is actually set by its frequency, which means that the visual *hue*, or impression of color, is not modified underwater by the refractive effects. However, it *is* changed by wavelength-selective attenuation, as will be discussed below, so that objects underwater take on bluish or greenish tints as a result of the altered color of the subsurface illumination.

Smooth Surface Reflectivities

Since seawater is only slightly absorbing when measured in terms of its fractional absorption per wavelength, the reflection coefficients can in the main be calculated without consideration of the imaginary part of the index of refraction. Because of this, we can derive somewhat more convenient expressions for those coefficients, as well as for the reflection law, Snell's law, and Brewster's law. It is easy to manipulate the formulas in Chapter 8 (Eqs. 8.62 and 8.63) to obtain:

$$r_{\parallel} = r_v = \frac{\tan^2 (\theta_i - \theta_t)}{\tan^2 (\theta_i + \theta_t)} , \tag{9.19}$$

$$r_{\perp} = r_h = \frac{\sin^2 (\theta_i - \theta_t)}{\sin^2 (\theta_i + \theta_t)} , \tag{9.20}$$

$$\sin \theta_i = \sin \theta_r , \tag{9.21}$$

$$\sin \theta_i = n \sin \theta_t , \tag{9.22}$$

and

$$\tan \theta_B = n . \tag{9.23}$$

Now the direct solar radiation is unpolarized, so that the effective reflectance for sunlight may be taken as the mean value of r_h and r_v:

$$r_{sun} = (r_h + r_v)/2 . \tag{9.24}$$

Figure 9.9 is a plot of the reflection coefficients for horizontal, vertical, and unpolarized light; it may be seen that only the vertically polarized component shows a minimum at the Brewster's angle of $\theta_B = 53.3°$ for seawater.

Reflection of the diffuse, partially polarized *skylight* is considerably more complicated than for direct sunlight. If $L(\theta,\phi)$ is the sky radiance (as shown in Fig. 9.8, for example), an average *diffuse reflection coefficient* for smooth water may be defined via

$$r_{dif} = \frac{1}{\langle E_d \rangle} \int_0^{2\pi} \int_0^{\pi/2} r(\theta) \, L(\theta,\phi) \cos \theta \sin \theta \, d\theta \, d\phi , \tag{9.25}$$

where the normalizing downward irradiance is

$$\langle E_d \rangle = \int_0^{2\pi} \int_0^{\pi/2} L(\theta,\phi) \cos \theta \sin \theta \, d\theta \, d\phi . \tag{9.26}$$

If L were unpolarized and uniform in all directions, the value of the diffuse radiation obtained from Eq. 9.25 would be $r_{dif} = 0.066$; if it had the cardioidal distribution of Eq. 9.18, the integral is equal to 0.052. Other models of illumination give roughly the same values, so that one may say that on the

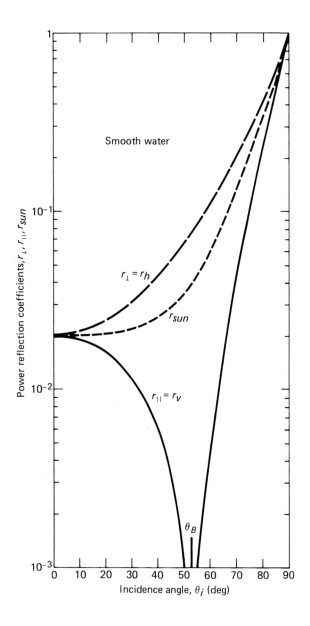

Fig. 9.9 Reflection coefficients for a smooth ocean surface for horizontally, vertically, and unpolarized light versus incidence angle. The minimum in vertically polarized intensity occurs at the Brewster angle of 53.3°.

average, a smooth sea surface reflects approximately 6% (and hence transmits 94%) of the incident skylight. Some fraction of the latter is returned across the surface as upwelling light.

The division between direct sunlight and skylight will obviously depend on the atmospheric Rayleigh and aerosol scattering, as well as the fractional cloud cover existing at the time and place of interest, so that the total reflectance can only be specified parametrically. If F is the ratio of sky radiation to total radiation, then the total reflectance from the smooth sea is given by

$$r_{smooth} = E_r/E_i = r_{sun}(1 - F) + r_{dif} F, \qquad (9.27)$$

where E_r and E_i are the reflected and incident radiances, respectively. Figure 9.10 illustrates the reflected irradiance as a function of *solar zenith angle,*

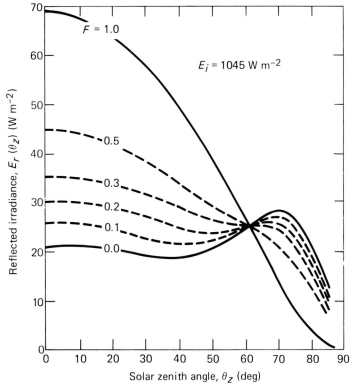

Fig. 9.10 Irradiance reflected from a smooth sea under varying ratios of sky radiation to total radiation (sky plus sun), as a function of solar zenith angle. Pure solar radiation corresponds to $F = 0$, and has a maximum near 70° zenith. Pure skylight ($F = 1$) varies approximately sinusoidally. [Adapted from Neumann, G., and R. Hollman, Union Geod. Geophys. Int. Monogr. (1961).]

θ_z (i.e., the angle of the sun from the local vertical); the parameter is F. An incident solar irradiance of 1045 W m^{-2} has been assumed, with the reflection coefficient for sunlight as given by Eq. 9.24, and that for diffuse skylight as $r_{dif} = 0.066$. The curve for solar radiation has a maximum near 70° due to a combination of increased reflectance and reduced radiance as θ_z tends toward 90°; an intersection of all the parametric curves occurs at $\theta_z = 61°$, where $r_{sun} = r_{dif} = 0.066$.

Rough Surface Reflectivities

The effect of wind and waves in modifying the smooth surface reflectances at near-vertical angles is small; however, there is a considerable increase from the smooth surface reflectivity as the wind speed increases for solar angles near the horizontal. An empirical result for reflected sun and sky radiances (normalized to the zenith sky radiance) is shown in Fig. 9.11 as a function of viewing angle measured from nadir, with cases illustrated for calm and rough seas (Beaufort Force 4 winds; i.e., $u_w \approx 5.5$ to 7.9 m s^{-1}), and for clear and overcast skies. The increase of intensity toward the horizon is due to the general variation of reflectivity with incidence angle that characterizes any dielectric. From Eqs. 9.19, 9.20, and 9.24, r ranges from 0.02 at the zenith to 1.00 at the horizon. This variation implies that the smooth sea would appear to be very dark near nadir (except for upwelling light from beneath the surface), but would totally reflect the sky at the horizon, which would therefore theoretically be invisible. Any wave action rapidly reduces the reflectivity at large incidence angles, as Fig. 9.11 suggests, thereby making the roughened sea appear much darker toward the horizon than the smooth one. Good marine artists intuitively recognize the visual consequences of these facts, if not the physics, and incorporate them into their renditions of the seascape.

A more detailed consideration of reflection and refraction at the sea surface is shown in Fig. 9.12, which illustrates a small section of the surface on a scale equivalent to the dimensions of short gravity and capillary waves. Two light rays are incident on the wavy surface at point B from points A and A'; these are partially refracted to points D and D', and partially reflected to C and C'. Ray $A'-B-C'$ is singly reflected, while Ray $A-B-C$ is reflected again from the surface at C toward C'', a process called double surface scattering or reflection. The doubly scattered ray also has a refracted component, ray $C-E$, which undergoes total internal reflection at point E. The underwater rays at D, D', and F suffer fates that will be discussed in the next section. The results of a statistical "Monte Carlo" calculation of the reflection coefficient from a nonisotropic, Gaussian, random sea surface are shown in Fig. 9.13 as a function of wind speed, u_w, with incidence angle, azimuth angle

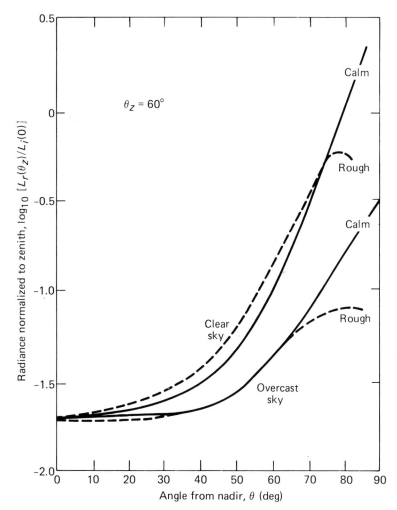

Fig. 9.11 Radiance reflected by the sea surface at various viewing angles, normalized to zenith sky radiance. Clear sky conditions produce higher fractional radiances, while overcast sky yields more uniform reflected energy. Rough conditions change radiance mainly closer to the horizon. [From Cox, C., and W. Munk, *J. Opt. Soc. Am.* (1954).]

(with respect to crosswind and alongwind directions), and multiplicity of scattering as parameters. The graph shows the azimuthal asymmetry in scattering, with crosswind conditions significantly increasing the reflection coefficient; this effect enters because of the upwind–crosswind asymmetry in the surface

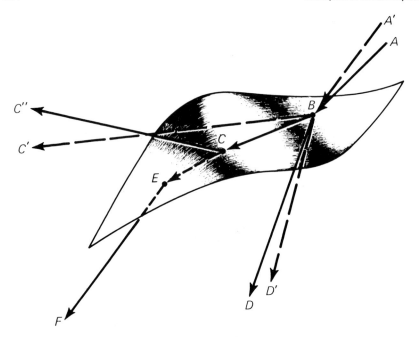

Fig. 9.12 Section of a short surface wave, illuminated from A and A′, and with light rays incident at B having reflected and refracted components. Reflected component undergoes a second scattering at C, and its refracted component is totally reflected internally at E. [Adapted from Preisendorfer, R. W., and C. D. Mobley, *J. Phys. Oceanogr.* (1986).]

slope probability distribution (see Eqs. 8.118 and 9.28 below). It also demonstrates the importance of multiple scattering in establishing the albedo. A crossplot of the total scattering coefficient of Fig. 9.13 is shown in Fig. 9.14 for the alongwind direction, and clearly illustrates the reduced albedo above some 50° as the wind speed increases. Such an approach yields a fairly realistic evaluation of the scattering coefficients for arbitrary geometries and surfaces.

The reflected glitter pattern, when viewed in forward scatter, is variably polarized, and depends on both the solar zenith angle, θ_z, and on the direction of reflection. In the direction of maximum radiance (the solar specular point), the reflected energy is unpolarized at $\theta_z = 0°$, is highly polarized in the range of 35° to 45°, and falls again to about 20% at $\theta_z = 90°$.

Oil and other surfactants modify the surface roughness, reflectivity, and polarization significantly. Under light wind conditions, surfactants readily dampen capillary and ultragravity waves, reducing their average slopes; at larger incidence angles, the reflectivity and percentage polarization of the surface

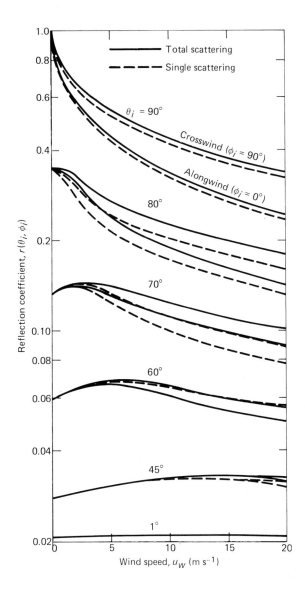

Fig. 9.13 Reflection coefficient for a Gaussian, random sea as a function of wind speed, with incidence angle, θ_i, wind azimuth angle, ϕ_i, and multiplicity of scattering as parameters. Scattering at larger incidence angles is more sensitive to wind speed (see Fig. 9.11), and is larger in crosswind direction. [Adapted from Preisendorfer, R. W., and C. D. Mobley, *J. Phys. Oceanogr.* (1986).]

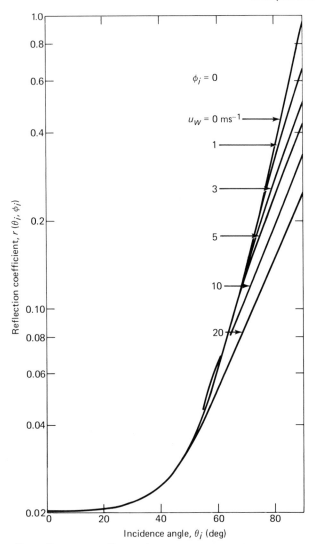

Fig. 9.14 The reflection coefficient for a nonisotropic, Gaussian, random sea as a function of incidence angle, with wind speed as a parameter; the azimuth is the along-wind direction. [Adapted from Preisendorfer, R. W., and C. D. Mobley, *J. Phys. Oceanogr.* (1986).]

are thereby increased. As the wind speed rises, the surfactants are broken up and mixed down into the water column (unless they consist of heavy, viscous oils or similar materials), so that their presence on the surface is not readily detected by reflectivity changes.

The statistical description of reflection from the sea surface uses the probability distribution function for surface slopes; this quantity was cited in Sections 7.9 and 8.6. At optical, infrared, and short radar wavelengths, the *Cox and Munk* distribution appears most appropriate. Measurements of the glitter pattern of the sea surface under light to moderate wind conditions yield the result that the probability distribution of the surface slopes, $p(\zeta_x, \zeta_y)$, is a *Gram–Charlier* function with a variance that depends on wind speed and direction (see Eq. 8.118). Thus we may write for that distribution

$$p(\zeta_x, \zeta_y) \simeq \frac{1}{2\pi\sigma_u\sigma_c}\{\exp[-\tfrac{1}{2}(s_u^2 + s_c^2)]\}\{1 - (1/2)c_{21}(s_c^2 - 1)$$

$$- (1/6)c_{03}(s_u^3 - 3s_u) + (1/4)c_{22}(s_c^2 - 1)(s_u^2 - 1)$$

$$+ (1/24)[c_{40}(s_c^4 - 6s_c^2 + 3) + c_{04}(s_u^4 - 6s_u^2 + 3)]\}\,,$$

$$(9.28a)$$

where the normalized upwind slope is

$$s_u^2 = \zeta_x^2/\sigma_u^2 \qquad (9.28b)$$

and the normalized crosswind slope is

$$s_c^2 = \zeta_y^2/\sigma_c^2\,. \qquad (9.28c)$$

The slope distributions are plotted in Fig. 9.15, with the crosswind component being symmetrical about zero slope and the upwind–downwind component being skewed, and displaced from zero. This is due to the fact that the longer gravity waves tend to have a higher spectral density of short waves on their downwind faces than in their troughs or upwind faces. The asymmetry is described by the quantity in the curved braces. The variances are given by

$$\sigma_u^2 = a + bu_w \qquad (9.29a)$$

and

$$\sigma_c^2 = c + du_w \qquad (9.29b)$$

and thus increase linearly with wind speed. The *skewness coefficients* are given by

$$c_{21} = e - fu_w \qquad (9.30a)$$

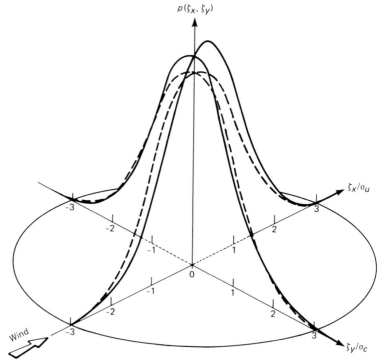

Fig. 9.15 The probability distribution function for sea surface slopes as a function of direction. The alongwind direction is $+\zeta_x$, with positive slopes in the downwind direction. The peak of distribution in direction of wind is shifted toward the downwind side by several degrees. The dashed curves represent a Gaussian distribution. [Adapted from Phillips, O. M., *The Dynamics of the Upper Ocean* (1977). Originally due to Cox, C., and W. Munk, *J. Opt. Soc. Am.* (1954).]

and

$$c_{03} = g - hu_w .\qquad(9.30\text{b})$$

Values for the coefficients in these relationships are given in Table 9.3.

The increase in the widths of the slope distributions with u_w explains how the image of the sun's disk reflected from the mirrorlike surface of the calm sea evolves from a spot that subtends an angle of approximately 0.5° at zero wind speed, to one that spreads out diffusely over angles of 20 to 30° at higher speeds. Plate 2 (Fig. 5.32) shows such diffuse surface reflection in the region southeast of the Strait of Gibraltar.

TABLE 9.3

Parameters in Cox and Munk Slope Probability Distribution
(multiply all quantities by 10^{-3})

Parameter	Symbol	Clean Surface	Slick Surface
Upwind coefficients	a	0.0	5.0
	b	3.16	0.78
Crosswind coefficients	c	3.0	3.0
	d	1.92	0.84
Skewness coefficients	e	10.0	0.0
	f	8.6	0.0
	g	40.0	20.0
	h	33.0	0.0
Peakedness coefficients	c_{40}	400	360
	c_{22}	120	100
	c_{04}	230	260

[Adapted from Cox, C., and W. Munk, *J. Opt. Soc. Amer.* (1954).]

Optical Albedo

The overall albedo of the sea surface is the ratio of the total irradiance leaving the sea surface, both reflected and upwelled, to that entering it, and is an important quantity in the study of solar heating and subsurface illumination. The average albedo, $\langle \alpha \rangle$, is defined as

$$\langle \alpha \rangle \equiv E_{u_{air}} / E_{d_{air}} , \qquad (9.31)$$

where the quantities on the right are the downward irradiance, $E_{d_{air}}$, and the upward irradiance, $E_{u_{air}}$, measured just above the surface; similar quantities may be defined for the irradiances just below the water surface. By continuity, the irradiance immediately above and below the surface must be the same, so that

$$E(0) = E_{d_{air}} - E_{u_{air}} = E_{d_{water}} - E_{u_{water}} . \qquad (9.32)$$

The upward irradiance in air is composed of the surface reflected irradiance and the escaping subsurface upwelling irradiance. Let r_s be the reflectivity of the upper surface of the water, and r_w be that for the under surface; then the upwelling light in the air has the form

$$E_{u_{air}} = r_s E_{d_{air}} + (1 - r_w) E_{u_{water}} . \tag{9.33}$$

From Eqs. 9.31 to 9.33, the albedo is then

$$\langle \alpha \rangle = r_s + (1 - r_w) E_{u_{water}} / E_{d_{air}} , \tag{9.34}$$

which is clearly greater than the surface reflectivity, r_s. An application of the diffuse reflectivity formula of Eq. 9.25 to the reflection of the undersurface upwelling irradiance yields $r_w \simeq 0.46$ to 0.48, assuming that the upwelling light is unpolarized and the upwelling radiance distribution is uniform. The fraction of the total insolation (both direct sunlight and diffuse skylight) absorbed by the ocean is then

$$1 - \langle \alpha \rangle = 1 - r_s + (1 - r_w) E_{u_{water}} / E_{d_{air}} . \tag{9.35}$$

It is this energy that enters into the thermodynamic and photosynthetic processes occurring beneath the sea surface.

Experimental values suggest that to a first approximation, the average albedo is only weakly dependent on atmospheric turbidity; it is somewhat more strongly dependent on surface wind speed; and is most sensitively dependent on solar zenith angle. Clouds reduce this incident energy significantly (see Fig. 9.6, for example). Overall experimentally derived values for $\langle \alpha \rangle$ versus solar zenith angle for combined clear-sky sunlight and unpolarized skylight are shown in Fig. 9.16. It is seen that as the sun approaches low elevation (large zenith) angles, the albedo increases rapidly toward unity.

In clear oceanic waters, the ratio of $E_{u_{water}}$ to $E_{d_{air}}$ may be as great as 0.03 to 0.04; using the value of $r_w = 0.48$ and the average reflectivity from Fig. 9.16 of 0.03, we obtain from Eq. 9.34 an overall albedo of 0.045 for the sun near its zenith, which means that 95.5% of the incident sunlight is absorbed in the ocean. At larger angles, say at some 70°, $\langle \alpha \rangle = 0.12 + 0.015 = 0.135$, and thus less than approximately 86.5% of the incident light is absorbed. This decrease in the absorbed light toward the horizon, when coupled with the decrease in the level of illumination that occurs with lower sun angles (see Figs. 9.7 and 9.8), largely accounts for the daily, seasonal, and latitudinal changes in sunlight absorbed beneath the surface. Greatly reduced amounts of sunlight therefore enter subpolar and polar waters during the winter, with the amount absorbed poleward of the Arctic and Antarctic Circles being essentially zero at the winter solstice.

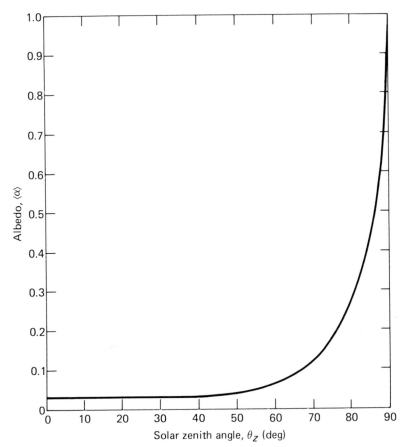

Fig. 9.16 The albedo of a smooth water surface as a function of the solar zenith angle, for a mixture of direct solar and clear-sky radiances. Low sun angles result in large reflected and small transmitted fractions.

Refraction

Snell's law, as given by Eq. 9.22, implies that when $\theta_i = 90°$, the refracted angle, θ_t, is given by $\theta_t = 48.3°$, or that a light ray at grazing incidence on a smooth water surface will refract into the sea at a *critical transmitted angle*, θ_c, with respect to the vertical. This means that all of the downwelling light incident on a smooth surface will be mapped into a cone of full angle 96.6°, the *Snell cone*. Figure 9.17 shows schematically how such refraction appears to an underwater detector located at the apex of the cone. Light re-

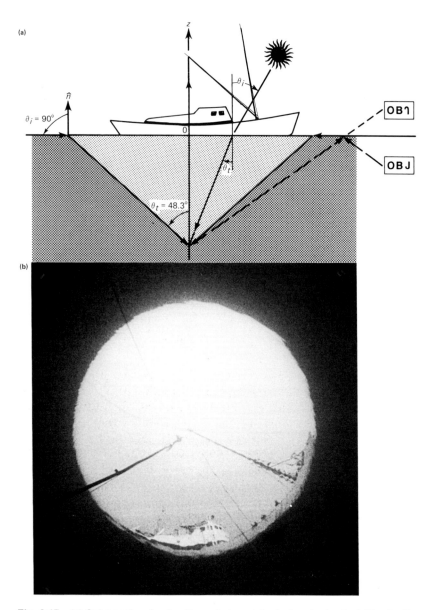

Fig. 9.17 (a) Schematic of refraction of atmospheric hemispherical illumination into an underwater cone of half-angle 48.3°. Beneath the surface, images seen at angles greater than the critical angle must be from underwater objects. Waves diffuse the cone angle somewhat but do not change imaging properties significantly. [Adapted from Preisendorfer, R. W., *Hydrologic Optics*, Vol. I (1976).] (b) Subsurface image at $z = -1.75$ m, illustrating mapping of objects in upper hemisphere into Snell cone. Radiance camera suspended from side of ship seen on the horizon. [Photo courtesy of R. C. Smith, from *Optical Aspects of Oceanography* (1974).]

ceived at the detector from angles greater than θ_c must have come from objects *beneath* the surface, and these will appear as inverted and perverted images. The presence of waves diffuses the cone angle into a distributed geometry, but the critical angle effect is nevertheless dominant. Thus the entire sky hemisphere is contained within an illuminated circle located vertically above the detector.

Other consequences of the refractive index are to make underwater objects appear to be closer, larger, and moving more rapidly than they actually are. When viewed directly from above, a submerged object appears closer by the factor $1/n \simeq 0.746$. Also, a rod or string inserted at a nonzero incidence angle appears to be bent toward the vertical by an amount shown in Eq. 9.22. Figure 9.18 shows the magnification effect, whereby a target at A appears to be at A' and to subtend a larger angle as seen by the observer in air than its underwater geometry dictates. The magnification factor, M, is the ratio of the subtended incidence angle to the subtended refracted angle, or

$$M \equiv \frac{\delta\theta_i}{\delta\theta_t} = n \left[\frac{1 - \sin^2 \theta_t}{1 - n^2 \sin^2 \theta_t} \right]^{\frac{1}{2}}. \tag{9.36}$$

Here θ_t is the mean angle of refraction. It may be seen from Eq. 9.36 that the magnification becomes quite large near the critical angle. The velocity variation may also be illustrated with Fig. 9.18; for example, during the time that the object travels from A to B, the image appears to go from A' to B', which is clearly larger than AB. As it nears the critical cone angle, its apparent velocity becomes very large, indeed. In practice, the ability to see beneath the surface becomes poor as these large angles are reached, but the effects are amply visible over the nominal range. They may very well be partially responsible for the apparently exaggerated claims as to the size and speed of various kinds of fish and marine mammals observed at sea.

There is also a refraction-induced concentration of radiance in crossing the surface. If, in air, $\delta\Omega_a$ is the unit solid angle subtended by a small target of area dA, then refraction maps it into a smaller solid angle, $\delta\Omega_w$, in water, where

$$\delta\Omega_w = \frac{1}{n^2} \, \delta\Omega_a . \tag{9.37}$$

This result follows from the fact that a solid angle is essentially the square of a linear angle, plus an application of the law of refraction, along with the definition of radiance as power per unit solid angle. The same mapping carries the 180° air hemisphere into the 96.6° full-angle cone beneath the surface. If $L_{d_{air}}$ is the downwelling radiance in air, then reflection reduces that

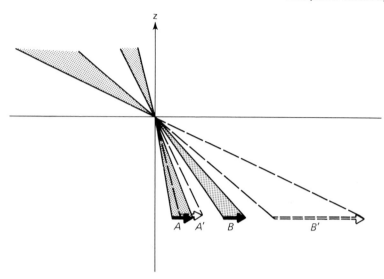

Fig. 9.18 Distortion and magnification of submerged objects by refraction. Targets underwater appear closer, larger, and faster-moving. [Adapted from Williams, J., *Optical Properties of the Sea* (1970).]

value to $(1 - r_a)L_{d_{air}}$; however, just beneath the surface the radiance underwater is increased by the compression of the solid angle, so that we have

$$L_{d_{water}} = n^2 (1 - r_a)L_{d_{air}} . \qquad (9.38)$$

Further small-scale fluctuations in the subsurface radiance result from short, highly curved surface waves, which act much as cylindrical lenses in focusing sunlight. The illumination of the sun, when viewed from just beneath the undulating surface, disintegrates into a myriad of streaked, glittering filaments under the rolling action of the wavelets. The refracted, focused images subtend smaller angles than the solar object, with the consequence that the underwater radiance may be magnified as much as 1000 times, and can approach levels damaging to the human eye.

9.5 Subsurface Interactions: Pure Seawater

Absorption and scattering in *pure* seawater are established by the molecular constituents listed in Table 4.1, with the main contributions being from H_2O, but with a nominal increase in scattering due to the ionic components. The diffuse attenuation coefficient for irradiance, $K_\lambda(sw)$, and the beam

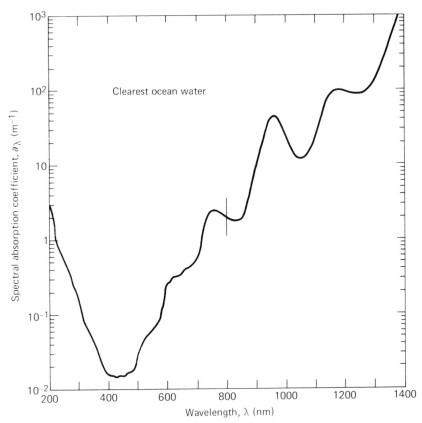

Fig. 9.19 The beam absorption coefficient for irradiance in clearest seawater versus wavelength. [Adapted from Smith, R. C., and K. S. Baker, *Appl. Optics* (1981) (200 to 800 nm), and from Curcio, J. A., and C. C. Petty, *J. Opt. Soc. Am.* (1951) (800 to 1400 nm).]

scattering coefficient, $b_\lambda (sw)$ for the clearest ocean waters are listed in Table 9.4 for the range of $200 \le \lambda \le 800$ nm; the seawater beam absorption is plotted in Fig. 9.19 for an even greater range of wavelengths extending into the near-IR. For comparison's sake, also listed in Table 9.4 are the beam absorption coefficient, $a_\lambda (fw)$, and the beam scattering coefficient, $b_\lambda (fw)$, both for pure fresh water. The scattering in both salt and fresh water is due to fluctuations (see below), with the salt further increasing the coefficient by roughly 30% over the fresh water values. For absorption, there is no significant difference between the fresh water and the clearest seawater values. It may be seen that scattering in clear water is much less than absorption (although this is not true in turbid waters).

TABLE 9.4
Diffuse Attenuation Coefficient for Irradiance in Clearest Ocean Waters,
and Beam Absorption and Scattering for Clear Seawater and Fresh Water

Wavelength, λ (nm)	Diffuse Attenuation, K_λ (sw) (m^{-1})	Beam Absorption, a_λ (fw) (m^{-1})	Beam Scattering, b_λ (sw) (m^{-1})	Beam Scattering, b_λ (fw) (m^{-1})
200	3.14	3.07	0.151	0.116
250	0.588	0.559	0.0575	0.0443
300	0.154	0.141	0.0262	0.0201
350	0.0530	0.0463	0.0134	0.0103
400	0.0209	0.0171	0.0076	0.0058
450	0.0168	0.0145	0.0045	0.0035
500	0.0271	0.0257	0.0029	0.0022
550	0.0648	0.0638	0.0019	0.0015
600	0.245	0.244	0.0014	0.0011
650	0.350	0.349	0.0010	0.0007
700	0.650	0.650	0.0007	0.0005
750	2.47	2.47	0.0005	0.0004
800	2.07	2.07	0.0004	0.0003

[Adapted from Smith, R. C., and K. S. Baker, *Appl. Optics* (1981).]

Except in the backward direction, scattering by very pure fresh water is relatively weak when compared with scattering by even the most transparent of natural waters, because of significant concentrations of particulate matter in the latter. In order to arrive at reliable measurements of pure water scattering, repeated distillation *in vacuo* is required. The scattering is dominated by small fluctuations in the index of refraction (and hence the dielectric function) due to fluctuations in the liquid density caused by thermal motions. Calculations of the volume scattering function for purified water, $\beta_0(\theta)$, are of the same general type as those for the scattering cross section from a random sea surface. They yield a functional dependence on wave number similar to that for Rayleigh scattering, but with temperature and polarization dependencies as well:

$$\beta_{0\lambda}(\theta) = \frac{k^4}{(2\pi)^2} k_B T \, a_p^{-1} \left(n \frac{\partial n}{\partial p} \right)^2 \frac{3(1+\delta)}{6-7\delta}$$

$$\times \left[1 + \left(\frac{1-\delta}{1+\delta} \right) \cos^2 \theta \right] . \tag{9.39}$$

Here k_B is Boltzmann's constant, T is the absolute temperature, a_p is the isothermal compressibility, $\partial n/\partial p$ is the change in index of refraction with pressure, and δ is the so-called polarization defect:

$$\delta \equiv \frac{I_v(\pi/2)}{I_h(\pi/2)} . \tag{9.40}$$

The intensities I_h and I_v are those of the horizontally and vertically polarized light, respectively, scattered at $90°$ to the beam direction. For fluctuations in seawater, it is found experimentally that $(1 - \delta)/(1 + \delta) \simeq 0.835$ at $\lambda = 550$ nm. A plot of $\beta_0(\theta)$ versus θ is shown in Fig. 9.20, and it exhibits the $\cos^2\theta$ behavior characteristic of Rayleigh scattering, with equal amounts of forward scatter and backscatter. An integration over solid angle gives for the beam scattering coefficient of very pure water:

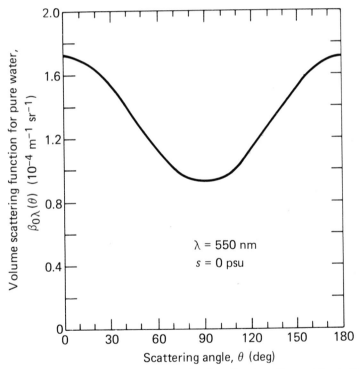

Fig. 9.20 The theoretical volume scattering function from thermal fluctuations in pure fresh water, for $\lambda = 550$ nm. Forward and backward scatter are equal in this case. Fluctuations contribute a major portion of the backscatter in seawater; compare with Fig. 9.26. [Adapted from Kirk, J. T. O., *Light and Photosynthesis in Aquatic Systems* (1983), following Morel, A., in *Optical Aspects of Oceanography* (1974).]

$$b_{0\lambda} = \frac{8\pi}{3} \beta_{0\lambda}(\pi/2) \frac{2 + \delta}{1 + \delta} . \tag{9.41}$$

The addition of salts to pure water increases the scatterance by approximately one-third. Table 9.5 lists values of $\beta_{0\lambda}(\pi/2)$ for pure fresh and salt water over a range of visible wavelengths.

TABLE 9.5
Volume Scattering Function at 90° for Pure Fresh and Salt Water
(s = 35 to 39 psu)

Wavelength, λ (nm)	$\beta_{0\lambda}(\pi/2)(fw)$ $(10^{-4} \text{ m}^{-1} \text{ sr}^{-1})$	$\beta_{0\lambda}(\pi/2)(sw)$ $(10^{-4} \text{ m}^{-1} \text{ sr}^{-1})$
350	6.47	8.41
400	3.63	4.72
450	2.18	2.84
500	1.38	1.80
550	0.93	1.21
600	0.68	0.88

[Adapted from Morel, A., in *Optical Aspects of Oceanography* (1974).]

From Eq. 9.39, the theoretical k dependence is k^4; however, because of the additional variations of the refractive index and its pressure derivative with wavelength (see Table 9.2), the actual dependence is found to be

$$\beta_{0\lambda}(\theta) \sim k^{4.32} . \tag{9.42}$$

This variation, together with the more complicated form of spectral absorption, work together to establish the color of pure water. The strong dependencies give a generally blue color to water, which, because of the relative weakness of the total attenuation coefficient, $c_{0\lambda} = a_{0\lambda} + b_{0\lambda}$, is only visible in large volumes of fluid. In the sea, however, the presence of other constituents in the water modifies the color to a greater or lesser extent, as we shall see after we discuss the absorption and scattering properties of natural oceanic waters.

9.6 Subsurface Interactions: Particles, Plankton, and Gelbstoffe

Beyond the absorptive and scattering properties of pure water, the natural ingredients of seawater increase the light attenuation coefficients by very large but variable factors and make the prediction of light levels beneath the sea a difficult undertaking without detailed knowledge of the properties and distributions of the attenuators. There are three main sources of the latter: inorganic particulates, plankton and its derivatives, and yellow substance (gelbstoffe). Near river outfalls, tannins and similar materials make additional contributions to the concentration of yellow substance.

Particulate Scattering

Scattering in the ocean is due to both inorganic and organic particulates, with the latter including both living cells and detrital material of biological origin. Organic particles are the dominant scatterers in the open ocean, whereas on the continental shelf and in coastal waters, inorganic components constitute perhaps 40 to 80% of the total scatterers, depending on location, with land runoff and winds being the main causative agents. The injection of inorganic materials can take place at long distances from the source, with aeolian Sahara dust, for example, being deposited in the ocean thousands of kilometers from the desert region. Thus inorganic materials are not restricted to near-shore regions. SiO_2 in quartz therefore becomes the most important inorganic constituent of the particulate population. Organic materials derive from disintegrated phytoplankton cells and skeletons of zooplankton; the latter make $CaCO_3$ a chemical of great importance in the marine ecosystem.

Scattering from particulates depends on (1) the degree of external reflection and diffraction by their geometrical form, and (2) on internal refraction and reflection and hence on the index of refraction of the particles, n_p. Figure 9.21 suggests these processes in schematic form; the totality of their effects establishes the important theoretical parameters to be the particle radius, r, and its index of refraction, n', relative to water. Here the relative index for a particle of absolute index, n_p, is

$$n' = n_p/n . \tag{9.43}$$

In addition, a quantity termed the *cumulative number distribution* is needed to characterize the scattering from an ensemble of particles. Such a distribution is generally a power law of the form

$$N(d) = n_0(d/d_0)^{-\gamma} , \tag{9.44}$$

where $N(d)$ is the number of particles having diameters greater than d, and n_0 and d_0 establish scales for the distribution and diameter. This type of distribution is typical of *grinding laws* that describe the size distributions of

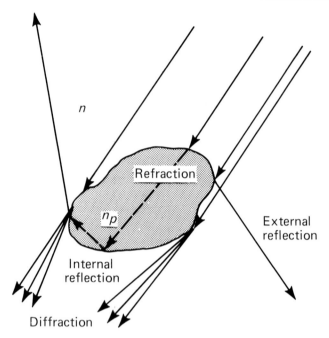

Fig. 9.21   Schematic of scattering by a particle of index of refraction n_p. External reflection and diffraction and internal refraction and reflection establish the overall scattering.

both terrestrial and extraterrestrial particles. Figure 9.22 shows several size distributions of oceanic particles for different water masses, with the power law exponent, γ, ranging from 0.7 to 6.0; an average value of $\gamma \simeq 2.5$ appears reasonable.

The shapes of the particles are not overly important in scattering as long as their aspect ratios are not too great; when low aspect ratio particles are the case, *Mie scattering theory* may be applied to a statistical collection of nonspherical scatterers and reasonable results obtained, because of the random orientations of the particles that occur in the aggregate. A variant of this theory was met in Chapter 7 in the context of acoustic scattering (see Figs. 7.28 and 7.29). Mie theory is quite complicated, and analytical solutions are available for only a few simple shapes, such as spheres and cylinders. The oscillations apparent in the cross section of Fig. 7.28 in the resonant region are due to constructive and destructive interference among diffracted edge waves traversing the perimeter of the particle, and these are smoothed out in the case of nonuniform geometric shapes such as are found in natural particulates. What can be said is that for particles small compared with the

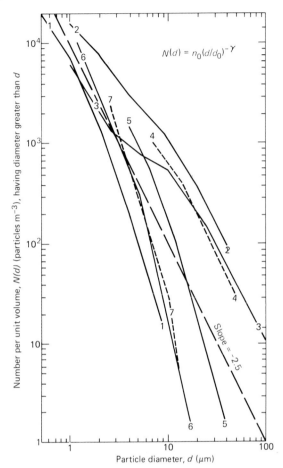

$$N(d) = n_0(d/d_0)^{-\gamma}$$

Number per unit volume, $N(d)$ (particles m^{-3}), having diameter greater than d

Particle diameter, d (μm)

Slope = -2.5

Fig. 9.22 Cumulative size distributions of oceanic particles. (1) Pacific deep water; (2) Swedish fjord; (3) Mediterranean (microscopic); (4) mean North Atlantic; (5) western North Atlantic; (6) Mediterranean (particle counter); (7) subtropical western Pacific. Average exponent for power law is $\gamma = 2.5$. [Adapted from Jerlov, N. G., *Marine Optics* (1976).]

wavelength of light, the scattering will behave approximately as k^4; for wavelengths much less than the particle size, the cross section is essentially geometrical and is nearly constant.

When a variable particle index of refraction is taken into account, the problem becomes more complicated. A useful summary diagram showing the effects of r and n' is given in Fig. 9.23, where contours of constant total cross section are shown, normalized to the geometrical cross section of πr^2 (a quantity termed σ^0 in Chapter 8, and which is also called the *efficiency fac-*

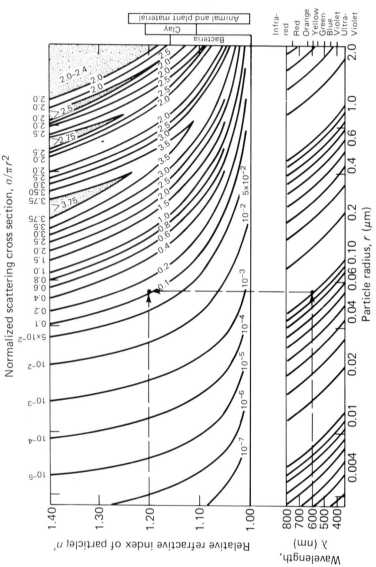

Fig. 9.23 Contours of constant normalized scattering cross section for micrometer-sized particles of relative index of refraction, n'. [Adapted from Burt, W. V., *J. Mar. Res.* (1956).]

tor). Contours of constant cross section are plotted versus particle radius and relative index of refraction. To use the diagram, enter the lower graph with the wavelength of light and the particle radius, and next project on the upper graph to the relative refractive index; then interpolate between the upper curves to obtain the scattering efficiency, or cross section. A cross-plot of Fig. 9.23 is shown in Fig. 9.24, for λ = 550 nm and n' = 1.17 (or n_p = 1.57); the Rayleigh, resonance, and geometrical regimes are indicated. One sees that the total cross section tends asymptotically toward 2 for large particles; this is due to the fact that there are two sources of the scattered field, one the directly transmitted component and the other the diffracted component, each equal in magnitude at θ = 0. A comparison with the acoustic scattering case of Fig. 7.28 shows many similarities and a few interesting differences from the optical case.

Measurements of particulate scattering without obfuscation by the pure water contribution are difficult if not impossible to carry out, and are usually made by taking the difference between observations of the total and pure-water scatterances. Figure 9.25 illustrates the volume scattering function for the forward direction for several different water masses, and suggests that the particulate scattering varies by nearly two orders of magnitude from turbid coastal waters to the clearest ocean waters such as the Sargasso Sea. It appears that scattering in the forward hemisphere is set mainly by the parti-

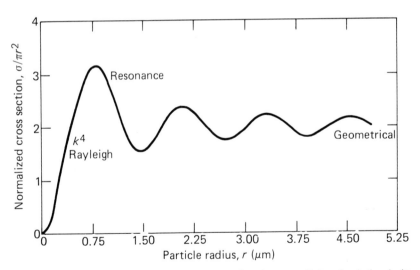

Fig. 9.24 Normalized scattering cross section for a particle of relative index, n' = 1.17, illuminated by light of λ = 550 nm. Rayleigh, resonance, and geometrical regimes are indicated. Compare with Fig. 7.28 for the acoustic case. [Adapted from Kirk, J. T. O., *Light and Photosynthesis in Aquatic Ecosystems* (1983).]

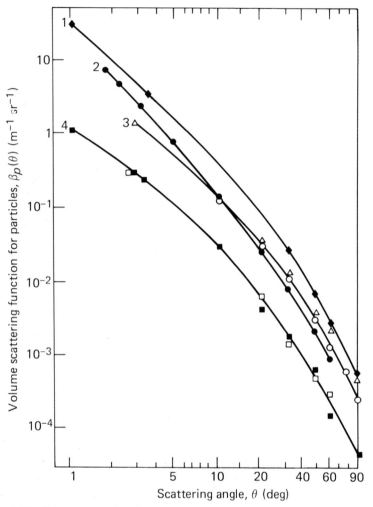

Fig. 9.25 Volume scattering function for particulates, as obtained by subtracting theoretical scattering from measured values. (1) Baltic surface water; (2) Mediterranean Sea; (3) Atlantic and Gibraltar near-surface water; (4) Subsurface Sargasso and Western Mediterranean water. [Adapted from Jerlov, N. G., *Marine Optics* (1976). Data originally due to Kullenberg, Bauer, Morel, Jerlov, and Olsen.]

cle size, somewhat less by their index of refraction, and very little by their shape; the latter accounts for the efficacy of Mie scattering theory in so many applications.

The data in Fig. 9.26 represent measurements of the total volume scattering function for several different water types, with the upper curves being

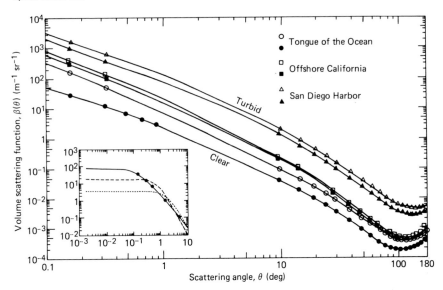

Fig. 9.26 Volume scattering function for various water types, with an emphasis on small-angle scattering; inset suggests that the volume scattering approaches a constant as $\theta \rightarrow 0$. [Adapted from Petzold, T. J., *SIO Report 72-78* (1972). Insert adapted from Blizard, M. A., "Ocean Optics: Introduction and Overview," (1986); originally due to G. W. Kattawar.]

for turbid water in San Diego Harbor, and the lower ones for quite clear water in the Bahamas. Forward scattering strongly dominates the total scattering coefficient, but the insert suggests that $\beta(\theta)$ approaches a constant as θ approaches 0, with a beam pattern that is strongly peaked in that direction, as Fig. 7.29f illustrates for the acoustic case. The scatter at 90° for the clearest water is quite close to that listed in Table 9.5, which implies that *backscatter* levels are set by fluctuational scattering. Even in rather turbid waters, thermal fluctuations are responsible for 10 to 30% of the scattering at 180°, and thus contribute significantly to the *upwelling* light. However, an integration of these data over the entire solid angle yields the result that only 1% or so of the *total* beam scattering is due to fluctuations, with the vast majority coming from particulates or other suspended material in the water.

Phytoplankton Pigment Absorption

The constituents of seawater causing the greatest variability of optical properties in both space and time are *phytoplankton*, which are unicellular plants found throughout the oceans in the euphotic layer. Equipped with a variety of light-absorbing pigments, they interact with radiant energy across the wavelength range from approximately 350 to 700 nm, and convert that energy to organic carbohydrates via the process of *photosynthesis*. Species on higher trophic (nutrient) levels of the oceanic food chain, e.g., zooplankton, bacteria, and ichthyoplankton (larval fish) graze on the phytoplankton, thereby passing on stored solar energy to larger animals.

Light, nutrients, and carbon dioxide are the essential elements required for phytoplankton growth. The *euphotic zone* is the region of the upper ocean illuminated by sunlight, and is conventionally defined as reaching to the depth, $z_{0.01}$, at which depth the photon flux is 1% of the value immediately beneath the surface. Elevated levels of nutrients are usually located in deeper waters or near land, and are brought to the surface through upwelling or vertical mixing processes. Thus the most highly productive areas are those well-fertilized regions where vertical motions of the water can bring nutrients up into the euphotic zone from greater depths. These are generally found on the continental shelves, shoal banks, and in areas of strong upwelling, with the latter occurring most frequently along the west coasts of continents and in the equatorial regions, because of the character of Ekman pumping there (see Section 6.2).

During the winter, the deep vertical mixing by winds places much of the phytoplankton population below the euphotic zone and thus inhibits cell growth. Toward the end of winter, the march of the seasons, the associated reduction in average wind stress, and the increases in sunlight and upper-ocean temperatures re-establish the stratified upper levels. These factors initiate a spring plankton bloom that leads to rapid increases in plant and animal life in regions of ample nutrition. This greening of the ocean may proceed almost exponentially in its early stages, but approaches saturation in its mature phases for a number of reasons. First, the rapid growth depletes the local supply of nutrients unless further mixing occurs. Secondly, at higher concentrations, the strong absorption by a thick layer of phytoplankton leads to a partial extinction of the light at greater depths and to a reduction of the growth rates. Thirdly, during the warm, calm summer months, the stratified upper layer becomes isolated from the nutrient-rich waters below the thermocline. Also, higher levels of sunlight inhibit the growth of many species. These factors result in a decrease in the phytoplankton concentrations near the surface and commonly, the development of a higher subsurface population at deeper levels, which is thought to occur where the balance of nutrients and insolation is the best obtainable. Summer storms can sometimes reintroduce nutrients from beneath the thermocline and stimulate growth for a few days. In the autumn, episodic blooms occur as the seasonal stratification breaks down and nutrients again come within reach. Thus the concentrations of phytoplankton are variable on time scales ranging from a very few days to many months.

The vertically integrated *carbon fixation rate*, with units of kilograms of organic carbon per square meter of water column per year, is an important measure of the annual primary production rate of phytoplankton in the ocean. Typical values are 0.15 to 0.25 kg m^{-2} yr^{-1} on the continental shelves, 0.32 in the Antarctic, and as great as 1.50 in the Peruvian upwelling. In the open ocean, primary productivity is more nearly 0.05 to 0.10 kg m^{-2} yr^{-1}, except in the equatorial upwelling, where indications are that it is somewhat closer to continental shelf values. Thus a great deal of the ocean is relatively low in biomass production, with the result that most fisheries are found close to shore; the major exceptions are the deep sea *pelagic* (far from land) species such as tuna, albacore, and whales.

Chlorophyll-a is the major pigment present in all photosynthetic plankton and thus can be used as a convenient measure of the phytoplankton biomass. In addition to it, the plants contain a number of other so-called *accessory pigments* that also absorb light, but pass the energy on to the chlorophyll-a. The absorption characteristics of these pigments are highly specialized, each capturing light in a distinct spectral region, as illustrated in Fig. 9.27. However, the absorption of the chlorophyll by itself shows only two main peaks, as illustrated in Fig. 9.28. This graph gives the *specific absorption spectrum* (spectral absorption coefficient per unit concentration) for in vivo oceanic phytoplankton, and shows the two main, relatively broad chlorophyll absorption peaks near 440 and 670 nm. However, the true absorption coefficient per molecule is actually much more peaked than in vivo spectra; examples of absorption by chlorophyll-a and -b in dilute solution in vitro are also shown in Fig. 9.28. The broadening is due to the tendency for the living plant cellular material containing the absorbing pigments to coalesce into discrete units, which in turn lessens the effectiveness with which they collect photons and thereby reduces the absorption. The complement of accessory pigments depends on the species assemblage, nutrients, and light, and this variability ultimately limits the precision with which one can predict the absorption, even when the concentration of chlorophyll-a is known.

Photosynthesis fixes carbon into organic carbohydrates and releases molecular oxygen according to the general equation

$$CO_2 + 2\,H_2O \xrightarrow{8\hbar\omega} (CH_2O) + H_2O + O_2 , \qquad (9.45)$$

which requires a minimum of 8 quanta of radiation energetically capable of the molecular conversion of carbon dioxide to the simple carbohydrate CH_2O. Since more complex carbohydrates and other organic compounds

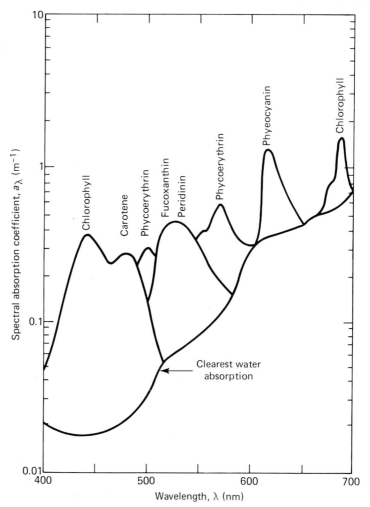

Fig. 9.27 Absorption spectra of photosynthetic algal pigments superimposed on that for clear water. The main chlorophyll peaks are near 440 and 680 nm. [Adapted from Yentsch, C. M., and C. S. Yentsch, *Oceanogr. Mar. Biol. Ann. Rev:* (1984).]

comprise the phytoplankton biomass, a more practical requirement is 10 to 12 quanta. The inverse process (respiration) consumes carbohydrates, other nutrients, and oxygen, and releases heat and CO_2.

Models developed to describe the attenuation of light by phytoplankton generally represent attenuation coefficients as linearly dependent on chlorophyll concentration, C. For values of that quantity below perhaps 1 mg

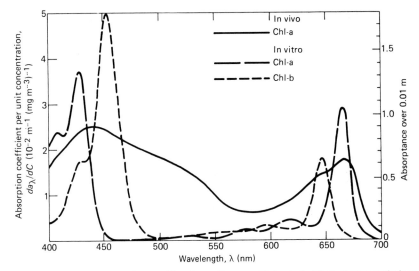

Fig. 9.28 Monochromatic specific absorption coefficient of chlorophyll-a and phae-ophytin-a for in vivo oceanic phytoplankton (solid line and left-hand scale) and for a dilute diethyl ether solution of chlorophyll-a and -b in vitro at a concentration of 10 μg ml^{-1} over a 0.01 m path (dashed lines and right-hand scale). [Adapted from Morel, A., and L. Prieur, *Limnol. Oceanogr.* (1977), and from Kirk, J. T. O., *Light and Photosynthesis in Aquatic Ecosystems* (1983).]

m^{-3}, one may respresent the spectral diffuse attenuation coefficient for irradiance, $K_{T\lambda}$, as

$$K_{T\lambda} = K_{w\lambda} + [dK/dC]_\lambda \, C + K_{x\lambda} . \tag{9.46}$$

Here $K_{w\lambda}$ is the spectral diffuse attenuation coefficient of pure seawater (see Fig. 9.19), $[dK/dC]_\lambda$ is the chlorophyll specific diffuse attenuation coefficient, and $K_{x\lambda}$ is a component attributed to substances other than phytoplankton. Each of these terms can be determined empirically by fitting multiple linear regression models to measurements of $K_{T\lambda}$. Observations of the total coefficient are shown in Fig. 9.29 as solid lines, and calculations using data in a model of the form of Eq. 9.46 are shown as dashed lines; the agreement is considered good. Like the absorption spectra of Fig. 9.28, the specific attenuation coefficient has strong absorption lines at blue wavelengths, with a secondary line in the red. Its effect on the total diffuse attenuation of Fig. 9.29 is to shift the wavelengths of minimum absorption from the blue (near 460 nm for clear water) to approximately 570 nm (in the green) for chlorophyll concentrations near 10 mg m^{-3}. Thus, the spectral absorption characteristics of the chlorophyll partially determine the col-

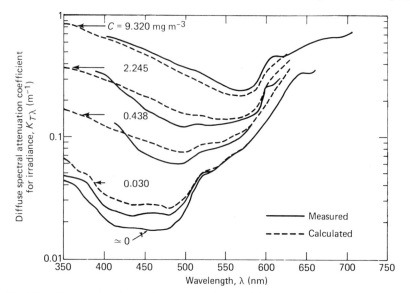

Fig. 9.29 Measured and calculated values for the diffuse spectral attenuation coefficient, for several ocean waters varying in chlorophyll-like pigment concentrations. The clear seawater values are limiting at $C = 0$. [Adapted from Smith, R. C., and K. S. Baker, SIO Ref. 77-4 (1977).]

or of water at depth (as well as at the surface), and its vertical distribution often controls the absolute intensity.

In studies of photosynthesis, a quantity termed *photosynthetically available radiation*, or PAR, is used as a measure of total available light; because of the quantized interaction between photons and the molecular species of interest, PAR is defined in terms of a flux of quanta rather than energy. It is generally considered as the total photon flux lying between 400 and 700 nm (although on occasion the lower limit is taken as 350 nm), and constitutes about 38% of the extraterrestrial solar irradiance. The units of PAR are einsteins per square meter per second (E m^{-2} s^{-1}), where one einstein is 6.022×10^{23} quanta, or one mole of photons. It is related to the spectrally integrated irradiance via the energy per quantum, $\mathcal{E} = \hbar\omega = hc/\lambda$. Letting N_0 be Avogadro's number, PAR(z) is

$$\text{PAR}(z) = \frac{1}{N_0} \int_{400}^{700} E_{\lambda 0}(z) \ (\lambda/hc) \ d\lambda , \qquad (9.47)$$

where Planck's constant, h, and the speed of light, c, are defined in Chapter 8. An appreciation of the numerical value of such radiation may be gained from the observation that underwater, approximately 2.5×10^{18} quanta per second per square meter in this wavelength range are equivalent to 1 W m^{-2} (to within 10%). To convert one mole of CO_2 to its carbohydrate equivalent through photosynthesis requires not less than 8 molar equivalents of light, or 8 E; as mentioned above, the true minimum requirement is more nearly 10 to 12 E.

An instrument called a *quantum meter* is used to measure underwater PAR directly, and from it, one may estimate another diffuse attenuation coefficient, this one for the total photosynthetically available light, termed K_{par}. Because we want to consider the euphotic zone in terms of photosynthesis rather than simply the total light intensity, we define the euphotic depth, $z_{0.01}$, as the level at which the PAR has fallen to 1% of its value immediately beneath the surface. This depth is at 4.61 attenuation lengths beneath the surface, since $\exp(-4.61) = 0.01$:

$$K_{par} = 4.61/z_{0.01} \ . \tag{9.48}$$

As we shall see later, approximately 90% of the upwelled energy received by remote sensing instruments comes from within 1 attenuation depth of the surface; the effective depth to which this sensor sees is then

$$z_k = 1/K_{par} \ , \tag{9.49}$$

which is about 20% of the depth of the euphotic zone. Thus values for pigment deduced from remote sensing devices represent the near-surface concentrations only; these may be even in greater error if the phenomenon of subsurface maxima is present.

There is often a high degree of correlation between the surface value of chlorophyll, $C(0)$, and the total concentration, C_T, as averaged over the euphotic zone. Then the average total concentration over the euphotic zone is

$$C_T = \frac{K_T}{4.61} \int_{-4.61/K_T}^{0} C(z) \ dz \ . \tag{9.50}$$

More recent work, however, has raised some doubt as to the general validity of the assumption of a relationship between the surface and integrated concentrations. There usually exists a subsurface *chlorophyll maximum* at depths that depend on latitude and time of year. Figure 9.30 shows the locus and width of such a late-summer maximum in the Pacific, and demonstrates that the layer is near 120 m depth in the lower latitudes of the subtropical gyres, but comes close to the surface as the sub-Arctic waters of the Gulf of Alaska are approached. The maximum is thought to be due to the sequence of mixing and stratification processes discussed earlier. Figure 9.31 shows subsur-

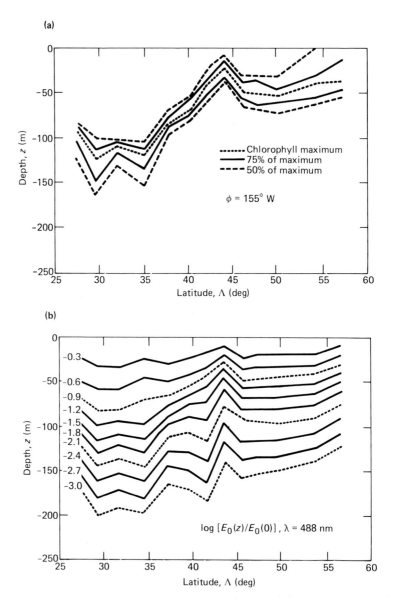

Fig. 9.30 (a) Depth of subsurface chlorophyll maximum along a summertime transect in the North Pacific. (b) Contours of constant scalar irradiance at $\lambda = 488$ nm, in logarithmic units, normalized to the near-surface values. [Adapted from Pak, H., "Cruise Data Report, R/V *Discoverer*, 28 June–23 July 1985," (1985).]

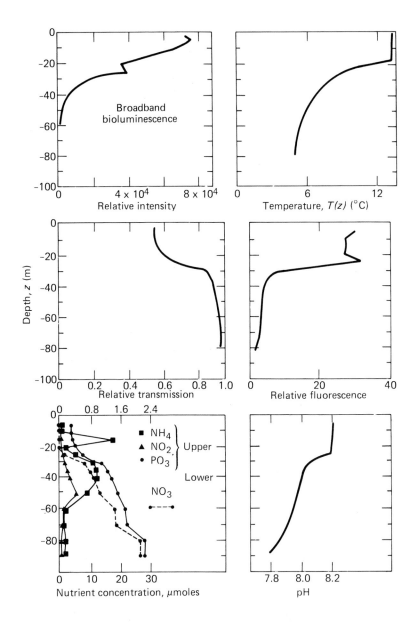

Fig. 9.31 Vertical profiles of optically relevant quantities for the summertime Gulf of Alaska. Maxima occur in several quantities at the bottom of the mixed layer. [Adapted from Losee, J., et al., *Ocean Optics VII* (1984).]

face profiles for several quantities of optical relevance in the summertime Gulf of Alaska, and illustrates the shallow maxima in nutrients, fluorescence, and bioluminescence at the bottom of the mixed layer.

Absorption by Gelbstoffe

Dissolved organic compounds collectively known as *yellow substance* (German: gelbstoffe) or *gilvin*, constitute an optically important class of chemically complex material generally present in oceanic waters at low concentrations and in coastal waters at somewhat higher levels. They derive from the decay of organic matter under microbial action and are loosely classified as *humic* materials. Such substances are generally in solution in the sea and their effects are not readily visible to the eye in the coloration of oceanic waters. However, near river discharges, concentrations are often high enough to be apparent as brownish or yellowish hues imparted to those waters.

The spectral absorption coefficient for yellow substance shows no structure in the optical and near-ultraviolet region other than an exponential behavior with wavelength. If $a_{\lambda 0}$ is the beam absorption at an arbitrary wavelength, λ_0, then the coefficient at other wavelengths behaves as

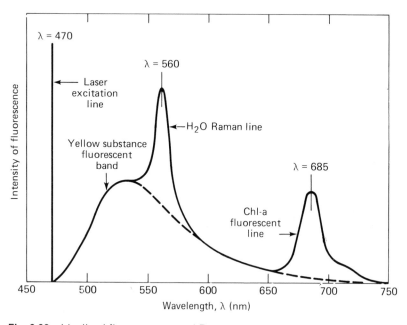

Fig. 9.32 Idealized fluorescence and Raman spectra resulting from excitation of natural waters by a laser at 470 nm. [From Bristow, M., et al., *Appl. Optics* (1981).]

$$a_\lambda = a_{\lambda 0} \exp[-0.014(\lambda - \lambda_0)] . \qquad (9.51)$$

Table 9.6 lists a few values for $a_{\lambda 0}$ at $\lambda_0 = 440$ nm, for oceanic and coastal waters. From the values cited and from Fig. 9.32, for example, it may be seen that the attenuation due to yellow substance (there is no apparent scattering from this material) is generally smaller than that from plankton in the open ocean, but that in coastal waters, it may contribute significantly to the overall diffuse attenuation in the blue region. At longer wavelengths, its effects are negligible compared with those of clear seawater (see Fig. 9.19).

TABLE 9.6
Absorption Coefficients at $\lambda_0 = 440$ nm
from Yellow Substance

Location	$a_{\lambda 0}$ (m^{-1})
Sargasso Sea	0
Off Bermuda	0.01
North Atlantic	0.02
Galapagos Islands	0.02
Mauritanian Upwelling	0.034–0.075
Off Peru	0.05
Villefranche Bay	0.060–0.161
Marseilles drainage	0.073–0.646
Baltic Sea	0.24
Gulf of Bothnia	0.41

[Adapted from Kirk, J. T. O., *Light and Photosynthesis in Aquatic Ecosystems* (1983).]

9.7 Fluorescence and Bioluminescence

Marine organisms (and to some extent, their decay products) exhibit two important optical properties, fluorescence and bioluminescence, both of which are sources of light in the sea.

Fluorescence

Both in vivo phytoplankton and dissolved yellow substances contribute to *fluorescence*, i.e., the re-emission of light energy at a lower frequency by an absorber illuminated with optical energy. The fluorescent response is immediate, occurring within nanoseconds of the excitation; the signal is typically very weak, being of order 1 to 3% of the incident intensity, and displaced from the exciting wavelength toward longer wavelengths by tens to hundreds of nanometers.

Fluorescent substances found in seawater have characteristic emission spectra that serve to identify them. Figure 9.32 illustrates an idealized fluorescence emission spectrum resulting from excitation by a laser source with $\lambda = 470$ nm; the three components are (1) a broad band of emission from dissolved yellow substances, (2) Raman emission (here centered at 560 nm) resulting from vibrational/stretching modes of the H_2O molecule (not part of the fluorescence process), and (3) more narrowband fluorescence at 685 nm due to living chlorophyll-a. Figure 9.33 illustrates emission spectra in distilled water and in various natural waters; the peak for pure water is the Raman line, which must be subtracted from the other measurements to obtain the contributions from gelbstoffe and plankton. Figure 9.34 shows a depth profile of fluorescence that exhibits a more or less monotonic increase with depth, presumably from detritus in the deeper regions.

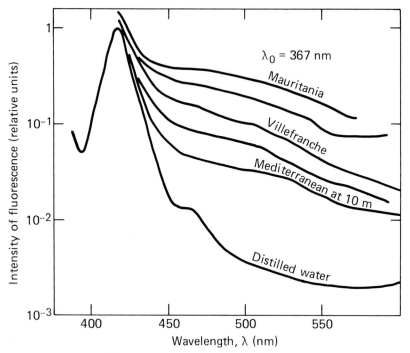

Fig. 9.33 Fluorescent emission spectra for distilled water and various natural waters. The exciting line is at $\lambda = 376$ nm, and the peak near 420 nm is due to Raman emission from vibrational states of the water molecule. [From Jerlov, N. G., *Marine Optics* (1976). Originally due to A. Morel (1972).]

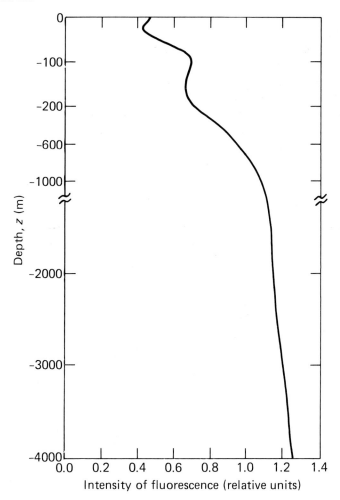

Fig. 9.34 Subsurface distribution of fluorescence in the Northeastern Atlantic. [From Ivanoff, A., in *Optics of the Sea: AGARD Lect. Ser. 61* (1973).]

The emission peak for chlorophyll-a is centered at about 685 nm, as Fig. 9.32 shows, while that for another major pigment, phycoerythrin, falls at approximately 580 nm. Because of the nature of such stimulated molecular emission, the locations of the peaks are independent of the wavelength of the excitation. However, the intensity of the peak is strongly dependent on the wavelength of the incident light, which must fall where absorption occurs. It is possible to map the absorption spectrum of phytoplankton by varying the excitation wavelength and measuring the fluorescent intensity at a fixed wavelength (e.g., 685 nm).

It is also possible to use the intensity of the characteristic line to determine the concentration of fluorescent pigments, since, for a fixed choice of excitation and observation, the intensity is generally found to be linearly dependent on the concentration. This is the basis for rapid fluorometer measurements of chlorophyll concentration routinely used at sea; such measurements, however, must be periodically calibrated against concentrations determined by more direct methods.

Fluorescence is also the basis for airborne remote measurements of the near-surface concentrations of chlorophyll-a and phycoerytherin. A pulsed laser system termed a *lidar* (for light detection and ranging) uses an intense laser source for excitation, and records the entire emission spectrum received from the upper levels of the ocean. The fluorescence peaks are then normalized by the Raman peak to correct for variations in the depth of penetration of the laser beam and the path length of the fluorescent emission, under the assumption that the intensity of the Raman line is proportional to the penetration depth.

The use of fluorescent dyes such as fluorescein and rhodamine-B is common in optical experiments at sea, with processes such as advection and diffusion being studied over limited time and space scales using such techniques.

Bioluminescence

Bioluminescence is the emission of light by living marine organisms, and is a characteristic of a wide variety of species, particularly when agitated. The wake of a ship at night often contains a myriad of small, phosphorescent light sources. Certain animals apparently use light as an attractor or as a guide, perhaps for mating or for other reasons known only to themselves. Such creatures are, at various times and places, found at all levels of the water column, including in benthic communities in very deep water. In the aggregate, their visual effect as seen from submersibles has been described as "marine snow," having something of the appearance of an underwater snowstorm.

The distribution of bioluminescent sources in the sea is not random but is correlated with regions of enhanced biomass production; fronts, upwelling areas, and cold, productive polar waters appear to be especially rich areas. Within them, zooplankton species such as dinoflagellates, cocepods, and euphausids constitute the dominant animals. Their light emission generally occurs in discreet flashes having durations of order 10 ms. At deeper levels, say below 200 m, larger species become more important but less frequent.

The spectral distribution of the emitted light is generally broadband and centered in the blue-green, as Fig. 9.35 illustrates. It is possible that higher spectral resolution would show narrowband structure superimposed on the broad spectrum of Fig. 9.35. Intensity levels of the aggregated organisms, when viewed over an extended area, are considerable.

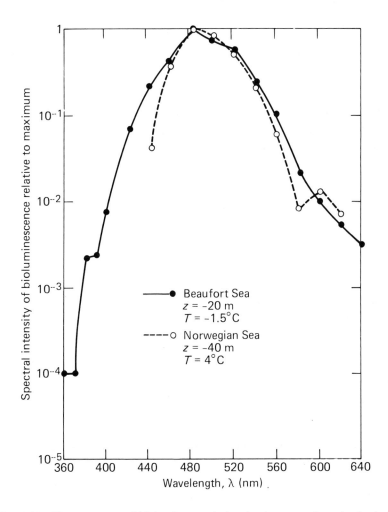

Fig. 9.35 The spectrum of bioluminescent signals at near-surface depths in two subpolar water masses. The peak is in the blue-green. [Adapted from Losee, J., et al., in *Ocean Optics VII* (1984).]

9.8   Radiative Transfer Beneath the Sea

We are now ready to assemble the information from the preceding sections into a discussion of propagation of the optical light field beneath the sea surface. In addition to the source intensity, the radiance level is governed by the processes of absorption, scattering into and out of the beam and, to a small extent, emission by bioluminescent and fluorescent sources. The relationship that governs the radiance level is termed the *radiative transfer equation*, and we shall derive it below after a discussion of some examples of observational data on subsurface irradiances.

Subsurface Irradiances

The surface solar spectral irradiance shown in Fig. 9.5 is attenuated with increasing depth much as shown in Fig. 9.36. The infrared and ultraviolet wavelengths are most rapidly eroded, so that when $z \leq -1$ m, essentially only visible radiation remains. By 100 m, blue-green wavelengths dominate, with the least rapidly attenuated spectral components being those with a blue wavelength of approximately 462 nm in clear oceanic waters (see Section 9.9). At even greater depths, this "monochrometer effect" (narrowband filtering of the incident light) is even more pronounced, so that by $z = -400$ m, the irradiance has a bandwidth of only 10 or 20 nm, and at this asymptotic wave-

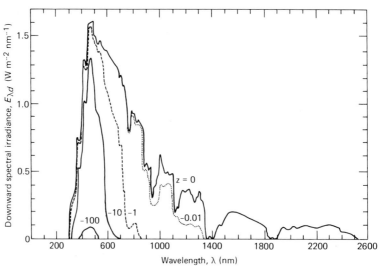

Fig. 9.36   The variation of downwelling solar spectral irradiance for clear ocean waters. The tendency for clearest seawater to filter light at depth toward an asymptotic wavelength of $\lambda = 462$ nm is apparent. [From Jerlov, N. G., *Marine Optics* (1976).]

length, the irradiance is 10^{-6} that at the surface (see Fig. 9.37). The bottom of the euphotic zone in this water lies slightly below -100 m.

Another method of displaying the same kind of data is shown for the Caribbean Sea in Fig. 9.38, which illustrates the attenuation of surface irradiance at fixed wavelengths. The semilog plots reveal the essentially exponential character of the attenuation; however, the slopes of these curves, which are proportional to the diffuse attenuation coefficients (see Eq. 9.10), are not constant. This could be due to varying plankton distributions at shallow depths, or to the change from single to multiple scattering in a homogeneous medium. Clearly blue wavelengths near 475 nm are attenuated the least, while ultraviolet and, to a much greater extent, infrared wavelengths are rapidly reduced to low levels. In more turbid waters such as the Baltic Sea, the asymptotic wavelength is in the green near 550 nm as a result of the minimum in the absorption bands of chlorophyll there (see Figs. 9.27 and 9.28).

A more broadband plot of subsurface downwelling irradiance is shown in Fig. 9.39. On the left are data for the Caribbean, which is clear (but not as clear as the Sargasso); on the right, Sargasso Sea water is compared with the West African upwelling off Senegal, a biologically active region. The shift

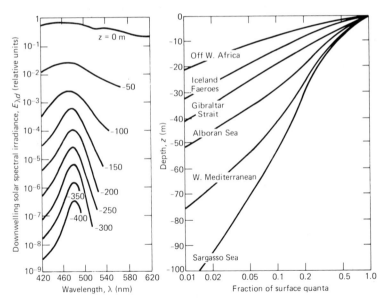

Fig. 9.37 The decrease of solar spectral irradiance with depth over a 200 nm spectral width, plotted on a logarithmic scale. At $z = -400$ m, the "monochrometer" or filtering effect of clear seawater has narrowed the radiation to a bandwidth of 10 to 15 nm centered near 462 nm. [Adapted from Kampa, E. M., Union Geod. Geophys. Int. Monogr. 10 (1961).]

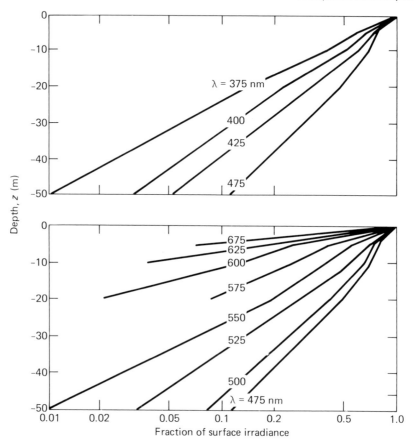

Fig. 9.38 The attenuation of surface irradiance with depth, at varying wavelengths, for high solar elevations. The nonlinear behavior is evidence of the layering of absorbers, or perhaps a change from a single- to a multiple-scattering medium. Data are for the Caribbean Sea. [Adapted from Jerlov, N. G., *Marine Optics* (1976).]

in spectral maxima and the increased attenuation between the clear and chlorophyll-laden waters is striking.

For purposes of determining the net energy absorbed in the water, the up-welling irradiance must be measured as well. Figure 9.40 shows downward and upward spectral irradiances for two water types falling at opposite ends of the turbidity range for the ocean, those being the Sargasso Sea and the Baltic Sea. According to Eq. 9.8, the vertical derivative of the net irradiance gives the absorbed power density, so that a differencing and a differentiation of the data of Fig. 9.40 will yield the spectral components of the absorbed power per unit volume as a function of depth, and the integral over

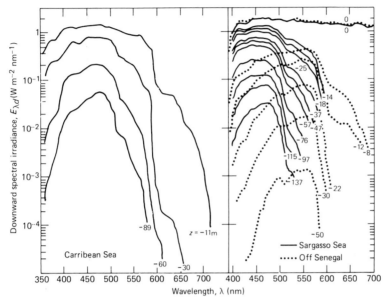

Fig. 9.39 Downward spectral irradiances: (a) The Caribbean Sea. [Adapted from Smith, R. C., *Data Report "Discovery" Expedition, 3 May–5 June, 1970* (1973).] (b) The Sargasso Sea and the West African upwelling. [Adapted from Morel, A., and L. Caloumenos, *Upwelling Ecosystem Analysis Conf. Proc., Marseille, 1973* (1974).]

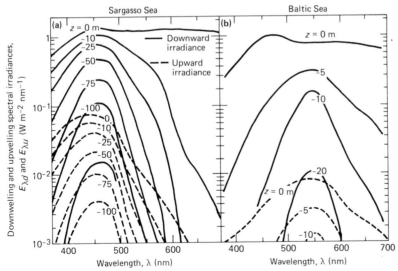

Fig. 9.40 Downward and upward spectral irradiances for (a) the Sargasso Sea [Adapted from Lundgren, B., and N. K. Hojerslev, Univ. Copenhagen, Inst. Phys. Oceanogr. Rep. 14 (1971)] and (b) the Baltic Sea [Adapted from Ahlquist, C. D., unpublished (1965)].

all wavelengths will give the total power deposited in a unit volume of water. Such elaborate data are not required for simple observations of total absorbed power, but the more detailed descriptions shown also give information on the characteristics of the absorption.

The Radiative Transfer Equation

The theory describing the behavior of the data of the preceding section is complex and we shall not attempt its general development; we shall instead confine ourselves to consideration of the steady state variations of radiance and irradiance in the vertical coordinate. These variations are influenced by absorption and scattering into and out of the beam.

Consider now a plane, parallel ocean that is horizontally homogeneous but which has varying properties in the vertical, and which is illuminated from above. In Fig. 9.41, the monochromatic radiance at level z' contained in a solid angle $d\Omega$ centered at (θ,ϕ) is termed $L_\lambda(z',\theta,\phi)$; this radiance is due to a source, L_0, located at $z = z_0$. The processes that have worked to alter the radiance from its source intensity are described by: (1) the monochromatic beam absorption coefficient, $a_\lambda(z')$; (2) the beam-scattering coefficient, $b_\lambda(z')$; and (3) the integral of the volume scattering function, β_λ, where

$$\beta_\lambda = \beta_\lambda(z', \theta, \phi, \theta', \phi') , \qquad (9.52)$$

which describes scattering *into* the angles θ and ϕ from all other angles, θ' and ϕ'. The first two of these processes are sinks and the last is a source for radiance, and are considered to be operative at some intermediate level, z', when the observation or field point is z.

In what follows, we shall temporarily suppress the wavelength and angular dependencies of the field quantities and beam coefficients, and indicate those arguments implicitly by primes where no ambiguity results. After the derivations are complete, the suppressed arguments will be restored.

We next assume a steady flux, and consider the total change in the monochromatic radiance at level z' after having traversed a slant path, ds', passing through a small volume, $d^3\mathbf{x}'$, where the path length is

$$ds' = \sec\theta \, dz' \qquad (9.53)$$

and the volume element is

$$d^3\mathbf{x}' = dx' \, dy' \, dz' . \qquad (9.54)$$

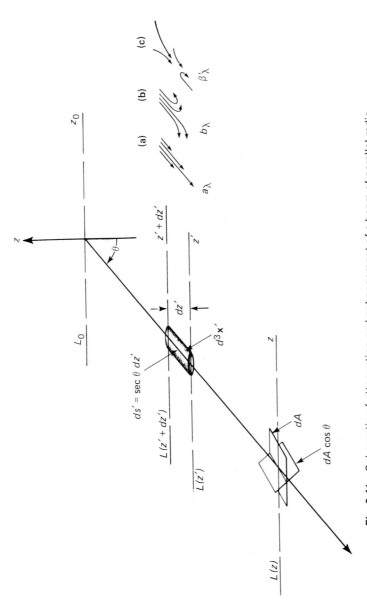

Fig. 9.41 Schematic of attenuation and enhancement of a beam of parallel radiation traversing a vertically inhomogeneous ocean. The processes of (a) absorption, (b) scattering out of, and (c) scattering into the beam are illustrated on the right.

The change in radiance may be written as

$$dL(z') = -aL(z')\,ds' - bL(z')\,ds' + [dL^*/ds]_{in}\,ds' \, . \quad (9.55)$$

On the right-hand side, the first term represents the decrease in radiance in traversing ds' due to absorption, while the second is the decrement arising from scattering out of the beam. The third, as given by the quantity $[dL^*/ds]_{in}$, is the change in radiance per unit slant path due to scattering into the beam from all other angles in the sphere surrounding the scattering element. From the definition of the volume-scattering function (see Eqs. 9.15 and 9.17), the gain in radiance from scattering is

$$[dL_\lambda^*(z',\theta,\phi)/ds]_{in} = \int_0^{2\pi} \int_0^\pi \beta_\lambda(z',\theta,\phi,\theta'',\phi'')\,L_\lambda(z',\theta'',\phi'')$$

$$\times \sin\theta''\,d\theta''\,d\phi''$$

$$\equiv \int_0^{4\pi} \beta_\lambda L_\lambda\,d\Omega'' \, . \quad (9.56)$$

With the definitions and assumptions above, the rate of change of radiance may be written in our truncated notation as

$$\frac{dL(z')}{ds'} = -(a+b)\,L(z') + \int_0^{4\pi} \beta L(z'')\,d\Omega'' \, . \quad (9.57)$$

This is an integro-differential equation for the spectral radiance that may be solved by standard (or if necessary, numerical) techniques, provided that the coefficients are known along the path from source point to field point.

To gain an appreciation for the physics embodied in Eq. 9.57, we will consider a few special cases. For the first case, assume a nonscattering medium, in which event, $b = \beta = 0$, leaving absorption as the only active process. The radiative transfer equation then reduces to

$$\frac{dL(z')}{ds'} = -a(z')\,L(z') \, , \quad (9.58)$$

with the solution termed *Beer's* (or *Bouguer's*) *Law*:

$$L(z,\theta) = L_0(z_0,\theta) \exp\left[-\int_z^{z_0} a(z') \sec\theta\, dz'\right]. \tag{9.59}$$

Here the source radiance at z_0 is $L_0(z_0,\theta)$, and the integration is over the entire slant range separating source and field. Thus Eq. 9.59 describes exponential absorption of radiation while traversing the path. The integral in Eq. 9.59 is termed the *optical thickness*, $\tau(z_0,z,\theta)$:

$$\tau(z_0,z,\theta) = \int_z^{z_0} a(z') \sec\theta\, dz'. \tag{9.60}$$

Without the $\sec\theta$ term, this quantity is called the *optical depth*. The optical thickness measures the number of exponentiation lengths that exist between source and receiver, while the optical depth gives the same number, but assumes a vertical separation only. Using the optical thickness, Beer's Law in simplified notation is written

$$L = L_0\, e^{-\tau}. \tag{9.61}$$

With this assistance as to its interpretation, the more general statement of Eq. 9.57 may be seen to describe the effects of scattering as well as absorption.

A different manipulation of Eq. 9.57 yields a result obtained earlier for the radiative flux divergence, $\nabla\cdot\mathbf{E}_{rad}$. We integrate the equation over a sphere surrounding an observation point at z' to obtain the *net irradiance* (as defined by Eq. 9.7) flowing from above and below. Thus

$$\int_0^{4\pi} \frac{dL_\lambda(z')}{ds'}\, d\Omega = \int_0^{4\pi} \frac{dL_\lambda(z')}{dz'} \cos\theta\, d\Omega$$

$$= \frac{dE_{\lambda_{net}}}{dz'}$$

$$= -\int_0^{4\pi} c_\lambda L_\lambda\, d\Omega + \int_0^{4\pi}\int_0^{4\pi} \beta_\lambda L_\lambda\, d\Omega\, d\Omega''$$

$$= -c_\lambda E_{\lambda 0} + b_\lambda E_{\lambda 0} = -a_\lambda E_{\lambda 0} . \tag{9.62}$$

Here $E_{\lambda 0}(z')$ is the monochromatic scalar irradiance of Eq. 9.3, and $b_\lambda(z')$ is the beam-scattering coefficient, as defined by Eq. 9.16. This result is called *Gershun's Law*, and relates the divergence of the *net* spectral irradiance to the absorption of the omnidirectional *scalar* spectral irradiance. Thus

$$\frac{dE_{\lambda net}}{dz'} = - a_\lambda (z') E_{\lambda 0}(z') . \tag{9.63}$$

A further integration over wavelength gives the desired item, the divergence of the radiative flux, or Poynting vector (see Eq. 8.177):

$$\nabla \cdot \langle S \rangle \equiv \frac{dE_{net}}{dz'} = - \int_0^\infty a_\lambda (z') E_{\lambda 0}(z') \, d\lambda , \tag{9.64}$$

which is the quantity describing solar heating of the upper water column (neglecting chemical processes), and which can in principle be obtained, for example, from a spectral integration of the data of Fig. 9.40.

A further item that can be derived from the radiative transfer equation is the relationship between diffuse attenuation and beam absorption coefficients, a topic heretofore neglected. From Eq. 9.63 and the definition of the diffuse attenuation coefficient for net irradiance (Eq. 9.10),

$$a_\lambda = - \frac{1}{E_{\lambda 0}} \frac{dE_{\lambda net}}{dz} \tag{9.65}$$

and

$$K_{\lambda net} = - \frac{1}{E_{\lambda net}} \frac{dE_{\lambda net}}{dz} , \tag{9.66}$$

so that

$$a_\lambda = K_{\lambda net} \frac{\int L_\lambda \cos \theta \, d\Omega}{\int L_\lambda \, d\Omega} = K_{\lambda net} \langle \cos \theta \rangle . \tag{9.67}$$

Here the quantity $\langle \cos \theta \rangle$ is the ratio of the spherical average of $\cos \theta$ weighted by the radiance of Eq. 9.7. Thus the beam absorption coefficient is somewhat less than the diffuse attenuation coefficient for net irradiance, since $\langle \cos \theta \rangle \leq 1$. The quantity $\langle \cos \theta \rangle$ is termed the *average cosine*. While the beam attenuation coefficient was introduced to circumvent the problem of the dependence of the diffuse attenuation on the source function, it may be seen that the difference between them is not great.

Solutions to the radiative transfer equation are in general complicated, but for the case of constant coefficients, a formal analytical solution is available. In an optically uniform medium, the solution is

$$L_\lambda(z,\theta,\phi) = L_{\lambda 0}(z_0,\theta,\phi) \exp[c_\lambda(z - z_0) \sec \theta] \qquad (9.68)$$

$$+ \int_z^{z_0} [dL_\lambda^*(z',\theta,\phi)/ds]_{in} \exp[c_\lambda(z - z') \sec \theta] \sec \theta \, dz' \, .$$

Here we have restored the implicit dependencies and have written the integration to be over z' rather than s'. It may be seen that the radiance at depth is the sum of the directly transmitted radiance over the path, plus a path radiance giving the flux scattered into the beam between the source and the scattering points, as attenuated during the remainder of the transit to the observation point. Remembering that z is negative beneath the sea surface, we see that the exponentially decaying behavior of the solution is in accord with the data presented above, especially in Fig. 9.38.

From the form of the path integral of Eq. 9.56 (which contains the radiance function itself), a reasonable first approximation to $[dL_\lambda^*/ds]_{in}$ is to assume that the in-scattering also falls off exponentially in z (not in s) with the scale of the diffuse attenuation coefficient, K_λ; thus we write

$$[dL_\lambda^*/ds]_{in} = [dL^*/ds]_0 \, e^{K_\lambda(z - z_0)} \, . \qquad (9.69)$$

This form allows the path integral to be evaluated explicitly, so that for the case of an ocean with constant coefficients and this form of path function, the subsurface radiance becomes

$$L_\lambda(z,\theta) = L_{\lambda 0}(z_0,\theta) \exp[c_\lambda(z - z_0) \sec \theta]$$

$$+ \frac{[dL^*/ds]_0 \, \exp[(K_\lambda(z - z_0)]}{(c_\lambda - K_\lambda \cos \theta)}$$

$$\times \{1 - \exp[(c_\lambda - K_\lambda \cos \theta)(z - z_0) \sec \theta]\} \, . \qquad (9.70)$$

Since $\cos \theta \geq 0$ in the downward direction, it is readily seen that the integral describing scattering into the beam decays exponentially with an attenuation coefficient that is less than that for the directly transmitted beam, because scattering into the beam represents a source of energy in the equation. In reality, Eq. 9.70 is not fully determined, since the function $[dL^*/ds]_{in}$ must be specified either empirically or numerically. However, computations using this formula agree well with observations made in the uniformly mixed waters of Lake Pend Oreille, Idaho.

In more realistic cases of variable attenuation and scattering coefficients, numerical integration of Eq. 9.57 or its equivalents is readily carried out. Increased complications from the inclusion of other processes such as fluorescence may be handled in the same way.

Jerlov Water Types

Jerlov has compiled and classified the optical properties of various waters according to a scheme that divides them into oceanic (Types I to III) and coastal (Types 1 to 9) origins, based on their irradiance transmissivity in the upper 10 m . This arrangement is a convenient if not entirely precise method of characterizing water clarity, and has come into fairly standard use. Figure 9.42 illustrates the spectral transmittance for these water types, and demonstrates the shift in the transmittance peak toward longer wavelengths with increasing turbidity. A slightly different representation of these data is shown in Fig. 9.43, which gives the variation in the spectral diffuse attenuation coefficient for irradiance, K_λ, for the oceanic water types. It is convenient to subdivide the clearest waters into subclasses I, IA, and IB, as the figure shows. Table 9.7 lists recent values for the diffuse attenuation coefficient for these idealized types, which may be used in regression equations to derive K_λ for any real water for which a measurement of K_{λ_0} is available at only one wavelength, λ_0.

Data useful in assessing the PAR available in these water types are shown in Fig. 9.44, which gives the attenuation of spectrally integrated downward irradiance over the wavelength range of 350 to 700 nm. The 1% level on these graphs handily approximates the bottom of the euphotic zone, which is as shallow as 19 m in the Baltic (Type 3), and 6 m in very turbid coastal waters. The global distribution of water mass types based on this classification is shown in Fig. 9.45. It is seen that many oceanic areas have yet to be classified; however, remote measurement of the diffuse attenuation coefficient

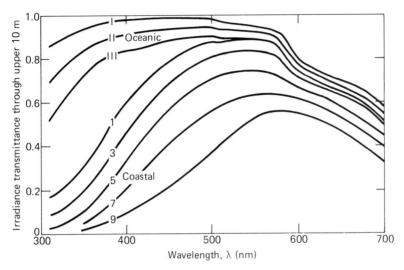

Fig. 9.42   The spectral transmittance over the upper 10 m of water for Jerlov water types I to III and 1 to 9. The migration of the peak in transmittance with increasing turbidity is apparent. [Adapted from Jerlov, N. G., *Marine Optics* (1976).]

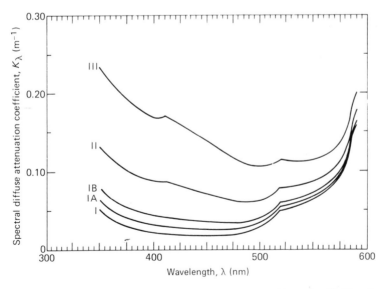

Fig. 9.43   Diffuse attenuation coefficient for Jerlov Water Types I to III. The clearest waters are subdivided into I, IA, and IB. [Adapted from Austin, R. W., and T. J. Petzold, in *Optical Engineering* (1986).]

TABLE 9.7

Spectral Downward Diffuse Attenuation Coefficient, K_λ (m^{-1}), for Jerlov Water Types I to III and 1

Water Type	Wavelength, λ (nm)														
	350	375	400	425	450	475	500	525	550	575	600	625	650	675	700
I	0.0510	0.0302	0.0217	0.0185	0.0176	0.0184	0.0280	0.0504	0.0640	0.0931	0.2408	0.3174	0.3559	0.4372	0.6513
IA	0.0632	0.0412	0.0316	0.0280	0.0257	0.0250	0.0332	0.0545	0.0674	0.0960	0.2437	0.3206	0.3601	0.4410	0.6530
IB	0.0782	0.0546	0.0438	0.0395	0.0355	0.0330	0.0396	0.0596	0.0715	0.0995	0.2471	0.3245	0.3652	0.4457	0.6550
II	0.1325	0.1031	0.0878	0.0814	0.0714	0.0620	0.0627	0.0779	0.0863	0.1122	0.2595	0.3389	0.3837	0.4626	0.6623
III	0.2335	0.1935	0.1697	0.1594	0.1381	0.1160	0.1056	0.1120	0.1139	0.1359	0.2826	0.3655	0.4181	0.4942	0.6760
1	0.3345	0.2839	0.2516	0.2374	0.2048	0.1700	0.1486	0.1461	0.1415	0.1596	0.3057	0.3922	0.4525	0.5257	0.6896

[From Austin, R. W., and T. J. Petzold, in *Optical Engineering* (1986).]

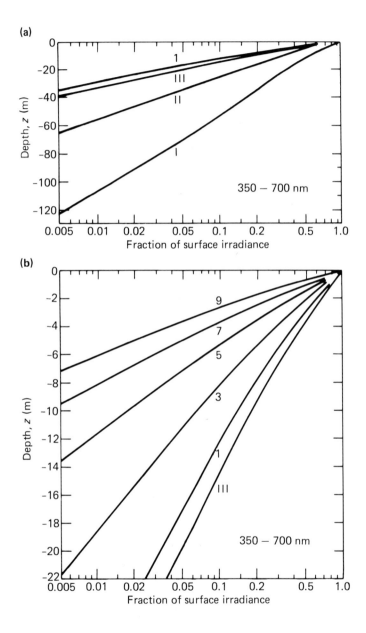

Fig. 9.44 Attenuation of surface irradiance with depth, for various Jerlov water types in (a) oceanic waters and (b) coastal waters. [From Jerlov, N. G., *Marine Optics* (1976).]

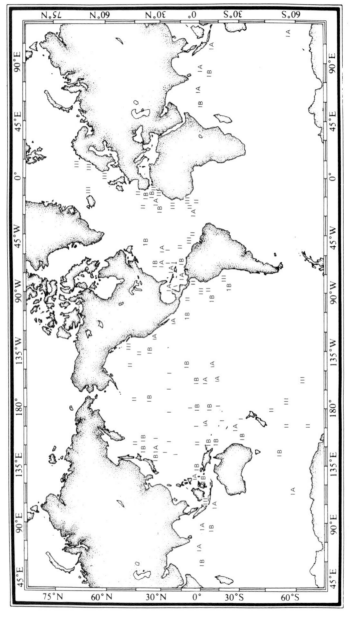

Fig. 9.45 The global distribution of Jerlov water types. The interiors of the subtropical gyres tend to be the clearest natural waters; upwelling regions off the west coasts of continents are much more turbid because of plankton production. [Adapted from Jerlov, N. G., *Marine Optics* (1976).]

and chlorophyll concentrations from satellites is correcting this deficiency rapidly (see Section 9.10).

9.9 Ocean Color; Underwater Imaging

The term "color" in the scientific sense means the visual response elicited in the human optical sensory system by a spectral radiance or irradiance, and as such, clearly involves the response of the human eye. Photometric workers have defined standard visual response curves, based on the *tristimulus color theory*, which holds that the perception of color is due to the response of three distinct sets of color receptors in the eye. These response curves (or filter functions), as defined by the *Commission Internationale de l'Eclairage*, are known as the C.I.E. tristimulus response functions, $\bar{x}_\lambda$, $\bar{y}_\lambda$, and $\bar{z}_\lambda$, and are plotted in Fig. 9.46. Any visual observation of an irradiance, E_λ, is

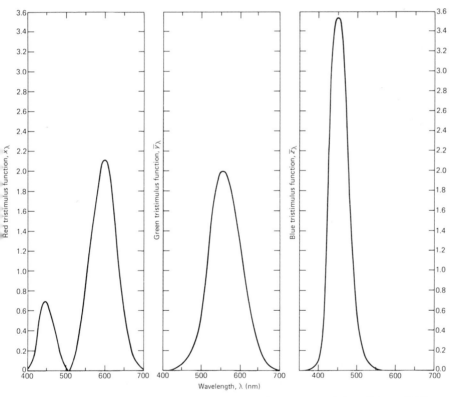

Fig. 9.46 The tristimulus response curves for human vision. [From C.I.E., *Vocabularie International de l'Eclairage* (1957).]

weighted by these functions according to the integrals of Eqs. 9.71a to 9.71c, to give the *visual chromaticities*, X, Y, and Z, which are the instrumental (eye) responses to the illumination field:

$$X = \int_{400}^{700} E_\lambda \bar{x}_\lambda \, d\lambda \, , \tag{9.71a}$$

$$Y = \int_{400}^{700} E_\lambda \bar{y}_\lambda \, d\lambda \, , \tag{9.71b}$$

and

$$Z = \int_{400}^{700} E_\lambda \bar{z}_\lambda \, d\lambda \, . \tag{9.71c}$$

The chromaticity values may then be normalized by dividing them by their sum, $X + Y + Z$, and these values then form the *chromaticity coordinates*, x, y, and z. Since by their definition, the sum of the coordinates is unity, only two of them are needed to represent a color, since the third may be calculated from

$$z = 1 - (x + y) \, . \tag{9.72}$$

Usually x and y are plotted in Cartesian coordinates; such a plot forms a *chromaticity diagram*, an example of which is shown in Fig. 9.47 for odd-numbered Jerlov water types from I to III and 1 to 9, as indicated by the encircled numbers. The outer boundary of the diagram is defined in terms of the red and green response curves, $(x,y) = (\bar{x}_\lambda, \bar{y}_\lambda)/(\bar{x}_\lambda + \bar{y}_\lambda + \bar{z}_\lambda)$, and the numbers along that curve are the wavelengths associated with $\bar{x}_\lambda$ and $\bar{y}_\lambda$. For a constant value of E_λ, the point $x = y = z = 0.333$ represents the "white point" S, on the diagram (although all neutral grays are also represented by it).

What is meant by "ocean color" is generally the specification of the chromaticity of the upward irradiance; it is this intrinsic character that establishes the *hue* (color) and the *chroma*, or strength, of the color of the sea. Measurements of the spectral radiance or irradiance made as a function of depth can be used to compute the three color coordinates, and the locus of points of x and y can be plotted on the diagram, as is shown in Fig. 9.47 for depths from 1 to 200 m, and for Jerlov water Types I to III and 1 to 9. A line drawn from the white point, S, through the intersection of the depth and the water-type curves at Q will give the dominant wavelength for that water type at that depth, when extended to the outward boundary at A. The

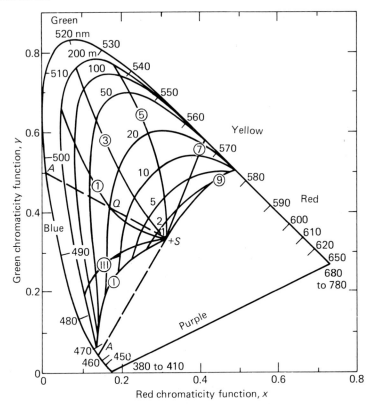

Fig. 9.47 A chromaticity diagram for color of Jerlov Water Types I, III, 1, 3, 5, 7, and 9 (as identified by the circled numbers) for depths from the surface to the asymptotic depth, wherein ocean color takes on its purist hue. The "white point" is at *S*, and the dominant color at great depths is obtained by drawing a straight line from *S* through the intersection, *Q*, of the depth and water type curves, and then extending it to the outer boundary, at *A*. The value of the wavelength at that boundary is the dominant wavelength at depth. [Adapted from Jerlov, N. G., *Marine Optics* (1976).]

ratio of *QS* to *AS* gives the purity of the radiance. An example is shown for water Type 1 at a depth of 20 m, and gives the dominant wavelength as approximately 499 nm, and the purity as approximately 48%. For Type I water, the line from *S* through the deep-water limit goes through the previously cited value of 462 nm as the asymptotic color of water at great depths, and also gives the purity as 100%. Thus the result of the color-selective processes at work in the water is to yield pure blue light at 462 nm coming from all directions surrounding the deep observation point; this wavelength is that of maximum transmittance.

From the earlier discussions of the mechanisms at work in modifying the subsurface light field, we can say that for pure water, the combination of selective absorption of light, along with fluctuational scattering, establishes the color of the sea. In more turbid waters, scattering by particulates, and absorption by both particles and yellow substance play important parts, and help to establish the shift of the wavelength of maximum transmission toward the red (see Fig. 9.42) as the dominant effect.

Above the surface, the state of overhead illumination often overwhelms the upwelling light. The sea surface takes on the color of the sky away from regions of solar specular reflection, and may be seen to reflect clouds, blue sky, and the entire panoply of the skylight in the upper hemisphere. In the shadow of a research vessel, the blockage of skylight and the reflection of the hull cause major modifications to the color locally. In regions of specular reflection, the surface takes on the color of the source of illumination, so that a copper sunset is mirrored in the reflected light. Little can be said about this in general, except that the laws of scattering and reflection from a rough surface hold under such conditions. For these reasons, remote sensing of the color of the sea must take place with carefully selected parameters if spurious color measurements are to be avoided. We will return to this later.

An approximate but useful method of determining the color of the sea is the *Forel Scale*, which consists of a set of 23 colored water standards enclosed in tubes, numbered from 0 (ultramarine blue) to 22 (yellow-green-brown). By holding the tube up against the sunlit sea in a region where the surface coloration and reflection are minimized, a subjective estimate of the water color is obtained. Since color, as defined, is a human sensory perception, such a scale is not without scientific validity in terms of colorimetry; however, it is clearly not a spectroradiometer.

Underwater Imaging

The visibility of objects underwater depends primarily on the scattering properties of the water between the object and the detector, and between the illumination source and the object. An underwater object that emits light cannot be imaged at arbitrarily large distances; beyond a certain distance, typically 15 to 20 attenuation lengths in darkened water, no image of the luminous object may be constructed, but only a bright, diffuse glow is seen, even in the clearest waters. (In this context, an attenuation length is defined as the reciprocal of the beam attenuation coefficient.) Effectively, the resolution has been destroyed by the following mechanism: The image is formed by those rays that leave the object and propagate to the detector without being scattered. However, in water having large scattering coefficients, light that has departed from the object at larger angles but which has then been scattered back into the detector acceptance angle will flood the image plane without contributing to the image formation. Since the image *contrast* depends on the difference between the radiance of the object, L_{obj}, and that of the background, L_b, according to

$$C_{image} = \frac{L_{obj} - L_b}{L_b}, \qquad (9.73)$$

any increase in scattered light that contributes to the background will serve to obscure the image. Small-angle scattering is the major cause of increased background light of self-luminous objects (see Fig. 9.26).

In the sea, with non-self-radiant objects having their source of illumination elsewhere (say, daylight), ordinary targets cannot be sighted at more than 3 to 6 attenuation lengths away. The exact distance depends on the position of the source of light but for natural light the figure given is nominal.

From the discussions of scattering given in Section 9.5, it might be suspected that thermal fluctuations that modify the index of refraction would be the major source of the loss of image resolution. However, while fluctuational scattering is important for backscattering, is is negligible for forward scattering, which is mainly set by the biological and particulate material in the water. Similar considerations also rule out image degradation due to salinity gradients. In the absence of biological or particulate scatterers, ocean water appears to be a better optical medium than air, provided that the loss of power due to absorption (rather than scattering) can be overcome.

An example of the degradation of the image of the sun when seen beneath the surface is presented in Fig. 9.48. These radiance measurements were made with a radiometer having an acceptance angle of $1.3°$, and show the direct image of the sun off nadir as a narrow-angle source at an apparent zenith angle of approximately $12°$ that is resolvable at depths between 1 and 25 m. This image then broadens out into a diffuse light source that appears to be nearly overhead at depths of 275 m. This result is also due to scattering and the accompanying diffusion of the source illumination. It is worth noting the sharp decrease in radiance at angles of $\pm 48.3°$ in the more shallow measurements; this is due to the effect illustrated in Fig. 9.17, wherein light from beyond the refraction cone comes mainly from secondary sources (e.g., scatterers or surface reflections) beneath the sea.

A simple instrument called the *Secchi disk* uses the loss of contrast beneath the surface to estimate the sum of the beam attenuation and diffuse attenuation coefficients. The Secchi disk is a round, white target having a diameter near 0.25 m that is lowered from a vessel and which is viewed from above the surface in full solar illumination, generally with a glass-bottomed bucket held in the water to reduce surface reflectance. The depth at which

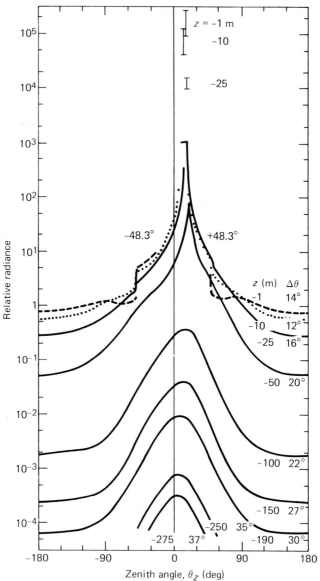

Fig. 9.48 The change of solar radiance with depth as observed in the vertical plane in clear, calm Mediterranean water. The instrument resolution is 1.3°. The solar zenith angle is $\Delta\theta$. For near-surface depths, the sharp falloff in radiance at ± 48.3° from the sun is clearly visible, as are the secondary maxima near ± 90°, which are due to light that has been totally reflected at the surface. The spreading of the solar beam is clear. [From Lundgren, B., "Radiance and Polarization Measurements in the Mediterranean" (1976).]

it disappears, i.e., where the contrast is zero, may be related to the total attenuation. The instrument gives very approximate but useful measurements of water turbidity.

9.10   Remote Sensing of Optical Parameters

The spectral components of the upwelling light emerging from beneath the sea surface have impressed on them, via absorption and scattering, information on the concentrations of the agents causing the attenuation. This light has sampled the upper reaches of the sea, and approximately 90% of it has been returned from within a depth of $1/K_\lambda$. Because of the character of spectral signatures of chlorophyll (see Fig. 9.28), and the proportionality between diffuse attenuation and chlorophyll concentration (Eq. 9.46), spectroradiometric measurements of the flux should allow estimates of the concentration to be made remotely. Scanning imaging radiometers on aircraft or satellites permit rapid measurements over large areas to be made in the form of multispectral images, whose scanning geometry is shown in Fig. 9.49.

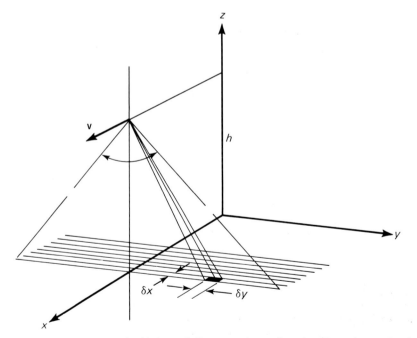

Fig.9.49   The geometry for high-resolution scanning radiometer. Image is constructed from pixels, or picture elements, defined by the instantaneous field of view through the scanning telescope, $\delta x\,\delta y$.

Earlier we presented examples of such spectral radiometric measurements at thermal-IR wavelengths (Figs 2.9 and 6.41); similar imagery may be obtained at optical wavelengths.

The water-leaving spectral radiance is not the same as the radiance immediately beneath the sea surface. Denote the radiance in air propagating upward across the surface at a refracted angle, θ, by $L_{\lambda ua}(\theta)$, and that just below the surface by $L_{\lambda uw}(\theta_i)$. Then, applying the same kind of reasoning that led to Eq. 9.38 for upward-propagating radiances, one obtains

$$L_{\lambda ua}(\theta) = L_{\lambda uw}(\theta_i) [1 - r_\lambda(\theta_i)]/n^2 . \qquad (9.74)$$

Here $r_\lambda(\theta_i)$ is the specular reflection coefficient at incidence angle θ_i; n is the index of refraction; and $\sin \theta_i = (1/n) \sin \theta$. The factor $[1 - r_\lambda(\theta_i)]/n^2$ is approximately 0.544 for angles near the vertical; thus the emergent radiation is about 55% of the radiance just beneath the surface. Surface waves and roughness increase this value only slightly for near-vertical incidence angles (see Figs. 9.13 and 9.14).

Because of the very small backscattering from the ocean near vertical incidence angles, the upwelling light from the sea is much less intense than the light scattered from the atmosphere. A radiometer that looks down through the atmosphere receives signals from all radiating objects within its field of view. These usually consist of (1) the upward radiance leaving the surface of the sea, (2) direct sun and sky radiance reflected from the surface, and (3) the scattered atmospheric radiance. This last is perhaps 5 times as intense as the surface-leaving flux. Therefore, it is essential that corrections for atmospheric effects be made in order to obtain quantitative measurements of the oceanic signal. However, we will not discuss these corrections here because of space limitations.

Figure 9.50 shows the spectral distribution of radiance leaving the sea surface, with chlorophyll concentration, C, as a parameter. With increasing concentrations, the signal in the blue drops rapidly, while that in the green increases; there is a narrow region of approximate invariance of the radiance with C, in the vicinity of 520 nm, that can be used to considerable advantage to normalize the observed signals. If the spectral radiance is sampled at several wavelengths with narrow-band filters (as suggested by the 20-nm-wide boxes along the top of Fig. 9.50), the general form of the curve may be recovered, and along with it, the concentration of phytoplankton pigments. These include not only chlorophyll and similar photosynthetic molecules, but their associated decay products, termed *phaeophytin*. Since the latter derives from the former, its concentration may be expected to covary very approximately linearly with the chlorophyll, which it does approximately at lower concentrations, say $C \leq 1$ mg m^{-3}.

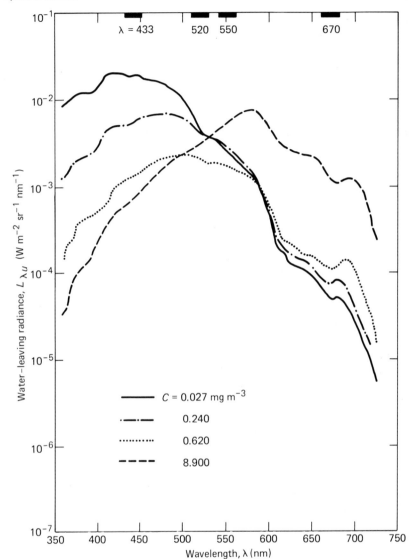

Fig. 9.50 Surface-leaving spectral radiance for varying concentrations of chlorophyll. Measurements have had atmospheric effects removed. [From Gordon, H. R., et al., *Advances in Geopysics,* Vol. 27 (1985).]

In the open ocean, it is generally true that water color is established by (1) the inherent optical properties of seawater, (2) living algal cells, (3) the associated debris products coming from grazing zooplankton and natural decay, and (4) dissolved organic matter liberated by algae, i.e., yellow substance. Waters for which these are the only determinants of color are termed *Case 1* waters. Closer to shore, or in offshore shallow regions, color is altered by the introduction of suspended sediments, terrigeneous particles from river and glacial runoff, yellow substance from land drainage, and influxes from anthropogenic sources. Waters containing any or all of these materials are termed *Case 2* waters.

For waters not bearing appreciable sediments, there exist empirically derived relationships between pigment concentrations and the upwelling radiances measured at various wavelengths (typically two). Denote the wavelengths of measurement by λ_i and λ_j; then the relationships are generally of the form

$$C = C_0 \, r_{ij}^{\eta} \, , \tag{9.75}$$

where C_0 is a coefficient setting the magnitude of the concentration, and η is the power to which the reflectance ratio, r_{ij}, is raised. This ratio is defined as

$$r_{ij} \equiv r_{\lambda_i} / r_{\lambda_j} \, , \tag{9.76}$$

where r_{λ_i} and r_{λ_j} are the *observed* spectral reflectivities at two distinct wavelengths, corrected for atmospheric effects. Alternatively, the reflectivities may be replaced by the irradiances, whose ratio does not differ appreciably from the ratio of the reflection coefficients.

Various approximations to the reflectances are available, the most important of which reduces to a simple relationship between the *beam backscatter* and *beam absorption coefficients,* $b_{\lambda u}$ and a_λ, viz:

$$r_\lambda \simeq \tfrac{1}{3} \, b_{\lambda u} / a_\lambda \, , \tag{9.77}$$

where the irradiance ratio, r_λ, is defined by

$$r_\lambda \equiv \frac{E_{u\lambda}}{E_{d\lambda}} \, , \tag{9.78}$$

as evaluated at the surface. The beam backscatter coefficient, $b_{\lambda u}$, is the integral of β_λ over the upper hemisphere:

$$b_{\lambda u} \equiv 2\pi \int_{\pi/2}^{\pi} \beta_\lambda(\theta) \sin \theta \, d\theta \, . \tag{9.79}$$

Thus spectral reflectivity is governed by the ratio of backscattering to absorption, a parameter of some significance.

For Case 1 waters, a calibrated, broadly applicable algorithm has been derived that is accurate to better than $\pm$ 40% in C when compared with in-situ measurements of pigment concentrations. Written in similar terms to the functional form of Eq. 9.75, it is given by

$$C = 1.71 \, r_{ij}^{-1.82} \, , \qquad 0.02 \leq C \leq 1.5 \text{ mg m}^{-3} \, . \qquad (9.80)$$

Here the wavelengths at which the radiances are evaluated are $\lambda_i = 440$ nm and $\lambda_j = 560$ nm, where the phytoplankton pigments are, respectively, strongly and weakly absorbing. The accuracy claimed is of the same order of magnitude (or even higher) than the natural fluctuations in chlorophyll concentrations occurring over short distances, and thus is within the biological "noise." This algorithm is plotted in Fig. 9.51 as a log-log plot spanning four decades in C, and correlates the data shown with a coefficient of 94.5%.

Figure 9.52a is an image from the Coastal Zone Color Scanner on the Nimbus-7 spacecraft showing the Middle Atlantic Bight. The algorithm used is approximately that of Eq. 9.80. Figure 9.52a shows the warm-core rings and eddies north of the Gulf Stream, and the pigment-rich waters of the continental shelf and George's Bank off the "bent arm" of Cape Cod, for which $C > 1$ mg m^{-3}. A simultaneous, co-registered image in the thermal infrared is shown in Fig. 9.52b. Here the tones represent uncalibrated temperatures. The correlation between color and temperature patterns is striking, especially in the region of the Gulf Stream and its rings and eddies, and implies interesting relationships between biological and physical processes. In this region, colder water seems to support higher pigment levels, and conversely, but this rule is not observed with perfection. On occasion, correlations between color and temperature are much lower than shown here; such appears to be the case in the Mediterranean in the summer.

Other optical quantities may also be derived from such multispectral imagery in their digital format. The diffuse attenuation coefficient for downwelling irradiance given by Eq. 9.46 is related to the concentration by

$$K_{\lambda d} = K_{\lambda w} + [dK/dC]_\lambda \, C \, , \qquad (9.81)$$

where $K_{\lambda w}$ is the value for clear ocean water, and $[dK/dC]_\lambda$ is the specific diffuse attenuation for phytoplankton pigments. A similar relationship for

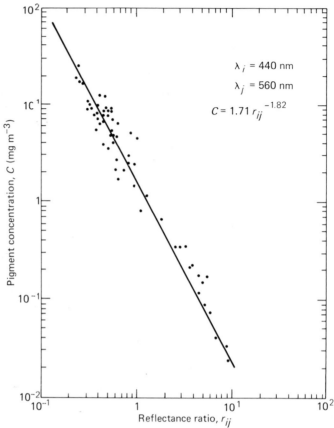

Fig. 9.51 The algorithm for relating pigment (chlorophyll + phaeophytin) concentration to ratio of irradiances at two different wavelengths (440 and 560 nm). The correlation coefficient for the straight regression line is greater than 94%. [From Gordon, H. R., and A. Y. Morel, *Remote Assessment of Ocean Color for Interpretation of Satellite Visible Imagery: A Review* (1983).]

upwelling irradiance holds, which, when combined with the concentration formula of Eq. 9.75, results in

$$K_{\lambda u} = K_{\lambda w} + K_0 \, r_{ij}^{\eta} \, , \tag{9.82}$$

where now r_{ij} is the ratio of radiances (or irradiances) rather than reflectances:

$$r_{ij} \equiv L_{\lambda_{iu}} / L_{\lambda_{ju}} \, . \tag{9.83}$$

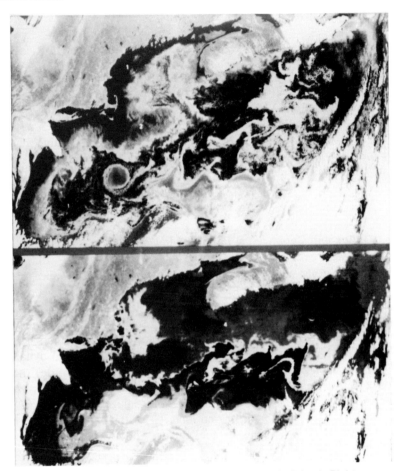

Fig. 9.52 Coastal Zone Color Scanner image of the Middle Atlantic Bight, processed to give quantitative values. (a) Pigment concentration derived using algorithm of Eq. 9.80, valid to within approximately a factor of two over the range $0.05 \leq C \leq 1.0$ mg m^{-3}. (b) Uncalibrated thermal-infrared radiances, adjusted to a temperature scale of $6 \leq T \leq 28$ °C. Much correlation exists between chlorophyll and temperature patterns. [From Gordon, H. R., and A. Y. Morel, *Remote Assessment of Ocean Color for Interpretation of Satellite Visible Imagery: A Review* (1983).]

Values for $K_{\lambda u}$ at $\lambda = 490$ and 520 nm are

$$K_{490} = 0.022 + 0.088 \, r_{ij}^{-1.491} \qquad (9.84a)$$

and

$$K_{520} = 0.044 + 0.066 \, r_{ij}^{-1.398} \,, \qquad\qquad (9.84b)$$

where all K's are in inverse meters, $\lambda_i = 443$, and $\lambda_j = 550$ nm. (The values for $K_{\lambda w}$ are obtained by interpolation in Table 9.4.) The accuracy of these algorithms is estimated to be about $\pm 25\%$ over the range of $K_{\lambda w}$ for that of clear seawater, to $K_{\lambda u} \simeq 0.3$ m^{-1}, corresponding to a pigment concentration of approximately 2.0 mg m^{-3}.

Thus at least two important optical parameters may be obtained via remote measurement: C and K; clearly, from the earlier discussion, these are not independent of each other.

9.11 Optics and Geophysical Fluid Dynamics

As with the other subdisciplines of acoustics and electromagnetics, the optics of the sea finds utility in the understanding of large-scale geophysical fluid dynamics, in addition to its intrinsic value in biological and thermodynamical studies. The scanner images of the preceding section are a case in point, and it is obvious from them that the color of surface waters helps to visualize flow, a requirement that is always a problem in fluid dynamics. For systems such as the Gulf Stream, where the water type varies sharply across a well-defined front, the dynamics are revealed in readily discernable color and temperature variations. Indeed, color images are more trustworthy tags for such flows than infrared images, for the latter come from the very thin IR skin depth at the surface, and are not always representative of the deeper upper ocean layers, while color observations sample the water column down to a depth of order $1/K_\lambda$. With a sequence of such images, the two-dimensional surface color field may often be derived as a function of time, within the limitations of color differentiability, cloud and satellite cover, and data availability.

In biologically productive areas, color is not a conserved quantity, because phytoplankton production may significantly change the spectrum of upwelling light in the course of a very few days. One then obtains from color measurements only the instantaneous spatial distribution of pigment, and even given a sequence of images, it is difficult to separate the local time derivative, $\partial C/\partial t$, from the advective term, $\mathbf{u} \cdot \nabla C$. Flow visualization is realized only by way of the advective term, if at all.

Plankton blooms often occur initially in regions of cold, nutrient-rich, upwelled water. As time goes on, this water warms and becomes depleted in nutrients as the plankton evolution proceeds; the color changes drastically during this process, first greening and then shifting back again toward blue. Thus there is sometimes a correlation between temperature and color for specific masses of water, but the character of the correlation changes with time. An inverse correlation between these two variables is shown in Fig. 9.52, which illustrates a warm-core ring that has been detached from the Gulf Stream; this ring is also seen in Fig. 9.53 to the south of Cape Cod (cf. Fig. 6.42). Here the warm, low-productivity Sargasso Sea water has been encased in a surrounding mass of cooler, higher-productivity slope water, and the thermal and biological fields reflect this.

A basin-wide view of ocean color variability is shown in Plate 4 (Fig. 9.54), which is a false-color map of the North Atlantic for the month of May 1979; the oceanic portions were constructed from Coastal Zone Color Scanner data, while the land portions used data from the Advanced Very High Resolution Radiometer. In the ocean, reddish colors signify high chlorophyll concentrations, while bluish tones denote low concentrations; white regions are clouds, while black areas are devoid of data. A spring plankton bloom extends across the North Atlantic and into the North Sea, as well as along the continental shelves and offshore banks. In the Sargasso Sea, however, the pigment concentrations are less than 0.05 mg m^{-3}. Such a map is a quantitative and detailed update to the chart of Jerlov water types shown in Fig. 9.45.

A more classical use of light measurements in geophysical fluid flow is illustrated in Fig. 9.55, which shows a subsurface section of relative light scatterance to the west of the Gibraltar sill, and demonstrates how the tongue of increased particulate scattering mimics the outflow and sinking of saline water shown in Fig. 2.24. The tongue is overlain by layers of alternating low and high scattering. Nearer to the surface, Fig. 9.56 shows the depth of the euphotic zone, $z_{0.01}$, in the region off Iberia and North Africa. Here clear surface water may be seen making its way in from the west to replace the saltier, deeper Mediterranean outflow, a pattern revealed by the topography of the bottom of the euphotic region.

In areas of riverine outflow, the distribution of suspended sediments, as revealed by light scattering, often maps the flow to a reasonable approximation. Figure 9.57 shows the coastal ocean immediately south of the mouth of the Columbia River in the Pacific Northwest, and illustrates near-surface contours of salinity, temperature, density anomaly, and light scattering during early summer, when river flow is moderately strong; scattering is measured by $\beta_{475}(\pi/4)$, and station locations are shown via the dots. The optical scattering is moderately well-correlated with the other physical parameters, but the different diffusivities for the various quantities shown imply that the contours should not necessarily be highly similar. When such in-situ measurements are used with quantitative remote imaging of the water-leaving spectral radiance, one may obtain fine-grained, calibrated maps of the diffuse attenuation coefficient, and even infer something of the sediment concentration, if the sediment properties such as index of refraction and size distribution are also known.

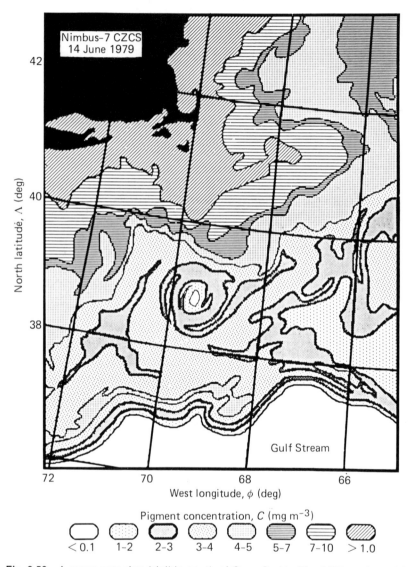

Fig. 9.53 A warm-core ring (visible south of Cape Cod in Fig. 9.52) contoured to show pigment concentration in mg m^{-3}. Data are from the Coastal Zone Color Scanner. [Adapted from Clark, D. K., and J. W. Sherman, *Marine Tech. Soc. J.* (1986).]

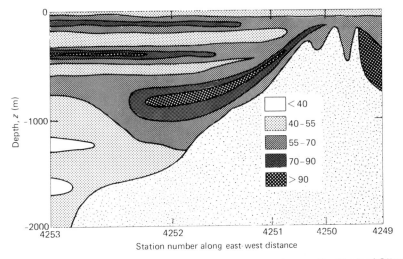

Fig. 9.55 Subsurface relative scatterance measurements near the Strait of Gibraltar, showing sinking of saline Mediterranean water to the west of the sill (cf. Fig. 2.24). [Adapted from Jerlov, N. G., Medd. Oceanogr. Inst. Göteborg (1961).]

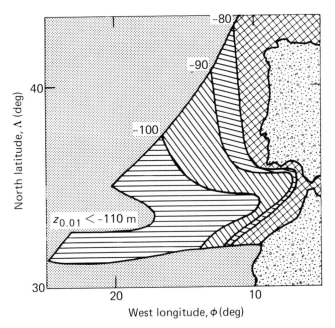

Fig. 9.56 Depth of the euphotic zone, $z_{0.01}$, in meters, at $\lambda = 465$ nm, to the west of Gibraltar. [Adapted from Jerlov, N. G., and K. Nygård, Medd. Oceanogr. Inst. Göteborg (1961).]

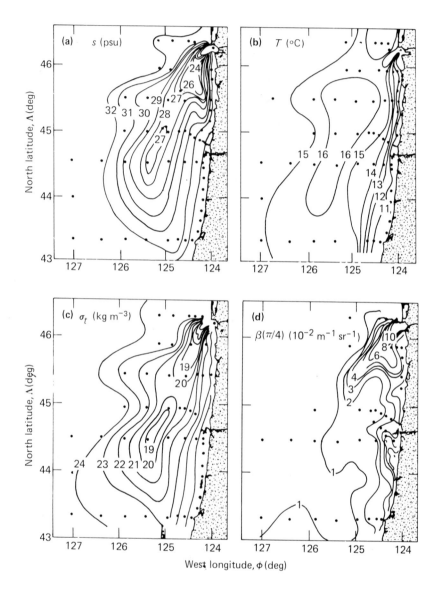

Fig. 9.57 Horizontal distributions of physical parameters in the coastal Pacific south of the Columbia River. (a) salinity, s, in psu; (b) temperature, T, in $^{\circ}C$; (c) density anomaly, σ_t, in kg m^{-3}; (d) volume scattering function, $\beta_\lambda(\pi/4)$, in 10^{-2} m^{-1} sr^{-1}. All measurements at $z = -10$ m and $\lambda = 475$ nm. Moderate correlation exists between quantities. [Adapted from Pak, H., et al., *J. Geophys. Res.* (1970).]

Bibliography

Books

Chandrasekhar, S., *Radiative Transfer,* Oxford, Clarendon Press, London, England (1950).

Gordon, H. R., R. W. Austin, D. K. Clark, W. A. Hovis and C. S. Yentsch, "Ocean Color Measurements," in *Advances in Geophysics: Satellite Oceanic Remote Sensing,* Vol. 27, B. Saltzman, Ed., Academic Press, Orlando, Fla. (1985)

Gordon, H. R., and A. Y. Morel, *Remote Assessment of Ocean Color for Interpretation of Satellite Visible Imagery: A Review,* Springer-Verlag, New York, N.Y. (1983).

Hodara, H., and W. H. Wells, *Optics of the Sea,* NATO/AGARD Lecture Series No. 61, Tetra Tech, Inc., Pasadena, Calif. (1973).

Jerlov, N. G., *Marine Optics,* Elsevier Scientific Publishing Co., Amsterdam, Netherlands (1976).

Jerlov, N. G., and E. S. Nielsen, Eds., *Optical Aspects of Oceanography,* Academic Press, London, England (1974).

Kirk, J. T. O., *Light and Photosynthesis in Aquatic Ecosystems,* Cambridge University Press, Cambridge, England (1983).

Phillips, O. M., *The Dynamics of the Upper Ocean,* 2nd ed., Cambridge University Press, Cambridge, England (1977).

Preisendorfer, R. W., *Hydrologic Optics,* Vols. I to VI, National Oceanic and Atmospheric Administration, Pacific Marine Environmental Laboratory, Honolulu, Hawaii (1976).

Williams, J., *Optical Properties of the Sea,* United States Naval Institute, Annapolis, Md. (1970).

Journal Articles and Reports

Austin, R. W., and G. Halikas, "The Index of Refraction of Seawater," Scripps Institution of Oceanography Report SIO Ref. 76-1 (1976).

Austin, R. W., and T. J. Petzold, "Spectral Dependence of the Diffuse Attenuation Coefficient of Light in Ocean Waters," *Optical Engr.* Vol. 25, p. 471 (1986).

Blizard, M. A., Ed., *Ocean Optics VII,* Proc. SPIE Vol. 489 (1984).

Blizard, M. A., "Ocean Optics: Introduction and Overview," U.S. Office of Naval Research, Arlington, Va. (1986).

Bristow, M., D. Nielsen, D. Bundy, and R. Furtek, "Use of Water Raman Emission to Correct Airborne Laser Fluorosensor Data for Effects of Water Optical Attenuation," *Appl. Optics,* Vol. 20, p. 2889 (1981).

Burt, W. V., "A Light Scattering Diagram," *J. Mar. Res.,* Vol. 15, p. 76 (1956).

Clark, D. K., and J. W. Sherman, "Nimbus-7 Coastal Zone Color Scanner: Ocean Color Applications," *Marine Tech. Soc. J.,* Vol. 20, p. 48 (1986).

Commission Internationale de l'Eclairage (C.I.E.), "Vocabulaire Internationale de l'Eclairage," 2nd ed., C.I.E., Paris, France (1957).

Cox, C., and W. Munk, "Measurement of the Roughness of the Sea Surface from Photographs of the Sun's Glitter," *J. Opt. Soc. Am.,* Vol. 44, p. 838 (1954).

Curcio, J. A., and C. C. Petty, "The Near Infrared Absorption Spectrum of Liquid Water," *J. Opt. Soc. Am.*, Vol. 41, p. 302 (1951).

Duntley, S. Q., "Underwater Visibility and Photography," in *Optical Aspects of Oceanography*, N. G. Jerlov and E. S. Nielsen, Eds., Academic Press, London, England (1974).

Duntley, S. Q., Ed., *Ocean Optics VI*, Proc. SPIE Vol. 208 (1980).

Esaias, W. E., G. C. Feldman, C. R. McClain, and J.A. Elrod, "Monthly Satellite-Derived Phytoplankton Pigment Distribution for the North Atlantic Ocean Basin," *Eos Trans., Am. Geophys. Union*, Vol. 67, p. 835 (1986).

Gordon, H. R., "Diffuse Reflectance of the Ocean: The Theory of Its Augmentation by Cholorphyll-a Fluorescence at 685 nm," *Appl. Optics* Vol. 18, p. 1161 (1979),

Ivanoff, A., "Facteurs Physiques, Chimiques et Biologiques Affectant la Propagation de la Lumière dans les Eaux de Mer," *Optics of the Sea*, AGARD Lect. Ser., No. 61 (1973).

Jerlov, N. G., and K. Nygård, "Measured Irradiance," in *Optical Measurements in the Eastern North Atlantic*, Medd. Oceanogr. Inst. Göteburg, Ser. B, Vol.8, p. 22 (1961).

Kampa, E. M., "Daylight Penetration Measurements in Three Oceans," Union Geod. Geophys. Int., Monogr. 10, p. 91 (1961).

Losee, J., D. Lapota, M. Geiger, and S. Lieberman, "Bioluminescence in the Marine Environment," in *Ocean Optics VII*, M. A. Blizard, Ed., Proc. SPIE Vol. 489, p. 77 (1984).

Lundgren, B., "Radiance and Polarization Measurements in the Mediterranean," Univ. Copenhagen Internal Report (1976).

Lundgren, B., and N. K. Højerslev, "Daylight Measurements in the Sargasso Sea. Results from the "Dana" Expedition, January–April 1966," Univ. Copenhagen, Inst. Phys. Oceanogr. Rep. 14 (1971).

Morel, A., and L. Caloumenos, "Variabilité de la Répartition Spectrale de l'Énergie Photosynthétique," in *Upwelling Ecosystems Analysis Conf. Proc., Marseille, 1973. Tethys*, Vol. 6, p. 93 (1974).

Morel, A., and L. Prieur, "Analysis of Variations in Ocean Colour," *Limnol. Oceanogr.* Vol. 22, p. 709 (1977).

Neumann, G., and R. Hollman, "On the Albedo of the Sea Surface," Union Geod. Geophys. Int., *Monogr.* 10, p. 72 (1961).

Pak, H., "Cruise Data Report, R/V *Discoverer*, 28 June–23 July, 1985" (1985).

Pak, H., G. F. Beardsley, Jr., and P. K. Pak, "The Columbia River as a Source of Marine Light Scattering Particles," *J. Geophys. Res.* Vol. 75, p. 4570 (1970).

Preisendorfer, R. W., and C. D. Mobley, "Albedos and Glitter Patterns of a Wind-Roughened Sea Surface, *J. Phys. Oceanogr.* Vol 16, p. 1293 (1986).

Smith, R. C., "Scripps Spectroradiometer Data," *Data Rep. "Discoverer" Expedition 3 May–5 June, 1970*, Vol. II, J. Tyler, Ed., SIO Ref. 73-16, p. 95 (1973).

Smith, R. C., "Structure of Solar Radiation in the Upper Layers of the Sea," in *Optical Aspects of Oceanography*, N. G. Jerlov and E. S. Nielsen, Eds., Academic Press, New York, N.Y. (1974).

Smith, R. C., and K. S. Baker, "Optical Classification of Natural Waters," SIO Ref. 77-4 (1977).

Smith, R. C., and K. S. Baker, "The Bio-Optical State of Ocean Waters and Remote Sensing," *Limnol., Oceanogr.* Vol. 23, p. 247 (1978).

Yentsch, C. M., and C. S. Yentsch, "Emergence of Optical Instrumentation for Measuring Biological Properties," *Oceanogr. Mar. Biol. Ann. Rev.* Vol. 22, p. 55 (1984).

Fundamental Physical Constants

Speed of light in free space	c_0	2.997924562×10^8	m s^{-1}
Planck constant	h	6.626176×10^{-34}	J Hz^{-1}
Elementary charge	e	$1.6021892 \times 10^{-19}$	C
Boltzmann constant	k_B	1.380662×10^{-23}	J K^{-1}
Stefan-Boltzmann constant	σ_{SB}	5.67032×10^{-8}	W m^{-2} K^{-4}
Wien displacement constant	α_r	2897.82	μm K
Avogadro's number	N_0	6.022045×10^{23}	mole^{-1}
Mechanical equivalent of heat	J	4185.80	J (kg cal)$^{-1}$
Temperature of absolute zero	T_0	-273.155	°C
Atomic mass unit	m_a	$1.6605655 \times 10^{-27}$	kg
Mass of proton	m_+	$1.6726485 \times 10^{-27}$	kg
Atomic mass, carbon-12 nucleus	m_{c12}	12 (exactly)	

Appendix Two

Astronomical and Geodetic Parameters

Parameter	Symbol	Value	Units
Newtonian gravitational constant	G	6.673×10^{-11}	$\mathrm{m^3\ s^{-2}\ kg^{-1}}$
Astronomical unit	R_{es}	$1.49597870 \times 10^{11}$	m
Mean solar radius	R_s	6.960×10^{8}	m
Mean solar constant	$\langle S \rangle$	1376	$\mathrm{W\ m^{-2}}$
Mean radius of earth	R_e	6370.949	km
Equatorial radius of earth	R_{ee}	6378.137	km
Mass of earth	m_e	5.9733×10^{24}	kg
Mean gravitational acceleration	g	9.7976	$\mathrm{m\ s^{-2}}$
Equatorial gravitational acceleration	g_e	9.78032	$\mathrm{m\ s^{-2}}$
Flattening of earth	$1/f$	298.257	—
Second degree harmonic coefficient	J_2	1.082629×10^{-3}	—
Magnetic dipole moment	M_e	8.02×10^{22}	$\mathrm{A\ m^2}$
Area of ocean	A_w	3.61254×10^{8}	$\mathrm{km^2}$
Mass of ocean	m_w	1.4×10^{21}	kg
Average depth of ocean	H	3800	m
Angular rotation rate of earth	Ω	7.292115×10^{-5}	$\mathrm{rad\ s^{-1}}$
Sidereal day	T_s	86,164.09	s
Solar day	T_{sun}	86,400.00	s
Length of year, solar days	T_y	365.24219878	d

Appendix Three

Representative Values for Oceanic and Atmospheric Parameters

Freezing point of seawater, $s = 24.7$ psu	T_f	-1.33	°C
Temperature of triple point (pure water)	T_3	0.0100	°C
Temperature of maximum density (pure water)	T_ρ	3.99	°C
Average surface temperature of earth	T_a	15	°C
Atmospheric surface pressure	p_a	1.013×10^5	Pa
Atmospheric surface density	ρ_a	1.225	kg m^{-3}
Atmospheric surface wind speed	u_w	5	m s^{-1}
Atmospheric drag coefficient (10 m) (neutral conditions)	c_d	0.95×10^{-3}	—
Atmospheric surface stress	τ_a	0.03	N m^{-2}
Specific heat—constant pressure and salinity $T = 20°C$, $s = 35$ psu	C_{ps}	3994	J (kg°C)$^{-1}$
Specific heat—constant volume and salinity $T = 20°C$, $s = 35$ psu	$C_{\alpha s}$	3939	J (kg°C)$^{-1}$
Ratio of specific heats, $C_{ps}/C_{\alpha s}$ $T = 20°C$, $s = 35$ psu	γ_h	1.014	—

Quantity	Symbol	Value	Units
Latent heat of fusion, $T = 0°C$	L_f	0.335×10^6	$J\ kg^{-1}$
Latent heat of vaporization, $T = 20°C$	L_v	2.453×10^6	$J\ kg^{-1}$
Isothermal compressibility coefficient $T = 20°C$, $s = 35$ psu	a_p	4.27×10^{-10}	$(Pa)^{-1}$
Thermal expansion coefficient $T = 20°C$, $s = 35$ psu	a_T	2.41×10^{-4}	$(°C)^{-1}$
Saline contraction coefficient $T = 20°C$, $s = 35$ psu	a_s	7.45×10^{-4}	$(psu)^{-1}$
Speed of sound $T = 20°C$, $s = 35$ psu, $z = 0$	c	1520	$m\ s^{-1}$
Density of seawater $T = 20°C$, $s = 35$ psu, $z = 0$	ρ	1024.8	$kg\ m^{-3}$
Salinity of seawater	s	35	10^{-3} (psu)
Scale height of ocean	H_s	230	km
Acoustic reference pressure	p_r	10^{-6}	$N\ m^{-2}$
Acoustic impedance	ρc	1.5×10^6	$kg\ m^{-2}\ s^{-1}$
Molecular mass, pure water	M_w	18.0160	—
Surface tension, clean water	τ_s	0.079	$N\ m^{-1}$

Quantity	Symbol	Value	Units
Molecular viscosity $T = 20°$, $s = 36$ psu	μ	1.075×10^{-3}	$\text{kg m}^{-1}\,\text{s}^{-1}$
Molecular kinematic viscosity $T = 20°$, $s = 36$ psu	ν_m	1.049×10^{-6}	$\text{m}^2\,\text{s}^{-1}$
Horizontal momentum diffusivity	K_h	$10^2 - 10^5$	$\text{m}^2\,\text{s}^{-1}$
Vertical momentum diffusivity	K_v	$3 \times 10^{-5} - 2 \times 10^{-2}$	$\text{m}^2\,\text{s}^{-1}$
Molecular salinity diffusivity	κ_s	1.5×10^{-9}	$\text{m}^2\,\text{s}^{-1}$
Horizontal salinity diffusivity	κ_{sh}	$1.5 \times 10^3 - 3 \times 10^3$	$\text{m}^2\,\text{s}^{-1}$
Vertical salinity diffusivity	κ_{sv}	$3 - 7 \times 10^{-5}$	$\text{m}^2\,\text{s}^{-1}$
Molecular thermal diffusivity	K_q	1.49×10^{-7}	$\text{m}^2\,\text{s}^{-1}$
Horizontal thermal diffusivity	K_{qh}	$10 - 10^5$	$\text{m}^2\,\text{s}^{-1}$
Vertical thermal diffusivity	K_{qv}	$2 \times 10^{-6} - 3 \times 10^{-2}$	$\text{m}^2\,\text{s}^{-1}$
Molecular thermal conductivity	κ_q	0.596	$\text{W m}^{-1}\,\text{K}^{-1}$
Minimum surface wave speed	c_{min}	0.23	m s^{-1}
Wavelength of slowest wave	λ_{min}	0.017	m
β term at 45° latitude	β	1.619×10^{-11}	$\text{m}^{-1}\,\text{s}^{-1}$
f term at 45° latitude	f_o	1.031×10^{-4}	s^{-1}
Internal Rossby radius at mid-latitudes	δ_R	40	km

Electrical Parameters

Permittivity of free space	ϵ_0	$8.85418782 \times 10^{-12}$	F m^{-1}
Permeability of free space	μ_0	$4\pi \times 10^{-7}$	H m^{-1}
DC dielectric constant	ϵ_r	81	—
DC conductivity	σ_s	4.29140	S m^{-1}
$T = 15.0°C$, $s = 35.0$ psu			
Optical index of refraction	n	1.34295	—
$T = 20°C$, $s = 35$ psu, $\lambda = 500$ nm			
Optical wavelength of max. penetration	λ_m	462	nm
(clearest seawater)			
Diffuse attenuation coefficient	K	0.02 – 0.04	m^{-1}
(clearest seawater)			
Brewster angle (microwave)	θ_B	83.7	deg
Brewster angle (optical)	θ_B	53.3	deg
Emissivity (microwave, vertical)	e	~0.4	—
Emissivity (thermal IR)	e	0.99	—
Reflectivity (optical)	r	~0.06	—
Dipole moment (pure water)	q	3.3357×10^{-30}	C m

Appendix Five

Dimensionless Numbers for Fluids

Characteristic scale	Velocity, U, ΔU	Length, L, ΔL	Radius, R	Time, L/U	Density, ρ, $\Delta\rho$	Pressure, P, ΔP
Momentum equation	$\dfrac{\partial \mathbf{u}/\partial t}{U^2/L}$	$\dfrac{\mathbf{u}\cdot\nabla\mathbf{u}}{U^2/L}$	$\dfrac{2\mathbf{\Omega}\times\mathbf{u}}{fU}$	$\dfrac{(1/\rho)\nabla p}{\Delta P/\rho L}$	$\dfrac{\nabla\Phi}{g,\ g'}$	$\dfrac{(A/\rho)\nabla^2\mathbf{u}}{2AU/\rho L^2}$
Heat flow equation	$\dfrac{\partial T/\partial t}{TU/L}$	$\dfrac{\mathbf{u}\cdot\nabla T}{TU/L}$	$\dfrac{K_q\nabla^2 T}{K_q T/L^2}$			

Name	Force ratio	Formulas	
Froude	$\dfrac{\text{inertial}}{\text{gravitational}}$	$\mathrm{Fr} = \dfrac{U^2/L}{g}$	$= \dfrac{U^2}{gL}$
Internal Froude	$\dfrac{\text{inertial}}{\text{buoyancy}}$	$\mathrm{Fr} = \dfrac{U^2/L}{g\Delta\rho/\rho}$	$= \dfrac{U^2}{g'L}$
Euler	$\dfrac{\text{pressure}}{\text{inertial}}$	$\mathrm{Eu} = \dfrac{\Delta P/\rho L}{U^2/L}$	$= \dfrac{\Delta P}{\rho U^2}$
Richardson	$\dfrac{\text{buoyancy}}{\text{inertial}}$	$\mathrm{Ri} = \dfrac{g\Delta\rho/\rho}{U^2/L}$	$= \dfrac{N^2}{(U/L)^2}$
Gradient Richardson	$\dfrac{\text{buoyancy}}{\text{shear}}$	$\mathrm{Ri} = \dfrac{g\Delta\rho/\rho\Delta L}{(\Delta U/\Delta L)^2}$	$= \dfrac{N^2}{(\Delta U/\Delta L)^2}$
Rossby	$\dfrac{\text{centrifugal}}{\text{Coriolis}}$	$\mathrm{Ro} = \dfrac{U^2/R}{fU}$	$= \dfrac{U}{fR}$
Ekman (horizontal)	$\dfrac{\text{horizontal viscous}}{\text{Coriolis}}$	$\mathrm{Eh} = \dfrac{2A_h U/\rho L^2}{fU}$	$= \dfrac{2K_h}{fL^2}$
Ekman (vertical)	$\dfrac{\text{vertical viscous}}{\text{Coriolis}}$	$\mathrm{Ev} = \dfrac{2A_v U/\rho\delta_E^2}{fU}$	$= \dfrac{2K_v}{f\delta_E^2}$

Name	Force ratio	Formulas
Reynolds	$\dfrac{\text{inertial}}{\text{viscous}}$	$\mathrm{Re} = \dfrac{U^2/L}{\nu_m U/L^2} = \dfrac{UL}{\nu_m}$
Reynolds (horizontal)	$\dfrac{\text{inertial}}{\text{horizontal viscous}}$	$\mathrm{Re} = \dfrac{U^2/L}{A_h U/L^2 \rho} = \dfrac{UL}{K_h}$
Stokes	$\dfrac{\text{viscous}}{\text{gravitational}}$	$\mathrm{St} = \dfrac{\nu_m U/R^2}{g\Delta\rho/\rho} = \dfrac{\mu U}{(\rho_s - \rho)gR^2}$
Prandtl	$\dfrac{\text{viscous diffusive}}{\text{thermal diffusive}}$	$\mathrm{Pr} = \dfrac{\nu_m}{K_q} = \dfrac{\mu C_\alpha}{\kappa_q}$
Péclet	$\dfrac{\text{forced convective}}{\text{thermal diffusive}}$	$\mathrm{Pe}' = \dfrac{U/L}{K_q/L^2} = \dfrac{UL}{K_q}$
Grashof	$\dfrac{\text{free convective}}{\text{viscous}}$	$\mathrm{Gr} = \dfrac{g\Delta\rho/\rho}{\nu_m^2/L^3} = \dfrac{g\Delta\rho L^3}{\rho\nu_m^2}$

Index

International Geophysics Series

EDITED BY

RENATA DMOWSKA
Division of Applied Science
Harvard University

JAMES R. HOLTON
Department of Atmospheric Sciences
University of Washington
Seattle, Washington

Volume 1 BENO GUTENBERG. Physics of the Earth's Interior. 1959*

Volume 2 JOSEPH W. CHAMBERLAIN. Physics of the Aurora and Airglow. 1961*

Volume 3 S. K. RUNCORN (ed.). Continental Drift. 1962*

Volume 4 C. E. JUNGE. Air Chemistry and Radioactivity. 1963*

Volume 5 ROBERT G. FLEAGLE AND JOOST A. BUSINGER. An introduction to Atmospheric Physics. 1963*

Volume 6 L. DUFOUR AND R. DEFAY. Thermodynamics of Clouds. 1963*

Volume 7 H. U. ROLL. Physics of the Marine Atmosphere. 1965*

Volume 8 RICHARD A. CRAIG. The Upper Atmosphere: Meteorology and Physics. 1965*

Volume 9 WILLIS L. WEBB. Structure of the Stratosphere and Mesosphere. 1966*

Volume 10 MICHELE CAPUTO. The Gravity Field of the Earth from Classical and Modern Methods. 1967*

Volume 11 S. MATSUSHITA AND WALLACE H. CAMPBELL (eds.). Physics of Geomagnetic Phenomena. (In two volumes.) 1967*

Volume 12 K. YA. KONDRATYEV. Radiation in the Atmosphere. 1969

Volume 13 E. PALMEN AND C. W. NEWTON. Atmospheric Circulation Systems: Their Structure and Physical Interpretation. 1969

Volume 14 HENRY RISHBETH AND OWEN K. GARRIOTT. Introduction to Ionospheric Physics. 1969*

Volume 15 C. S. RAMAGE. Monsoon Meteorology. 1971*

Volume 16 JAMES R. HOLTON. An Introduction to Dynamic Meteorology. 1972*

Volume 17 K. C. YEH AND C. H. LIU. Theory of Ionospheric Waves. 1972

Volume 18 M. I. BUDYKO. Climate and Life. 1974

Volume 19 MELVIN E. STERN. Ocean Circulation Physics. 1975

Volume 20 J. A. JACOBS. The Earth's Core. 1975*

* Out of Print